P9-EDK-055

THE BIOLOGY OF TREMATODES

THE BIOLOGY OF TREMATODES

David A. Erasmus, Ph.D.,
Reader in Parasitology, Department of Zoology,
University College, Cardiff

EDWARD ARNOLD

First published 1972
by Edward Arnold (Publishers) Limited,
25 Hill Street,
London, WIX 8LL

ISBN: 0 7131 2326 5

Printed in Great Britain at The Universities Press, Belfast

PREFACE

The experimental investigation of the biology of the trematodes has always been a particularly challenging task. The establishment of a trematode species in the laboratory is dependent on a supply of parasite-free hosts and, because of the systematic diversity of the host animals sometimes required, culture and maintenance procedures become very elaborate. The use of natural infections from the field is often unsatisfactory for many purposes because of seasonal variation in availability and because the material is ill-defined in terms of age and host/parasite strains. The situation is rendered more difficult by the relatively short life of many of the larvae.

In all stages, the isolation of organ systems is not easily accomplished because of the absence of a body cavity and the intimate association of the structures with the surrounding parenchyma. Thus, trematodes do not lend themselves readily to the isolation and fractionization techniques favoured by biochemists. In spite of these difficulties, the introduction of methods for the maintenance of species such as *Schistosoma* and *Fasciola* on a large scale in the laboratory, and the application of histochemistry, autoradiography, electron microscopy and *in vitro* culture techniques has enabled considerable progress to be made in our understanding of these parasites over the last two decades.

The aim of this text is to introduce final honours and first year postgraduate students, who are already familiar with the basic characteristics of the group, to some of the more recent discoveries and ideas developed through the use of modern techniques. The choice of the title, which might concern the more pedantic reader, is justified by the following reasons. The treatment in this book is largely non-systematic and the term 'Biology' has been chosen because its inclusive nature permits the correlation of information from a wide variety of disciplines. The much criticised systematic unit 'Trematoda' has been retained because it enables me to consider both 'monogenean' and 'digenean' trematodes in a comparative fashion. In spite of the controversy concerning the systematic status of the units, both groups become involved in similar host-parasite situations and possess systems which are similar in structure. Thus both the oncomiracidium and the miracidium are free swimming ciliated stages and have similar problems of host location and attachment to overcome. Adults of both groups have an external cytoplasmic tegument which is the 'buffer' between the parasite and its external environment and is a surface which is involved in secretion and absorption. The alimentary tract and the excretory system of both groups show considerable morphological similarity. Thus in order to discuss the essentially nonsystematic but largely biological aspects of these parasites the older, and perhaps more familiar, systematic unit has been retained. A final point is that the emphasis in this book is on larval stages, which have so often been given inadequate coverage in the past.

My interest in the trematodes has been fostered and encouraged by Professor J. Brough, D.Sc., and I will always be indebted to him for the facilities which have enabled my interests to be fulfilled. This book would not have been completed but for the magnificent way in which Mrs. R. Hyde accepted the challenge of endless notes and alterations, and translated them into readable text. I am sincerely grateful to her for her considerable labour and to my colleague, Dr. R. Hammond, who read the text in its early stage and made many valuable suggestions.

Several of my postgraduate students have generously allowed me the use of their photographs and I thank Dr. A. J. Probert, Dr. J. Jenkinson, Mrs. L. Gregory and Mr. R. Robson for this. I am also grateful to Dr. R. A. Wilson who sent me his manuscript on the body wall of *Fasciola* miracidium and on whose work my text-fig. 32 is based. Finally, but by no means least I will always be grateful to my wife, Sylvia, for her understanding and encouragement, without which this book would not have been possible.

This text contains many illustrations taken from other publications and I gratefully acknowledge the permission of Professor B. Dawes and the editors of the American Journal of Tropical Medicine and Hygiene, Blackwell's Scientific Publications, Ltd., Comparative Biochemistry and Physiology, Det Kongelige Danske Videnskabernes Selskab, Experimental Parasitology, Journal of Helminthology, University of Illinois Press, Japanese Journal of Medical Science and Biology, Journal of the Marine Biological Association, U.K., Journal of Parasitology, University of Neuchatel, Oliver and Boyd, Parasitology, Proceedings of the Helminthological Society of Washington, the Royal Society of Tropical Medicine and Hygiene, Journal of Shanghai Science, Sobretire del Libro Homenge al Dr. Eduardo Caballero y Caballero, T. Cheng and W. B. Saunders Co., Transactions of the American Microscopical Society, Zeitschrift fur Parasitenkunde and the Zoological Society of London to reproduce them. Figures 49a and b are reproduced by permission of the National Research Council of Canada from the Canadian Journal of Zoology, 34, 295–386 (1956) and figure 7 from 'The Invertebrata, vol. II, Platyhelminthes and Rhynchocoela' by L. H. Hyman, copyright 1951 held by the McGraw-Hill Book Company Inc., and used with the permission of McGraw-Hill Book Company Inc. Specific author acknowledgements are made alongside each figure.

Cardiff, 1972 D. A. E.

CONTENTS

I General Features

All trematodes are parasitic and may live on the skin, fins, gills, buccal and cloacal surfaces of aquatic hosts, in the alimentary tract and its accessory structures, or in the reproductive, excretory, respiratory, blood and nervous systems. In fact there are very few organ systems which have not become invaded by trematodes. The majority of trematodes are parasitic in chordate hosts although the wide distribution of larval stages in the invertebrate fauna results in the association of most animal taxa with the biology of the trematodes.

Attachment to the host is usually by means of suckers of one sort or another and the evidence suggests that attachment is not permanent but that the parasites are able to move around and reattach themselves within the selected environment. The selection of certain organs, e.g. the brain or eye as habitats often necessitates considerable migration within the host body by some parasite species. The life-cycle of trematode parasites may be simple in possessing a single larval stage with no multiplieation so that one egg is potentially capable of giving rise to one adult. In other cycles the development is complex in that several larval stages may occur involving a succession of different intermediate hosts as well as multiplication so that a single egg is potentially capable of giving rise to hundreds or thousands of adults, and this compensates to a considerable extent for the hazards which are encountered during the developmental period.

The Trematoda represents a somewhat heterogeneous group of parasites which, because of the nature of their life-cycle, exhibit a strong and persistent association with aquatic habitats. The appearance and survival of these parasites is therefore influenced by the general biotic factors associated with aquatic environments as well as by the more intimate physiological and immunological interchanges present in any host-parasite relationship. In this way the survival of both adult and larval trematodes is particularly susceptible to climatic and seasonal changes in the environment and to fluctuations in the free-living fauna also resulting from such changes. The freshwater habitat for example is considerably affected by seasonal changes so that the trematode fauna associated with such environments is in a continual state of flux.

The complexity of the life-cycle, particularly in the case of the digeneans, is reflected in the involvement in it of many representatives, both permanent and temporary, of the free-living fauna. Thus, trematode parasites can be regarded as a component of the fauna associated with aquatic environments although this aspect is generally overlooked until epidemic disease makes the association inescapable. Water is essential to all animals and many essentially terrestrial animals have to maintain permanent or temporary contact with natural bodies of water. The evolution of the parasites of these terrestrial hosts has occurred in such a way that this period of contact with water permits the establishment of parasitic stages in host animals which lead a largely terrestrial existence.

The association of adult trematodes with clinical disease is relatively rare in natural populations although several genera (e.g. *Fasciola*, *Schistosoma*, *Clonorchis* and *Paragonimus*) are notorious in producing severe disease and mortality in Man and domestic animals.

ADULT MORPHOLOGY

The majority of trematodes fall into the range of 2–15 mm but some may reach a length of 80 mm (*Fasciola gigantica* and *Fasciolopsis buski*). NOBLE (1967) records a forty-foot fluke! They are usually cream or white in colour although the presence of large numbers of eggs in the uterus, or host blood or other tissues in the gut, may give a brown or reddish tinge to the body. Larval stages may possess pigment, which is intrinsic or derived from host food, so that these stages may be green, yellow or brown in colour, but the majority are translucent. The body usually has the form of an elongated oval in outline although many of the strigeoid trematodes possess a transverse constriction dividing the body into anterior and posterior regions. The conspicuous posterior attachment organs of the monogeneans may be correlated also with a posterior demarkation of the body. The body is dorso-ventrally flattened but variations on this may result in spherical, cup-shaped or cylindrical forms appearing. The absence of a rigid skeleton and the presence of an extensive musculature allow continual changes of body shape to occur, much to the dismay of many systematists. It is now realized that the trematodes (data not available yet on the Aspidobothria) are bounded externally by a highly modified epidermis in the form of a cytoplasmic tegument. Associated with the tegument are hooks and spines of various types and these are usually distributed in a manner constant and characteristic for the species. Below the tegument lies the basement layer which has, as its major component, a fibrous layer and it is this which serves as a skeleton for the attachment of muscles as well as limiting the extent by which the body changes shape.

Immediately below the basement layer lie the muscles arranged with circular muscles outermost and longitudinal innermost. The muscles (as seen in the strigeoid genera *Apatemon*, *Diplostomum* and *Cyathocotyle*) (Plate 1-1) resemble very closely these described by MORITA (1965) from the planarian *Dugesia dorotocephala*. The sarcoplasm encloses two types of myofilaments—a central one approximately 0·02 μ in diameter and this is surrounded by a ring of smaller myofilaments approximately 70 Å in diameter. The sarcoplasmic extension of the muscle contains the nucleus and patches of dense endoplasmic reticulum and mitochondria. Each muscle cell is enclosed in a sheath of connective tissue. The characteristic features of the muscle seem to be the lack of bands or striations, the parallel arrangement of the fibres (not helical as in the molluscs) and the presence of two types of myofilaments. The fibres of the basement layer merge into the interstitial fibre material (GRESSON and THREADGOLD, 1964; THREADGOLD and GALLAGHER, 1966) and these fibres become inserted into conical depressions in the connective tissue coat of the muscle so that the muscle becomes attached to the fibrous layers. At the point of insertion dense material is usually apparent. This arrangement is illustrated by the electron micrographs of the oral sucker of *Diplostomum* (Plate 1-2). Within the suckers the muscles are arranged in a radial fashion.

The basic connective tissue of the trematodes is the parenchyma and the ultrastructure of this tissue has been studied in *Fasciola hepatica* (THREADGOLD and GALLAGHER, 1966) as well as in several turbellarians. It appears that the connective tissue consists of parenchymal cells which lie in contact with the cells of other tissues and organs or may be separated by fibrous, interstitial material. A similar pattern exists in certain strigeoid trematodes studied by the author and there seems little evidence to support the concept of large fluid filled spaces described by earlier workers. The parenchymal cell is large and polymorphic with a finely granular cytoplasm containing a nucleus, numerous mitochondria and relatively sparse endoplasmic reticulum.

The cytoplasm is generally rich in glycogen but rarely contains lipid droplets (THREADGOLD and GALLAGHER, 1966).

Extending between the parenchymal cells are several, fairly elaborate, organ systems. The alimentary tract is well developed and possesses a mouth, pharynx and two intestinal caeca although the system is variously modified in different trematodes. The excretory system is basically a protonephridial one consisting of flame-cells, capillaries, large collecting ducts, a bladder and an excretory pore. The pore may be double and anterior or single and posterior. The system is extensively modified in different trematodes, and the reader is referred to DAWES (1946 and 1953), HYMAN (1951) and GRASSÉ (1961) for further details. In some digenean trematodes a lymphatic system exists containing fluid and cells and recent descriptions of this system in Paramphistomes have been made by TANDON (1960a,b) and LOWE (1966).

The nervous system (Fig. 7) consists of cerebral ganglia in the vicinity of the pharynx and from this extends a number of longitudinal trunks joined in some cases by transverse commissures. From this bilaterally symmetrical system arise branches which innervate the suckers and the organ systems. One of the best descriptions is still that by BETTENDORF (1897) although recent accounts have been published by REISINGER and GRAAK (1962). Cholinesterases have been recorded from the nervous system of adult *Fasciola hepatica* by HALTON (1967c). The ultrastructure of the nervous system has been described for the cercarial stage of *F.hepatica* (DIXON and MERCER, 1965) and these authors, as well as UDE (1962) and ROHDE (1965), suggest the presence of a neurosecretory system in the digeneans. The publication by MORITA and BEST (1965) and that of other authors reviewed by them indicate the presence of a neurosecretory system in the turbellarians and it seems likely that many of their comments will apply to the trematodes. Several sense organs have been described in ultrastructural studies and these are considered in more detail in chapter 9. There seems no justification for perpetuating the idea implying a degeneration of sensory and nervous systems in the trematodes.

The majority of trematodes are hermaphrodite and possess a fairly elaborate reproductive system (Fig. 1). The ovary is usually single and there may be one to many testes. As in all platyhelminthes the yolk is produced by a separate structure the vitelline gland. The uterus is long and accommodates a large number of eggs. The female system may possess a single or a double copulatory canal and the male system terminates in a cirrus. The female system possesses a characteristic chamber—the ootype which has associated with it a distinct gland, usually referred to as Mehlis' gland, and the entire structure has considerable significance in the assemblage of the ovum and yolk cells and also in the formation of the egg shell. In some trematodes e.g. *Schistosoma* and *Didymozoidae* the sexes are separate or exhibit protandry. The variation in the structure of the reproductive system is considerable and for further details the reader is referred to BYCHOWSKY (1957) and to the general texts mentioned earlier. Ultrastructural observations on the flagellum of the spermatozoa of the digeneans *Haematoloechus* (SHAPIRO *et al*, 1961), *Fasciola* (GRESSON *et al*, 1961), and *Gorgodera amplicava* (HERSHENOV *et al*, 1966) have shown that it consists of a double filament in its central and proximal regions and that each filament contains nine peripheral fibres encircling a central single one. The pattern of the axial filament complex is therefore nine plus one. In addition to the axial filament, a number of single micro-tubules lining the sheath are present. This arrangement has also been observed in *Cyathocotyle bushiensis* (see Plate 2-3). The nature of the digenean spermatozoa and the process of spermatogenesis has been reviewed by FRANZEN (1956) and HENDELBERG (1962) and the similarities existing between these structures and processes of the Digenea and Cestoda has been commented on by RYBICKA (1966). In certain monogeneans (LLEWELLYN and EUZET (1964);

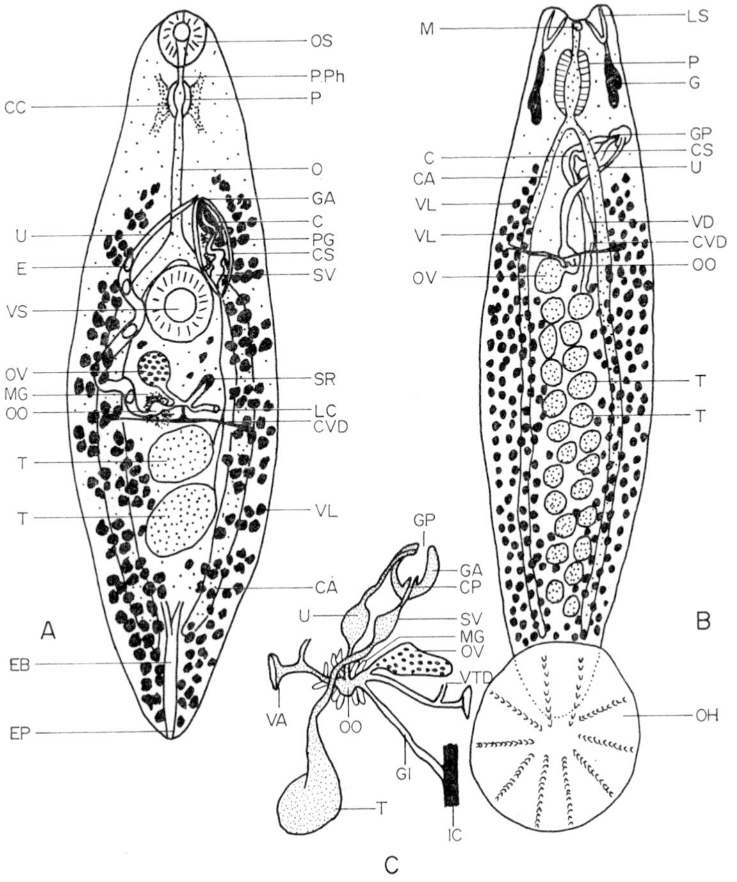

FIG. 1 The generalized structure of monogenean and digenean trematodes. **A** Digenean adult; **B** Monogenean adult; **C** schematized representation of the monogenean reproductive system. (C: cirrus; CA: caecum; CC: cerebral complex; CP: cirrus papilla; CS: cirrus sac; CVD: common vitelline duct; E: eggs; EB: excretory bladder; EP: excretory pore; G: glands; GA: genital atrium; GI: gastrointestinal canal; GP: genital pore; IC: intestinal caecum; LC: Laurer's canal; LS: lateral suckers; M: mouth; MG: Mehlis' Gland; O: oesophagus; OH: opisthaptor; OO: ootype; OS: oral sucker; OV: ovary; P: pharynx; PG: prostate glands; P.Ph: prepharynx; SR: seminal receptable; SV seminal vesicle; T: testes; U: uterus; VA: vagina; VD: vas deferens; VL: vitelline lobes; VTD: duct from vitelline gland; VS: ventral sucker.) All figures original.

PALING (1966)) a specialization in the form of spermatophore production occurs. The gross morphology of the ovary in *F.hepatica* has been studied frequently but a reappraisal of its structure was published by BJÖRKMAN and THORSELL (1964b). The ovary is surrounded by an outer capsule containing muscles. Within this is the basement membrane 80–130 mμ thick, and on the inner side occurs a continuous layer of peripheral cells. The germ cells adjacent to this layer are in intimate contact with it, with processes from the germ cells entering the peripheral layer. The authors suggested that the layer may function as a zone of nurse cells. The cytoplasm of these cells was dense, contained granular endoplasmic reticulum, mitochondria and a nucleus. The germ cells adjacent to this nurse layer were small (oogonia) and were roughly ovoid in outline. The larger germ cells (oocytes) in the centre of the ovary were polyhedral in outline and possessed microvilli facing the intercellular spaces. The cytoplasm of these germ cells was rich in ribosomes although most of them were free and not bound to a membrane system. The cytoplasm also contained dense spherical granules the nature and function of which was not apparent. The process of oogenesis as it occurs in the digeneans has been reviewed by GRESSON (1964).

CLASSIFICATION

The system of classification used in this text is the conventional one as it appears in DAWES (1946; 1953) and YAMAGUTI (1958; 1963). In this system the Trematoda constitutes a Class of the Phylum Platyhelminthes and contains three orders, the Monogenea, the Aspidogastrea and the Digenea. The distinctive features of these systematic units are given below:

Phylum platyhelminthes

Basically bilaterally symmetrical and usually dorsoventrally flattened animals. The body is acoelomate and without anus, or distinct respiratory and circulatory systems. There is a characteristic connective tissue referred to as parenchyma and a protonephridial flame-cell system. Hermaphrodite.

CLASS TURBELLARIA

Usually free-living. The body is covered with a simple cellular or syncytial epidermis and generally ciliated to some extent. The epidermis contains rhabdites which can be extruded to the exterior. The body is undivided and with intestinal caeca except in the Acoela. The life-cycles are simple and species are usually free-living in an aquatic environment although some are commensal and parasitic in their habit. Hermaphrodite.

CLASS CESTODA

Endoparasitic platyhelminthes, divided into segments and without an alimentary tract. The body is covered with a specialized epidermis in the form of a cytoplasmic tegument possessing microtriches. Anterior end specialized for attachment. Life-cycle complex involving a hooked embryo, a variety of larval stages and two or more hosts. Adult stage usually in the intestine of vertebrates. Hermaphrodite.

CLASS TREMATODA

Ecto- or endoparasitic platyhelminthes with an undivided body possessing a mouth, pharynx and intestinal caeca. The body is covered with a specialized epidermis in the form of a cytoplasmic tegument bounded on the outside by a plasma membrane not elevated into microtriches. Life-cycles simple, or complex involving one or more larval stages. Attachment structures in the form of suckers. Hermaphrodite (Fig. 1).

Order 1. Monogenea Main attachment organ at the posterior end of the body, usually consisting of adhesive structures and hooks and termed the opisthaptor (Plates 4-2 to 4-5). Oral sucker poorly developed or absent but the anterior end possessing some type of attachment

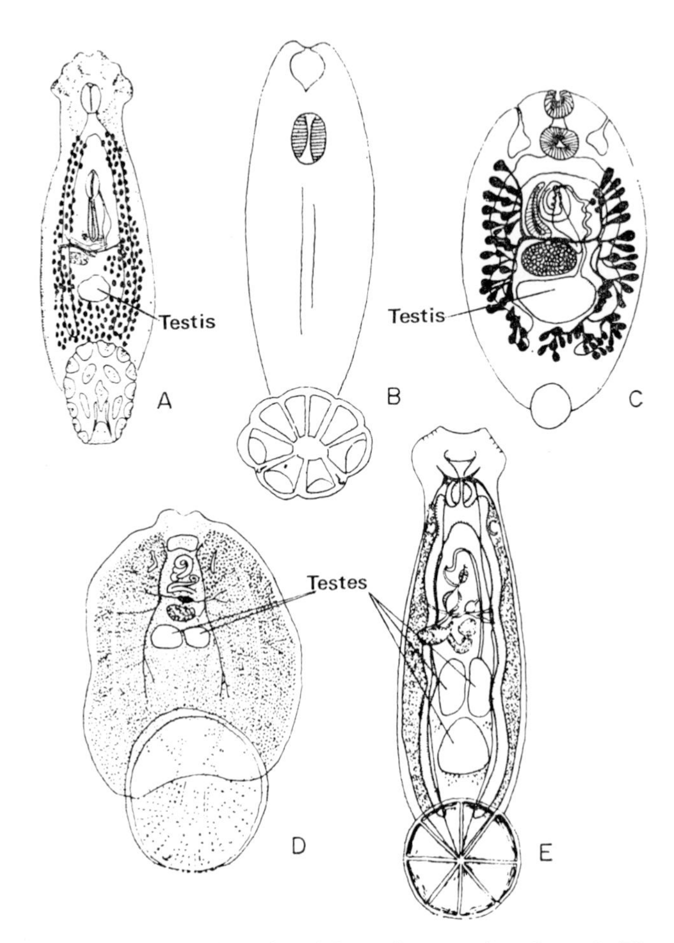

FIG. 2 Examples of monogenetic trematodes, viewed from the ventral surface. **A** *Thaumatacotyle dasybatis;* **B** *Hetercotyle pastinacae;* **C** *Leptocotyle minor;* **D** *Entobdella hippoglossi;* **E** *Monocotyle sp.* Note the relatively simple form of the opisthaptor in these Monopisthocotyleans. Redrawn from various sources from CHENG, 1964.

structure. The pores of the protonephridial system are paired, anterior and dorsal. Endo- or ectoparasitic and usually with a simple life cycle with no alternation of hosts. Hermaphrodite with one ovary and one to many testes. Uterus short, containing relatively few eggs which may possess filaments and may or may not have an operculum. Usually oviparous but sometimes viviparous. The larva is ciliated and possesses a posterior attachment disc and is described as an 'oncomiracidium'. Monogenea are usually parasitic on or in cold blooded vertebrates, sometimes on aquatic mammals, cephalopods and occasionally on parasitic crustacea from fish. (Figs. 2 and 3).

Generally divided into two suborders: 1. Monopisthocotylea · · · with the opisthaptor as a single structure. Genito-intestinal canal absent. 2. Polyopisthocotylea · · · with a complex opisthaptor consisting of several suckers or clamps. Gastro-intestinal canal present.

Order 2. Aspidogastrea With a distinctive attachment organ consisting of a large ventral sucker subdivided into loculi or of a row of suckers. Protonephridial system with a single posterior, terminal or dorsoterminal pore. Life-cycle simple with a larva possessing suckers and

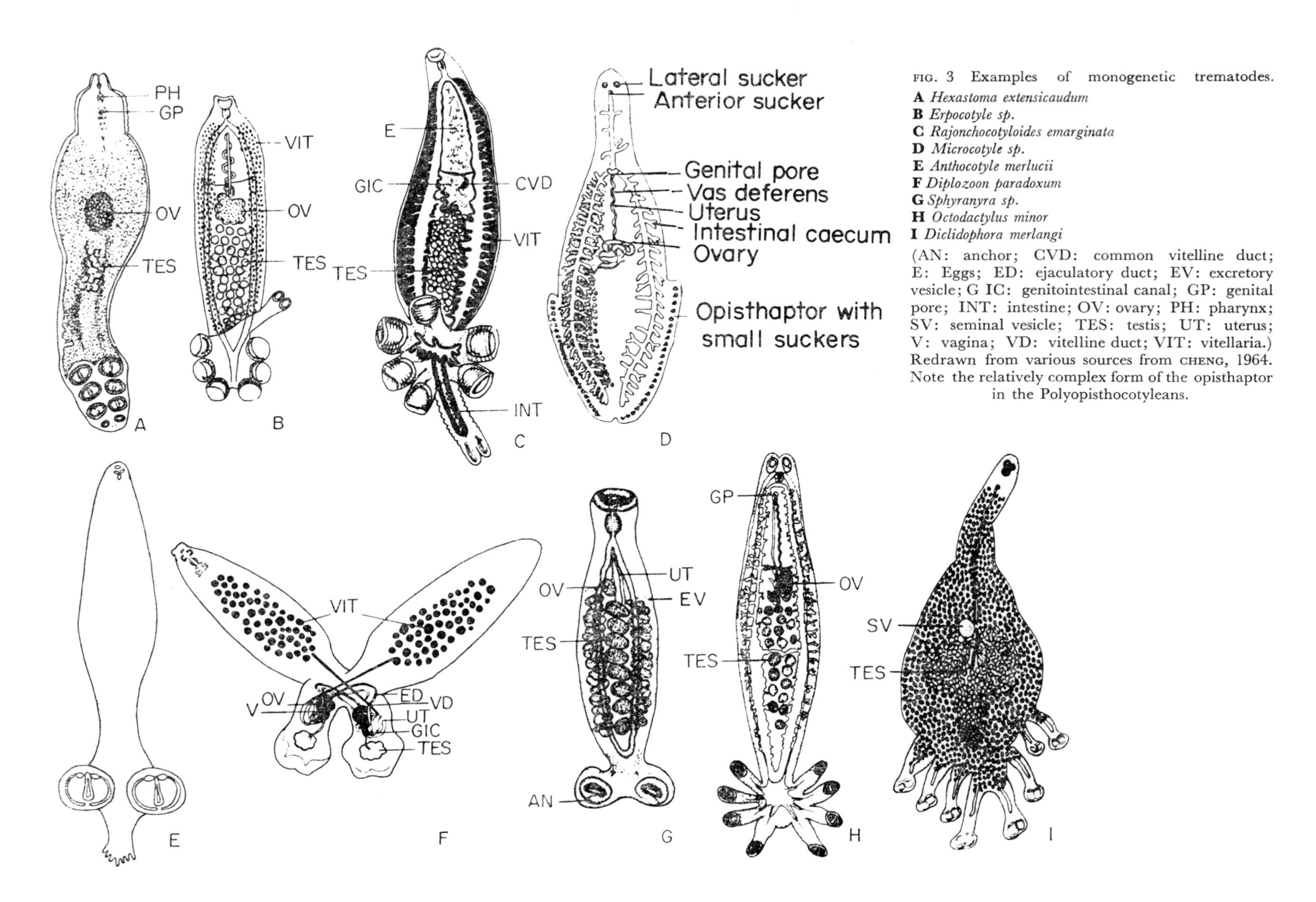

FIG. 3 Examples of monogenetic trematodes.
A *Hexastoma extensicaudum*
B *Erpocotyle sp.*
C *Rajonchocotyloides emarginata*
D *Microcotyle sp.*
E *Anthocotyle merlucii*
F *Diplozoon paradoxum*
G *Sphyranyra sp.*
H *Octodactylus minor*
I *Diclidophora merlangi*
(AN: anchor; CVD: common vitelline duct; E: Eggs; ED: ejaculatory duct; EV: excretory vesicle; G IC: genitointestinal canal; GP: genital pore; INT: intestine; OV: ovary; PH: pharynx; SV: seminal vesicle; TES: testis; UT: uterus; V: vagina; VD: vitelline duct; VIT: vitellaria.) Redrawn from various sources from CHENG, 1964. Note the relatively complex form of the opisthaptor in the Polyopisthocotyleans.

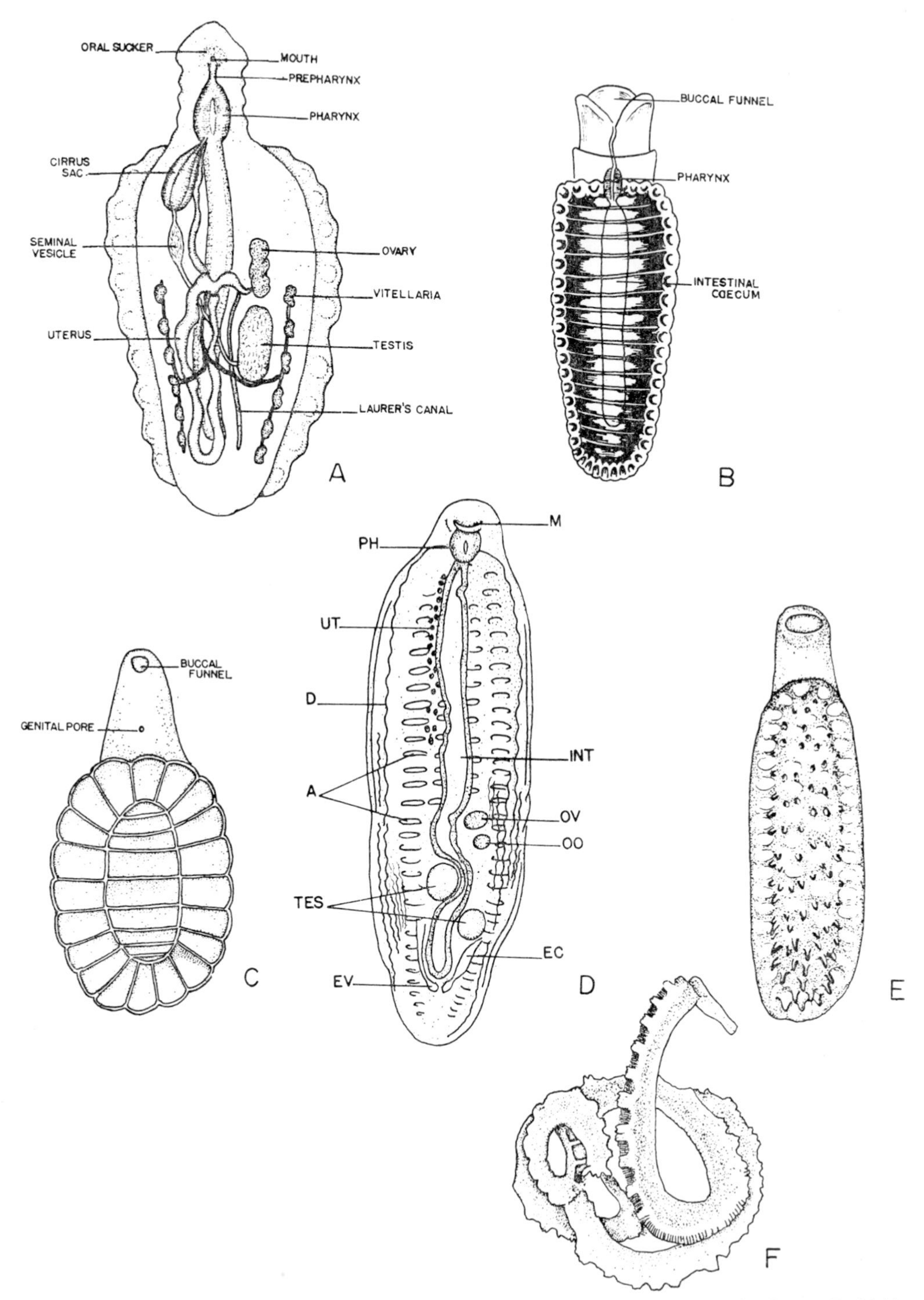

FIG. 4 Examples of Aspidogastreans. **A** *Aspidogaster conchicola;* **B** *Cotylogaster sp.;* **C** *Cotylaspis sp.;* **D** *Multicotyle purvisi* (A: Alveoli: D: Adhesive disc; EC: excretory canal; EV: excretory vesicle; INT: intestine; M: mouth; OO: ootype; OV: ovary; PH: pharynx; TES: testis; UT: uterus) E *Lophotaspis sp.*; **F** *Stichocotyle cristata*. Redrawn from various sources from CHENG, 1964.

with or without cilia. Described as a 'cotylocidium' (WOOTTON, 1966a). Generally endoparasitic in molluscs and cold blooded vertebrates (Fig. 4).

Order 3. Digenea Usually with two suckers—oral and ventral. Protonephridial pore single and posterior. Uterus long and containing many eggs, usually without filaments and operculate. Complex life-cycle with several larval stages and an alternation of hosts. Eggs hatch to produce, in most cases a ciliated larva 'miracidium' (Figs. 5 and 6).

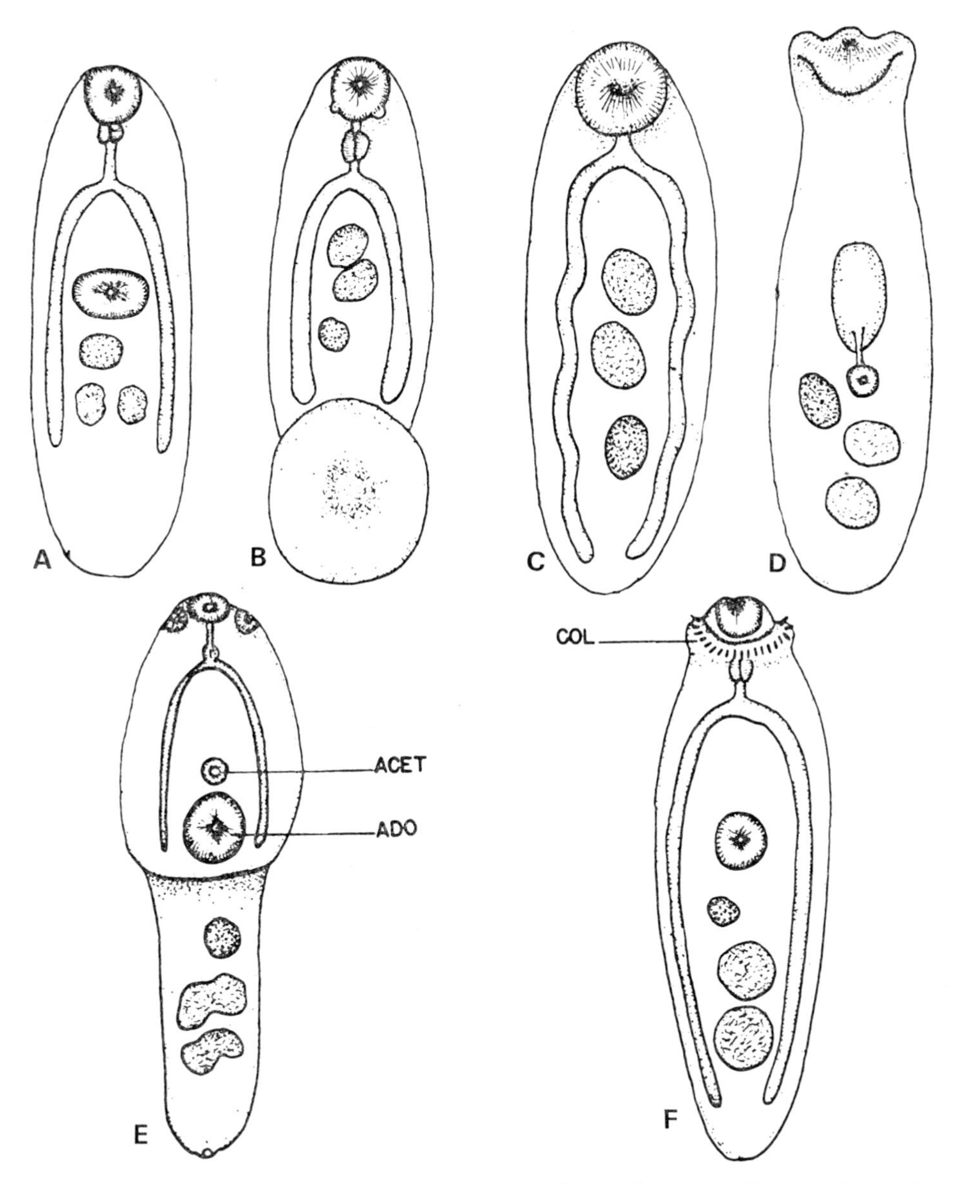

FIG. 5 Examples of the basic gross variations in body form within the digenea. **A** Distome; **B** Amphistome; **C** Monostome; **D** Gasterostome; **E** Holostome; **F** Echinostome. (Acet: acetabulum; Ado: adhesive organ; Col: collar of spines.) From CHENG, 1964.

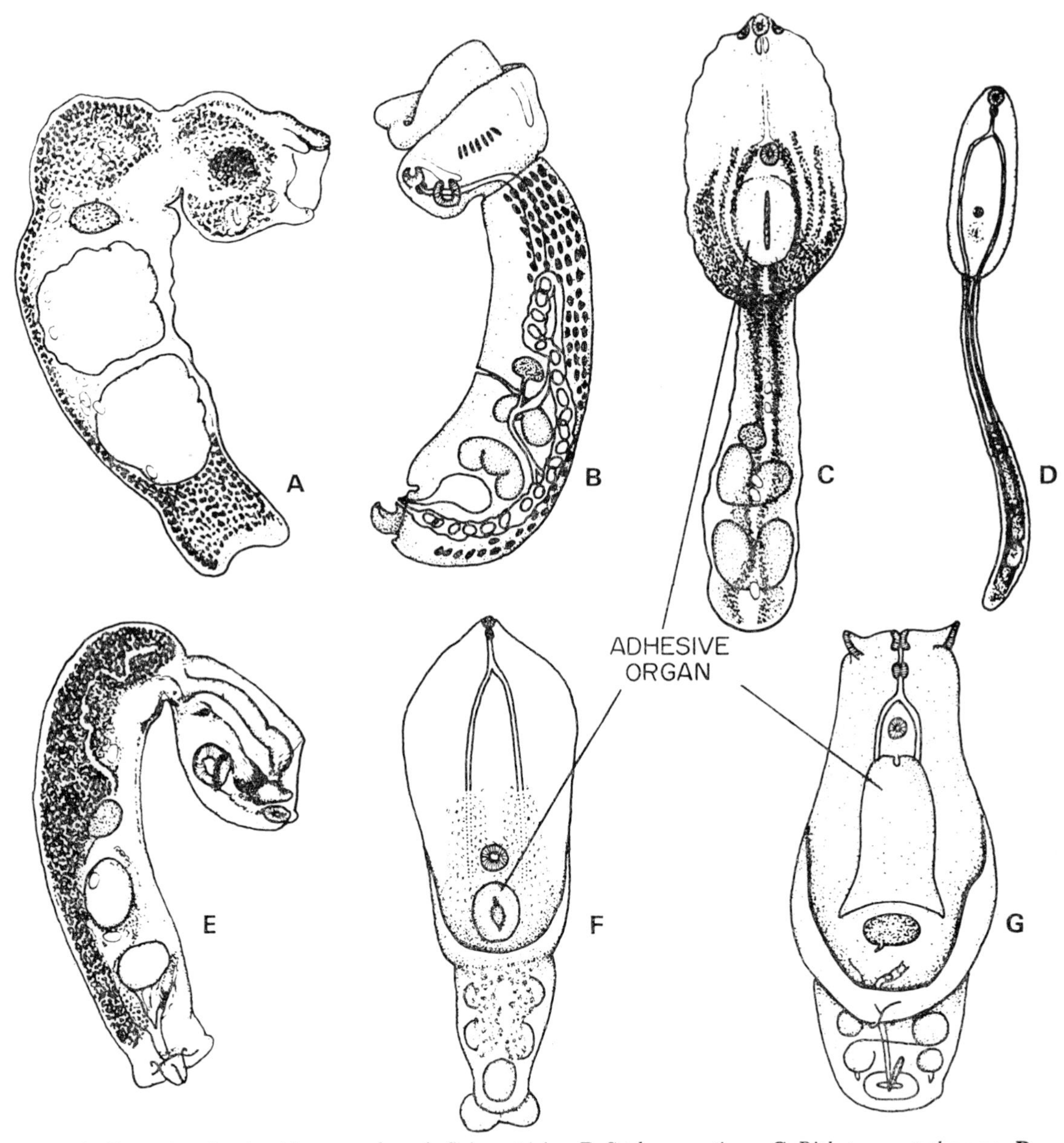

FIG. 6 Examples of strigeoid trematodes. **A** *Strigea strigis;* **B** *Cotylurus erraticus;* **C** *Diplostomum spathaceum;* **D** *Uvulifer gracilis;* **E** *Apatemon gracilis;* **F** *Neodiplostomum ochilongum.* **G** *Alaria alata.* Redrawn from various sources from CHENG, 1964.

The taxonomic treatment of the Trematoda has been frequently reassessed as new information becomes available, particularly that concerned with life-cycles and larval stages. The scheme outlined above is in general usage at present but is unsatisfactory in several ways and new approaches have been discussed and reviewed by LLEWELLYN (1965), STUNKARD (1946, 1963), LA RUE (1938, 1957), and BYCHOWSKY (1957). The main bases for disagreement are (a) the status of the Monogenea; (b) the status and position of the Aspidogastrea; (c) the basis

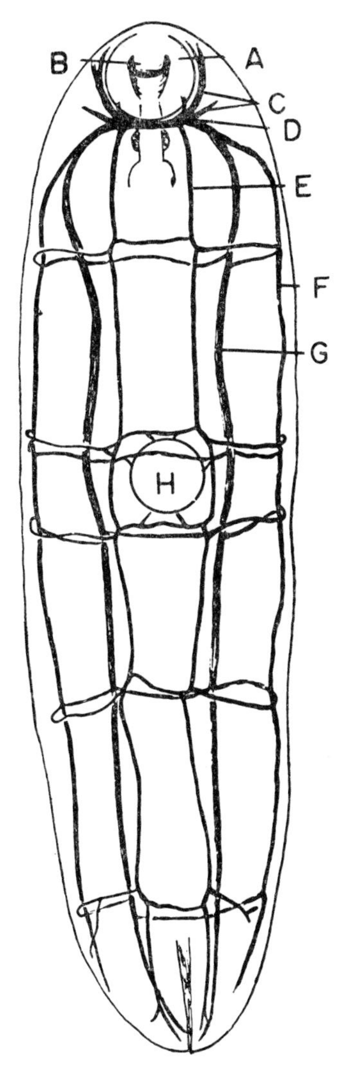

FIG. 7 Semidiagrammatic representation of the digenean nervous system of an adult parasite. (A: oral sucker; B: mouth; C: anterior nerves; D: cerebral complex; E: ventral nerve cords; F: lateral nerve cords; G: dorsal nerve cords; H: ventral sucker.) After LOOSS, 1894 from HYMAN, 1951.

for the subdivision of the Digenea. Most of the evidence which has initiated a change in ideas has taken the form of the discovery and detailed descriptions of larval stages previously little known. These main features will be discussed briefly below but for the details upon which the arguments rest the reader is referred to the reviews listed above.

(a) THE STATUS OF THE MONOGENEA Investigations over the last decade have improved considerably our knowledge of monogenean larvae and their development. This information has shown that little similarity exists between the monogenean larva and the digenean miracidium and aspidobothridian cotylocidium but in contrast the greatest resemblance exists between the decacanth lycophore of the gyrocotylideans and amphilinideans and the hexacanth or coracidium of cestodes and the monogenean larva (BYCHOWSKY, 1957; LLEWELLYN, 1963,

1965). The differences between the monogeneans and digeneans at larval and adult levels (described earlier in the classification section on pages 5, 9) as well as the 'monogenetic' nature of the life-cycle in the former, has prompted these authors to propose that the monogeneans should be separated from the digeneans and given the status of a class—the Class Monogenoidea BYCHOWSKY, 1957. This classification has been adopted by BAER and EUZET (1961). The Class Trematoda is then represented by three sub-classes—the Sub-class Aspidogastrea FAUST and TANG, 1935, the Sub-Class Digenea Van Beneden, 1858 and the Sub-Class Didymozoides BAER and JOYEUX, 1961.

This rearrangement has been criticized by STUNKARD (1963) who has opposed the removal of the monogeneans from the Trematoda. He states that not all the monogeneans are 'monogenetic' and gives as exceptions the 'viviparous' species of *Gyrodactylus* and the complicated development of *Polystoma spp.* in which a type of alternation of generations exists. LLEWELLYN (1965) states however, that the assessment of monogenean-digenean affinities should not be based on such isolated and specialized occurrences and supports Bychowsky in the differentiation of this group from the Digenea, and from the Trematoda.

(b) THE STATUS OF THE ASPIDOGASTREA The appreciation of the true status of this group is rendered particularly difficult because of the lack of information concerning larvae and development. The adult morphology exhibits some quite distinctive features which separates this group from both the monogeneans and the digeneans. The life-cycle is closely linked with molluscs in which the adult stage occurs, although this is capable of survival in reptiles and fish if ingested. STUNKARD (1963) describes them as having strong affinities with the digeneans and considers them an aberrant group of parasites of molluscs and of lower vertebrates which feed on such molluscs and which have not acquired the feature of asexual multiplication in the life-cycle. LA RUE (1957) regards them as quite distinct and places them in the Class Aspidogastrea. BAER and JOYEUX (1961) have associated this group with the digeneans as a sub-class as mentioned above and STUNKARD (1963) has also linked them with the digeneans as can be seen from his scheme given below:

CLASS TREMATODA

Subclass Pectobothridia BURMEISTER, 1856 (= Monogenea)

Firm hard suckers, generally ectoparasitic on aquatic vertebrates and monogenetic.

Subclass Malacobothria BURMEISTER, 1856.

Soft, flexible suckers, generally endoparasitic in invertebrates and vertebrates; begin life-cycles in molluscs; with 1, 2, 3 or 4 hosts.

Order Aspidobothrea BURMEISTER, 1856 (= Aspidogastrea FAUST and TANG, 1936)

Order Digenea VAN BENEDEN, 1858.

It is obvious that opinion regarding the major systematic units is in a state of flux and LLEWELLYN (1965) raises the point as to whether or not there is now any justification in retaining the term 'Trematoda' solely to link two groups as diverse as the Aspidogastrea and Digenea. In spite of the divergence between these groups as implied in the proposals described above, most authors agree that all have a common ancestry in a primitive type of rhabdocoel turbellarian stock.

(c) SUBDIVISION OF THE DIGENEA The general sequence of events which was associated with the progression of ideas concerning the major systematic units applies also to digenean classification. Originally the morphology of the adult stage was the sole basis for classification but as details of life-cycles and larval stages appeared a new basis for assessment was obtained. The complexity of the life-cycle within the Digenea and the great variation in developmental patterns and larval structures has confused the picture. The existence of an organ system which persists through the entire life-cycle and which alters sequentially in a precise way has been exploited by different workers. This system—the 'protonephridial' or 'excretory system' has been used by LA RUE (1957) in the most recent subdivision of the Digenea, as well as incorporating as much life-history data as was possible. In this scheme the Subclass Digenea is divided into two superorders based mainly on the mode of development of the excretory bladder and its structure.

Superorder Anepitheliocystidia LA RUE. Primitive bladder retained i.e. not replaced by cells from mesoderm, hence definitive excretory bladder not epithelial; Cercariae with forked or single tails; caudal excretory vessels present in developing cercariae (except perhaps in certain species of Renicolidae); stylet always absent. *Superorder Epitheliocystidia* LA RUE. Primitive excretory bladder surrounded by, and then replaced by a layer of cells derived from mesoderm, hence definitive bladder thick-walled and epithelial: Cercarial tail single, reduced in size, or lacking; miracidium with one pair of flame-cells.

Further details of this classification are given in the Appendix. The concept of a non-epithelial, non-cellular excretory bladder is one which might have to be revised in the light of current ultrastructural studies. Included in this category by La Rue is the Strigeoidea and studies (ERASMUS, 1967b and page 222 in this text) have indicated already that the reserve bladder system in this group of trematodes is definitely cellular in a number of genera. It seems probable that it is the developmental aspect involving the incorporation of additional cells which is the more significant feature of the classification submitted by LA RUE (1957).

A difficulty which has concerned many systematists is the appropriate systematic position for the didymozoid trematodes. The group is a small one and unfortunately little information is available on life-history and larval forms. BAER and JOYEUX (1961) have isolated them into a distinct subclass but this has been criticized by STUNKARD (1963) who retains them within the Digenea. The miracidium, where known, is non-ciliated and has a spinose anterior tip which suggests a link with hemiuroid trematodes. LA RUE (1957) has included this family in the Superfamily Hemiuroidea FAUST, 1929 along with families such as Hemiuridea and Halipegidea. Until further information is available this seems the most reasonable step to take.

THE LIFE-CYCLE

The considerable amount of information which is now available seems to support the basic divergence in life-histories evident between the monogeneans and the digeneans. Although the majority of the monogeneans exhibit the typical 'monogenetic' cycle, a few exceptions do occur as mentioned earlier.

The digeneans exhibit many variations of life-cycle pattern and the evolution of these cycles has been discussed by HEYNEMAN (1960). The typical digenetic cycle consists of a molluscan primary host in which multiplication occurs, an intermediate transport host, and a vertebrate final host (Fig. 8).

Infection of the molluscan primary host results from the penetration of the mollusc by a free-swimming miracidium or by the ingestion and subsequent hatching of eggs containing a fully-developed miracidium. Within the snail there develop mother and then daughter

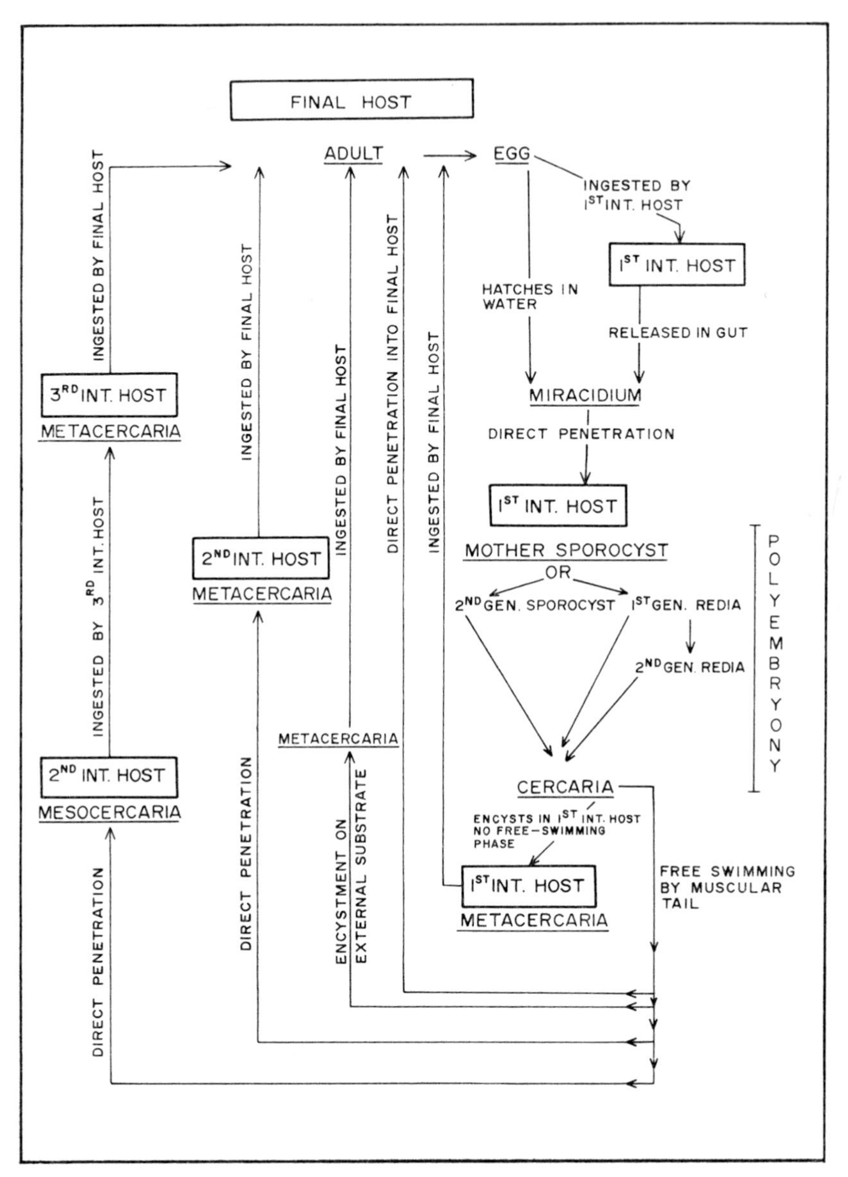

FIG. 8 Diagram illustrating the various types of digenean life-cycles.

sporocyst generations. In other life-cycles, the mother sporocyst gives rise to rediae which themselves are able to produce a generation of daughter rediae under certain circumstances. Daughter sporocysts and rediae appear to be mutually exclusive and both do not occur within the same cycle. These generations usually inhabit the digestive gland of the mollusc and exhibit polyembryony. The next stage—the cercaria—may arise from daughter sporocyst or rediae depending on the cycle. The cercaria is a disseminating and distributive phase and possesses specializations such as gland cells for penetration or encystment, eyespots and tactile sense organs as well as a muscular tail used in swimming. There is great variation in the morphology of this stage and this is discussed in a later chapter. In the majority of cycles, the cercaria gives rise to the metacercaria which usually represents the cercarial body with the addition of the progenitors of the adult reproductive system to a greater or lesser extent, as well as specialized features such as the adhesive organ characteristic of the strigeoid adult. The metacercariae are usually encysted and may lie free on vegetation or other external substrates or may be parasitic within a second intermediate host. The life-span of the cercarial stage is brief and the metacercarial stage may be regarded as a resistant stage which permits the infectivity of the life-cycle to be extended over a longer period of time. The life-cycle is completed in most cases by the development of the metacercaria into the adult, after the ingestion of the infested vegetation or the second intermediate host by the final host. The life-cycle may be considered as a sequence of phases exhibiting different specializations. Mobility, host detection, and penetration are the characteristics of the miracidial and cercarial stages, multiplication is a feature of the intramolluscan phases, and, at the metacercarial stage, resistance to adverse agents, prolonged survival with the pre-development of adult structures occur. The adult becomes specialized for maintaining its position in a microhabitat which provides favourable physiological and immunological conditions as well as permitting the passage of eggs to the exterior. The reproductive system is highly developed for the production of large numbers of eggs which are capable of surviving the passage to the exterior through the organ systems of the host and although very susceptible to desiccation, sufficiently resistant to adverse agents to protect the developing embryo. Development is associated with the differentiation of somatic and germinal cells with the latter persisting through the life-cycle from one phase to the next. This is considered in more detail in the next chapter.

The most complex variation on this life-cycle pattern appears in certain strigeoid trematodes e.g. *Alaria spp.* in which four hosts are present and an additional post-cercarial and pre-metacercarial stage—the mesocercaria. The extra host in this cycle functions as a transfer host. The life-cycle is less complex in other digeneans in that only two hosts are involved, and this may result from two possible modifications of the common pattern. The metacercarial stage may be missing so that the cercaria penetrates directly the final host and then develops into the adult. This cycle is seen in the schistosomes and is regarded by some authors as indicating the precocious maturation of a larva in its intermediate host which thus becomes the final host in the cycle. The cycle may become reduced to two hosts also by the loss of the intermediate host, and this sequence is exhibited in cycles where the cercariae encyst on vegetation or other external surfaces, or within the primary host, after emergence from the prior generation or without emergence and within its own sporocysts. The ultimate specialization is where the one host life-cycle appears. In these examples the redia produces progenetic metacercariae which mature and produce viable eggs capable of infecting other snails. This type of cycle involving a hemiurid digenean was described by SZIDAT (1956) for *Genarchella genarchella.* Although this species matures normally in the fish *Salminus maxillosus*, progenesis and maturation can take place in the molluscan intermediate host *Littoridina australis.* A one host cycle involving progenesis has also been described for a monorchiid *Asymphylodora progenetica* by SERKOVA and BYCHOWSKY (1940).

Progenesis (i.e. advanced development of genitalia in a larval form) has been described on several occasions at the metacercarial stage and it generally occurs within the second intermediate host. Examples frequently given in the literature are: *Pleurogenes medians* metacercaria in *Agricon sp.*; *Coitocaecum anaspides* in the shrimp *Anaspides tasmaniae*; *Asymphylodora progeneticum* metacercaria in *Bithynia tentaculata*; *Phyllodistomum lesteri* metacercaria in fresh-water shrimps; *Proctoeces subtenuis* in *Scrobicularia* and in the case of *Proterometra sagittaria* eggs appear within the uterus in the cercarial stage. The appearance of miracidia in cyathocotylid sporocysts has been described by SEWELL (1922) and PREMVATI (1955). These examples and their significance have been discussed by DAWES (1946, 53), LA RUE (1951), HEYNEMAN (1960) and BAER and JOYEUX (1961).

The digenetic cycle therefore exhibits considerable plasticity not only in its general pattern but also in the sequence and extent of development of particular stages. Such plasticity is an obvious advantage in the evolution of the group, and modifications are not confined to the latter stages of the cycle as illustrated by the occurrence of progenesis but may also occur at the miracidial stage. In a few species, occurring in different families, the miracidial stage contains a fully developed redia which is released after invasion of the first intermediate host. This specialization has been observed in the Echinostomatidae (*Parorchis acanthus*), the Cyclocoelidae

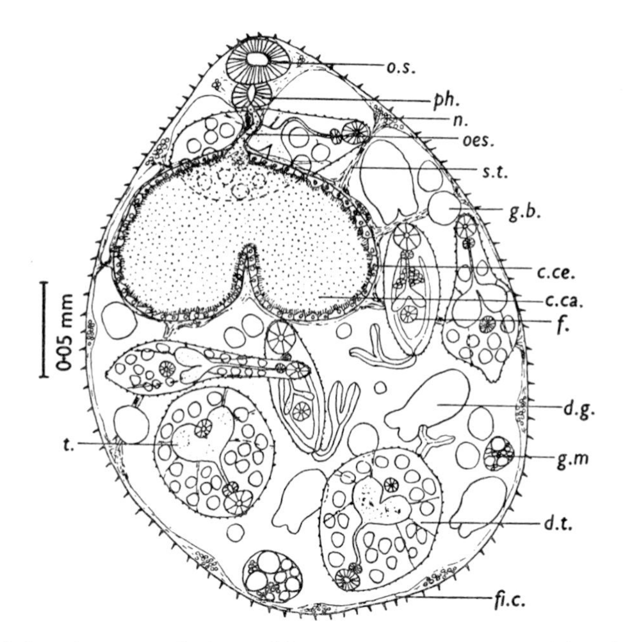

FIG. 9 Ventral view of the primary germinal sac of *Parvatrema homoeotecnum* containing daughter germinal sacs in various stages of development. (c.ca: caecal cavity; c.ce: caecal cells with microvilli; d.g.: developing daughter germinal sac; d.t.: developing daughter germinal sac with degenerating tail; f: furcocercous stage of developing daughter germinal sac; fi.c.: fibrous protoplasmic layer; g.b.: germinal ball; g.m.: germinal mass; oes: oesophagus; o.s: oral sucker; n: group of dividing cells; ph: pharynx; s.t.: tissue strand supporting caecum; t: earliest tailless stage of daughter germinal sac.) From JAMES, 1964.

(*Typhlocoelium cymbium*) and the Paramphistomidae (*Stichorchis subtriquetus*). A very unusual divergence from the general pattern has been described in the life cycle of *Parvatrema homoeotecnum* JAMES (1964). In this life-cycle the usual sequence of stages having the morphology of the typical sporocyst and redia cannot be clearly defined and are replaced by 'germinal sacs' which resemble the cercaria-metacercaria-adult generations. The first stage described by James, from the haemocoelic spaces of the digestive gland and gonad of *Littorina saxatilis tenebrosa*, is referred to as the primary germinal sac (Fig. 9) and is enclosed in a thin, 'cuticular' envelope associated with an inner protoplasmic layer. This outer covering is thought by James to represent the miracidium-sporocyst generation of this cycle. Within this covering lies the oval 'primary germinal sac' with its own cuticle and with an oral sucker, pharynx, oesophagus and two caeca as well as a ventral sucker. The excretory vesicle or bladder is present and associated flame cells. The primary sacs eventually produce germinal balls which come to lie in the parenchyma. The germinal sac becomes distended and ruptures the enclosing cuticular sheath so that the primary germinal sacs come to lie free in the tissues of the snail. The largest germinal sacs are spherical and approx. 0·16–0·5 mm in diameter and may contain five to twenty daughter germinal sacs.

The germinal balls from which the daughter sacs arise first develop into a tailed stage, resembling the furcocercous cercaria which develops later in the cycle, but differs in that it is unable to swim. The tail has a tail stem and furcae and the oval body possesses mouth, oral sucker, pharynx and caeca. The excretory system is well developed. The body of this daughter germinal sac increases in size and contains germinal balls. At this stage the tail degenerates so that a tailless, almost spherical daughter germinal sac results (Fig. 10). The germinal masses within the daughter sac, develop to form furcocercous cercariae which have a mobile body and a contractile tail so that they are able to swim feebly if released into sea water. The cercaria is unusual in not possessing papillae nor setae, eyespots nor penetration gland cells. In view of this it is surprising to read that the cercarial tail degenerates while still within the daughter sac but the body persists and changes into the metacercarial stage. Although this metacercaria does not increase in size to any great extent, which is probably related to its retention within the daughter sac, considerable organogenesis occurs so that the fully differentiated metacercaria contains cephalic glands, a well developed gut and excretory system and a reproductive system which reaches an advanced state of development. The genital pore, vesicula seminalis, testes, ovary, oviduct, uterus and vitelline glands are all present so that the metacercaria is infective to the final host when the parasitized *Littorina* are ingested. The adult parasite occurs in the intestine and rectum of the oystercatcher, *Haematopus ostralegus occidentalis* Neumann.

Such an ususual cycle as this needs some explanation which would relate it to the more typical digenetic cycle. JAMES (1964) submits two interpretations. The closely related species in this genus possess the typical sporocyst stage and follows the usual digenean sequence of development. James suggests that the characteristic asexual reproductive phase has been transferred in this unusual species to the cercaria-metacercaria generation so that the miracidium-mother and daughter sporocyst sequence has become suppressed, and the early primary germinal sac becomes homologous with the metacercarial stage of the related *P. borinqueñae* and *P. borealis*. The alternative proposal is that the germinal sacs could be regarded as representing the reappearance of an ancestral condition existing prior to the evolution of rediae and sporocysts. The fact that reproduction occurs within the mollusc in a stage with the characters of the cercaria-metacercaria-adult phase does seem to support the concept proposed by some (e.g. BAER and JOYEUX, 1961) that the ancestral digenean once present in the molluscan host resembles the cercaria-metacercaria-adult generation of present day forms. The homologous nature of all these stages is also supported by examples quoted by James where sporocyst stages have features of later generations i.e. the sporocyst of *Cercaria caribbea XLV* LE ZOTTE (1954) which has a forked tail and the sporocyst of *Cercaria lintoni* MILLER (1926b) which has an oral sucker. The suggestion that all generations in the life-cycle are fundamentally the same has been discussed in some detail by CABLE (1965), who also suggests the need for a reappraisal of existing concepts of the cycle as a whole as well as a reassessment of the type of reproduction occurring in germinal sacs. A review of the varied interpretations of the digenetic life-cycle has been made by JAMES and BOWERS (1967c). These authors suggest that true polyembryony i.e. larval multiplication does not generally occur in the Digenea and that the process of multiplication should be referred to as apomictic parthenogenesis. They suggest that the germinal sacs have evolved from sexual

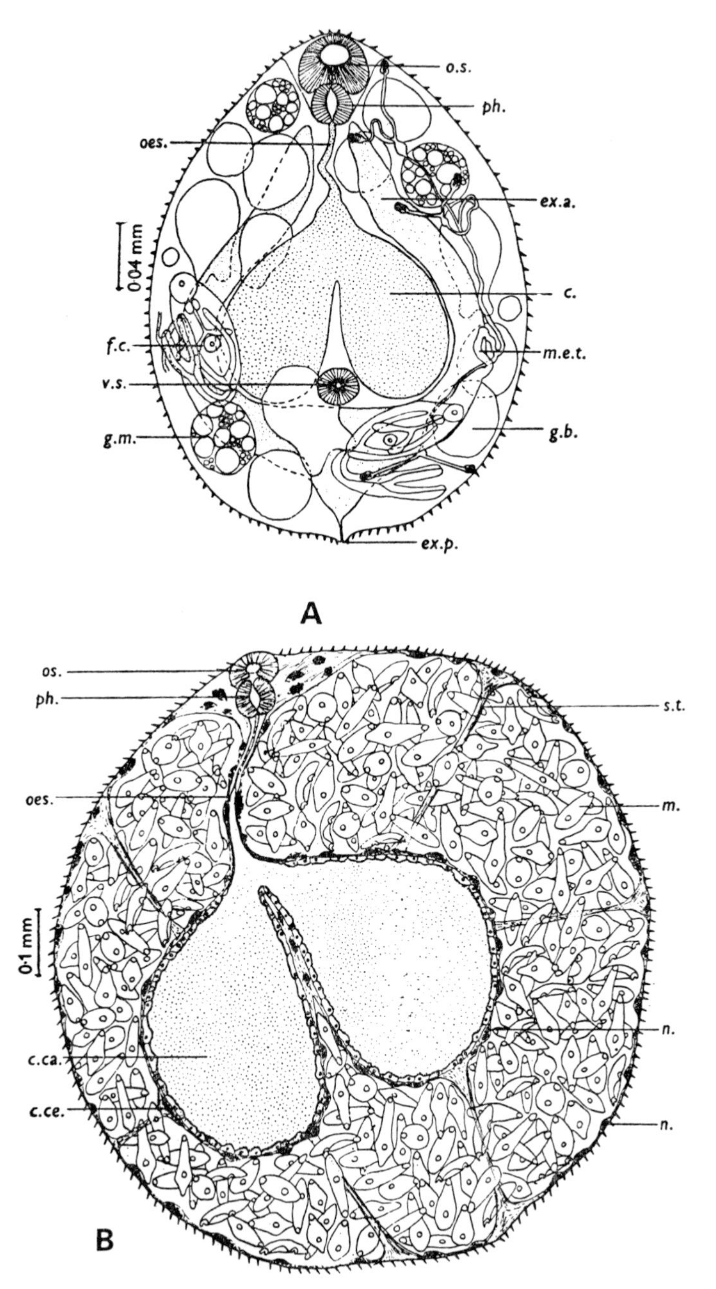

FIG. 10 **A** Daughter germinal sac of *P. homoeotecnum*; young specimen containing developing cercariae, ventral view. **B** Ventral view of fully formed daughter germinal sac containing metacercariae. (c: caecum; ex.a: arm of excretory vesicle; ex.p: excretory pore; f.c.: furcocercous cercaria; m: metacercaria: met: main excretory tubule; v.s: ventral sucker; Other abbreviations as in Fig. 9). From JAMES, 1964.

ancestors and that the mode of reproduction is parthenogenesis without meiosis. It is suggested therefore that the miracidium is a modified relic of the original rhabdoceole-like ancestor and on becoming parasitic, sexual reproduction was replaced by apomictic parthenogenesis. The sporocyst-redial phases could therefore be regarded as paedogenetic embryos in which development of the soma is retarded, associated with a precocious division and development of the germinal cells. Apomictic parthenogenesis represents the primitive condition but in certain circumstances true larval multiplication may occur secondarily. Thus germinal continuity is preserved throughout the life-cycles which could be regarded as a cyclical alternation of three or more homologous generations.

The idea of an ancestral adult generation parasitic in snails is criticised by HEYNEMAN (1960) who favours an evolutionary sequence from larval forms in a specific snail with free-living adult worms to a phase containing worms maturing in the gut of a vertebrate host resulting from the ingestion of overwintering encysted metacercariae.

In addition to these major variations on the digenetic pattern, developmental abnormalities also appear in cycles progressing along well established sequences. Twinning in miracidia rediae and cercariae have been described, as well as the appearance of trifurcate forms, individuals with bifurcate tails with furcae on each branch, and doubling of ventral sucker and stylets in the cercarial stage. In the adult, monorchism and lack of the ovary have been described. The reader is referred to the review of KUNTZ (1948) and the paper by SCHELL (1960) for details of some of these abnormalities.

Information on the chromosome numbers of adult trematodes has been reviewed by WALTON (1959) and their role in aiding taxonomy discussed. The presence of heterochromosomes seems to be restricted to the blood flukes where the heterogamic condition is related to the female of the species. In a general consideration of the digenetic cycle, GINETZINSKAJA (1965) describes the cycle as representing an alternation between hermaphrodite and parthenogenetic generations, the latter being phylogenetically older and showing more regression in morphology and physiology. Both generations are diploid. The significance of the hermaphrodite nature of the adult has been discussed briefly by SMYTH (1966a) who suggests that as mutant genes could appear simultaneously in both male and female germ cells homozygous double recessives may appear as a result of self-fertilization. In this way 'instant mutants' could appear.

The maturation and growth rates of adult trematodes are aspects of biology which have received little extensive study. A major contribution to the study of maturation has been the introduction by BELL and SMYTH (1958) of defined stages in the development of strigeoid trematodes and this pattern of terminology has been adopted in the study of other trematodes. SMYTH (1966b) lists seven stages: cell multiplication; body shaping; organogeny; early gametogeny; late gametogeny; egg-shell formation and vitellogenisis; oviposition. These stages can be readily determined by a combination of simple histochemistry, cell squashes with chromosome counts and morphological observations.

The rate of development varies considerably depending on the parasite species and host studied. Strigeoid trematodes such as *Diplostomum phoxini* and *Apatemon gracilis minor* reach maturity and become egglaying within 3 to 5 days after infection of the final host. In contrast, *Fasciola hepatica* takes considerably longer to reach the same stage—5–6 weeks in the mouse and 10–14 weeks in the sheep (PANTELOURIS, 1965b). Little precise data is available for the Monogenea and with these parasites the rate of development after the oncomiracidium is markedly affected by environmental temperature and may become seasonal in its occurrence. In most experimental studies growth has been demonstrated by linear measurements and not weight.

The accurate measurement of an adult digenean is difficult to accomplish because of the absence of a rigid skeleton and the continual changes in shape which occur so that mean figures of length and breadth have a value which is difficult to assess. In general there is, in digeneans, a brief initial lag followed by a rapid increase in size (as indicated by length and breadth) which gradually falls off after the egglaying stage is reached although increase in size may continue slowly for some time. SENGER (1954) found that although egglaying occurred in *Echinostoma revolutum* 6 days after infection, maximum length was not attained until after 28 days in the chick. Most workers comment on the lack of uniformity of size within a population which is of the same age. There is also variation in the time taken by different individuals within the same population to reach the various stages in development described above.

The various parts of the body of a digenean trematode do not grow at the same rate. In the strigeoid species development into the mature adult is characterized by a tremendous increase in the size of the hind-body, which contains the reproductive system, relative to that shown by the fore-body. DAWES (1962a) in his study of the growth of *F. hepatica* in mice showed that the greatest increments in size are added to length rather than to breadth, the ventral sucker increased in size to a greater extent than the oral sucker and that the major increase occurred in the body posterior to the ventral sucker.

Increase in size is also influenced by the host selected for infection. BERRIE (1960) found that in adult *Diplostomum phoxini* the relative proportion of fore- to hind-body differed between various hosts. The hind-body reached its maximum size in the herring-gull and its minimum size in the mouse. SENGER (1954) infected rats and chicks with *Echinostoma revolutum* and observed that the maximum size of parasites from the rat was 2·6 mm whereas, in parasites of the same age, in the chick the maximum size reached was 11 mm. These, and similar experiments, suggest that specific diagnoses based on measurements of parasites from experimental and possibly abnormal hosts must be treated with caution.

The ecology of the life-cycle

The brief account, given above, of the variations in the digenetic and monogenetic life cycles indicates that the establishment of a trematode cycle within a given environment is subject to considerable hazards. It is unfortunate that the recognized method of illustrating a life-cycle in the form of a diagram existing only in the plane of the page does not indicate very clearly the relationship between the trematode cycle and external ecological factors such as food chains, seasonal variations, migratory and crowding behaviour patterns on the part of the hosts and the physico-chemical characteristics of the environment. In many ways the completion of the cycle is the result of a succession of lucky accidents although the element of chance is reduced to some extent by the distinctive behaviour patterns exhibited by the trematode larval stages as well as by the various hosts. The fecundity of the parasite also provides it with a considerable advantage.

The consideration of the trematode life-cycle against the background of a complex aquatic environment with its characteristic flora and fauna and seasonal variation is a very difficult task and outstanding pioneer investigations in this field have been made by Dogiel and his associates. As an introduction to these extensive studies of the ecology of the parasites of fish and birds the reader is referred to DOGIEL (1962), BYCHOWSKY (1957), DOGIEL, PETRUSHEVSKI and POLYANSKI (1958), BAUER (1959), BYKHOVSKAJA-PAVLOVSKAJA (1964a; 1964b).

THE DIGENETIC CYCLE—LARVAL FORMS

With the exception of those stages parasitic in homiothermic animals, the entire life-cycle is considerably influenced by the temperature of the external environment. The activity and life-span of the free swimming and infective cercarial and miracidial stages, as well as their ability to emerge from the molluscan host or hatch from the egg, as the case may be, is considerably affected by temperature. The continuation of the developmental cycles within the molluscan host is also susceptible to temperature variation. Examples of this relationship will be discussed in more detail in the relevant chapters, and it is sufficient to point out that, in spite of the coincidence and availability of the appropriate hosts, the progression of the cycle may be interrupted by a drop of external temperature. In addition, the free-swimming stages are relatively feeble swimmers and the potential of their orientation patterns may be easily dissipated by an increase in the turbulence of the water. The relationship between the ability of the cercariae of *Schistosoma mansoni* to infect mice and the rate of water flow of the environment has been investigated by RADKE *et al* (1961) who found that infectivity was inversely proportional to water velocity.

In view of the fact that these infective stages (cercariae and miracidia) have to swim to contact their hosts it seems likely that some relationship must exist between the density of susceptible hosts and the level of infection exhibited by those hosts. EWERS (1964b) has examined the influence of the density of snails on the incidence of larval trematodes in *Velacumantus australis* from Botany Bay. The larval trematodes found belonged to the Heterophidae, Philophthalmidae and Schistosomatidae and the snails were considered under two categories—juvenile and adult. It was found that the incidence of trematode infection was highest (25 per cent) when the average density of snails was 15 snails per sq ft and further analysis of the data indicated that the variation observed was due almost entirely to the juvenile part of the snail population. In this case it was found that the infection rate of the low-density group (less than 20 snails per sq ft) was approximately twice that of the high density group (more than 20 snails per sq ft). These results suggest that the small number of parasites seeking hosts is the limiting factor in this population and that an excess of susceptible hosts produced a dilution of the incidence rate recorded.

The level of infection existing in the fauna associated with a particular aquatic environment, exhibits fluctuations which are related to a wide variety of factors. The fresh-water habitat is a relatively unstable environment which is markedly affected by seasonal changes. These changes have an effect on most of the fauna associated with the environment so that the parasite population is subjected to varying conditions in two ways. The parasites will be affected directly by environmental changes in temperature, turbulence and oxygen tension and indirectly by changes, also associated with alterations in the external environment, in host activity, abundance, feeding patterns, migratory and swarming behaviour. These direct and indirect effects interact in a complex manner and to this pattern must be added geographical variations in host distribution as well as variations in susceptibility to infection which might be related to age or to the genetic constitution of the strain. Because of the obvious susceptibility of the digenetic cycle to this wide range of agents, it becomes difficult to predict in detail the variety and incidence level of trematode parasites in a particular environment. In the case of *Fasciola hepatica* disease an attempt has been made to predict the appearance of a heavy 'fluckey' year based on the relationship between climatic conditions and the availability of *Lymnaea truncatula*. After a series of surveys it was suggested that when rainfall exceeded evaporation for 3 months within the period May to October in regions where good soil favoured the survival of the snail, a heavy flukey year could be predicted (OLLERENSHAW, 1958a and b; 1959; OLLERENSHAW and ROWLANDS, 1959).

The temperature during this period is suitable for the development of both snail and parasite so that humidity becomes one of the linking factors.

Certain aspects of the ecology of larval digeneans in freshwater habitats have been studied in some detail, and a number of general conclusions may be drawn.

Many workers, e.g. MANSON-BAHR and FAIRLEY, 1920; SEWELL, 1922; MILLER and NORTHUP, 1926; DUBOIS, 1929; REES, 1932; RANKIN, 1939; ROTHSCHILD, 1941a and b; SINGH, 1959; OLLERENSHAW, 1964; PROBERT, 1966b; JAMES, 1968a and b) have referred to seasonal variation in the level of infection by cercarial stages in both marine and freshwater molluscs. In general two periods of high incidence occur, one at late Spring (May) and the other in late Summer (September to October). In some cases the peaks are later and correspond to June and December. This major fluctuation may be due to a number of reasons, but in some cases and particularly in those species recorded from European countries, the peaks coincide with changes within the molluscan fauna. In British freshwaters some of the most frequently parasitized molluscs have a relatively short life-span of one or two years (HUNTER, 1961). The death of the older molluscs generally takes place after the egg laying season in May or June. In this way the initial peak of cercarial infection corresponds to the infection rate in the older molluscs and the one year old individuals infected in the previous Summer–Autumn. In July and August, mortality of the oldest individuals takes place with a subsequent drop in the level of infection. The degree by which this falls depends on the life-cycle of the particular mollusc. In those species which live for one year (e.g. *Lymnaea pereger*) there will be a considerable reduction corresponding to the complete replacement of the mollusc population by very young individuals which will either be uninfected at this time or may contain very young developing stages such as sporocysts or rediae. Other molluscs (*Lymnaea stagnalis* and *Bithynia tentaculata*) have a longer life span so that although mortality following egg deposition will occur in part of the population, there will not be a complete disappearance of adult molluscs carrying cercarial infections. In this way there will be a fall in incidence but not so marked as in annual molluscan species. The cercarial incidence in Great Britain gradually falls as winter approaches and does not rise again until the spring and the associated increase in environmental temperatures. This corresponds to the reduction of environmental temperatures below the threshold for larval multiplication so that infections recently acquired are prevented from reaching maturity and those already existing and producing cercariae will have their productivity reduced. It seems likely that trematode infections acquired in late summer are able to overwinter and complete their development with the release of cercariae in the spring as the temperature rises. Eggs deposited by adult parasites in the final host in autumn might also overwinter so that hatching and release of miracidia takes place in late spring and early summer. Thus, the release of infective miracidia and cercariae at maximum concentration corresponds to the appearance of young, uninfected molluscs of the new generation.

Seasonal variation is not so evident with metacercarial stages (PIKE, 1965) and this is correlated presumably with the relative longevity of this stage compared to that of the miracidia and cercaria. The initial level of infection in an uninfected population will be related to the degree of cercarial production within the molluscan hosts and the proximity of the primary and secondary hosts. The incidence will rise and reach a level which stays fairly constant over the winter but which rises again in the summer after exposure to a second cercarial population. The ultimate levels will be affected by the life-span of the second intermediate host particularly if it is an annual species, in the same way as was cercarial incidence. The ability of secondary hosts to resist subsequent infection by cercariae so that the metacercarial level remains fairly constant will be associated with the development of immunity. This may be the case for homiotherm hosts but seems unlikely for poikilotherms, such as molluscs, as PROBERT (1963, 1966b) has shown that some molluscan species, e.g. *Bithynia tentaculata* are able to harbour simultaneously metacercariae of several trematode species.

The maximum level of infection appearing in the molluscan fauna appears to be related to some extent to the type of freshwater habitat. WESENBURG-LUND (1934) in his study of larval digenea from Danish freshwaters found that in general the percentage infection of mollusca was higher in ponds (50–100 per cent) than in lakes (3–4 per cent). ROTHSCHILD (1941b) found the infection rate of *Peringia ulvae* in small pools to vary between 11 and 33 per cent depending upon the time of the year. REES (1932) studying infections in pools and marshy habitats records infection rates varying between 4·5–22·6 per cent and KHAN (1960a,b) found infection rates of 10–50 per cent in pools in Kent, Sussex and Essex. The low rates associated with large bodies of water have also been recorded by LARSON (1961) from Lake Itasca, Minnesota (although the rates varied from year to year (6·4 per cent in 1958 to 12·5 per cent in 1959). A general level of infection corresponding to 25–30 per cent has been indicated by DUBOIS (1929) from a variety of habitats in Switzerland, by KUPRIIANOVA-SHAKHMATOVA (1957) in Poland and by LLEWELYN (1957) from an artificial lake in S. Wales (Great Britain). In contrast, PROBERT (1966b), describing the fauna parasitizing molluscs in the small lake (Llangorse Lake) in mid-Wales, reports a general incidence of 51 per cent over a three year study. This habitat is relatively small and shallow and this must be one of the reasons for the high incidence. The relative infection rates of still and running waters was considered by DEFOREST (1957) who found that seepage ponds and lakes had the highest percentage infection and irrigation canals the lowest. WIKGREN (1956) comparing the levels of infection in molluscs from pools and the shore of regions in the Finnish Archipelago found rates varying from 40 to 58 per cent with the highest infection rate in the pools. In contrast in the largely terrestrial cycle of *Eurytrema pancreaticum* involving the land snail *Bradybaena similaris* and the grasshopper *Conocephalus maculatus* BASCH (1966) records a 6 per cent infection rate in the snail and a 3·8 per cent rate in the insect secondary intermediate host.

The figures given above represent the general rate for the molluscan population as a whole and it is not surprising to discover that the rate for different molluscan species varies considerably (Table 1) although their availability and abundance within the habitat may be similar. In the digenetic cycle the molluscan host may act as a primary host harbouring sporocyst, redial and cercarial stages or as a secondary intermediate host for metacercariae of the same or different species of trematode. It is possible that some molluscs within the habitat function mainly as primary hosts, others mainly as secondary hosts with a few occupying both roles simultaneously. EWERS (1964a) in a numerical analysis of the data available in YAMAGUTI (1958) found that in the life-cycles of 279 species of digenean trematodes the molluscan families Lymnaeidae, Physidae, Planorbidae, Hydrobiidae and Thiaridae were by far the most important hosts particularly for those species parasitizing birds and mammals. The data obtained by PROBERT (1966b) illustrates some of these concepts. He found that out of 11 molluscan species examined the most heavily parasitized were *Lymnaea pereger* and *Bithynia tentaculata* (total percentage infection 60 and 64 per cent respectively). These snails were important as primary and secondary hosts in that 25 per cent contained redia, sporocyst or cercarial stages and 39 per cent metacercariae in the case of *L. pereger* with 45 per cent and 37 per cent being the respective figures for *B. tentaculata*. In contrast to the dual role played by these gastropod species, out of three species of bivalve present in the habitat (*Sphaerium corneum*, *Unio pictorum* and *Anodonta cygnea*) only one harboured parasites and that species (*S. corneum*) functioned solely as a secondary intermediate host in this habitat and contained metacercarial stages only. In a similar manner *Planorbis corneus* contained a high level of infection (58 per cent) but this was due entirely to metacercarial stages, so that this species also was acting mainly as a secondary intermediate host. In this environment *Lymnaea stagnalis* contained a low level of infection (8 per cent) but was involved equally as a primary and secondary intermediate host.

It is apparent that the role played by individual species of the molluscan fauna will be related to the population of adult parasites harboured by the final hosts associated with that particular habitat. In some cases, habitats will have a fairly permanent fauna of final hosts, but others, particularly those in the path of migratory birds, will have a varying range of final hosts. Thus diverse digenean cycles will be introduced involving the molluscan fauna in different ways. In contrast with the observations of PROBERT (1966b) it was found by LLEWELYN (1957) that, in one of the habitats studied by her, *Planorbis corneus* was one of the most significant final hosts, and in the habitats examined by DUBOIS (1929) in Switzerland, *Lymnaea stagnalis* was the most significant mollusc. WIKGREN (1956) also found that in certain fresh water habitats in the Finnish Archipelago the majority of cercarial species were recorded from *L. stagnalis*. The relative significance of the various molluscan species in the establishment of digenean cycles in a few freshwater habitats in Europe is indicated in very general terms by Fig. 11. From this it can be seen that *Bithynia tentaculata* is capable of harbouring the greatest range of species with *Lymnaea stagnalis* or *L. pereger* coming second depending on the habitat. Other significant species are *Lymnaea palustris*, *Planorbis corneus* and *Physa fontinalis*. The bivalves seem to play a relatively small role in the establishment of digenean cycles. The importance of *Bithynia tentaculata* in this respect has also been indicated by WIŚNIEWSKI (1958a) who found it to contain the greatest number of cercarial species. This prosobranch species is remarkable in its susceptibility to molluscan infections and this is further emphasized by the fact that the related species *Bithynia leachii* was recorded as harbouring only two cercarial species by WESENBERG-LUND (1934) and WIŚNIEWSKI (1958a). Other prosobranch genera such as *Hydrobia jenkinsi* and *Valvata spp* seem to have little significance in the establishment of a cercarial fauna (WESENBERG-LUND, 1934; REES, 1932).

Although molluscs may harbour simultaneously cercarial and metacercarial stages of different species quite frequently, the incidence of multiple simultaneous infections of different cercarial species is comparatively rare. Most cercarial species are restricted to one molluscan host and this is related presumably to the host-specificity and selection mechanism exhibited by the miracidium (see chapters 2, 6). PROBERT (1966b) found that out of 21 cercarial species only two were recorded from more than one host. The xiphidiocercaria *C. microcaeca* was described from *Lymnaea pereger* and the closely related *L. auricularia*. The forked tail species *Cercaria Apatemon gracilis minor* occurred in both *Lymnaea palustris* and *L. pereger*. DUBOIS (1929) described 43 cercarial species and all, except for 12, were restricted to a single molluscan host. Although this group of twelve exhibiting a wide host range was confined to pulmonate molluscs and often to species of the same genus, others were recorded from widely differing pulmonate genera such as Lymnaea and Planorbis. This information suggests that simultaneous multiple infections will not be frequent and this is borne out by the field data. DUBOIS (1929) found that out of 2400 molluscs examined, 691 contained larval stages and of these 32 were simultaneous multiple infections. *Lymnaea stagnalis* contained 25 double infections and three treble infections and *Bithynia tentaculata* four double infections. REES (1932) after examining 5372 snails found 1420 infected and out of these only one case of double infection. The pattern of these double

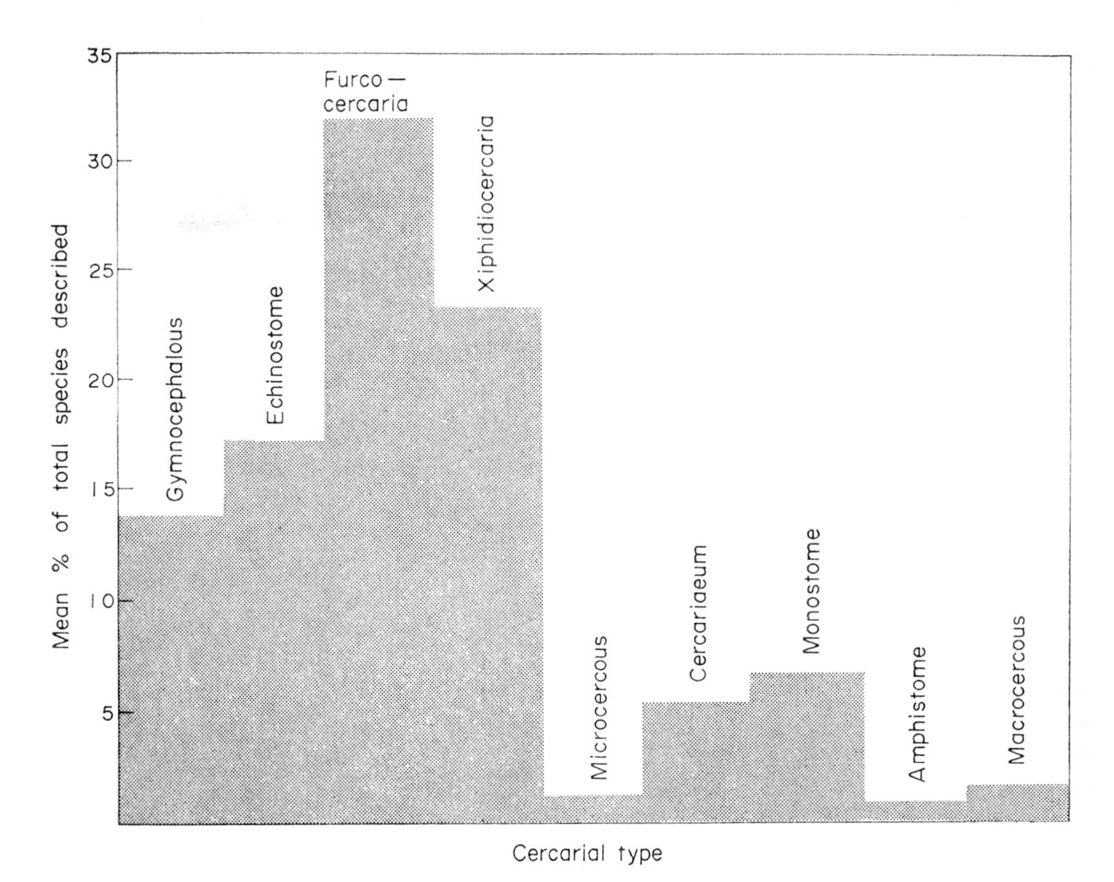

FIG. 11 The relative abundance of cercarial types recorded from molluscs of fresh-water habitats in Europe. The figure is based on data from WESENBURG-LUND, 1934; DUBOIS, 1929; LLEWELYN, 1957; REES, 1932; PROBERT, 1966b; WIKREN, 1956.

infections is as follows: Combinations of Xiphidiocercariae with other xiphidio species or with furcocercariae, echinostome, cercariae, monostome or gymnocephalous have been described by DUBOIS and REES (loc. cit.) as well as by CORT (1915), FAUST (1917) and SEWELL (1922). Other combinations include echinostome cercariae with furcocercariae and monostome with monostome or with furcocercariae. The morphological characteristics of these various cercarial types are given in the chapter on the Cercaria. It is apparent from this that most combinations seem to be possible although double echinostome infections have not been reported. The relative infrequency of these multiple infections is attributed either to the specificity of the miracidial stage or to the development of immunity subsequent to the establishment of the first infection. The idea that competition between two cercarial species might exist within the snail has been developed by LIE (1966a), LIE *et al* (1965, 66), BASCH *et al* (1966a,b) and they suggest that active foraging by the echinostome rediae within the host tissues may be the reason for the inability of other species to compete with and survive associated echinostome infections. These aspects are discussed in more detail in chapter 8.

The variation in morphology exhibited by the cercarial stage fall into distinctive patterns which have been defined as cercarial types (i.e. Monostome, xiphidiocercaria, gymnocephalous echinostome, furcocercaria, cercariae, and microcercous forms) and to some extent these larval

Table 1 The number of cercarial 'species' recorded from certain fresh-water molluscs.*

Mollusc	Number of cercarial 'species'
Gastropoda	
Pulmonata	
Lymnaea stagnalis	29
,, *palustris*	17
,, *truncatula*	3
,, *ovata*	7
,, *auricularia*	6
,, *pereger*	20
Physa fontinalis	1
Planorbis corneus	10
,, *carinatus*	4
,, *vortex*	1
,, *planorbis*	2
Prosobranchia	
Bithynia tentaculata	51
Valvata piscinalis	4
Bivalvia	
Sphaerium corneum	2
Anodonta sp.	1

* Data based on the figures of: WESENBURG-LUND, 1934; DUBOIS, 1929; REES, 1932; PROBERT, 1966; LLEWELYN, 1957. The names of the molluscs are those used by the original authors.

categories have been associated with systematic units at the adult stage. As a result of the many incidence studies carried out on the freshwater fauna it is apparent that the cercarial types are not represented equally in the various environments. The many surveys published suggest that the xiphidio and furcocercarial types are by far the most abundant (Fig. 11). In the larger bodies of water furcocercariae predominate, while in the smaller environments xiphidiocercariae are the most numerous. Gymnocephalous and echinostome cercariae are roughly equally abundant and are roughly half as frequent in occurrence as the first two types. All the other categories occur infrequently.

It is apparent that the incidence and type of larval trematodes present in a particular environment are ultimately governed by local conditions so that high incidence may occur in one habitat with a very low one in adjacent environments. This latter feature has been commented on by HUNTER and BIRKENHOLZ (1961) and in some environments must be related to the behaviour of the final host HOFF (1941). WIKGREN (1956) comparing the molluscan fauna of pools and the shore line of various islands of the Finnish Archipelago found that the rate of infection with larval trematodes was highest in the pools. He suggested that the degree of infection on the shore depended on (a) the density of hatched miracidia, (b) the density of the snail population, and (c) the water movements such as waves and currents. Inshore the infection rate varied from 0 to 100 per cent between different pools. The rate of infection could be related to a variety of factors. The extent of the bird fauna on a particular island influenced greatly the level of infection in the mollusc population. The relation of the surrounding rocks to the pool was considered significant as an inclination of the rock towards the pool favoured the washing into the pool of helminth eggs present in the bird's droppings. In those cases where vegetation was associated with the pool the infection rate was high and this could be correlated with the fact that the birds favoured pools with vegetation.

The restriction of larval trematodes to certain molluscan hosts may be related to the feeding behaviour of the final host and partly to the ecology of the intermediate host (JAMES, 1968a). The larval stage of *Parvatrema homoeotecnum* occurs only in *Littorina saxatilis tenebrosa* var *similis* although other subspecies were also present in certain habitats. The final host of this species is the Oystercatcher and James explains the restricted molluscan host range by the fact that only the subsp. *tenebrosa* has a shell thin enough to be crushed by the bird's gizzard. The subsp. *jugosa* and *rudis* have thick shells which cannot be crushed and are therefore not normally eaten by the Oystercatcher.

THE ECOLOGY OF THE DEVELOPMENT CYCLE—ADULT TREMATODES

In view of the direct relationship between larval and adult forms it is obvious that the adult trematode fauna present in a final host will exhibit variations in intensity and composition also but the factors affecting the adult populations will not necessarily be the same as those producing change in the larval fauna.

The adult fauna will be influenced by the behaviour of the host in relation to the aquatic environment. The frequency of visits and the duration of contact with the particular environment as well as the behaviour of the host within the environment will all affect the internal fauna. In this way the migratory, feeding and swarming patterns of the host will influence the degree and duration of the exposure to infective stages.

One of the most basic factors in the establishment of both adult and larval populations of parasites in hosts associated with fresh-water is the characteristic nature of the environment as

Table 2 The composition of the parasite fauna of the fishes of Llyn Tegid and Lake Drużno.

	Species of parasite							
	Llyn Tegid				*Lake Drużno*			
	Mature forms		Larval forms		Mature forms		Larval forms	
Parasite group	No.	Percentage	No.	Percentage	No.	Percentage	No.	Percentage
Myxosporidia	2	7·1	—	—	4	7·4	—	—
Monogenea	3	10·7	—	—	4	7·4	—	—
Digenea	3	10·7	0	0	10	15·1	12	22·2
Cestoda	5	17·8	4	14·2	7	12·9	4	7·4
Nematoda	6	21·4	0	0	3	5·5	1	1·5
Acanthocephala	3	10·7	0	0	4	7·4	0	0
Hirudinea	0	0	—	—	1	1.5	—	—
Copepoda	2	7·1	0	0	3	5·5	0	0
Lamellibranchiata	—	—	0	0	—	—	1	1·5
Totals	24	85·7	4	14·2	36	66·6	18	33·3
Total no. spp			28				54	
No. fish sp. examined			8				19	
Data from			Chubb		WISNIEWSKI, 1958; KOZICKA, 1959.			

(From CHUBB, 1963).

expressed in terms of oligotrophic and eutrophic conditions. It is obvious that the latter environment will possess a more varied fauna and flora and will favour particularly the survival of many molluscan species as well as attracting a greater variety of birds. The fact that there exists a relationship between the parasite fauna of an environment and the character of that body of water was suggested by WIŚNIEWSKI (1958a) who, with his associates, made a very detailed investigation of Drużno Lake in Poland (JARECKA, 1958; SULGOSTOWSKA, 1958; RYBICKA, 1958; DOBROWOLSKI, 1958; STYCZYŃSKA, 1958, a,b; KOZICKA, 1958, 1959). In this country CHUBB (1963, 1964) has characterized the parasite fauna of the fish from two types of oligotrophic environments and described differences between these environments and the data obtained from the Polish eutrophic lake (Table 2). The species typical of Salmonoidei and thus representing oligotrophic habitat (Llyn Tegid) were not found in the eutrophic lake, and the number of species in Pike, Roach and Perch was less in the oligotrophic environment. With reference to the trematode parasites the percentage of adult monogenea in the fish was approximately the same in both environments, whereas the percentage of adult and especially the larval digenea was much higher in the eutrophic lake than in the oligotrophic environment. It is tempting to relate these differences to the fact that the digenean cycle is completely dependant on intermediate hosts, particularly molluscs, which would not be very abundant in the oligotrophic environment as compared with the eutrophic habitat. Unfortunately the relationship between the nature of the environment and the incidence of adult and larval digenea is not as simple as this as CHUBB (1964), comparing a strictly oligotrophic lake (Llyn Padarn) with the late oligotrophic lake (Llyn Tegid) found a greater percentage of adult and larval digenea in the fish from the former environment. It seems that although the principle may be correct in broad

Table 3 Trematode infestations of birds of different ecological groups (From DOGIEL (1962) after BYKHOVSKAJA-PAVLOVSKAJA).

Groups of Birds	Species	% Infestation	No. of Trematode species	Feeding Grounds
Aquatic	Rails	70·3	20	Surface layers, mid-water and bottom of aquatic environments
	Gulls	60·4	22	
	Geese	48·6	34	
Swamp Dwellers	Limicoles	37·5	37	Swamps, marshy soils, mud, rushes, etc.
Terrestrial	Passerines	17·1	12	Dry land, air etc.

terms, much more information is needed on the factors which will produce variations within this main context.

The factors affecting the parasite fauna of birds have been discussed by BYKHOVSKAJA-PAVLOVSKAJA (1964a). One of the major features influencing the trematode parasites of birds is the nature of the food and the method of feeding of adult birds, with a particularly high incidence among those birds associated with an aquatic habitat (Table 3). Within the USSR 400 species of trematodes from 25 families were reported from predatory and omnivorous birds and only 120 species from 15 families in granivorous and insectivorous hosts. The highest intensity of infection is associated with those birds which feed in deep water and from the lake bottom; next were those feeding on swampy shores and marshes; then those which took food from the ground, with the lowest levels of infection occurring in those birds feeding on insects whilst in flight. It is apparent from this that birds with similar food preferences i.e. fish, aquatic insects, aquatic molluscs will tend to be associated with the same parasite species. The extent of contact with aquatic habitats will also influence the trematode fauna of amphibians and PAUL (1934) has described a higher infection with trematodes in those host species most closely associated with aquatic habitats. DUBOIS (1953), discussing the host specificity of the Strigeoidea, suggests that although the ecological preferences of the host and the nature of its food plays a part in determining the host distribution of a parasite species, physiological adaptation and specialization may be more significant in some cases.

The association between a host and its food and feeding pattern is particularly important in digenean development where infections occur frequently via the mouth and by the ingestion of intermediate hosts. The development of monogenean populations on a host are not greatly influenced by such factors and it is the gross movement of the host population which is more important. The initial infection by monogeneans may not be acquired immediately after the hatching of the fish host but may occur later when the young fish move from the spawning grounds to other waters. LLEWELLYN (1962a) in a study of the population dynamics of the monogenean parasites on the gills of *Trachurus trachurus* (Scad) found that the pelagic larvae of the fish less than 2 cm long were parasite free. In late September or early October the young fish descend to the sea bottom and it is here and at this time that the fish become infected by oncomiracidia. The parasites mature slowly and reach sexual maturity by the following summer. Eggs will be deposited at this time and these will sink to deeper waters, develop and produce a high density of infective oncomiracidia by the following October. After feeding in the surface waters for several months the fish return to deeper waters and will again be exposed to infection by oncomiracidia. Llewellyn suggests that the parasite *Gastrocotyle* survives for one year so that by late autumn the parasite population of the one year old scad has been replaced by juvenile forms from the second infection. The fish return to surface waters in the second summer,

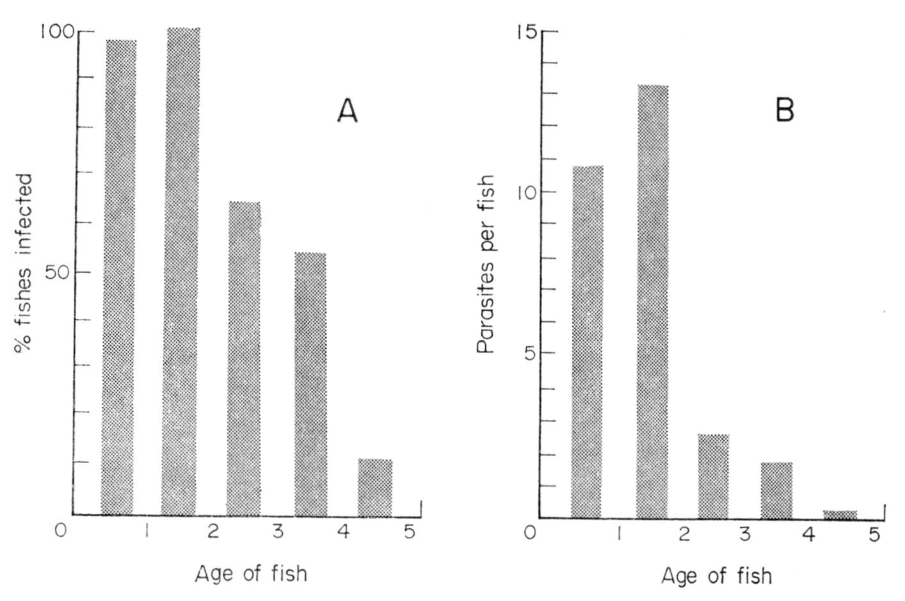

FIG. 12 The occurrence of *Gastrocotyle trachuri* on *Trachurus trachurus* of various ages. **A** percentage of *Trachurus* infected; **B** degree of infestation in Parasites per fish. From LLEWELLYN, 1962a.

are now sexually mature and breed, and also will reinfest the deeper waters with monogenean eggs. After this second season the scad disperse to deeper waters and eventually lose their *Gastrocotyle* infection (Fig. 12). This interesting study illustrates how the acquisition of a population of monogenean parasites is closely associated with the seasonal movements of the fish population.

The population dynamics of a monogenean parasitic in freshwater fish has been considered by PALING (1965). The hosts in this case were the brown trout (*Salmo trutta*) and the charr (*Salvelinus alpinus*). In the case of the trout spawning occurs in the streams so that the fish leave the lake (Windermere) over the period October to December. The young fish, after hatching, remain in the streams for approximately two years and then migrate to the lake in spring and summer. Paling found that the parasite *Discocotyle sagitta* invades its host predominantly in July and August and that this occurs in the deeper waters of the lake. In this way the young trout in the streams are largely free of *Discocotyle sagitta*. The parasites attain maturity approximately two months after infection of the host and may live for three or four years. The fish within the lake are therefore exposed to reinfection and it was found that the older fish on average harboured a larger number of *Discocotyle* with the parasite population reaching saturation at five to six years (Fig. 13). Thus no age immunity was demonstrated but a difference in the

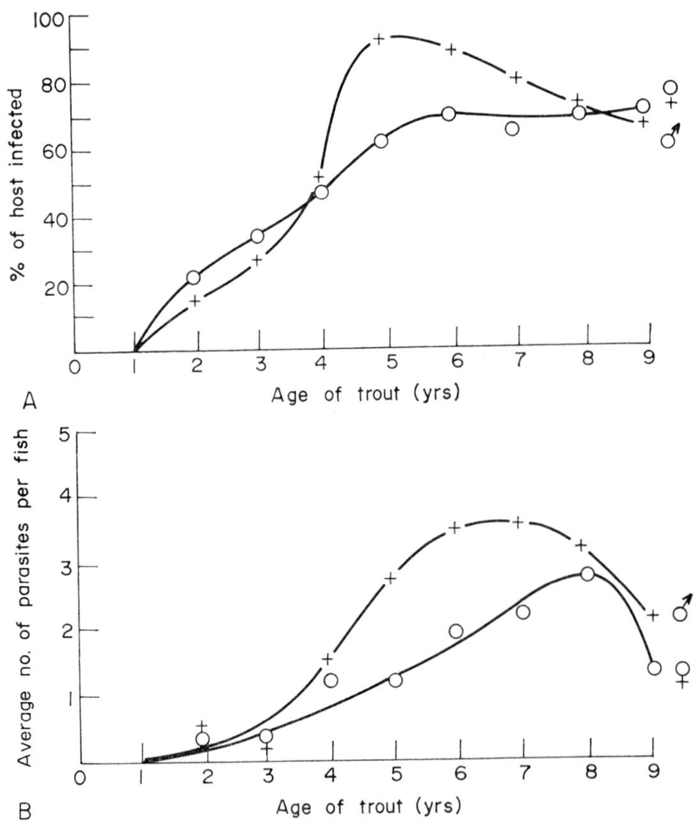

FIG. 13 The relationship between the sex of Windermere trout of different ages and the degree of infestation with *Discocotyle sagittata*. **A** percentage of hosts infected; **B** average number of parasites per fish for each host sex. From PALING, 1965.

numbers of parasite carried by each sex was found. Male trout between the ages of five and seven years carried significantly more *Discocotyle* than female fish of the same age.

These two studies have indicated the significance of host movements within fairly restricted areas. It is to be expected therefore that the extensive migrations exhibited by certain birds and catadromous and anadromous fish will influence considerably the parasite populations of those hosts. The extensive Russian literature on this subject has been discussed in some detail by DOGIEL *et al*, (1958) and DOGIEL (1962) and only brief comment will be made here. The anadromous migration of the Salmon provides an excellent example of the effect of host movements on the parasite population harboured by that host (Fig. 14). A general picture emerges which reveals the acquisition of an essentially freshwater parasite population in its early stages, and a gradual replacement of this population by a marine one, as the fish enters and establishes itself in this new environment. The parasite population is still not stable and will alter as the fish migrates from a marine to a freshwater habitat in relation to its breeding behaviour, so that the marine parasite population varies in its predominance depending on the habitat of the host. The two extremes are represented by those fish which are permanently freshwater in their habit. Population patterns of various degrees of heterogenicity exist depending on the periodicity and duration of the movement of the host from one habitat to another (Fig. 15).

In a similar way the parasite population of migratory birds is in a state of flux, the particular phase existing at any one time being dependent on the geographical location of the host (Fig. 16). The variation encountered, during migration, in diet and consequently the types of intermediate host ingested with its related larval stages will be considerable and will be reflected in changes in the parasite fauna. In general it is thought that the number of parasite species and the intensity of infestation decreases during migration and in the winter locality of the host and that maximum infestation occurs in the breeding grounds. The infestation of the nestlings has been commented on by BYKHOVSKAJA-PAVLOVSKAJA (1964a) who found that waterfowl became infested in the following order: cestodes, acanthocephalans, nematodes and finally trematodes. The process is very rapid and within one and a half to two weeks all groups of helminths are present.

Table 4 Incidence of trematodes in *Rana temporaria* from Britain (After LEES, 1962).

Parasite	No. of individuals parasitized	Parasitization Rate (%)	Mean no. of parasites per host
Polystoma intergerrimum	302	12·5	2·0
Gorgoderina vitelliloba	283	11·8	2·2
Haplometra cylindracea	1308	54·2	5·2
Dolichosaccus rastellus	173	8·0	3·0

The adult population of the host will also vary seasonally and in trematodes will be related to the availability of infective stages, such as cercariae, and the abundance of intermediate hosts and metacercarial stages. Seasonal incidence may also be related to host behaviour and migrations, as discussed above. The observations of HOLL (1932), BRANDT (1936) and RANKIN (1937) indicate that the helminth population of various amphibia is susceptible to seasonal

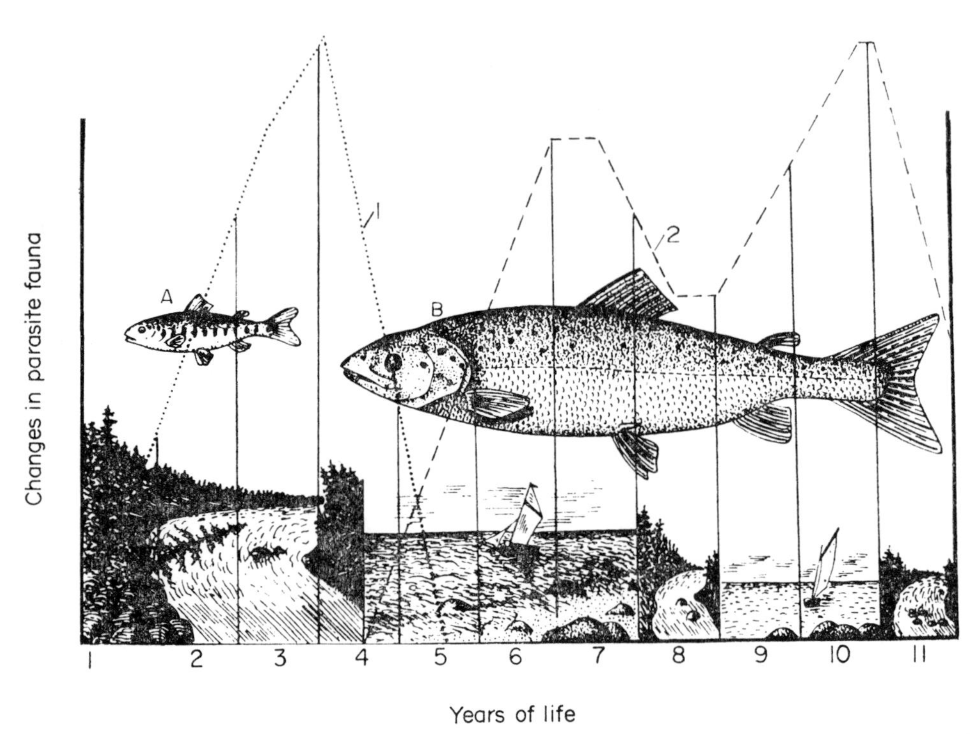

FIG. 14 The changes occurring in the parasite fauna of the Salmon in relation to host migration. **A** young salmon; **B** adult salmon. 1: changes in the freshwater parasite fauna; 2: changes in the marine parasite fauna. From DOGIEL, 1962 (1964 translation).

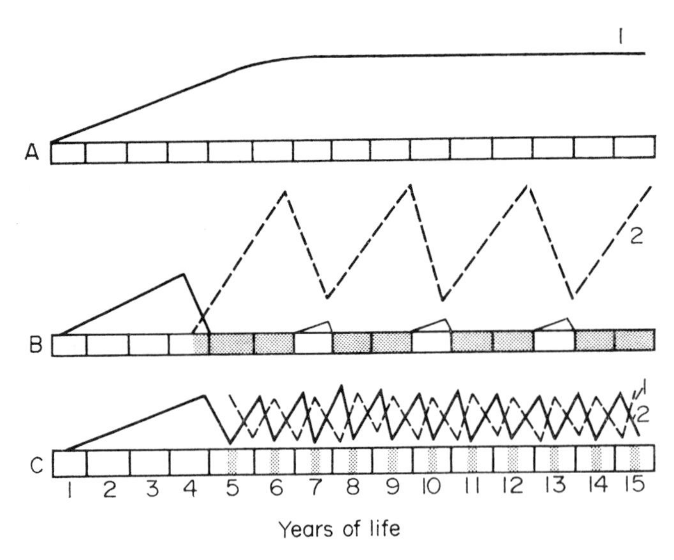

FIG. 15 The fluctuation in the parasite fauna of fish with different modes of life. 1: fluctuations in the freshwater parasite fauna; 2: fluctuations of marine parasites. The shaded areas represents time spent in the sea. **A** *Thymallus thymallus*—a freshwater fish; **B** *Salmo salar*—a fish spending long periods in the sea; **C** *Salvelinus alpinus*—as an adult spends nine months in lakes and three months in the sea. From DOGIEL, 1962 (1964 translation).

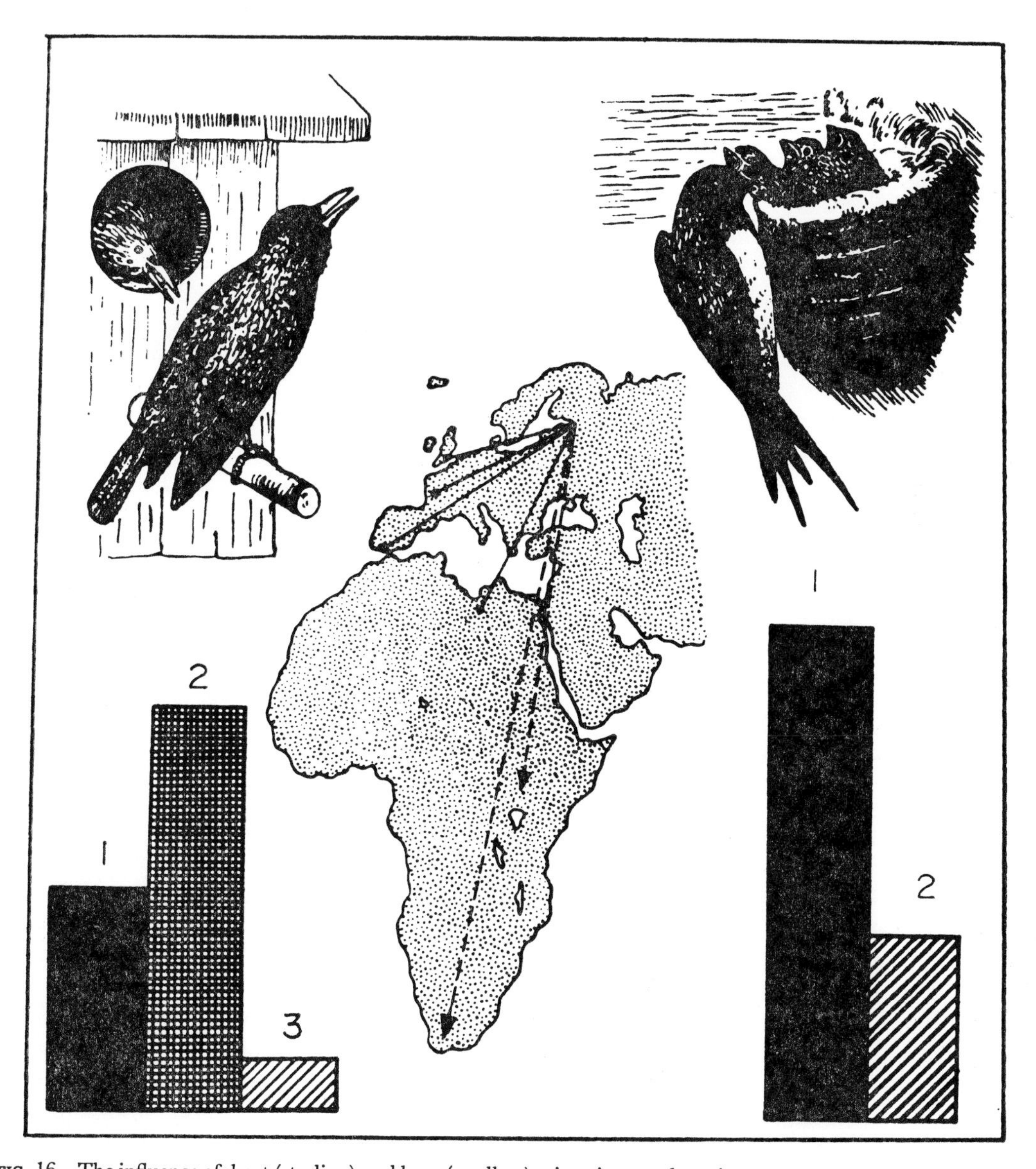

FIG. 16 The influence of short (starling) and long (swallow) migrations on the avian parasite population. 1: number of northern parasite species; 2: number of ubiquitous species; 3: number of southern parasite species. In long migrations ubiquitous parasites are absent and the number of southern species is small. From DOGIEL, 1962 (1964 translation).

changes. This variation was apparent in the trematode fauna of *Rana temporaria* from Great Britain examined by LEES (1962). The overall incidence varied depending on the particular parasite (Table 4) but the highest incidence and intensity of infection occurred in the autumn. The incidence varied between sexes (Table 5). Generally the highest levels of infection were present in the male frogs and this difference, particularly in the case of the trematode fauna, was exaggerated seasonally. In winter and spring the trematode fauna of female frogs fell to a much lower level of incidence than in the male frogs and it was suggested that, although the autumn maximum was correlated with the abundance of caddis and stone flies, the exaggeration of the seasonal difference between the sexes was related to the changes in the sex hormones apparent at this time of the year. Differences between the parasite incidence of male and female fish have also been described by PALING (1965) and THOMAS (1964). In the case of *Discocotyle sagittata* on trout (PALING, 1965) it was found that the male fish carried a significantly heavier parasite burden. The situation was not so clear in the populations of trout studied by Thomas. In young trout (one to two years) hardly any significant difference could be detected although in the older fish (3 years and over) the females tended to be more heavily infested with *Neoechino-*

Table 5 Seasonal incidence of helminth parasites in male and female frogs (After LEES, 1962).

Parasite	Host Sex	Percentage of hosts parasitized			
		Autumn	Winter	Spring	Summer
Polystoma	♂	18·3	14·4	16·3	16·2
intergerrimum	♀	15·6	8·2	17·8	15·7
Gorgoderina	♂	15·2	10·8	11·6	12·8
vitelliloba	♀	12·4	5·4	8·5	9·0
Haplometra	♂	62·8	50·1	53·8	54·2
cylindracea	♀	54·3	41·6	44·4	52·6
Dolichosaccus	♂	12·3	10·2	6·2	8·3
rastellus	♀	14·4	10·5	5·9	8·6

rhyncus rutili (Acanthocephala), *Phyllodistomum simile* and *Discocotyle sagittata* than the males, and this difference existed during and after the spawning season. Thomas mentions that mature male fish eat more food than the females during the period October to December and therefore would be expected to harbour the heavier parasite burden. However, the female fish must be under greater stress, and consequently more susceptible, because they contribute relatively more reproductive material, expend energy in cutting the redd and also eat less than the males. The situation is complex and the explanation is not simple nor obvious.

The life-cycle of the digenetic trematode can be very complex involving a succession of different hosts. The hosts may occur in widely differing systematic groups, may have entirely different habitats, and may exhibit great divergence in their food and feeding habits. The various stages of the parasite exhibit differences in the optimal conditions needed for survival and in their behaviour patterns. The interaction of all these features makes the assessment of the ecological nature of the life-cycle particularly difficult and the brief consideration given above illustrates some of the complexities associated with the biology of both adult and larval stages. In this kaleidoscope of relationships and requirements one cannot but be impressed by the way that the developmental cycle has evolved so that contact is established between host and parasite allowing the perpetuation of the parasite species.

It is obvious from this section that the study of the trematode life-cycle in relation to the ecology of the environment has been grossly neglected. Much of the existing work is descriptive and although very valuable provides limited information on the changes within a parasite population occurring in a given environment. There is considerable need for an exhaustive ecological study of selected habitats investigating the seasonal fluctuations in the entire fauna and their relationship to changes in the parasite population. There is at present little information on the way in which parasites become established in a freshwater environment and the relationship between the survival of trematode species and the biology of permanent and transitory host species.

2 The Egg: Oncomiracidium: Miracidium

One of the major biological characteristics attributed to parasites is their fecundity and this is regarded as a partial insurance against the many hazards likely to be encountered before the life-cycle is completed. Very few trematodes are viviparous and the oviparous forms vary considerably in the numbers of eggs which are produced. Some e.g. *Fasciola hepatica* lay thousands of eggs, but even so, it is unlikely that the number of eggs deposited by any trematode is comparable with that produced by certain parasitic nematodes e.g. *Ascaris*. As far as is known, none of the eggs of trematodes are able to resist desiccation to any great extent so that the life-cycle, from its earliest stages, is committed to an aquatic environment.

Because of the many morphological differences between the egg and its ensuing larval stage it seems convenient to consider these phases separately under the appropriate systematic grouping.

Monogenea

THE EGG

For many years in the teaching of Parasitology it has been traditional to contrast the monogenean egg with that of the digenean emphasizing, as a major point of distinction, the fact that monogenean eggs are non-operculate. Recent studies by BYCHOWSKY (1957) and LLEWELLYN (1957a) have indicated that monogenean eggs are generally operculate. At one of both ends, the eggs bear filaments which may be very long or short, straight or much coiled (Plate 3-2). BYCHOWSKY (1957) differentiates these filaments into the 'foot' extending from the non-operculate end and the 'filament' at the operculate end of the egg (Fig. 17). The eggs range from 0·02–0·18 mm in length, excluding the filaments which may be many times the length of the egg. The colour varies from bright yellow to dark brown and generally changes to a darker shade as the egg develops. This colour change suggests that quinone-tanning of the egg shell takes place in the manner suggested for digenean eggs. The only histochemical observations available are those of RENNISON (1953) and SMYTH and CLEGG (1959) on the egg of *Diclidophora merlangi*, FREEMAN and LLEWELLYN (1958) on *Gastrocotyle trachuri* and *Protoeces subtenuis* and LLEWELLYN (1965) on *Entobdella soleae* and *Dichidophora luscae*. The presence of proteins, phenols and a phenolase in the newly formed egg shell of *Diclidophora* and *Gastrocotyle* further supports the idea of quinone tanning in the monogenean egg although *Protoeces* was found to be phenolase negative. Observations on the formation of this type of egg shell are very few indeed. DAWES (1940) describing the formation of the egg capsule of *Hexacotyle extensicauda* from the Tunny, states that secretory material is discharged from the gland cells of the ootype into the lumen. This secretion, he suggested, formed a thin layer on the outside of the rudimentary capsule and then deposition on its inner surface of other droplets from the yolk cells took place. Fertilization of the ovum may occur as a result of self or cross fertilization and copulation may take place via the vagina or the uterus. The nature of the copulatory organ varies considerably. In *Diplozoon*

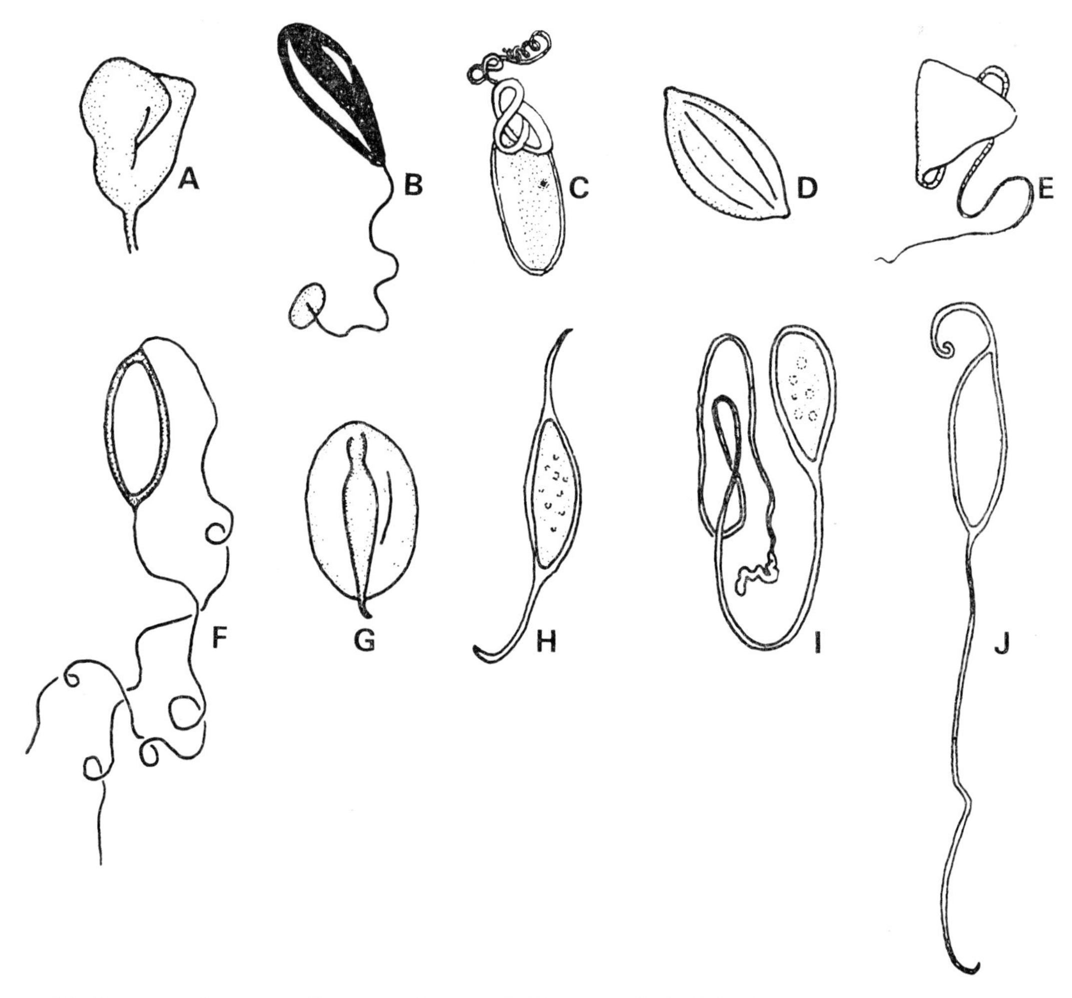

FIG. 17 Examples of monogenetic trematode eggs. **A** *Thaumatocotyle dasybatis;* **B** *Udonella caligorum;* **C** *Diplozoon paradoxum;* **D** *Rajonchocotyle alba;* **E** *Nitzschia monticelli;* **F** *Squalonchocotyle apiculatum;* **G** *Microbothrium apiculatum;* **H** *Erpocotyle laevis;* **I** *Hexabothrium canicula;* **J** *Diclidophora denticulata.* Redrawn from various sources from CHENG, 1964.

there is no copulatory structure as the terminal parts of the reproductive system of both individuals fuse. In other species there is generally no muscular eversible cirrus but a distinct, non-eversible 'penis' which may bear at its free end a corona of spines or hooks. A variation of this type of structure is represented by those specimens in which the ejaculatory duct opens into a long chitinous tube which may be twisted or coiled into a variety of shapes. For further morphological details the reader is referred to BYCHOWSKY (1957).

The deposition of egg capsules by the fluke takes place very slowly at 13°C with little or no production below 8°C (LLEWELLYN, 1957a). At 18°C the rate of production was 3–4 capsules per day. ALVEY (1936), describing the development of *Sphyranura oligorchis* reported an egg production rate of one every three hours. Individuals of *Polystoma intergerrimum* produced a total of from 1074–2290 egg capsules at 14°–16°C over a period of 18 days (BYCHOWSKY, 1957).

The hatching of the egg capsule is again very dependent on temperature and presumably oxygen tension. LLEWELLYN (1957a) reports the following development times at 20°C. *Diplectanum* five days; *Plectanocotyle* 8–11; *Acanthocotyle* 12; *Entobdella, Microcotyle labracis, Gastrocotyle* and *Pseudaxine* 14–16; *Rajonchocotyle* 25 and *Diclidophora merlangi* 27 days. ALVEY (1936) records the hatching of *Sphyranura oligorchis* capsules after 28–32 days at room temperature. Several workers, FRANKLAND (1955), BYCHOWSKY (1957) LLEWELLYN (1957a) describe the hatching of the larva being brought about by repeated pressure against the operculum by the active larva. Hatching seems to be affected by temperature and in some cases by light, but no detailed study has been made.

THE ONCOMIRACIDIUM

The egg hatches to release a very characteristic larva termed (by LLEWELLYN, 1957a and 1963) the oncomiracidium. The structure of this larval stage will vary in its details but generally speaking the following description contains the basic characters (Fig. 18). The larvae are not flattened but are approximately cylindrical in form between 100–300 μ long and 30–100 μ in diameter. Both anterior and posterior ends are slightly tapered, with the posterior end terminating in an adhesive organ in some species. The larvae have a ciliated epidermis except in such genera as *Gyrodactylus, Udonella* and *Acanthocotyle.* The extent of ciliation varies, exhibiting a general and continuous distribution in some, but in others a restriction of the cilia to three zones. In the anterior region of the body there are one or two pairs of eyes. These consist of pigment cups which may contain permanent crystalline lenses, e.g. *Diplectanum* and *Entobdella,* or oil droplets which may act as lenses as in some diclidophoroidean polyopisthocotylineans. The mouth may be anterior and terminal in its position or subterminal, opening at the level of or just in front of the anterior pair of eyes. There is a well developed muscular pharynx which leads into a simple sac-like gut. BYCHOWSKY (1957) refers to the intestine as circular, with the genital rudiments lying in the centre of this intestinal ring. The body also contains a simple nervous system (cerebral complex) and excretory system which opens out to the exterior on either side at the anterior end of the body. Two other morphological features are particularly characteristic of this larval stage. Opening out to the exterior at the anterior end of the body are two groups of gland cells. The function of these is obscure, but they may secrete substances which aid in attachment to the host. At the posterior end of the larva is the attachment organ. This consists of a terminal or subterminal disc or cup, directed ventrally and bearing marginal hooks, median hooks and accessory sclerites. The details of the armature vary considerably between different larvae. The systematic significance of larval structure in general and the opisthaptor morphology in particular has been discussed by LLEWELLYN (1963).

Aspidogastrea

As far as is known (WILLIAMS, 1942; WHARTON, 1939) the eggs of aspidogastriids are oval and operculate. In *Aspidogaster conchicola* the eggs (126–134 μ × 48–50 μ) contain active larvae. The eggs hatch to produce a non-ciliated larval stage 130–150 μ long. The terminal mouth is surrounded by a large oral sucker, followed by a muscular pharynx and then a sac-like intestine. At the posterior end of the larva is the invaginated ventral sucker. This posterior sucker develops to form the large attachment apparatus on the ventral surface. The larvae, when free, move about with leech-like motion.

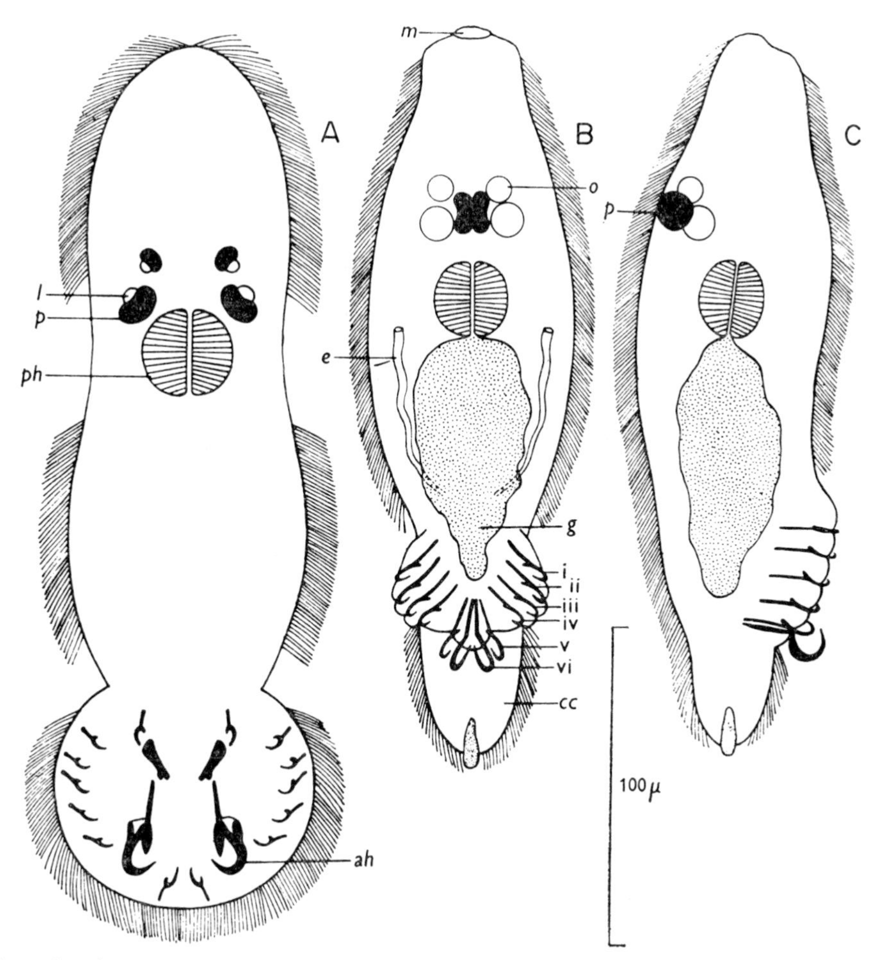

FIG. 18 Examples of monogenean larvae. **A** *Entobdella soleae;* **B** schematized larva of the superfamily Diclidophoroidea ventral view. **C** as for **B** but lateral view. (ah: primordia of adult hooks; cc: ciliated cone; e: excretory canal; g: gut; l: lens of eye; m: mouth; o: oil droplet; p: pigment cup of eye; ph: pharynx; i–iv,: lateral hooks; v: postero-lateral hook; vi: posterior hook.) From LLEWELLYN, 1957a.

The larva of *Lophotaspis vallei* (WHARTON, 1939) is similar to the above description but it also possesses eye spots and three patches of cilia.

Digenea

THE EGG

The digenean egg capsule is usually oval in shape and light to dark brown in colour (Plates 3-1, 3-3 and 3-6). At one end is the operculum attached to the capsule by a substance generally referred to as 'cement' although its true nature is unknown. The outstanding exception to this description is the egg capsule of the schistosomes which is non-operculate and which bears a large lateral spine. A study of these eggs with the scanning electron microscope (HOCKLEY, 1968) has shown that in addition to the large spine (15–25 μ long) the shells of *S. mansoni* and *S. haematobium* are covered with small spines 0·3 μ long. These small spines cover the entire outer surface of the egg including the large spine and Hockley estimates that there are approximately 120 spines/μ^2 of shell surface. However, egg capsules of most other species appear to be simple although filaments do occur in some non-schistosome species. The entry of sperm into the female reproductive system may take place via Laurer's canal (sometimes absent) or through the genital aperture and the uterus. As in the Monogenea self fertilization may take place although most workers suggest that cross fertilization is the general rule. It must be admitted however, that the number of observations on living trematode relating to this aspect are very few. NOLLEN (1968a) has investigated the relative frequency of occurrence of cross and self fertilization by introducing adult *Philophthalmus* labelled with ^{3}H-thymidine into a population of non-labelled adults. Using autoradiographic techniques, the appearance of labelled sperm in the non-labelled adults could be detected and therefore indicated the frequency of cross fertilization. He found that of 33 labelled adults only one inseminated itself, whereas the 33 inseminated 47 out of the 61 non-labelled adults available. The extent of cross-insemination varied with the age of the parasites and was less frequent with older worms. Thus in this species, and given the opportunity, *P. megalurus* is more likely to inseminate other individuals than itself. He was also able to demonstrate that in this species entry of sperm into the female reproductive system did not take

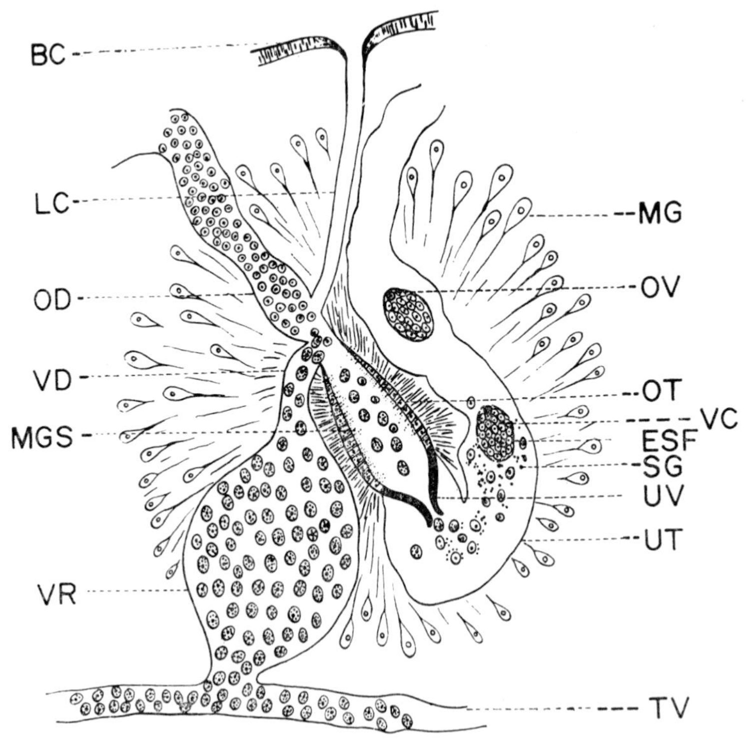

FIG. 19 Diagrammatic representation of the Mehlis' gland complex and associated ducts of *Fasciola hepatica*. (BC: external tegument; ESF: egg capsule in process of formation; LC: Laurer's canal; MG: Mehlis' gland cells; MGS: Mehlis' gland secretion; OD: oviduct; OT: ootype; OV: ovum; SG: shell granule; TV: transverse vitelline duct; UT: uterus; UV: uterine valve; VC: vitelline cells; VD: median vitelline duct; VR: vitelline reservoir.) From RAO, 1959.

place via Laurer's canal. The sperm are generally filiform, free or enclosed in spermatophores as in *Steringopharus*, *Haplocladus* and *Prosorhyncus*, or may be globular as in *Asymphylodora*.

Egg formation occurs as a result of a co-ordinated sequence of contractions in the proximal ducts of the female reproductive system (Fig. 19). The following account is based on the description by GUILDFORD (1961). Contractions of the ovarian wall push mature oocytes into the proximal part of the oviduct. This region possesses a sphincter which relaxes periodically releasing an oocyte into the more distal portions of the duct. The peristaltic contractions move the oocyte along and, at the opening of the seminal receptacle, sperm and fluid are discharged behind the oocyte. Movement of the oocyte and sperm continues until the junction of the oviduct with the vitelline duct is reached. Contractions of the vitelline ducts force vitelline cells into the vitelline reservoir which also contracts pushing groups of 6–9 vitelline cells into the ovivitelline duct. The contents of this duct pass into the ootype when a sudden release of globules from the vitelline cells takes place. These globules form the shell and movements of the ootype distribute the discharged material around the vitelline cells and ootype (Fig. 20). Although this

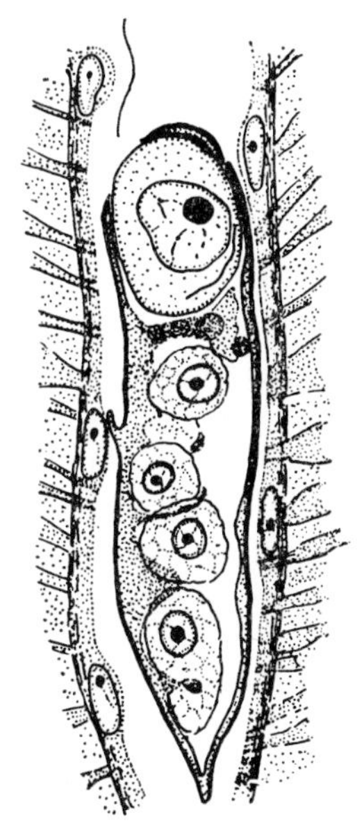

FIG. 20 Egg capsule formation in the ootype of *Halipegus eccentricus*. The uneven egg capsule encloses a primary oocyte with a sperm nucleus and four vitelline cells. From GUILFORD, 1961.

description represents the sequence of events in *Halipegus eccentricus* it is likely that it is fairly representative of what occurs in other digenea. In *F. hepatica* GÖNNERT (1962) also described the release of material from the vitelline cells whilst within the ootype and the subsequent formation of a thin shell within the ootype. GRESSON (1964) reviews oogenesis in digenetic trematodes and in this paper he lists data from 66 species. The site of fertilization is not given for 16 species, it is reported as occurring in the uterus for 41 species; in the oviduct for 5 species and in the ootype for 4 species. BURTON (1967) has recorded some ultrastructural observations on the process of fertilization of *Haematoloechus medioplexus* ova. He mentions that in the proximal uterus of this species, newly formed eggshells contain spermatozoa in the process of fertilization. The plasma membrane of the spermatozoon (approx, 400 μ long)

Table 6 The occurrence of the components of a quinone-tanning system in the vitellaria and newly formed egg shells.

Helminth	Phenols	Proteins	Phenolase	Author
Monogenea				
Diclidophora merlangi	+	+	+	Smyth and Clegg, 1959
Gastrocotyle trachuri	+	+	+	Freeman and Llewellyn, 1958
Protoeces subtenuis	+	+	0	
Entobdella soleae	+	?	+	Llewellyn, 1965
Diclidophora luscae	+	?	+	do.
Digenea				
Fasciola hepatica	+	+	+	Smyth and Clegg, 1959
Haplometra cylindracea	+	+	+	do.
Haematoloechus medioplexus	+	+	+	Burton, 1963
Clonorchis sinensis	?	?	+	Ma, 1963
Syncoelium spathulatum	+ uterine epithelium	+	?	Coil and Kuntz, 1963
Diplostomum phoxini	+	?	+	Smyth, 1959
Diplostomum spathacum	+	+	+	Erasmus, unpublished observations
Apatemon gracilis minor	+	+	+	
Holostephanus lühei	+	+	+	
Cyathocotyle bushiensis	+	+	+	
Gorgodera vitelliloba	+	?	0	Llewellyn, 1965

fuses with that of the ovum (approx. 8 μ in diam.) at many points along its length, so that the plasma membrane of both gametes becomes continuous. In contrast there is much evidence to support the view that in unisexual female infections of Schistosome species, eggs, capable of producing viable miracidia, are developed parthenogenetically. These miracidia are infective and ultimately result in cercaria producing infections in snails (SHORT and MENZEL, 1959). The time taken between the release of an oocyte from the sphincter of the ovary to the entry of the egg from the ootype into the uterus took from 20–30 seconds in compressed living specimens of *Halipegus eccentricus*. In its passage along the uterus the wall of the egg capsule becomes thicker and darker in colour. The number of eggs present in the uterus varies considerably from species to species. There may be only 5–6 in *Cyathocotyle bushiensis* but in *Fasciola hepatica* hundreds may occur.

CAPSULE FORMATION

The function and role of the ootype, Mehlis' gland and the vitelline gland cells in the formation of the egg capsule wall has attracted considerable interest. A review is not relevant here as this has been ably covered by BURTON (1963), SMYTH and CLEGG (1959) and EBRAHIMZADEH (1966). The majority of studies so far suggest very strongly that in some trematodes the egg shell is composed of sclerotin type proteins and that some form of tanning process occurs. In some species (Table 6) it seems that quinone tanning takes place, but in others where phenolase activity has not been demonstrated, another mechanism for protein hardening must be present. The precursors of the quinone-tanning system, protein, phenols and phenolase can be demonstrated histochemically in the globules of the vitelline cells. These globules may be released either in the ootype or just after the cell clusters have emerged from the ootype, although in an isolated electron microscope study by TULLOCH and SHAPIRO (1957) the authors described the release of globules while the cells were still in the vitelline gland lobules. BURTON (1963) records the uptake of tritium labelled tyrosine by cells in the vitelline follicles and its appearance in the young capsules in the proximal part of the uterus. On this and other evidence BURTON (1963) suggests that the tanning of capsules in *Haematoloechus medioplexus* involves the alteration of —OH and —NH_2 radicals in a tyrosine rich protein. The tyrosine residues could be enzymatically oxidized to DOPA and then to quinone. Further reactions involving the formation of stable linkages between quinone and amine groups of adjacent protein chains would be necessary. The evidence in the literature is not completely complementary and it is possible that several biochemical routes for tanning occur.

The role of Mehlis' gland is still uncertain, and interpretation is hindered by its indefinite histochemical characteristics. Under the electron microscope it may be seen that the wall of the ootype is cellular and that the luminal plasma membrane is thrown into a series of microvilli. The ducts from the gland cells penetrate this lining so that the secretory products of Mehlis' gland are discharged into the lumen. Functions suggested for the secretion range from lubricatory (KOURI and NAUSS, 1938), for the activation of spermatozoa (STEPHENSON, 1947c), causing release of shell globules from the vitelline cells (UJIIE, 1936), controlling or initiating the quinone tanning process is some way (UJIIE, 1936), to providing a membrane within which the secretory droplets accumulate forming the egg capsule (DAWES, 1946). Gland cells of the Mehlis' complex are diastase resistant PAS positive and in *Haplometra medioplexus* (BURTON, 1960) and *F. hepatica* (HANUMANTHA-RAO, 1959) contain phospholipid. BURTON (1963) was able to show that isolated Mehlis' gland from *H. medioplexus* was able to produce distinct membranes *in vitro*. This evidence thus supports DAWES' (1940) suggestion that the secretion from Mehlis' gland may form a membrane on the inner surface of which the secretion forming globules coalesce. More recent observations by CLEGG (1965) and CLEGG and MORGAN (1966) have provided further support for the secretion of lipoprotein-like material by Mehlis' gland in *F. hepatica*. Histochemical and biochemical analyses have shown that phospholipids and glycolipids were present in the extracts of Mehlis' gland as well as in extracts of fragments of egg-shell. These authors describe the release of lipoprotein granules which spread over the inner surface of the epithelium lining the ootype and the upper region of the uterus. The granules contributed to membranes covering the outer surface of the egg and also lining its inner surface. It was suggested that the lipoprotein layers might have considerable significance in affecting the permeability of the egg shell. WILSON (1967b) has shown in *F. hepatica*, using the electron microscope, that the egg shell consisted of a fine reticulum of fibres with no evidence of lamellae. Lying beneath the shell is a layer of featureless material 1000 Å thick and described as the peri-vitelline material, and below this, two unit membranes which are equated by Wilson to the 'vitelline membrane' of earlier workers. Wilson suggested that the secretions described by Clegg (loc. cit.) contributed to the peri-vitelline material and that the two unit membranes are cellular in origin. The membranes are separated by a gap of 300 Å which is filled with debris and inclusions. The viscous cushion described by ROWAN (1956) was laid down, at the end of development, between these two membranes. Ultrastructural investigations on the reproductive system of *Haematoloechus medioplexus* by BURTON (1967) have shown that Mehlis' gland complex contains three basic cell types. One type is interstitial and non-secretory whereas the remaining two are secretory, producing morphologically differing forms of secretion. One cell type produces a dense body enclosed by a plasma membrane whereas the other produces membrane bound, flattened 'golgi-like' vesicles. Both types of cell possess long tapering cytoplasmic processes which pass through the epithelium of the ootype and discharge their secretion into the lumen. The terminal regions of these processes are supported by a peripheral ring of microtubules about 200 Å in diameter. This arrangement of microtubules seems to be a common feature associated with certain types of secretory cells in the digeneans. They have been observed in strigeoid lappet gland cells, in 'penetration' gland cells of miracidia and cercaria as well as in similar cells comprising the Mehlis' gland of *Apatemon* (see Plate 2-2). Within the lumen of the ootype the dense secretory body seems to disintegrate and no obvious function can be attributed to it. Burton suggests a lubricatory function. The membraneous products from the other secretion become associated with the vitelline globules and with the shell in its early stages of formation. The eggshell itself is formed largely by the fusion of vitelline droplets and Burton was unable to detect a shell membrane such as was suggested by CLEGG (1965) and WILSON (1967b) might be present in *Fasciola hepatica*. Burton was also unable to confirm the hypothesis of DAWES (1940) that a membrane is formed first and that it is upon this that the vitelline globules coalesce.

A contrast in interpretation is provided by the observations of COIL and KUNZ (1963) on *Syncoelium spathulatum*. In this fluke the vitelline cells did not contain demonstrable quantities of phenolic substances nor proteins. The epithelium of the proximal uterus was positive for these substances and it seems that in this case shell formation might be largely a function of the uterus. In the majority of studies the role of the uterus in egg-shell formation is not regarded as being very significant. COIL (1965 and 1966), describing egg-shell formation in certain hemiurid and notocotylid trematodes, observed that in *Hydrophitrema gigantica* the uterus could be divided into three main regions differing significantly in their histology. The proximal region was thin-walled and seemed to possess a brush border on its inner surface, the middle uterus possessed unicellular glands which were PAS positive and the distal part appeared to be non-glandular. As the eggs passed through the middle portion they became coated with the PAS positive secretion, so that the outer coat present on the egg was derived from this region. In *Ogmocotyle indica* it was the distal part of the uterus which was glandular and which contributed to the formation of the thick outer shell. In all cases however the initial shell was formed in the ootype. As a result of his observation Coil suggested that the uterus of certain trematodes could be the site of operculum and filament formation and could also play a part in the formation of outer coats to the egg-shell. The physiologically active nature of the uterus is suggested by the observations of HALTON (1967a,c) who found alkaline phosphatase activity in the uterus of all seven monogenean species tested and in six out of eight digeneans. Acid phosphatase activity was absent from the monogenean uterus but was present in the uterus of two out of the eight digeneans (see Tables 47, 48). Non-specific esterase was also demonstrated in the uterus of adult *Fasciola hepatica* as well as cholinesterase in the more muscular portions of the cirrus sac and ejaculatory duct. All these observations suggest that the uterus is much more than a 'cuticular tube' carrying eggs to the exterior.

In addition to the need for a wider spectrum of observations a number of outstanding problems remain. The nature of the mechanism preventing premature reaction of components of the tanning system, the stimulus initiating the reaction, the way in which the final, external and very constant shape and size of the egg is provided and how regional differentiation into capsule, operculum and cementing substance takes place, all remain to be determined. The fracture zone between the egg capsule and the operculum (GÖNNERT, 1962) is said to be produced by the ovum extending a pseudopodium which interrupts the formation of the capsule at this point and so induces a weakened zone which could eventually break to release the miracidium.

DEVELOPMENT AND HATCHING

The first division of the fertilized ovum is unequal, producing a macromere termed the 'ectodermal cell' and the micromere described as the 'propagatory cell'. Detailed cytological studies by ISHII (1934), CHEN (1937), REES (1940), PIEPER (1953) and VAN DER WOUDE (1954) have shown that the germinal cells are derived solely from the propagatory cells whereas the somatic cells may be derived from both the propagatory and the ectodermal cell. The germinal cells do not undergo further differentiation at this stage but persist as a germinal mass, whereas the ectodermal cell gives rise to the majority of the somatic structures.

This is the general pattern although some variation exists depending on the species being studied. The observations suggest and support the concept of the germinal lineage throughout the life-cycle associated with multiplication at certain phases. In this way the germinal cells at each stage (i.e. miracidium, sporocyst generations, redia) remain distinct from the soma and form a germinal mass which gives rise to the successive stage in the cycle by mitotic division. The various hypotheses which have attempted to explain the nature of the digenetic life-cycle have been reviewed by BROOKS (1930), CABLE (1934) and DAWES (1946) and more recently by HEYNEMAN (1960) and JAMES and BOWERS (1967c). The ectoderm cells divide to form most of the miracidial body. From the cellular mass, certain cells—the vitelline membrane cells—migrate to the periphery and fuse to form the vitelline membrane. Epidermal and subepidermal cells become differentiated and then nervous and excretory systems and the gland cell mass appear.

Most workers suggest that the majority of somatic structures are derived from the ectodermal cell although GUILFORD (1958) thought that parenchymal and excretory structures in *Heronimus chelydrae* originated from the propagatory cell. VAN DER WOUDE (1954) also postulated the origin of gland, excretory and parenchymal structures from this cell. It is apparent that there is no formation of germ layers in development and in this respect the embryology resembles that of the ectolecithal Turbellaria.

In this context it is unfortunate that the term 'ectoderm' cell is used to describe one of the products of unequal cleavage. A more suitable term would be somatic as used by HYMAN (1951) so that the first division produces a somatic and a propagatory cell.

The embryo within the egg capsule emerges from the host in a very varied state of development depending on the particular genus and species. In *Schistosoma* spp. and *Halipegus eccentricus* for example, the egg capsule contains a fully developed, active miracidium when it reaches the outside world. In others e.g. *Fasciola hepatica* (Plates 3-3 and 3-4) and *Diplostomum spathaceum* the egg capsule contents are little differentiated and require a developmental period in the external environment before miracidial formation is completed. In those species which possess a completely differentiated miracidium within the egg capsule, hatching takes place very quickly. DUTT and SRIVASTAVA (1962a) give the following figures for the hatching of the egg capsules of *Orientobilharzia dattai* at a temperature of 30–32°C. Five minutes after the start of incubation in water, 20 per cent hatched, after 15 min 71·8 per cent, after 30 min 94·5 per cent, after 60 min 98·8, and after 120 min 0·9 per cent hatched. There is therefore a distinct peak of hatching between 30 and 60 min after the eggs reach water. The egg capsules of *Haplometrana intestinalis* have to be ingested by the molluscan intermediate host before they hatch (SCHELL, 1961). In this case hatching takes place in the snail stomach within 1 hour after the eggs are ingested. Egg capsules which are undifferentiated when deposited, generally take from 10 days to three weeks to mature depending on temperature and the particular species. MATHIAS (1925) observed that the hatching of the egg capsules of *Strigea tarda* varies considerably with temperature, maximum hatching taking place within 8 days at 27°C but 20–21 days at 20°C. DUBOIS (1929) constructed a graph using Mathias' figures and from this deduced that the temperature relationships of hatching followed the law of van't Hoff and Arrhenius and that the Q10 at 17–37°C was 4 and between 16–26°C was 5, so that an increase of ten degrees of the environmental temperature would shorten the incubation period by a factor of 4 or 5. Below 10°C little development took place and this feature may be utilized in the laboratory for the storage of egg capsules for long periods. PEARSON (1961) has maintained the egg capsules of *Neodiplostomum intermedium* for 3 months at 5°C before allowing them to hatch at normal temperatures and FRIEDL (1961a) refers to this storage of egg capsules of *Fascioloides magna* for 2 years at 5–10°C before inducing development and hatching. These observations also suggest that egg capsules may over-winter at low temperatures and then hatch in the spring when conditions are more favourable.

The development of egg capsules in water does not apparently require very specific conditions other than a sufficiently high temperature and an adequate oxygen supply. The mechanism of hatching and the factors affecting it seems to be more complex and has attracted much interest. The egg capsule of schistosome species is particularly interesting from two points of view. It does not possess an operculum and therefore hatching occurs by the shell splitting, thus allowing the miracidium to emerge. Secondly the eggs contain a fully developed miracidium whilst still within the gut or bladder of the warm-blooded final host and some mechanism must exist to prevent the premature hatching of the eggs. STANDEN (1951), MAGATH and MATHIESON (1946) and INGALLS *et al* (1949) provide some information on the factors affecting the hatching of *S. mansoni* and *S. japonicum* ova (Figs. 21, 22, 23, 24). Salinity of the surrounding environment constituted an important factor in hatching. Eggs of *S. japonicum* matured in 0·7 per cent sodium chloride did not hatch but when the saline was replaced by rain water hatching took place immediately. Similarly eggs present in raw stools maintained at 25–32°C did not hatch until dilution of the faeces with water took place. The maximum rate of hatch was between 25–30°C. STANDEN (1951) working with *S. mansoni* observed that the presence of light also increased the rate of hatching. SUGIURA *et al* (1954) found that although the eggs of *S. japonicum* hatched in total darkness, light produced a 60 per cent increase in the rate. Faecal decomposition tended to suppress hatching but the specific factor involved (low O_2 tension, products of decomposition) was not investigated. These experiments suggest that premature hatching of these eggs is prevented by high temperature (37°C), high osmotic pressure and the absence of light within the mammalian host. Once these conditions become altered by entry into an external environment exhibiting the appropriate characteristics, hatching takes place.

The hatching of operculate eggs must involve the release of the opercular cap (Plates 3-4, 3-5 and 3-6) and it is generally assumed that there is a cementing substance keeping the operculum closed and that this substance has to be removed in some way. So far there has been no real proof of the existence of this substance. In those life-cycles where the egg capsule is ingested by a mollusc and where hatching occurs in the molluscan gut, it has been suggested that the cementing substance becomes digested away by the secretions of the molluscan gut. BYRD and SCOFIELD (1952) obtained numerical data on the hatching of ochetosomatid eggs in physid snails. They estimate that 87·9 per cent of the eggs of *Neorenifer aniarum* hatched after ingestion and corresponding figures for other trematode species were 84 per cent for *N. georgianus* and 80·1 per cent for *Dasymetra conferta.* These authors were using experimental hosts and observed that 39·9 per cent of the released miracidium of *N. aniorum* passed out to the exterior with the faeces, 39 per cent of *N. georgianum* miracidia and 70 per cent of *D. conferta* larvae. Hatching

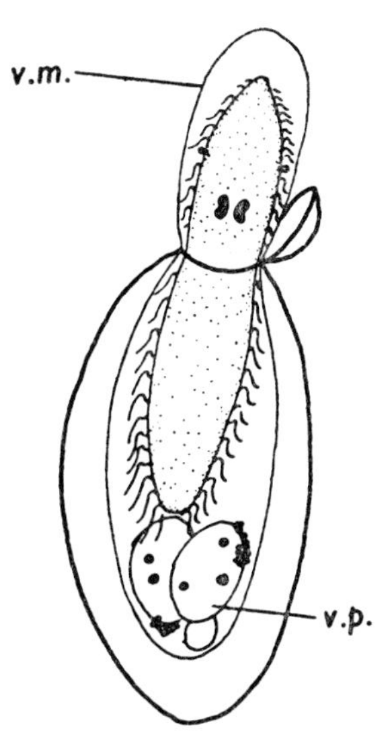

FIG. 21 The hatching of the egg capsule of *Neodiplostomum intermedium*. The operculum has opened and the miracidium is being ejected while still enclosed in the vitelline membrane (v.m.). Note the vitelline cells (vp) left in the capsule. From PEARSON, 1961.

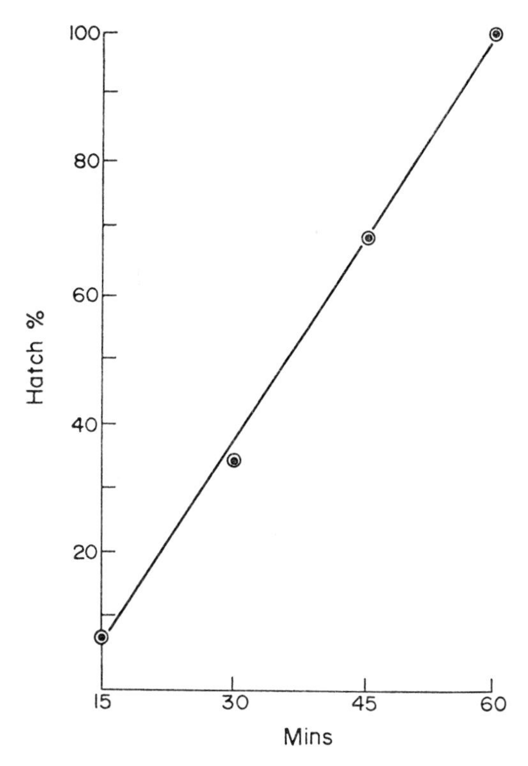

FIG. 22 The rate of hatching of the egg capsules of *Schistosoma mansoni* in fresh water during the first 60 min at 28°C From STANDEN, 1951.

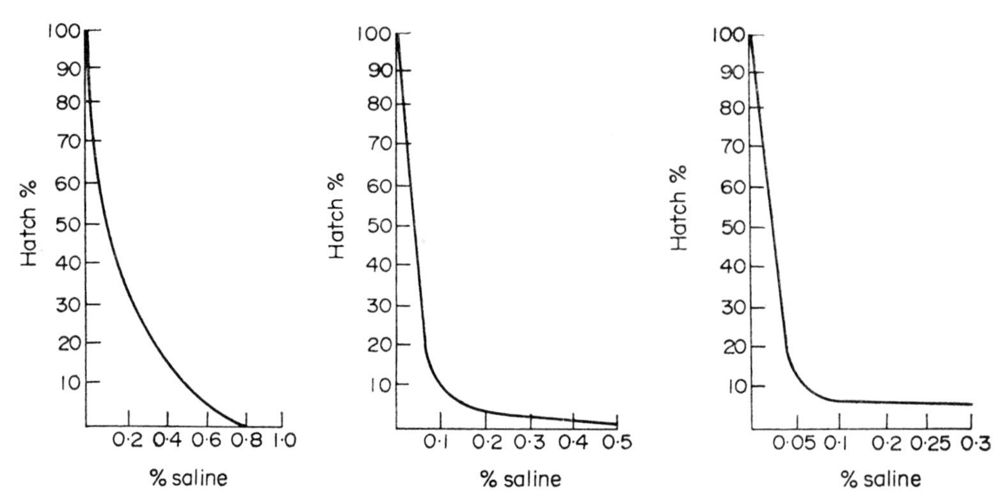

FIG. 23 The effect of variations in the salinity of the external environment on the hatching of the egg capsules of *Schistosoma mansoni*. From STANDEN, 1951.

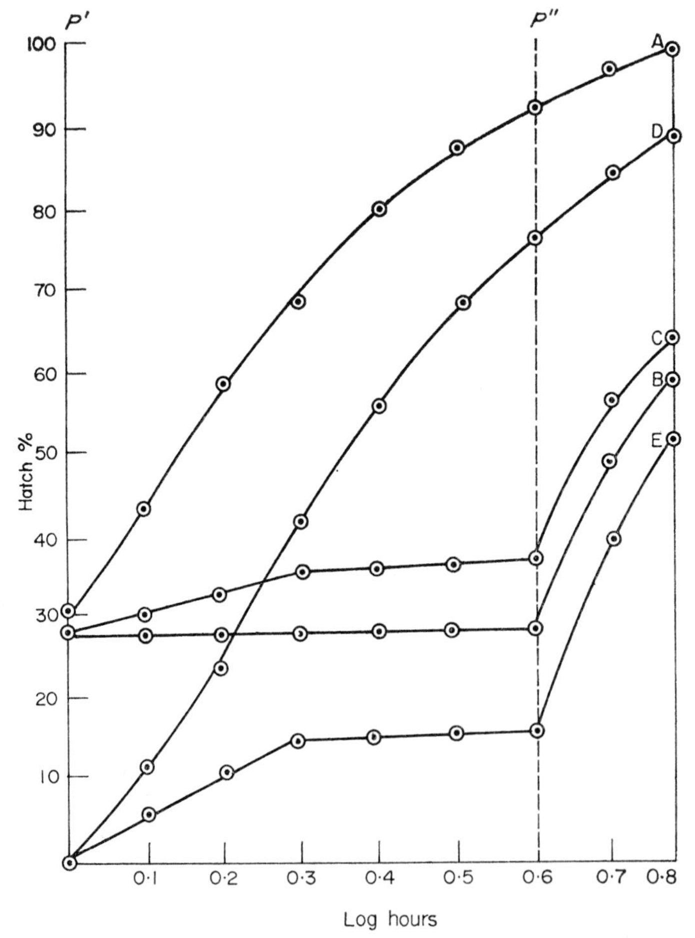

FIG. 24 The effect of temperature change (at points p' and pp'') of the external environment on the rate of hatching of the egg capsules of *Schistosoma mansoni*. Samples A, B, C, were at an initial temperature of 28°C (i.e. at point p'). Samples D and E were initially at 4°C. From STANDEN, 1951.

and penetration took place within 30 min and penetration of the gut by released larvae continued for several hours, but that miracidia which had not penetrated by this time were passed out with the faeces. These findings confirm that in the majority of ingested eggs hatching occurs rapidly and within the snail gut, but no explanation of the hatching mechanism was provided.

Further experimental data is provided by the observations of RUSSELL (1954) on the hatching of *Plagitura salamandra* eggs under *in vitro* conditions. Normally these eggs hatch only after ingestion by the molluscan host. Mature and motile miracidia emerged after 48 hours (18–20°C) incubation of the eggs in intestinal extracts of the intermediate host *Pseudosuccinea columella*. The time of incubation was reduced to 24 hours by the addition of sodium chloride to the extract. Extracts of other snails containing NaCl resulted in hatching but the emergent miracidia were non-motile. In other experiments hatching took place in mixtures consisting of 0·1–1·0 per cent NaCl or KCl maintained at a pH of 5·0–7·6 (NaCl) and pH 7·0–8·6 (KCl) with Clark and Lubs buffer at 18–21°C. However the released miracidia were not motile in any of the solutions.

Many observations suggest that light is an important factor influencing the hatching of operculate eggs in the external environment. In general terms it may be stated that fully developed eggs kept in the dark at 20–25°C either exhibit a very low hatching rate or do not hatch at all. Immediately they are brought into bright light mass hatching takes place. ROWAN (1956, 57) investigated this phenomenon in relation to the hatching of the eggs of *F. hepatica*. In a preliminary series of experiments he was able to show that diastase and hyaluronidase solutions had no effect on the opercular cement of intact and punctured eggs. Trypsin and pepsin preparations however, loosened the operculum of a relatively small number of intact eggs but when punctured eggs were exposed to the media the operculum was loosened in the majority of cases. From these experiments he suggested that the cementing and opercular bonding substance was proteinaceous in nature. In a second series of experiments eggs were exposed to water in which the metabolites from hatched eggs had accumulated. Few intact eggs hatched but the majority of punctured eggs did. These and other experiments suggested that under the influence of light (certain wavelengths being more effective than others) a 'hatching' enzyme is produced which attacks the inner surface of the opercular junction resulting in the opening of the operculum and the release of the miracidium.

In the egg of *F. hepatica* the vitelline membrane or membranes encloses the miracidium, the vitelline cells and a structure present at the opercular end of the egg known as the viscous cushion (Plate 3-4). This structure occupies about $\frac{1}{8}$th of the egg volume and is concave-convex in outline with the concave surface directed inwards. Hatching in this species follows the following pattern. The miracidium lies with its anterior end directed towards the concave surface of the cushion. As hatching commences the viscous cushion suddenly expands to approximately twice its size and 10–60 seconds later the operculum is released. The vitelline membrane ruptures, the cushion flows out, followed by the miracidium. Authors differ regarding the state of activity of the miracidium within the egg. Some maintain that it is active all the time whereas ROWAN (1956) states that although active initially it becomes quiescent at the moment of hatching and regains its activity as it passes out through the shell. He also suggests that during the actual ejection the muscular activity of the miracidium plays a secondary role. Some indication of the nature of the viscous cushion was given by ROWAN (1957). He described it as resembling a colloid containing some protein and that it changed from the contracted gel condition to an expanded sol as the ionic constitution of the fluid surrounding it changed. It was suggested that this expansion inside the egg may be a result of damage to the vitelline membrane or opercular seal as a result of the action of the hatching enzyme. MATTES (1926) confirms that the cushion expands immediately before the operculum opens, but suggested that the opening release of the operculum was the result of the increased pressure produced by the expansion of the cushion. WILSON (1968) has reinvestigated the hatching mechanism of the egg capsule of *F. hepatica* and has arrived at a number of conclusions which differed from those originally proposed by Rowan. He does not agree with the 'hatching enzyme' theory and proposes that light stimulated the miracidium to activity and that it is the miracidium itself which is responsible for altering the permeability of the membrane on the inner concave surface of the viscous cushion. Histochemical tests suggested that the cushion is a fibrillar mucoprotein complex which is normally in a dehydrated or semi-dehydrated state. The change in permeability of its enclosing membrane results in the hydration of this material and an increase in its volume. This produces an internal pressure culminating in the sudden rupture and release of the operculum and the resulting discharge of cushion and miracidium. In the hatching of *Spirorchis* eggs (ONORATO and STUNKARD, 1931) the viscous cushion forms an envelope protruding from the egg capsule but attached to it after the operculum has opened. They also suggest that the passage of the miracidium into this envelope was an essential intermediary step in the acclimation of the miracidium before it reached the hypotonic freshwater environment. These observations do not seem to apply to *F. hepatica* egg capsules. Other factors influencing the hatching of the eggs of these species have been mentioned by BEČEJAC and LUI (1959) who suggest that oxygen tensions below 0·2 mg/litre are lethal and by ROWCLIFFE and OLLERENSHAW (1960) who define three basic requirements: (a) the eggs must be free of faeces (b) they must be surrounded by a film of moisture and (c) the central critical temperature for development is 9·5°C.

Slightly different hatching mechanisms may be present in other eggs. In *Neodiplostomum intermedium* (PEARSON, 1961) the egg capsules do not contain a viscous cushion and the miracidium is extruded whilst still within the vitelline membrane (Fig. 25). A viscous cushion is also absent from the eggs of *Paramphistomum hiberniae* (WILLMOTT, 1952). Pearson states that the miracidium probed the vitelline membrane which eventually ruptured. The expulsion of a miracidium enclosed in a vitelline membrane has also been reported in certain spirorchid trematodes by WALL (1941a, b; 1951). In *Alaria arisaemoides* (PEARSON, 1956) the vitelline membrane was ruptured immediately after the release of the operculum so that the miracidium emerged from the egg without pause. HOFFMAN (1955) describing the life-cycle of *Fibricola craterna* observed that the fully developed eggs hatched in the dark as well as in the light. This observation suggests that the hatching enzyme, if present, may be produced in response to stimuli other than light. It may be relevant that, although the miracidia of *Paramphistomum hiberniae* do not possess eyespots, hatching can be induced by bringing the eggs from darkness into light.

Factors which influence hatching other than light and temperature are suggested by FRIEDL's observations (1961a,b) on the mass hatching of *Fascioloides magna* miracidia (Fig. 26). The exposure of the embryonated eggs to nitrogen results in a very rapid mass hatch of a large proportion of the eggs. The production of a partial vacuum above the culture medium also increased the hatching rate. Unfortunately no explanation of these results has been made. A rather unusual observation was recorded by TROMBA (1957) who reported hatching of the eggs of *Urotrema scrabridum* in 0·075 per cent formalin in 0·85 per cent saline. The miracidium emerged after 1 minute but none survived longer than 55 seconds after hatching! It seems likely that the hatching of these operculate eggs may be associated with differing mechanisms and that precisely defined environmental conditions are necessary for particular species.

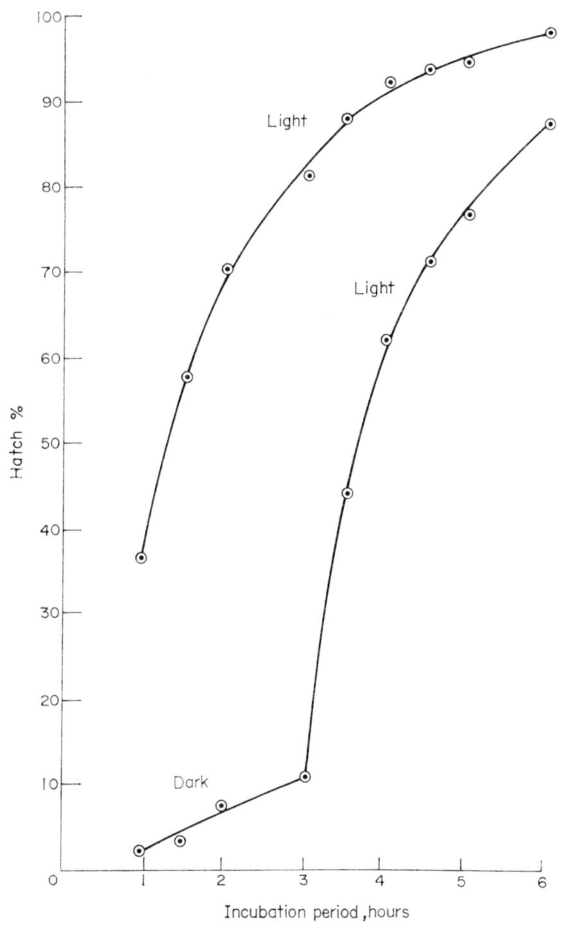

FIG. 25 The rate of hatching of *S. mansoni* egg capsules over a six hour period in the light (upper line) and the effect of darkness on hatching (lower line). From STANDEN, 1951.

The physiology of the egg developing in the external environment is poorly understood. A variety of enzymes has been described from the egg of *S. mansoni* (PEPLER, 1958; ANDRADE and BARKA 1962). The glycogen content of the egg decreases during development whereas the oxygen consumption rises (HORSTMANN, 1962) during this period. A histochemical study of the development of the ovum and vitelline cells has been made by RANZOLI (1956) but a more recent study by WILSON (1967c) of the development of the egg of *Fasciola hepatica* has involved a combination of histochemical and physiological techniques. The protein, carbohydrate, lipid and nucleic acid content of the egg changed in a regular manner during development and these findings are summarized in Fig. 27. The QO2 (μl of oxygen taken up/mg dry weight of tissue/hour) also altered during development, rising at first and then falling in the final phase. The idea that vitelline cells degenerated into a mass of yolk which was then utilized by the developing embryo is not favoured by Wilson. The vitelline cells themselves undergo an active and ordered metabolism of food stores and 'degeneration' of the cells occurs in the final stages only. He also suggested that the utilization of food reserves is organized by the embryo. The suggestion that the embryo undergoes a final, relatively quiescent phase (indicated largely by the fall in respiratory rate) is also supported by the fact that at this time two to five vitelline cells remain at the end

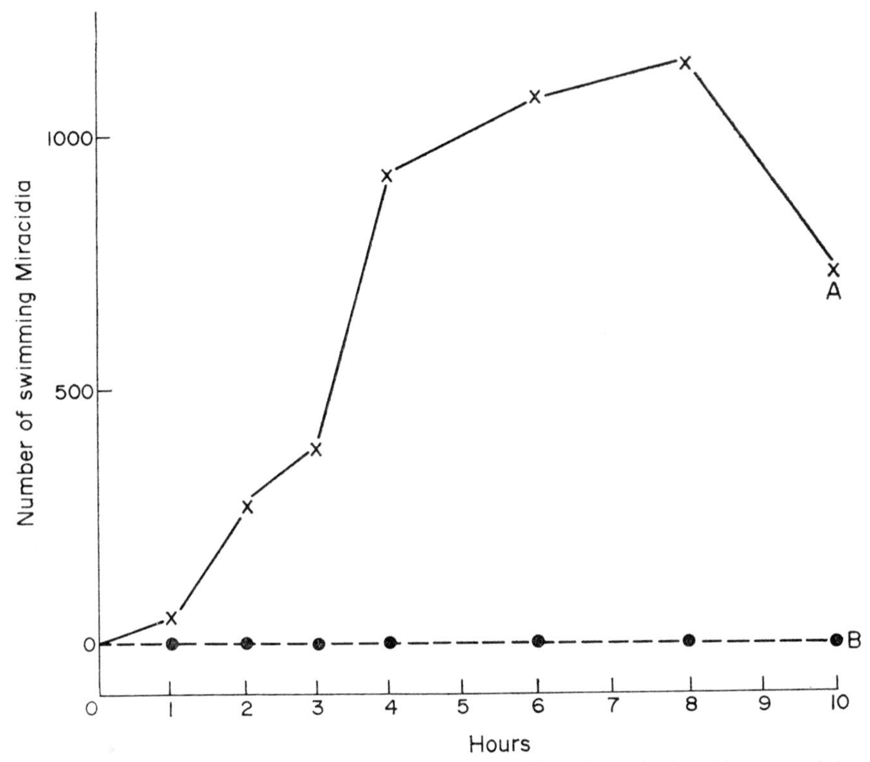

FIG. 26 The effect of exposure to an atmosphere of Nitrogen (line A) on the hatching rate of the egg capsules of *Fascioloides magna*. Line B represents the hatching rate when exposed to air. From FRIEDL, 1961.

of development and are not utilized. The activity of the protonephridial system varies during the developmental stage and may possibly reflect in a very indirect way changes in the metabolic activity of the embryo. WILSON (1967a) has recorded changes in the activity of the flame cells within the miracidium of *Fasciola hepatica* during development, hatching and free life (Table 7). The flame cells appear during the seventh day of development at a temperature of 25°C. The frequency of beat decreases as development proceeds so that the flame is almost stationary in

Table 7 Variations in flame cell activity correlated with size of developing miracidium (*F. hepatica*).

Size of embryo (μ)	Mean frequency of flame cells + S.E. (beats/sec)	No. of animals
60–69	6·0 ± 0·80	5
70–79	6·3 ± 0·65	11
80–89	6·2 ± 0·66	10
90–99	4·4 ± 0·56	16
100–109	3·2 ± 0·36	19
>110	2·8 ± 0·22	12

(From WILSON, 1967a).

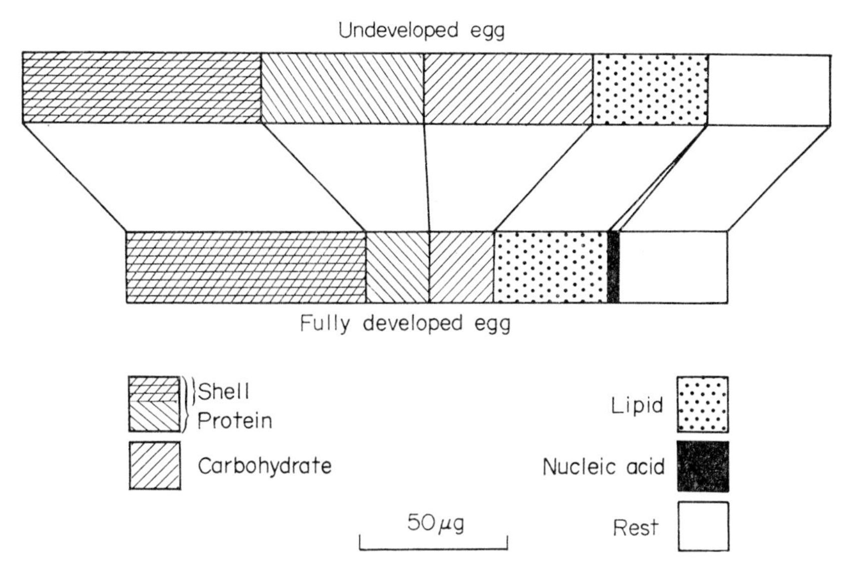

FIG. 27 A summary of the changes in chemical composition of 1000 eggs of *Fasciola hepatica* during development. The period of development represents 10 days at 25°C. From WILSON, 1967c.

the fully developed stage. As described earlier, exposure to light caused the eggs to hatch and during this period the flame cell activity increased rapidly. Light of high intensities had an inhibiting effect and the activation of the flame cell during hatching could be inhibited by a sharp increase in light intensity. This inhibition was reversible and the increased rate was resumed when the light intensity was reduced to normal levels. Wilson regards these changes as indicating that the control mechanism of flame cell activity involved the nervous system to some extent.

THE MIRACIDIUM

MORPHOLOGY

This is the first disseminating and infective stage in the digenetic cycle and a considerable quantity of literature is available concerning its morphology. Nearly all of this is the result of light microscope observations and there is no doubt that investigations with the electron microscope will reveal many details of considerable significance in the interpretation of the biology of this stage.

The body of an active miracidium is extremely variable in outline and contractions and extensions of the body occur almost continuously. Fixation with histological solutions generally produces a very distorted form and the most accurate record is obtained with electronic flash photography of swimming miracidia not subjected to coverslip pressures (Plates 3-7 and 3-8). Under such circumstances when the miracidium is swimming in a straight line the body outline is almost cylindrical with the anterior end slightly rounded and a tapering posterior extremity. In less active forms the body is shortened and more pear-shaped in outline with its greatest width just anterior to the eyespots and tapering to either extremity. The more rounded anterior generally bears at its centre a conical, non-ciliated projection termed the anterior papilla or rostrum, rostellum, terebratorium, tactile-organ or head papilla by various authors (see Plate 3-7). This anterior papilla may be very flexible and capable of introversion and eversion in some species (e.g. *F. hepatica*; *Neodiplostomum intermedium*) but in others seems to be fairly rigid e.g. that of *Parorchis acanthus* (REES, 1940). At its tip the apical papilla bears a median aperture,

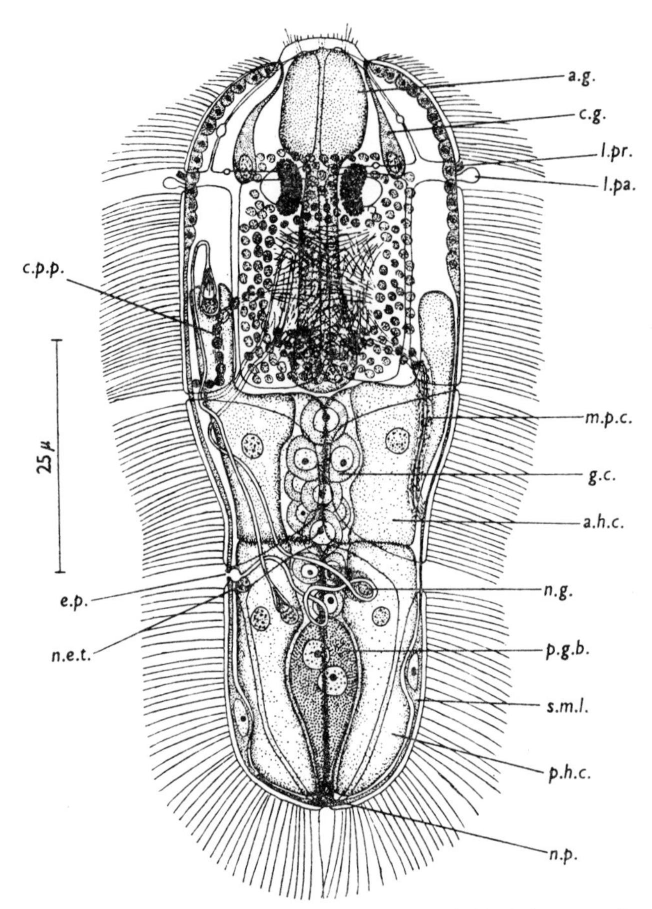

FIG. 28 Semi-diagrammatic representation of the miracidium of *Neodiplostomum intermedium* (dorsal view). (a.g.: apical gland; a.h.c.: anterior hyaline cell; c.g.: cephalic gland; c.p.p.: commissure to postero-lateral papilla; e.p.: excretory pore; g.c.: germinal mass; l.pa.: lateral papilla; l.pr.: lateral process; m.p.c.: main posterior commissure; n.e.t.: nucleus of excretory tubule; n.g.: nucleus of granular cell; n.p.: nucleus of posterior cell to which p.g.b. is attached; p.g.b.: posterior granular body; p.h.c.: posterior hyaline cell; s.m.l.: submuscular layer.) From PEARSON, 1961.

sometimes described as the mouth, and lateral pores representing the apertures of 'penetration' gland cells (Fig. 28).

The surface of the miracidium is covered by a regularly arranged series of epidermal plates. In the majority of miracidia these plates are ciliated but in a few unusual forms the plates may bear spines. In *Otodistomum cestoides* (MANTER, 1926) the spines or bristles are arranged in five rows radiating from the anterior tip of the body but in *Halipegus eccentricus* (THOMAS, 1939) the anterior end is crowned with eight pen-shaped spines 8 μ long and posterior to this zone there are alternate rows of smaller spines (Fig. 29).

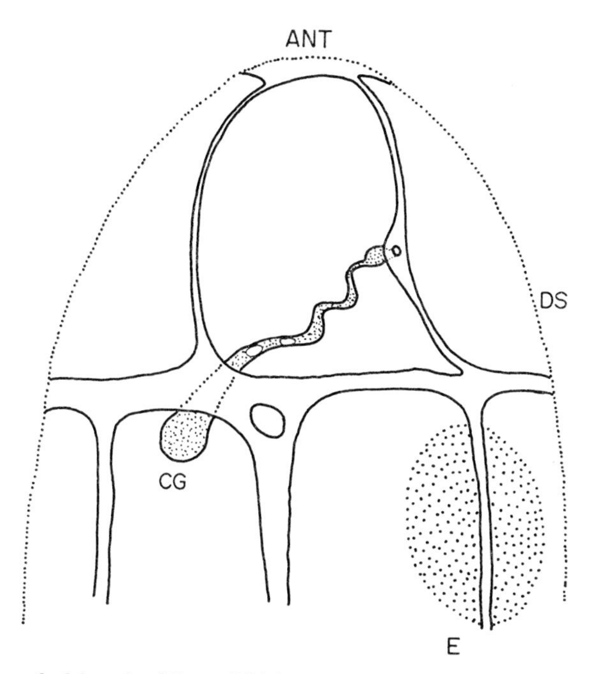

FIG. 29 Detail of anterior end of the miracidium of *Diplostomum* sp. The cephalic gland (cg) opens to the exterior at an intercellular junction. Ant: anterior; DS: dorsal surface; E: eyespot.

In the ciliated type of miracidium the cilia are generally shorter nearer the anterior end. Unusual arrangements occur in the Bucephalidae where the cilia are borne in tufts at the end of stalks and in *Leucochloridium morpha* where they appear as plumose structures. In the vicinity of the apical papilla of many miracidia are a number of long cilium-like structures which are immobile and probably represent sensory receptors of some kind (Fig. 31).

The ciliated plates or cells are unusual in that their edges are not in contact but are separated by extensions of the sub-epithelial layer (Figs. 30B & C; 30). These intercellular zones are particularly obvious after staining with silver nitrate solutions. Under the electron microscope it becomes apparent that these zones represent extensions of the cytoplasmic layer underlying the ciliated cells and do not represent intercellular cement as suggested by earlier authors. Observations of many workers indicate that the ciliated cells have a definite arrangement which may be characteristic to a limited extent of adult systematic units. The cells are arranged in 4 or 5 transverse rows and the total number ranges from 18–21. A list of described epidermal plate arrangements has been produced by BENNETT (1936). The arrangement is usually expressed as a formula, each figure representing the number of ciliated cells in a tier and the sequence of the numbers from left to right indicate the particular row working from anterior to posterior end of the body. The formula for many strigeoid trematodes is 6,8,4,3, and for *Fasciola hepatica* it is 6,6,3,4,2. The significance of these formulae in taxonomy has been critically discussed by LYNCH (1933) and DOBROVOLNY (1939).

The ciliated cells rest on a subepithelial layer of cytoplasm. This deeper layer extends outwards to form the intercellular zones described above and also contributes to the characteristic papillae particularly common in the anterior half of the body. The most commonly described projections are a pair of large clavate lateral papillae between the cells of the first and

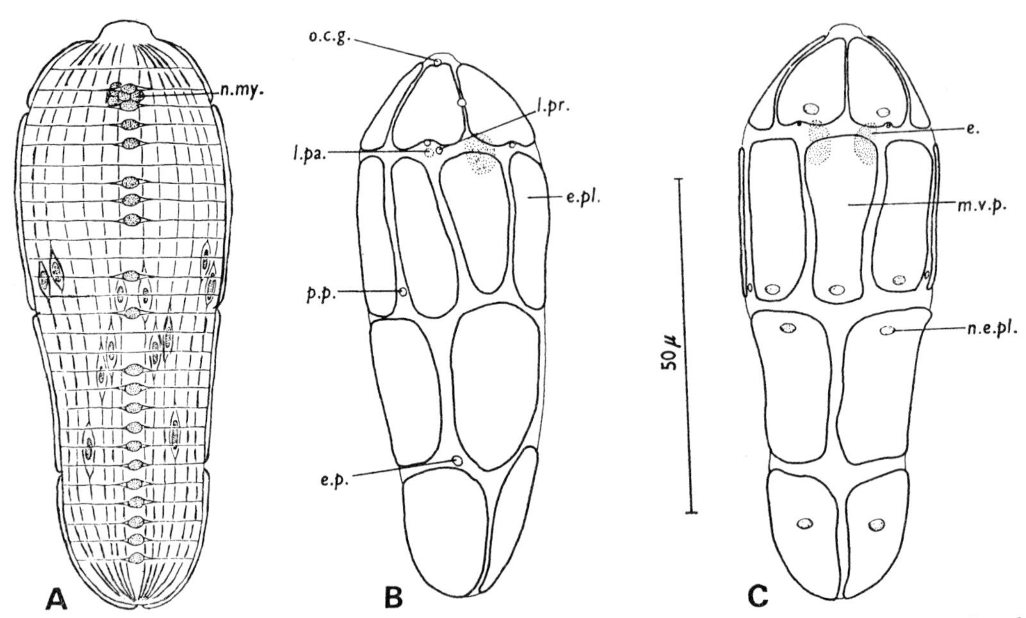

FIG. 30 The structure of the miracidium of *Neodiplostomum intermedium*. **A** Dorsal view of body musculature. **B** and **C** lateral and ventral aspects of the epidermal plates. (e: eyespot; e.p.: excretory pore; e.pl.: epidermal plate; l.pa.: lateral papilla; l.pr.: lateral process; m.v.p.: mid-ventral plate of second tier; n.e.p.: nucleus of epidermal plate; n.my.: nucleus of myoblast; o.c.g.: opening of cephalic gland; p.p.: postero-lateral papilla.) From PEARSON, 1961.

second tiers and closely associated with each of these papillae is a short slender process, slightly dorsal and anterior in position. In addition to these papillae and processes there may be other small projections and sometimes 'bristles' (DUTT and SRIVASTAVA, 1961). In *Parorchis acanthus* (REES, 1940) there is a circle of 24 such papillae between the first and second tiers of ciliated cells and a row of 24 between the third and fourth tiers. In some cases they may be less numerous than this and in *Neodiplostomum intermedium* (PEARSON, 1961) there is a row of six small processes associated with the posterior edge of the cells of the first tier. The function of these processes is not yet thoroughly understood. Some authors (CORT, 1919; FAUST and MELENEY, 1924; STUNKARD, 1923) have stated that substances were extruded from the lateral papillae. Others (e.g. PRICE, 1931; REES, 1940) describe the lateral processes as sensory and refer to nervous connections running from them to the brain. PEARSON (1961) however, was unable to detect such nervous connections in the miracidia studied by him. The functions of these papillae may therefore be sensory, responding to tactile or chemical stimuli and possibly facilitating host detection, or secretory.

Most authors describe a muscular layer immediately beneath the ciliated cells. In contrast to these observations PEARSON (1956, 1961) in a study of the miracidia of *Alaria arisaemoides* and *Neodiplostomum intermedium*, refers to a cuticle beneath the ciliated cells with the muscular layer internal to this. Electron microscope observations on the miracidium of *Fasciola hepatica* partly confirm Pearson's interpretation (Fig. 31). The ciliated cells rest on a cytoplasmic layer which corresponds to the 'cuticle' of Pearson and which may possibly represent the precursor of the cytoplasmic tegument of the succeeding sporocyst stage. This cytoplasmic layer rests on a basement membrane and it is below this that the muscles occur in this miracidium (Fig. 30A).

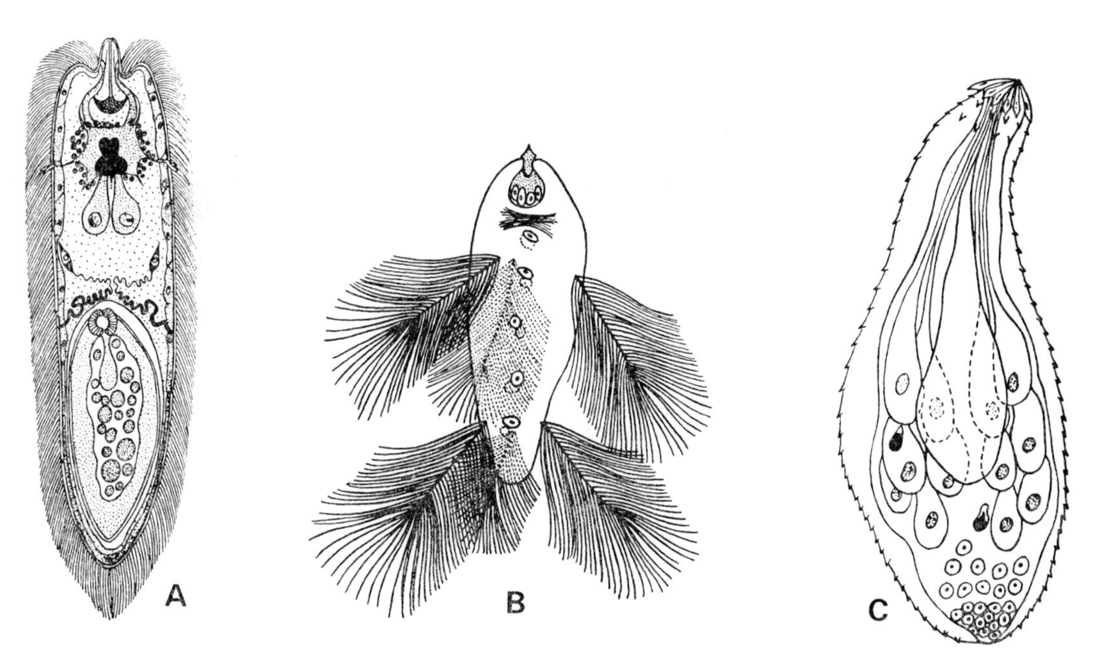

FIG. 31 Examples of miracidia. **A** *Parorchis acanthus* containing a completely formed redia; **B** *Leucochloridomorpha constantiae* with conspicuous 'plumose appendages'; **C** *Halipegus amherstensis* bearing spines and several gland cells. After various sources from BAER, 1952.

The muscles are arranged in an outer circular and inner longitudinal layer and the muscle fibre is spindle-shaped with a lateral extension containing the nucleus. Under the electron microscope the muscle fibre contains both large and small myofilaments. It is not clear whether or not the subepithelial layer is cellular or syncytial. The observations of WILSON (1967d) also confirm this arrangement in the case of the miracidium of *Fasciola hepatica.*

The most obvious structure within the anterior end of the body, lying in the mid-line is a granular mass, variable in shape and extending posteriorly to a position just posterior to the cerebral mass. The structure may contain up to four nuclei and lies near the posterior surface and tapers to a slender neck anteriorly. Earlier workers referred to this mass as a gut, 'primitive' or otherwise, but most observations now support the interpretation that it is a glandular mass of some kind—the apical gland. The exact nature of the secretion has not been determined, but it is suggested that the products may play a part in adhesion to and/or penetration into the molluscan tissues. On either side of this apical gland lie slender pear-shaped 'penetration' glands (Fig. 32). These open to the exterior on either side of the apical papilla and there is present in most cases one pair of cells. In some species e.g. *F. hepatica, Diplodiscus temperatus* and *Cotylophoron cotylophoron* the cells are small and difficult to distinguish. Early authors have referred to these cells as 'salivary' or 'pharyngeal' glands.

The majority of miracidia possess dorsally situated eyespots at a level corresponding to the posterior edge of the first tier of ciliated cells. There is usually a single pair of eyespots, each of which is roughly pear-shaped and composed of pigment granules partially enclosing a spherical lens. Although a single lens is usually present, VAN HAITSMA (1931b) described two lenses in each pigment cup of the miracidium of *Diplostomum flexicaudum.* Electron microscope studies by KUMMEL (1960b) and ISSEROF (1963, 64) have revealed additional details of the miracidial

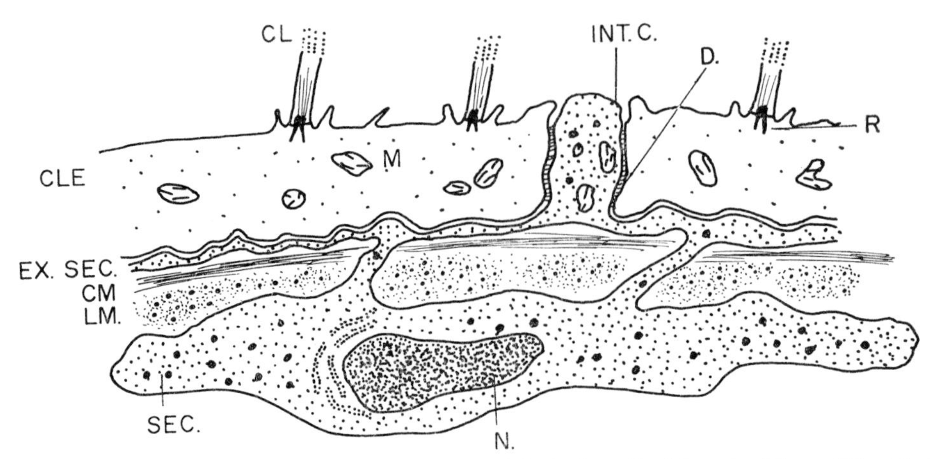

FIG. 32 Diagrammatic representation of the ultrastructure of the body wall of the miracidium of *Fasciola hepatica*. (CL: cilium; CLE: ciliated epithelial cell; CM: circular muscle; D: desmosome; EX.SEC: extension of the sub-epithelial cell; LM: longitudinal muscle; M: mitochondrion; N: nucleus; INT.C.: intercellular cytoplasmic extension between ciliated epithelial cells; R: rootlet system of cilium; SEC: sub-epithelial cell.) Redrawn with additions from an original drawing by Wilson, unpublished.

eyespots. In the miracidium of *Philophthalmus megalurus* the two pigment cups are closely opposed with the openings directed anterolaterally. The nucleus of each pigment cup is near its base and the cytoplasm contains a few mitochondria and a large number of pigment granules 1 μ in diameter. Each cup possesses two sensory cells which project beyond its edge. These cells are arranged anterior and posterior relative to one another and are separated at their base but not at their distal end by a septum derived from the pigment cup. The sensory cells are differentiated to form numerous microvilli 550 Å in diameter which extend towards the cup wall. This microvillous region is termed the rhabdome by ISSEROFF (1964). Outside the cup each sense cell becomes bulbous and it is this region which represents the lens. The sensory cells are also rich in mitochondria but the nucleus lies in the portion which continues towards the cerebral mass as a marrow extension outside and below the cup.

Beneath the eyespots lies the cerebral mass. This is approximately quadrangular in shape and from which run processes to the eyespots, the lateral papillae, in an anterior direction towards the apical papilla and posteriorly towards the posterior end of the miracidium. Most workers are unanimous in their interpretation of this structure but it is interesting to note that CHU and CUTRESS (1954) consider this structure to be neither nervous nor glandular in the miracidium of *Australobilharzia variglandis*. Under the light microscope the mass appears fibrous but the electron microscope reveals a structure similar to that described for the nervous system of *F. hepatica* cercaria by DIXON and MERCER (1965) (Plate 2-1).

The description of acetylcholinesterase activity in the cerebral mass of *F. hepatica* miracidium by PANITZ and KNAPP (1967) also confirms the nervous nature of this structure.

Posterior to the germinal mass lie the germinal cells which are large rounded cells with prominent nuclei. In some cases these cells may divide to form cell clusters described as germ balls.

In a number of strigeoid genera (*Alaria*, *Diplostomum* and *Neodiplostomum*) the posterior end of the body is occupied by a gland mass. This is approximately pear-shaped with the tapering posterior end opening out at the posterior extremity of the body. The cytoplasm of this mass may contain two to four nuclei. The function of this structure is unknown at present.

The protonephridial system at this stage consists of a bilaterally symmetrical arrangement of flame-cells and ducts opening out on each side of the body via a small aperture between the third and fourth tiers of ciliated cells. Each half of the system may possess a single or a pair of flame cells opening into a convoluted tubule which may dilate before the point of discharge into a small 'bladder'. The flame cells begin their activity within the developing embryo at quite an early stage as mentioned above. WILSON (1967a) found that the flame-cell activity within the miracidium of *F. hepatica* was not affected by changes in the external osmotic environment although the rate was directly proportional to the temperature. This author also suggested that osmo-regulation was not controlled by variations in flame activity but by variations in the tubule diameter. The activity of the flame produced a hydrostatic pressure which maintained the tubules in a dilated state. When flame activity ceased the tubule collapsed. In this way constriction of the tubule at some point would result in an increased hydrostatic pressure and consequently a reduction in the efficiency of the flame as an ultrafiltration mechanism.

Table 8 Per cent incorporation from [^{14}C] Glucose into the soluble Intermediates, excluding residual substrate.

Intermediate	Adults	Miracidia	
Hexose phosphate and phosphoenal-pyruvate	15	21	Associated with Glycolytic activity
Alanine	10	26	
Lactate	18	30	
Succinate	17	0	Associated with Krebs' cycle activity
Fumarate	1	0	
Malate	4	0	
Citrate	4	0	
Glutamate	10	0	
Y-Aminobutyrate	6	0	
Disaccharide	3	7	

(From BRYANT and WILLIAMS, 1962)

In some miracidia there may be precocious development of the propagatory cells. In *Parorchis acanthus* (REES, 1940) the cells differentiate to form a well developed redia which becomes released after penetration of the first intermediate host. Rediae have been recorded from the miracidia of *Stichorchis* (BENNETT and HUMES, 1939), *Typhlocoelium* (STUNKARD, 1934) and *Ophthalmophagus nasicola* (YAMAGUTI, 1940).

HISTOCHEMISTRY

Relatively few histo- and biochemical studies have been carried out on the miracidial stage. The presence of glycogen and fat has been demonstrated in miracidia by AXMANN (1947), GINETZINSKAJA and DOBROVOLSKII (1962) and BOGOMOLOVA (1957). PEARSON (1961) suggested that the four large hyaline cells at the posterior end of the miracidium of *Neodiplostomum intermedium* might represent cells rich in glycogen. It seems likely that the glycogen would function as an energy source during the free-swimming period as has been suggested by ERASMUS (1958) for certain strigeoid cercaria. BRYANT and WILLIAMS (1962) have shown that labelled C14 glucose becomes incorporated into glycolytic intermediates such as hexose-phosphatases, phosphoenol-pyruvic acid, alanine and lactic acid but no radio-activity was detected in acids of the Krebs' cycle suggesting very strongly the presence of active glycolytic pathways in the miracidium (Table 8).

LEWERT and LEE (1954) have described PAS positive contents in the lateral penetration gland cells and apical gland of *S. mansoni* miracidia. They observed that the penetration gland cells still retained their positive reaction after penetration and that droplets of PAS positive material emerged from the apical papilla presumably derived from the apical gland mass. DUSANIC (1959) records alkaline phosphatase activity in the apical gland, propagatory cells and nuclei and nucleoli of the penetration gland cells of *S. mansoni* although the cell cytoplasm was negative. These two sets of observations suggest that the term penetration gland may be a misnomer for these cells and that the histolytic aspects of penetration may be more closely associated with the apical gland.

BIOLOGY

The longevity of the miracidium is a vital aspect of the biology of this stage as the degree of dispersal of the species and continuation of the developmental cycle is directly related to the duration of the free-swimming period. A knowledge of the factors affecting this phase might be useful in determining control measures but in spite of this, little study has been made of this period. Data on non-schistosome species is relatively scarce and is usually given without detailed information on the characteristics of the surrounding environment. The miracidium of *Cotylophoron cotylophoron* (BENNETT and JENKINS, 1950) survives on average 10·4 hours and up to a maximum of 16 hours in water at a temperature of 23–27°C. WILLMOTT (1952), describing the miracidium of *Paramphistomum hiberniae*, stated that at temperatures between 20–22°C the miracidia swam actively for 6–8 hours. After this they moved slowly for another hour at the bottom of the container. Some were quite active after ten hours but all were dead after fifteen. The miracidia of lung-flukes may live for 24 hours at 25°C and longer at lower temperatures (YOKOGAWA *et al*, 1960). The observations on schistosome miracidia are more numerous (see OLIVER and SHORT, 1956) but there is considerable variation in the observations recorded. SCHREIBER and SCHUBERT (1949a), studying the survival of *S. mansoni* miracidia at 24–26°C, reported that approximately half were dead in 8 hours and all by 22 hours. In contrast, LAMPE (1927) stated that these miracidia survived for 40 hours at 33°C. Data on the survival of *Schistosomatium douthitti* miracidia has been published by OLIVER and SHORT (1956) and FARLEY (1962). At temperatures of 22·2–25·6°C Oliver and Short showed that 12·8 per cent of the miracidia had died in the first hour, 25·7 per cent by the end of the third hour; 49·2 per cent after 9 hours; a few survived 24 hours but all were dead after 25 hours. Graphs based on this experimental data indicated two death rates; the first is slower, and starts one hour after hatching and the second peak is more rapid and begins approximately 17–18 hours after hatching. These authors suggested that this initial rate may be related to and a consequence of the time the egg has been confined within the host tissues before deposition and hatching. It is possible that ageing of the miracidium begins while the egg is still within the host tissues and the more extended this phase is the higher will be the death rate of the miracidia soon after hatching. In addition it is known that schistosome egg capsules containing living miracidia within host tissues are antigenic and so senescence in miracidia at this stage may be the result of immunological responses by the host.

Farley (1962) also observed that miracidia hatching within the first hour had a half life of twelve hours (at 20°C; pH 7·5) whereas late hatchers had a half life of six hours. The second more rapid death rate is dependent upon and related to the exhaustion of the miracidial energy sources. Bennett and Jenkins (1950) found that the life-span of *C. cotylophoron* miracidia could be extended by the addition of glucose to the water. The survival of *F. hepatica* miracidia at 22°C was observed by Bryant and Williams (1962) under a variety of conditions. They found that the control survival times varied from 4–12 hours but that the survival period in the experimental solutions was always longer. Representing survival time in water as 100, survival in Hedon-Fleig saline containing 0·1 per cent glucose was 216 units; in aqueous 0·1 and 1·0 per cent succinate solution it was 172 units; in 0·1 per cent aqueous glucose solution it was 166 units and in 1·0 per cent aqueous glucose solution it was 144 units. These observations suggest strongly that under natural conditions survival is dependent upon the utilization of endogenous sources but that these can be supplemented from the external environment under appropriate conditions and the survival times extended.

The rate of utilization of endogenous sources must be linked closely to the environmental temperature. Standen (1952) has investigated this relationship in *S. mansoni* miracidia and Farley (1962) in the miracidia of *Schistosomatium douthitti*. Farley's figures show that the half-life figures of the miracidia varied from 1·5 hours at 35°C to 11 hours at 8°C (pH 7·3). Below 5°C the miracidia became moribund. He was also able to show that the pH of the surrounding environment played a significant role in survival. The optimum pH at 2°C seemed to be pH 7·5 and on either side of this the survival rate dropped rapidly. The results were similar in both buffered media and KOH solutions showing that the differences were due to changes in pH and not to salt concentrations in the buffers.

The mechanism by which miracidia locate the first intermediate host is still not known. Most observations are based on work with schistosome species and *Fasciola hepatica*. Unfortunately many of the observations are contradictory and factors influencing the behaviour pattern are not clear. It must be realized that the practical difficulties involved in observing and

Table 9 Effect of temperature and light intensity on phototaxis in miracidia of *Schistosoma japonicum*.

°C / Lux	15	18	20	22	23	24	25	26	28	30	34
4500	+	−	−	−	−	−	−	−	−	−	−
2500	+	±	−	−	−	−	−	−	−	−	−
2000	+	±	±	−	−	−	−	−	−	−	−
1000	+	+	±	±	±	±	±	±	−	−	−
500	+	+	+	+	+	+	±	±	−	−	−
250	+	+	+	+	+	+	+	+	±	−	−
100	+	+	+	+	+	+	+	+	+	−	−
50	+	+	+	+	+	+	+	+	+	±	−
25	+	+	+	+	+	+	+	+	+	+	−
10	+	+	+	+	+	+	+	+	+	+	±

+: positive phototaxis

−: negative phototaxis

±: some miracidia incubated in the vessel showed a positive response and others were negative in the same vessel.

(From Takahashi *et al*, 1961.)

recording the behaviour of fast moving miracidia are very great, although the use of the flying spot microscope as described by DAVENPORT *et al* (1962) would greatly facilitate observations. CHERNIN and DUNAVAN (1962) describe the ability of *S. mansoni* miracidia to infect *Australorbis glabratus* from 33 (vertical distance) to 86 cm (horizontal distance) away. The rate of movement immediately after hatching was approximately 1·3–2·8 mm per second. These authors concluded that (a) under ordinary field conditions water depth is not an important barrier to host-location by miracidia; (b) certain submerged 'margins' may represent the sites where miracidium-snail interactions are likeliest to occur; (c) in view of the marked scanning capacity of the miracidia the threshold below which snails may not become infected will vary according to the degree of miracidial pressure; and (d) in regions of high endemicity such density thresholds may be close to the extinction point of the snail population. In a study of the miracidial taxes negative geotactic as well as positive phototactic responses were demonstrated and it seemed as if the former behaviour pattern was dominant over the latter. A similar conclusion was reached by TAKAHASHI *et al* (1961) after studying the behaviour pattern of *S. japonicum* miracidia (Table 9). At temperatures below 20°C the negatively geotactic response was stronger than the positive phototactic behaviour although this relationship was disturbed when the light intensity rose above 5000 Lux. A particularly interesting comment by these authors was that the snail host *Oncomelania nosophora* tended to crawl out of the water within the temperature range 17–25°C. Thus both host and parasite exhibited similar responses under certain conditions and this parallelism would favour host-parasite contact. Between temperatures of 15–35°C the miracidia showed a positive phototactic response but this was related to some extent to light intensity. At 15°C they exhibited a positive response to any light intensity but at other temperatures the thresholds were as follows: 1000 Lux at 20°C, 500 Lux at 25°C and 100 Lux at 30°C. In all cases the phototactic response was stronger than the thermotactic one. At 10°C there was a lack of phototactic responses. Other observations on host-location by miracidia are discussed in the chapter 'Entry into the Host'.

Because the future of the entire life-cycle depends on the success of this first stage, it is interesting to consider the infectivity of miracidia and the numbers necessary to obtain a successful infection in the appropriate molluscan host. It is possible to obtain a successful infection after the exposure of a single snail to a single miracidium (CHERNIN and DUNAVAN, 1962) as long as host and parasite are compatible. In this experiment 95 per cent infection rate was obtained in 1·5 ml of water and this fell to 6 per cent when the experimental volume was increased to 700 ml. Increasing the number of miracidia per snail from one to five raised the infection rate from 42–87 per cent in 1·5 ml of water; from 21 to 79 per cent in 100 ml and from 6 to 29 per cent in 700 ml of water. Increasing the number of snails available to a single miracidium did not alter significantly the infection rate. Thus, under confined experimental conditions, one miracidium was just as likely to infect the only snail available as it was to infect any one snail if five were available. STANDEN (1952) reported the most successful infection of *Australorbis glabratus* with *S. mansoni* miracidia when single snails were exposed to 6 miracidia in 2 ml of water at 28°C. In a volume of water as large as 500 ml very low infection rates were obtained. From these observations it seems reasonable to assume that the dispersal of miracidia in large volumes of water would result in lower infection rates but MOORE (1964) has reported higher rates under experimental conditions of mass exposure in large volumes of water. When 300 snails were exposed to miracidia in tanks of three gallon capacity the percentage infection of the snails was greater than when the snails were infected individually. Other evidence from a slightly different viewpoint was provided by EWERS (1964b) after studying the incidence of larval trematodes in

Velacumantis australis in Botany Bay, Australia. Ewers found that the highest incidence of infection (25 per cent) occurred when the average density of susceptible snails was 15 snails/sq ft. It is possible that the relationship between infection rate and miracidium/snail density may not be as simple as is generally assumed.

The extent of the infection established in the mollusc will be related to the number of miracidia entering the snail. SCHREIBER and SCHUBERT (1949a) working with *Australorbis glabratus* and *S. mansoni* miracidia found that of snails exposed to 1 miracidium, 8 per cent produced cercariae by the tenth week after exposure. When each snail was exposed to 57 miracidia the percentage rose to 48–60 per cent and when 12 miracidia were used the number of snails shedding cercariae rose to 85 per cent. KENDALL (1949a,b) found that with single miracidial infections of *Limnaea truncatula* with *Fasciola hepatica,* approximately 30 per cent of the exposed snails became infected and generally not more than forty rediae developed in each snail.

The susceptibility of snails to infection with miracidia is governed by a number of other factors. It is apparent that as well as affecting the survival rate, temperature of the surrounding environment influences the infectivity of the miracida. STIREWALT (1954) has shown that in single miracidial infections of *A. glabratus* with *S. mansoni* the infection rate at 23–25°C was 9 per cent; in the range of 26–28°C the rate increased to 35 per cent. DEWITT (1955) observed that below 10°C no infection of *A. glabratus* with *S. mansoni* miracidia took place. Presumably a threshold of activity must be passed before the physiologically active processes of penetration and migration within the host tissues can take place and obviously the temperature of the surrounding water will decide whether or not the threshold is passed. The age of the snail will also influence the interaction between it and the miracidium. MCQUAY (1953) studying the susceptibility of *Tropicorbis havanensis* to the Puerto Rican strain of *S. mansoni* found that adults (6–9 mm diam.) and large juveniles (4–5 mm) were more susceptible to infection than median juveniles (203 mm) and small juveniles (1 mm or less). KENDALL (1950) was able to infect *Lymnaea stagnalis, palustris, glabra* and *pereger* with *F. hepatica* when exposed to miracidia within a few days of hatching. Beyond this period the snails were resistant to infection. He was able to infect 13 out of 101 *L. stagnalis* in this way and viable cercariae and metacercariae were produced which were infective to rabbits. Similarly 34 out of 101 *L. palustris* and 8 out of 46 *L. glabra* were infected and all provided metacercariae which were infective to mice. The most refractory snail was *L. pereger* and only 1 out of 108 snails became infected and then development continued only as far as the redial stage.

In non-susceptible strains of *A. glabratus* NEWTON (1952) observed that penetration by *S. mansoni* miracidia took place but that the parasites were destroyed by extensive cellular infiltration within 24–48°C hours after penetration. Thus in this case the barrier to infection was not represented by an inability to penetrate but by some factor or factors preventing development. This observation, along with that of BARBOSA and COELHA (1956), suggests an immunological response by the snail to the foreign miracidium. DUSANIC and LEWERT (1963) have demonstrated differences in the relative concentrations of the proteins and free amino acids of *A. glabratus* haemolymph one hour after exposure to *S. mansoni* miracidia and it is possible that such changes would initiate immunological responses in the snail tissues. The mobilization of actively phagocytizing amoebocytes in molluscs has been described by TRIPP (1961) in experimental procedures, and by PROBERT and ERASMUS (1965) against migrating cercariae. These phenomena are described in more detail in Chapter 8.

The ability of the miracidium to infect a mollusc will, in some cases, depend on the geographical race of the particular species. Many studies involving the susceptibility of geographical strains of *Australorbis glabratus* to *S. mansoni* infections have been reported. NEWTON's (1952) observations in this field have been mentioned already. In his experiments he found that a Puerto Rican strain of *A. glabratus* was very susceptible (95 per cent) to infection but that the Brazilian strain was completely refractory. Cross breeding experiments (NEWTON, 1953) showed that susceptible offspring were obtained suggesting that susceptibility was an inheritable factor. BARBOSA and BARRETO (1960) demonstrated that *A. glabratus* from Salvador (Bahia, Brazil) was highly resistent to infection (1·7 per cent) with the Pernambuco strain of *S. mansoni*. Infection experiments with the Pernambuco snails produced infection rates of 83·9 per cent. The tissue response of the Bahian snails was strong and most of the invading larvae in the snail tissues were destroyed between 36–96 hours after infection. No responses were observed in the Pernambucan snails. Similar observations have been made for *Schistosoma japonicum* infections in *Oncomelania spp.* (HSÜ and HSÜ, 1962). The relationship between schistosomes and their molluscan hosts in Africa has been discussed in general terms by WRIGHT (1966b).

The metamorphosis of a miracidium into the sporocyst stage represents the release of a new series of developmental processes. MUFTIC (1969) has isolated a crystalline substance extracted with ethanol from crushed *Australorbis glabratus* which seems to influence this metamorphosis. When this substance was added to cultures, in snail haemolymph, of the miracidia of *Schistosoma mansoni*, metamorphosis into sporocysts occurred and development proceeded via mother and daughter sporocysts to the stage at which release of actively swimming, infective cercariae took place. This substance, when injected into ligated house-fly larvae, also exhibited the biological activity of ecdyson and induced metamorphosis in the insect. Insect ecdyson did not however, induce metamorphosis of the miracidia.

It seems possible therefore that metamorphosis of larval trematodes (at all stages) might be influenced by steroids in a manner similar to that occurring in other animal groups. MUFTIC (1969) has suggested that parasites may use certain steroids which are produced by their hosts as factors controlling metamorphosis and has suggested the term 'morphogenetic vitamins' for such substances.

3 Sporocyst and Redia

Within the molluscan first intermediate host the miracidium gives rise to developmental stages which ultimately produce large numbers of cercariae. The sequence of stages varies slightly—there may be two generations of sporocysts (mother and daughter) or the initial sporocyst generation may produce rediae, which then produce daughter rediae. Daughter sporocysts and rediae seem to be mutually exclusive and do not occur in the same cycle. The more unusual types of development representing a telescoping of the stages and/or progenesis has been discussed in an earlier chapter. This developmental sequence in the mollusc represents a period of consolidation in the life-cycle and a phase where the successful host location and entry of a few miracidia culminates in the productions of hundreds, thousands, and even millions of an active disseminating stage—the cercaria. Because these intra-molluscan stages are non-disseminative in function their structure is much less complex than that of the preceding and successive stages in the life-cycle. In a most remarkable manner the potentiality and momentum of the life-cycle becomes temporarily directed towards multiplication.

The phylogenetic significance of the sporocyst and redial stages has aroused much speculation. Recent assessments of the various concepts have been made by HEYNEMAN (1960) LLEWELLYN (1965) and JAMES and BOWERS (1967c). Most authorities believe that the intra-molluscan phases of the life-cycle are the more ancient parts of the life-cycle and represent a period when the development did not extend beyond the mollusc or where the molluscan stages were the only parasitic ones and the adults were free living. There is a strong similarity between the redial stage and the Dalyelloid turbellarian and this has been regarded as significant in postulating the origin of the Trematoda. The presence of the contrasting miracidial and sporocyst stages which precede the redial stage in the life-cycle make this philosophical path a somewhat thorny one.

The sporocyst

In most digenetic life-cycles this represents the first stage arising directly from the miracidium. Exceptions do exist and in *Parorchis acanthus* (REES, 1940) for example, the miracidium contains a fully developed redia which becomes released after the entry into the host and subsequent death of the miracidium.

A gradual series of changes may be observed after penetration has occurred, representing the development of the mother sporocyst from the miracidium. DAWES (1960) is rather unusual in regarding the penetrating stage as a sporocyst and not a miracidium. However the structures characteristic of the miracidium do not disappear simultaneously with penetration. During penetrating and immediately after, the epidermal plates bearing cilia or spines are lost, and the various lateral papillae and processes associated with the epidermis also disappear very rapidly. Other structures such as the apical gland, eyespots, cerebral body and posterior gland mass may persist for 24–48 hours after penetration [e.g. *Schistosomatium douthitti*, CORT, AMEEL and OLIVIER (1944), *Alaria arisaemoides*, PEARSON (1956) and *Glypthelmins quieta*, SCHELL (1962)].

The mother sporocysts generally occur in a different site to that occupied by the daughter sporocysts. They are frequently found attached to the wall of the alimentary canal protruding into the perivisceral coelome and occur near the buccal mass or the anterior regions of the intestine (Fig. 33A). In other trematodes, they may be present attached to the kidney or within

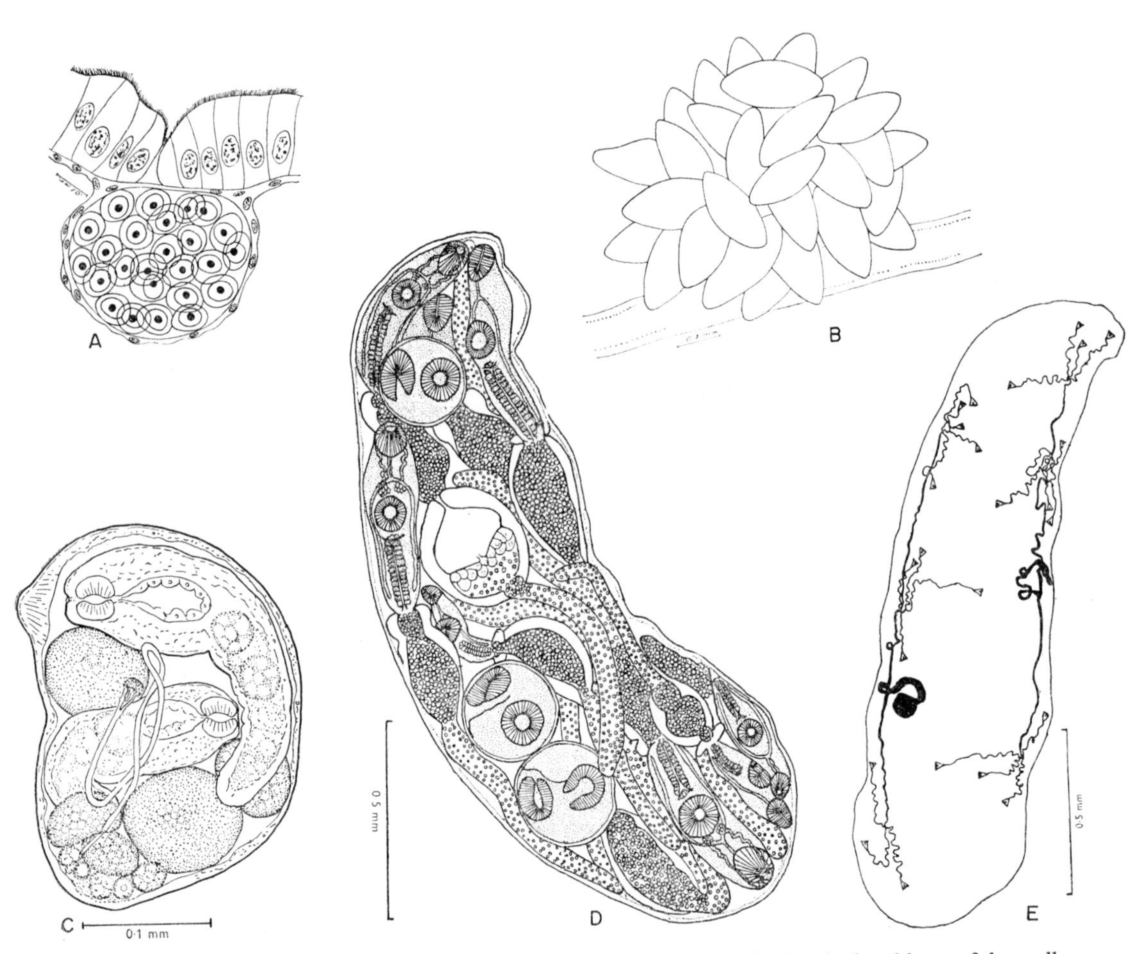

FIG. 33 **A** Spherical mother sporocyst, 7 days old, of *Glypthelmins quieta* attached to the basal layer of the molluscan gut. **B** Daughter sporocyst mass of *G. quieta* 23 days old attached to the snail host intestine. A and B from SCHELL, 1962. **C** Ten day old sporocyst of *Carmyerius exoporus*, from DINNIK and DINNIK, 1960. **D** Daughter sporocyst of *Phyllodistomum simile* containing cercariae and metacercariae. **E** The excretory system of the daughter sporocyst of *P. simile*. **D** and **E** from THOMAS, 1958.

the blood sinuses. This part of development is frequently overlooked in life-cycle studies, but the development of these early generations has received particular attention from Cort and fellow workers. Their studies have encompassed Hemiuroids (AMEEL, CORT, VAN DER WOUDE, 1949); Echinostomes and Psilostomes (CORT, AMEEL and VAN DER WOUDE, 1949); Plagiorchiids (CORT and AMEEL, 1944; CORT and OLIVIER, 1943a); Schistosomes (CORT, AMEEL and OLIVIER, 1944; CORT and OLIVIER, 1943b), Strigeids (CORT and OLIVIER, 1941) OLIVIER and MAO (1949), Fasciolatoidea (CORT, AMEEL and VAN DER WOUDE, 1948) and a general review (CORT, AMEEL and VAN

DER WOUDE, 1954). Observations on experimental infections are particularly valuable in that a time scale can be placed against the developmental sequence (e.g. CORT, AMEEL and OLIVIER, 1944; AMEEL, CORT and VAN DER WOUDE, 1949; SCHELL, 1962; PEARSON, 1956).

In the early development of *Schistosomatium douthitti* the youngest sporocysts were recovered from tissues near the origin of the oesophagus and near the cerebral ganglia, 65–69 hours after exposure of *Stagnicola palustris elodes* to miracidia. These young mother sporocysts were 0·12–0·2 mm long and capable of muscular wriggling. All traces of specialized miracidial structures had disappeared and the body cavity was traversed by cytoplasmic strands, extending from the body wall, which contained the germ cells. The number of germ cells ranged from 30–40 representing little increase on the original number present in the miracidium. As development proceeds the cytoplasmic strands tend to break up so that the germ cells come to lie against the body wall and this is accompanied by an increase in the number of germ cells (after 110–116 hours approximately 150 cells are present). At a later stage development continues with the formation of cellular masses termed 'germ balls' which protrude as spherical masses into the lumen of the sporocyst but remain attached to the wall by thin cytoplasmic strands. At the same time the mother sporocyst itself grows rapidly and although much variation exists, a 13 day old mother sporocyst was 4·7 mm long by 0·26 mm in diameter. Each germ ball consists of a closely packed mass of cells enclosed by a thin membrane containing a few flattened nuclei. It is from this outer membrane that the body wall of the daughter sporocyst develops. These spherical germ balls gradually elongate, become free and lie in the lumen of the mother sporocyst and eventually form the daughter sporocysts. Estimates of the number of daughter sporocysts in each mother sporocyst of *S. douthitti* ranged from 172 to 420.

This early development follows a similar pattern in strigeoid trematodes (CORT and OLIVIER, 1941; CORT, AMEEL and VAN DER WOUDE, 1951, 1954; PEARSON, 1956) but in *Diplostomum flexicaudum* CORT *et al* describe an increase in the number of germinal cells compared to that present in the miracidium. The miracidium contained about eight germinal cells whereas older mother sporocysts contained 20–24 germinal masses representing an increase of germinal cells prior to germ ball formation.

The mother sporocyst of strigeoid and schistosome cercariae are rather similar (Fig. 34). They are hollow, elongated structures with one end rounded and the other conical and more solid in appearance. The conical end frequently contains a narrow birth canal and birth pore. The outer surface is covered by a thin tegument, cytoplasmic in nature (see chapter nine), devoid of spines, and within which is a layer of circular and longitudinal muscles and finally a cellular epithelium lining the lumen. The excretory system is well developed and the number of flame cells increases on the miracidial number to an extent depending on the age of the sporocyst. Whether or not a fixed number of flame cells is reached is not really known, nor is the method by which multiplication of flame cells is achieved. The cytoplasm of the cells of the older mother sporocyst becomes filled with small refractile droplets giving a rather opaque appearance to the body wall.

The daughter sporocysts within the mother sporocyst are, at the later stages of development, quite active and well differentiated morphologically (Fig. 33, B, C, D, E). In *S. douthitti* they contained 127–238 germ cells and in *Cercaria stagnicolae* about 100 cells. The daughter sporocysts emerge through the birth pore of the mother sporocyst and then migrate through the molluscan tissues usually localizing in the digestive diverticulum (Plate 7-1). The effects of the presence of daughter sporocysts in this structure are extensive and have been reviewed by CHENG and SNYDER (1962a) and WRIGHT (1966a). The appearance of free daughter sporocysts within the

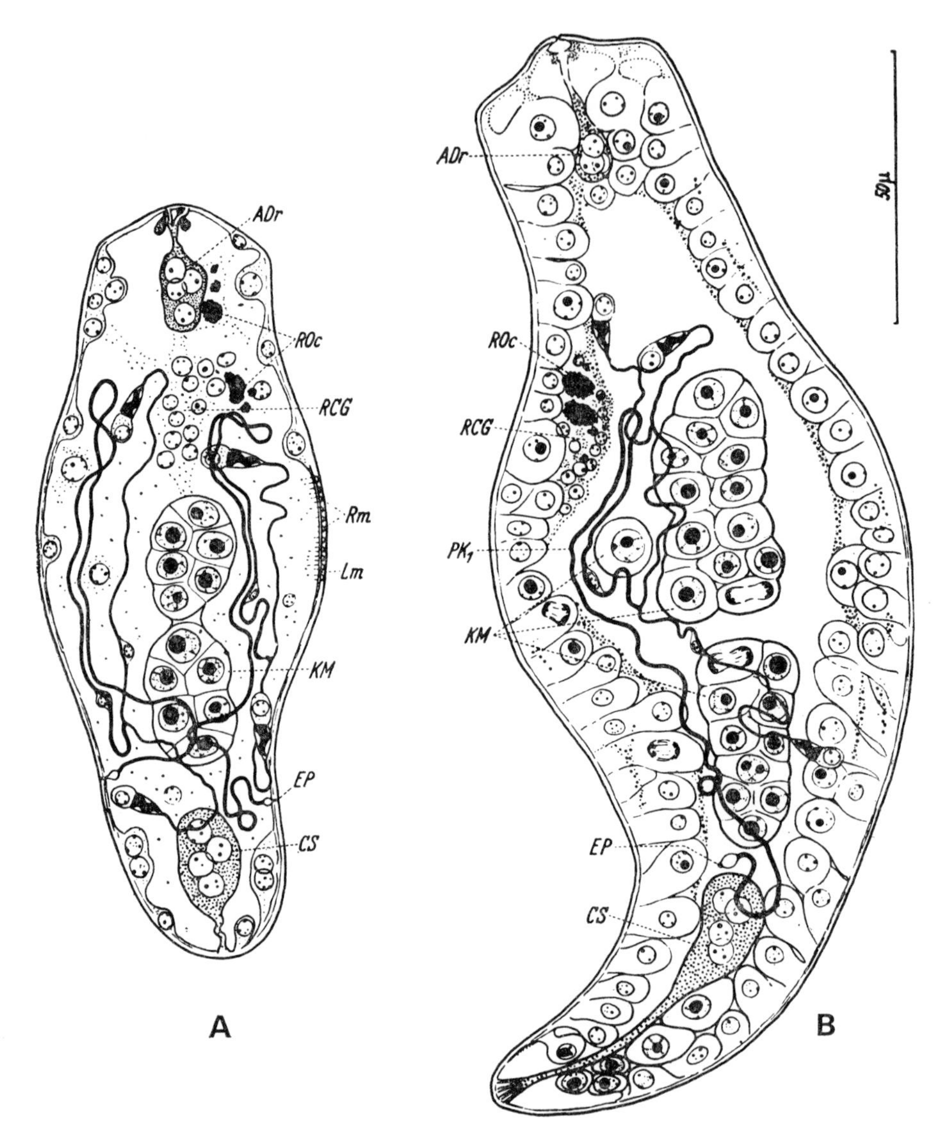

FIG. 34 Dorsal view of a four day old and lateral view of a seven day old mother sporocyst of *Posthodiplostomum cuticola*. (ADr: apical gland cells; CS: caudal sac; EP: excretory pore; KM: germinal mass; Lm: longitudinal muscle; PK: excretory canal; RCG: residual cerebral ganglion; Rm: circular muscle; ROc: residual eyespots.) From DÖNGES, 1964.

molluscan tissues may occur after 18 days (*Alaria arisaemoides*) or 21–22 days (*Schistosomatium douthitti*). They are elongate, hollow structures with a solid conical end containing a birth pore in most cases. The tegument may differ from that of the mother sporocyst in possessing fine spines at the conical end (*S. douthitti*; *Schistosoma mansoni*) or fine hair-like processes (*Alaria arisaemoides*). Within the digestive gland the daughter sporocysts increase in length often attaining a length of several millimetres. It is frequently difficult to dissect out from the molluscan tissues entire daughter sporocysts because they become very intimately bound up with the host tissue. Germinal development continues with the formation of cercarial embryos. The rate of development is obviously closely linked with temperature and in *S. douthitti* cercariae begin to emerge through the birth pores of the daughter sporocysts approximately three weeks after the release of the sporocysts from the mother sporocyst. The length of life of the sporocyst generations is difficult to assess except under experimental conditions. In *S. douthitti* mother sporocysts were observed up to 154 days after infection and in *Alaria arisaemoides* mother sporocysts were present after 13 months and persisted for the life span of the snail. The duration of the period over which cercariae are discharged will be related to the life span and productivity of the mother sporocyst and to the longevity and productivity of the daughter sporocysts. The extent of cercarial production will also be affected by the presence of birth pores in the sporocyst which would allow release of cercariae over long periods whereas if release is effected by rupture of the sporocyst wall cercarial production cannot be a continuous process. One of the most astonishing records of length of cercarial production is that by MEYERHOF and ROTHSCHILD (1940) who, after infecting *Littorina littorea* with a miracidium of *Cryptocotyle lingua* recorded cercarial production over a period of five years. No generalization can be made however, and the cercarial production of the molluscan phases will vary according to the parasite species and the features mentioned. Multiplication seems to occur in two phases. There may be a multiplication of the miracidial germ cells in the young mother sporocyst followed by a second increase in germ cell numbers in the daughter sporocysts with each germ cell giving rise to a daughter sporocyst or cercaria depending on the stage. In *Cercaria stagnicolae* there is a rapid multiplication of germ cells to form germinal masses which eventually divide and develop into daughter sporocysts. A similar pattern takes place in the daughter sporocyst. The strigeoid sporocyst on the other hand is characterized by a preponderance of germinal masses in both mother and daughter sporocysts and these provide a continuous supply of the succeeding stages by proliferation and breaking off of cellular masses.

The sporocyst generations associated with development in the Plagiorchiidae differ in a number of respects from those of schistosome and strigeoid cercariae. In the life cycles of *Dicrocoelium dendriticum*, *Platynosum fastosum* and *Lechriorchis primus* the young mother sporocyst is represented by a small sac containing germinal cells and which lies on the surface of the digestive gland. This sac eventually grows to form an irregular branching mass filling the space between the lobes of the digestive gland. The mother sporocyst of *Macroderoides typicus* also comprises three small string-like tubules radiating from a common centre (CORT, AMEEL, VAN DER WOUDE, 1954). In *Plagiorchis muris* the mother sporocyst becomes divided into compartments by ingrowths from the body wall. Each compartment contains germinal cells and as these divide and develop into embryos so they become enclosed by an ingrowth of the septal wall. CORT and OLIVIER (1943a) and CORT and AMEEL (1944) describe this outer covering of cells enclosing both mother and daughter sporocysts. This covering is referred to as the 'paletot' and both papers review the frequent early references to this structure. Although these workers stated that the paletot of the daughter sporocysts is derived from the epithelial lining of the mother sporocyst no suggestion was made for the origin of the paletot of the mother sporocyst. SCHELL (1962) on the other hand, describing the sporocyst generations of *Glypthelmins quieta* regarded the paletot enveloping both mother and daughter sporocysts as consisting of host basement membrane cells. This proliferation of the cells of the basement membrane of the molluscan gut wall is presumably a type of local host response to the presence of the parasite. The fact that in some cases the paletot becomes pigmented fits in with the concept of a host response.

The other interesting feature of this plagiorchiid type of sporocyst is the development of a large germinal mass in the daughter sporocyst which continually produces cercarial embryos.

The sporocysts of gymnocephalous and xiphidiocercaria are not so elongated as those of strigeoid and schistosome cercariae and may be almost spherical or ovoid in form. The branching of daughter sporocysts has been reported from the Bucephalidae and Brachylaemidae. ROBINSON (1949), describing the life-history of *Postharmostomum helicis*, observed the start of branching in daughter sporocysts 38 days after infection. The fully developed daughter sporocysts consist of a branched system of tubes with the lumen constricted at irregularly spaced intervals to form narrow channels.

The differential distribution of mother and daughter sporocysts represents their ability to migrate. This ability is particularly well developed in the sporocysts of *Plagioporus* and *Leucochloridium* (see Chapter on entry into host). The sporocysts of *Leucochloridium* are also brightly coloured with bands of green and brown pigment. Generally, sporocysts are pale cream or yellow in colour. NADAKAL (1960b) has investigated the nature of the pigments in a number of sporocysts and redia from the marine snail *Cerithidea californica*. The colour of the sporocyst and redia was usually orange or yellow although in some cases the redia was green. The main pigment isolated was B carotene and Nadakal suggested that the pigment was absorbed from a variety of carotenoids present in the snail. The snail itself derived them from the algal food. The B carotene appeared to be unaltered, but the chlorophyll, identified in a few cases, had a slightly different absorption spectrum from that in the green alga.

Very little is known concerning the physiology of the sporocyst stage and relatively few histochemical studies have been undertaken. Glycogen is present in the body wall of several sporocysts (ERASMUS, 1958; GINETZINSKAJA, 1960) but in others (AXMANN, 1947 and CHENG and SNYDER, 1962a) glycogen deposits have not been demonstrated. The presence of daughter sporocysts in molluscan digestive gland is generally associated with some degree of pathology. As well as changes in the cytological appearance of cells, CHENG and SNYDER (1962a) and CHENG (1963a) have observed a decrease in the endogenous glycogen of molluscan tissues parasitized by sporocyst stages. Similar observations have been made in other trematodes by FAUST (1920), HURST (1927), VON BRAND and FILES (1947) and JAMES and BOWERS (1967b). CHENG and SNYDER however observed that the increase in glycogen stored in the cercarial bodies corresponded with the decrease in glycogen within the digestive gland of the mollusc. In a later paper (CHENG and SNYDER, 1963) they reported the presence of glucose in the tissues of the sporocyst wall of *Glypthelmins pennsylvaniensis* as well as in the host tissue. These observations suggest that glucose, either obtained as such or derived from host glycogen, passes through the sporocyst wall and becomes stored as glycogen in the cercarial body. This concept may be supported by the demonstration by ERASMUS (1958) and DUSANIC (1959) of alkaline phosphatase activity in the sporocyst wall of a strigeoid and schistosome trematode respectively. The possible association of this type of enzyme activity with active transfer of materials through membranes has been commented upon frequently and the presence of activity in the wall of a stage without an alimentary tract is very suggestive. Similar observations have been made in other helminths without a gut (e.g. Acanthocephala: BULLOCK, 1949: and Cestodes: ERASMUS, 1957a,b,). These ideas regarding the uptake of carbohydrates via the sporocyst wall are supported also by the observations of JAMES and BOWERS (1967b). Even fewer observations are available with regard to lipoid metabolism. GINETZINSKAJA (1961) reported that the walls of sporocysts located near the host intestine were almost devoid of lipoids whereas those present in the digestive gland e.g. *Cotylurus brevis* and *Cercaria spinulosa* were rich in fats. Phospholipids, neutral fats and fatty acids have also been demonstrated in the tegument of the daughter sporocyst of *Cercaria bucephalopsis haimaena* (JAMES and BOWERS, 1967b).

CHENG and SNYDER (1962b) also reported the presence of fatty acids adhering to and on the interior of the sporocyst wall of *G. pennsylvaniensis*. Whether or not these observations represent fats passing through the sporocyst wall or whether they represent metabolites or metabolic products remains to be determined.

Observations on the amino acid and protein metabolism of sporocyst stages have been made by CHENG (1963b). In an analysis of the amino acid (bound and free) contents of hepatopancreas of *Physa gyrina* and the sporocysts of *Glypthelmins quieta*, Cheng was able to show that the amino acids identified in the sporocyst correspond with those removed from the digestive gland tissues. In addition the blood protein level of infected snails was significantly less than that of parasite free snails and these observations have been interpreted as representing the derivation of sporocyst amino acids from the molluscan digestive gland tissues and serum. In the bivalve *Musculium partumeium* infected with *Gorgodera amplicava* sporocysts, the bound amino acids in infected and uninfected sera corresponded but the content of free amino acids was nil in infected sera compared with 11 amino acids in uninfected sera. Similarly the mean protein content of serum from infected bivalves was 0·08 g/100 ml. The amino acids asparagine and proline were recovered from the serum of *M. partumeium* and the presence of both these substances in the sporocysts of *G. amplicava* suggests that some amino acids are obtained from the molluscan serum rather than from the tissues of the digestive gland. Considering the enormous rate of multiplication which takes place in the sporocyst stage it is not surprising that these observations suggest a high level of protein metabolism within these larval stages and it seems likely that the necessary amino acids are obtained from the molluscan digestive gland or serum or possibly from both. Histological observations by many workers reveal extensive cellular breakdown in the vicinity of sporocysts but the basis of this histolysis is not clear. It may be a compression

Table 9 Q_{10} values of trematode larval stages.

°C	*Z. rubellus* Sporocyst	*H. quissetensis* Redia	*Z. rubellus* Cercaria	*H. quissetensis* Cercaria
6–12	1·9	2·1	3·4	5·5
12–18	4·8	9·3	6·7	4·8
18–24	—	3·6	—	2·9
24–30	3·9	2·3	3·9	4·9
30–41	—	1·8	—	1·5

(After VERNBERG, 1961)

artefact, the result of the effects of the excretory products of the sporocyst on the host tissues or possibly the result of histolytic secretions produced by the sporocyst. Unfortunately the evidence available is insufficient to clarify this point.

In contrast to these observations NEGUS (1968) in a study of the nutrition of sporocysts of *Cercaria doricha* found no evidence of toxic or mechanical damage to the visceral hump of infected *Turritella*. The head-foot muscle of infected molluscs contained 7 per cent more free amino nitrogen than that of normal animals. The average concentration of soluble protein in the extra-pallial fluid of parasitized animals was higher than normal animals and there was no qualitative difference between the free purine content of normal and parasitized molluscs. The sporocysts of this species are found in close proximity to the gonadial tubules and Negus suggests that they derive their food supply directly from the gonad rather than from the digestive gland and other molluscan tissues. The wall of the sporocyst in contact with the gonad differed slightly in structure from the rest of the parasite body wall, and it was suggested that this region might correspond to a 'placenta-like' structure via which nutrients were obtained.

The respiratory rate of sporocysts has been investigated in *Zoogonus rubellus* by VERNBERG (1961). In this study it was found that the oxygen consumption rose with increasing temperature up to 18°C, maintained this rate, up to 24°C but at 30°C the rate rose sharply and then decreased at 36°C (Fig. 37). The sporocysts were unable to survive temperatures above 36°C and at 39°C were killed in 30 minutes. The QIO value between 12°–30°C was 3·9–4·8. The inability of this larva to survive temperatures above 36°C may be related to the fact that the adult of this stage occurs in marine fish (Table 9).

The redia

The mother sporocyst in many life-cycles gives rise to generations of rediae and not daughter sporocysts. Amphistome, monostome, gymocephalous, some cystocercous, echinostome, some microcercous and the cercariaea type of cercariae all arise from redia. In those cycles containing sporocysts only, there is one generation of daughter sporocysts but in redial development there may be one, two or even more generations of redia after the mother sporocyst stage, DINNICK and DINNICK (1957).

Redial morphology is slightly more complex than that of the sporocyst stage (Fig. 36). The body is elongate, generally tapering towards the posterior end and having a mouth at the more rounded anterior end. The oral aperture opens into a large, muscular pharynx which leads into a simple, sac-like caecum, which may be short or may extend almost the entire length of the body cavity. A short distance below the oral aperture the redia may possess a ridge-like collar and a little distance posterior to this may be the birth pore. At a point roughly two thirds the way along the length of the body there is generally a pair of lobe like lappets which are said to play a part in locomotion through the host tissues.

The external surface of the body is covered with a thin tegument (see chapter 9 and Plate 12-4) and beneath this lies the circular and longitudinal muscles and finally the cellular layer (Fig. 35). The interior of the redia is occupied by a large fluid filled body cavity into which protrude the germinal cells and germinal masses. Also floating in the fluid will be germinal cells, germinal masses and embryos. The excretory system is well developed consisting of flame cells opening into two trunks, one on either side of the body. These open to the exterior via two lateral pores. The daughter redia of *Paragonimus westermani* has the flame cell formula $2[(1+1+1+1)+(1+1+1+1)]=16$. (The derivation and explanation of the flame-cell formula is given on page 92 in chapter four.) In some redia (e.g. those of *F. hepatica*) there may be a cluster of small gland cells opening into the junction of caecum and pharynx. The body wall of the redia may be coloured as well as the gut within. The pigments described from some species (NADAKAL, 1960a,b,c; see reference in sporocyst section), is B carotene and chlorophyll. The pigment in the gut wall is in the form of spherical droplets whereas in the body wall it is restricted to stellate-shaped cells.

The basic pattern of development involving separate somatic and germinal lineages continues in the redial generation. In the Paramphistomes the germinal cells exhibit little

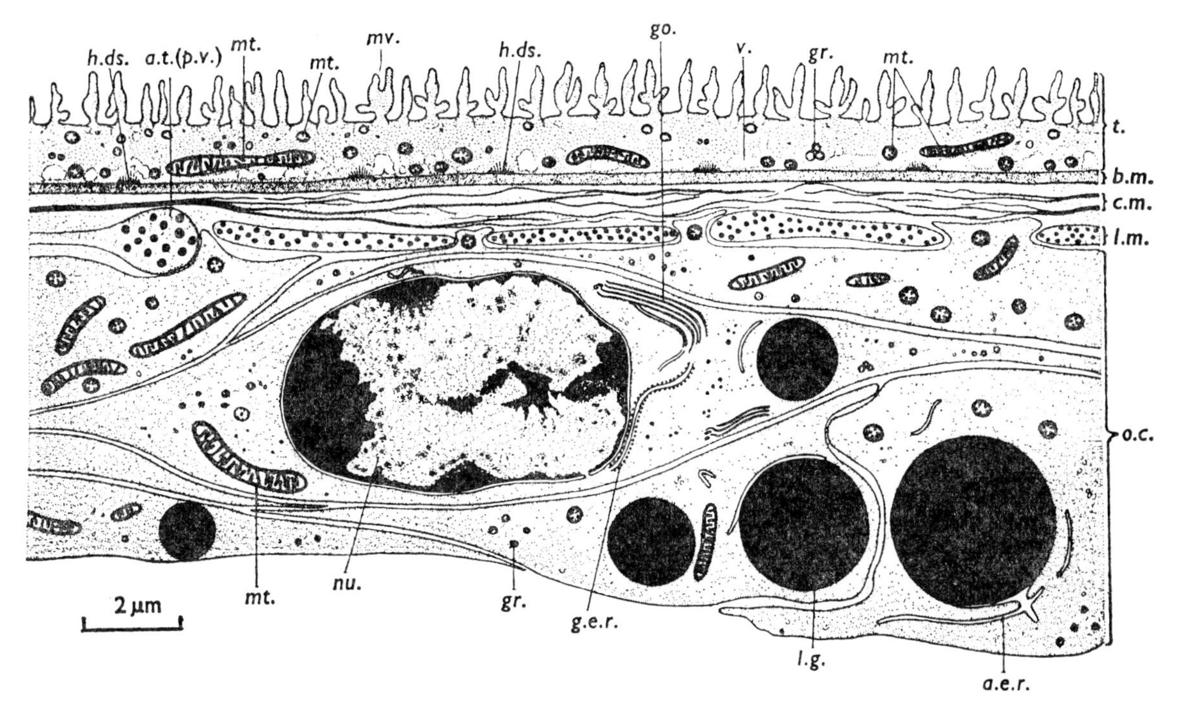

FIG. 35 Diagrammatic representation of the ultrastructure of the body wall of the redia of *Parorchis acanthus*. (a.e.r.: agranular endoplasmic reticulum; a.t. (p.v.) axon terminal containing synaptic vesicles; b.m.: basement membrane; c.m.: circular muscle; g.e.r.: granular endoplasmic reticulum; go.: Golgi complex; gr.: granule; h.ds.: half-desmosome; l.g.: lipid globule; l.m.: longitudinal muscle; mt: mitochondrion; mv.: microvillus; nu.: nucleus; o.c.: overlapping cell; t.: tegument; v.: vacuole.) From REES, 1966.

multiplication so that no true germinal masses are formed and there is no persistent centre of multiplication. The Echinostomes and Psilostomes however exhibit germinal multiplication with the formation of germinal masses which persist. These masses are irregular in shape and are generally tightly attached to the posterior end of the body cavity. Free embryos are produced from germinal cells which break off from the main germinal mass. These centres of multiplication persist in both generations of rediae so that large numbers of offspring may be produced. The degree of multiplication is taken even further in the Hemiuridae. In *Halipegus eccentricus* persistent germinal masses occur in both mother sporocysts and redia. The ultimate number of larvae produced is more a function of the size of the mollusc rather than of the reproductive capacity of the germinal masses. In seven large *Heliosoma trivolvis* (25 mm diameter) CORT *et al* (1948) recorded 558 to 3960 with an average of 1724 daughter redia of *Echinostoma revolutum*, whereas in snails half this size the numbers recorded ranged from 175 to 540 with an average of 360. WIŠNIEWSKI (1937) observed that the numbers of daughter rediae of *Parafasciolopsis fasciolaemorpha* were much higher (1700 to 8000, av: 4217) in snails 32 mm in diameter than in smaller snails 12–16 mm diameter (redial numbers 670–1800; av: 1058).

Some of the other factors which might influence the development of redial stages have been investigated by KENDALL (1949a). Artificially induced aestivation of snails which had just been exposed to miracidia did not inhibit the development of the redial and cercarial generations although the rate of development was slow and the numbers of rediae and cercariae produced

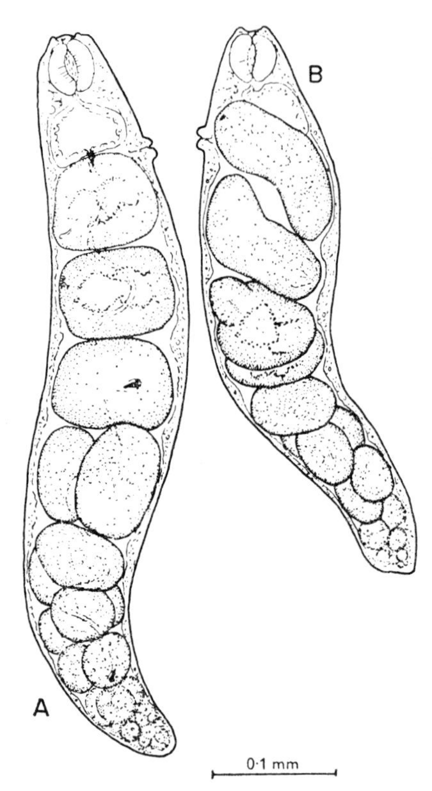

FIG. 36 Redial stages of *Carmyerius exoporus*. **A** First generation redia, 15 days old, containing cercarial embryos; **B** First generation redia containing daughter rediae and cercarial embryos, 25 days old. From DINNIK and DINNIK, 1960.

was reduced. In an already established infection aestivation retarded the rate of development of cercariae and also resulted in a loss of mature cercariae during this drought period. When aestivating snails were returned to water and provided with food, parasite development returned to normal providing that the environmental temperatures were suitable. Thus a basic level of host nutrition derived from outside is necessary for the establishment and development of a parasite infection. The size of the infection and the numbers of cercariae produced is however also affected by the size of the parasite population and the resulting competition for space and food within the host. The observations of CORT *et al* (1948) and WIŠNIEWSKY (1937) on the relationship between numbers of parasites developing and the size of the snail, complement these conclusions. In the development of *Fasciola hepatica* environmental temperature seems to have a more fundamental effect than simply affecting the rate of growth. KENDALL (1964) observed that in infections maintained at approximately 20·7°C no daughter rediae were produced, but in snails subjected to cooling at 4–5°C for 4½ hours daily, daughter rediae were produced. This plasticity in the type of progeny arising from mother rediae was first commented on by THOMAS (1883) and later confirmed by KRULL in 1933. Similar phenomena have been reported in *Clinostomum marginatum* by EDNEY (1950) and in *Fasciola gigantica* by DINNICK and DINNICK (1956).

Because of the more complex structure of the redial stage the acquisition of nutrients may occur via two routes. There is present the alimentary tract comprising mouth, an active muscular pharynx and caecum. Many histological observations (see review by CHENG and SNYDER,

1962a) indicate that cellular material of host origin is taken in through the mouth and presumably digested in the parasite caecum. The body wall of the redial stage resembles, under the optical and electron microscope, that of the sporocyst and there seems no reason to disagree with the idea that soluble nutrients may also pass in through it.

The experiments of MCDANIEL and DIXON (1967) have shown that redial tissues absorbed exogenous glucose and incorporated significant amounts into polysaccharide under *in vitro* conditions. The rate of glucose uptake was markedly depressed by phlorizin and was also depressed under atmospheres of nitrogen and those containing carbon dioxide. It has also been suggested by CHENG and YEE (1968) that the presence of leucine aminopeptidase activity in the body wall of the rediae of *Philophthalmus gralli* may contribute to the lysis of host cells and may facilitate migration through host tissues.

Histochemical tests have indicated alkaline phosphatase activity in the body wall of certain echinostome redia (CHENG, 1964) although PROBERT (1966a) detected acid phosphatase activity in the cuticle of *Echinoparyphium recurvatum* rediae. The redial gut exhibits various types of enzyme activity. The pharynx exhibits strong alkaline phosphatase activity (PROBERT, 1966a) and activity also occurs in the caecal cells (CHENG, 1964; PROBERT, 1966a). The caecal wall is also very rich in acid phosphatase and non-specific esterase activity is present in the pharynx wall and cuticle (PROBERT, 1966a). The contents of the caecal gut exhibit strong acid and alkaline phosphatase activity and the aggregates of cellular debris found in the digestive diverticulum of the host also exhibits alkaline phosphatase activity. These histochemical observations confirm the histological findings which suggest that the redial gut is an active structure and is used for the digestion of host cells and the absorption of the resultant nutrients although the phosphatases described may not be directly involved in absorption of nutrients (see page 246, chapter 9).

Fats have been demonstrated in redia by GINETZINSKAJA (1961) and the demonstration of esterases by Probert may be a significant association. The redial body wall and pharynx also contain glycogen (CHENG, 1963a) although in the case of *Echinoparyphium* parasitizing *Helisoma trivolvis* Cheng suggests that the glycogen is obtained by the ingestion of host digestive gland cells containing glycogen rather than by the absorption of glucose via the tegument as is postulated for sporocyst stages. The fact that rediae possess a gut does not however preclude the entry of nutrients through the tegument and the endogenous carbohydrate of the redia may be derived via several routes. Cheng also mentions that less glycogen (as indicated by histochemical tests) is removed from the digestive gland cells of *H. trivolvis* when infected with redia than when containing sporocysts.

Other indications of the nutritional requirements of rediae have been made by INGERSOLL (1956) and FRIEDL (1961a,b) who have attempted to maintain rediae alive in *in vitro* culture. FRIEDL (1961b) was able to show that the survival time of rediae in saline containing certain amino acids was greatly increased compared with that of rediae in simple Ringer solution. He found that the amino acids hydroxyproline and serine favoured survival the most, with alanine, glycine, leucine, phenyalanine, proline, threonine and the amide asparagine having a less pronounced and more variable effect. Isoleucine, methionine, tryptophan, valine and glutamine had no significant effect on prolonging survival times over that exhibited by the controls. Of five sugars tested only glucose and galactose produced any significant effect on survival times. There was no indication in the experiments whether the amino acids were taken in orally or through the body wall. INGERSOLL (1956) however, observed that the rediae of *Cyclocoelium microstomum* ingested some of the medium as well as occasional red blood corpuscles remaining in the serum incorporated in the culture medium, CHENG (1963b) found that the serum protein content of *Helisoma trivolvis* infected with *Echinoparyphium* sp. rediae ranged from 0·268 g/100 ml to 0·765 g/100 ml (mean 0·602 g/100 ml) compared with values of 0·098–3·380 g/100 ml (mean 1·70 g/100 ml) for sera from uninfected snails. He also observed that the sera of uninfected snails contained 17 bound amino acids and 12 free amino acids whereas the sera of infected snails contained 17 bound but only 1 free amino acid. The redia of *Echinoparyphium* contained all of 17 amino acids described from the host sera but the free amino acid composition of rediae was deficient in lysine, asparagine, serine and tyrosine. This evidence again suggests that nutrients may be taken in by ingesting host cells but also by absorption from the host serum through the body surface.

VERNBERG (1961) found that the QO_2 of rediae of *Himasthla quiessetensis* increased with each rise of temperature up to 41°C and remained fairly normal for periods at this temperature (Fig. 37). This ability to survive at high temperatures may be related to the fact that the adult

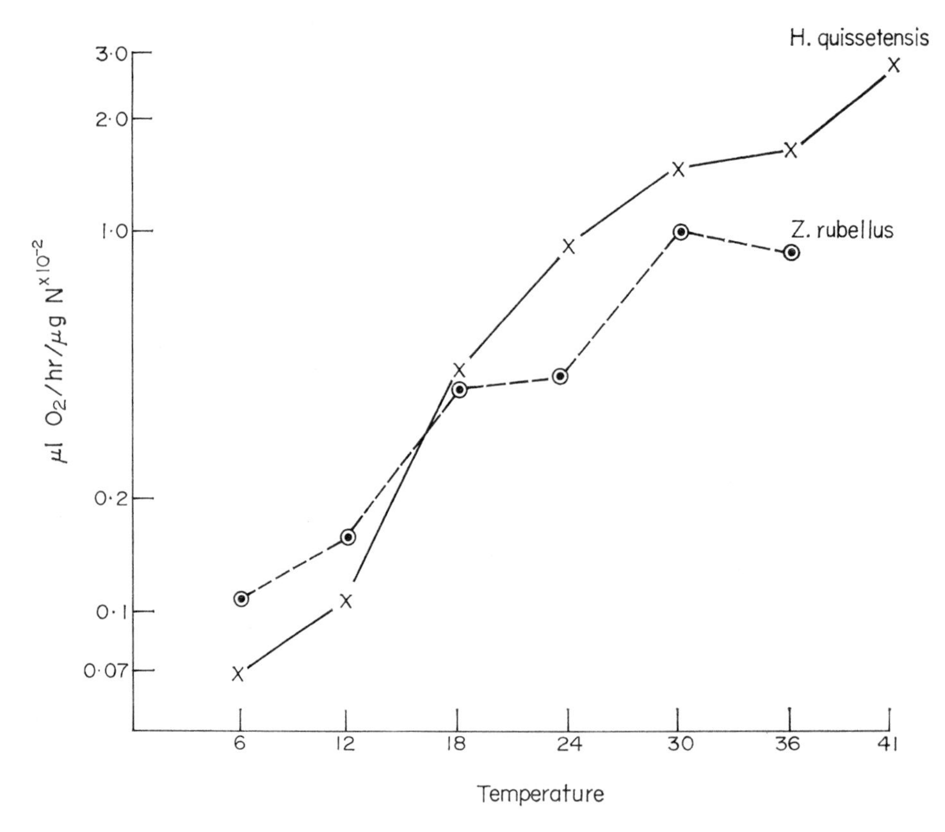

FIG. 37 The oxygen consumption in relation to nitrogen content of the sporocyst stage of *Zoogonus rubellus* and the rediae of *Himasthla quissetensis* at different temperatures. From VERNBERG, 1961.

of this stage is an avian parasite. The respiratory rate of these rediae was not closely dependent on oxygen tension, the QO_2 in 0·5 per cent oxygen was 61 per cent of the rate in 5 per cent oxygen and 39 per cent of the rate with atmospheric air as the gaseous phase (VERNBERG, 1963). This independence of the respiratory rate may be associated with the ability of redial stages to survive long periods of aestivation as described previously.

There is little doubt that the redial stage represents a more active phase than the sporocyst. This larval stage is capable of active movement and of browsing on the host cells using its mouth and pharynx. The small amount of experimental evidence available suggests that this stage is extremely tenacious and able to survive many of the adverse conditions to which its host may be exposed.

4 The Cercaria

The extensive sequence of multiplication within the molluscan first intermediate host culminates in the production of cercariae. The potentialities of the cycle are now transferred to this, usually free-swimming, stage and associated with this change of emphasis we see the appearance of new and highly significant morphological and behavioural features. The basic function of this stage is dissemination, although in many cycles the ability to enter and infect a host directly is combined with this function. The propulsive agent is the tail and this exhibits much variation in structure and will be described in detail later. The body proper has become very elaborate in structure relative to previous stage and possesses feeding, attachment, penetration and cyst forming structures (Fig. 38). Sensory apparatus in the form of tactile hairs and eyespots are also present and play a part in spatial orientation as well as in host detection and penetration. Although many of these structural features are adaptive and strictly related to the biological needs and function of the cercarial stage, some of the systems persist through the rest of the life-cycle and appear in the adult trematode. This aspect of morphological continuity through subsequent developmental phases has attracted the interest of systematists, and consequently the cercarial stage and its morphology plays a slightly more significant part in establishing phylogeny than do the preceding stages which exhibil a highly adaptive and non-persistent morphology.

The cercariae are differentiated from germinal masses arising in rediae or daughter sporocysts as described in the previous chapter. The cellular masses undergo complex development (see HUSSEY, 1941, 1943; KUNTZ, 1950, 1951, 1952; KOMIYA, 1938) eventually terminating in the fully developed cercaria, although in some cases, development may be completed outside the sporocyst and redia, within the snail tissues.

Cercarial types

There is a considerable range of cercarial morphology and the simplest expression of this variation is the classification proposed by LÜHE (1909) and presented below. The systematic value of this particular scheme is very small (CABLE, 1956; LA RUE, 1957; STUNKARD, 1963) and its usage in this text is solely as a descriptive scheme with no implication regarding adult systematics. The terminology therefore is descriptive and not systematic (after LÜHE, 1909) (Figs. 39 and 40).

MONOSTOME CERCARIAE

Tail simple but the body possesses an oral sucker only. The ventral sucker is absent. Cercariae developing in rediae.

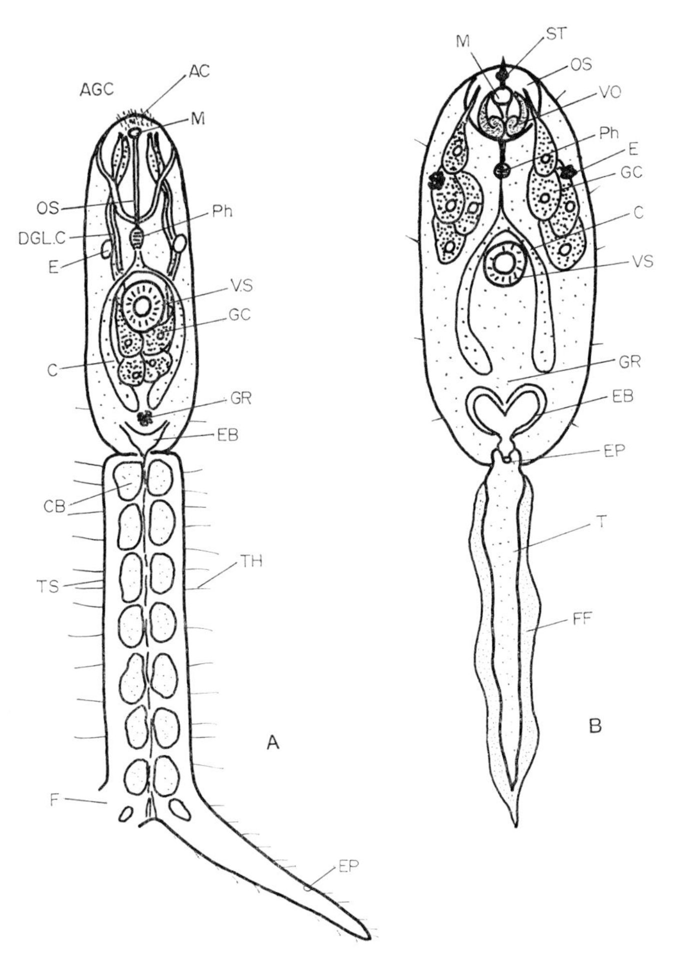

FIG. 38 The general morphology of the cercarial stage. **A** Ventral view of a furcocercaria of the strigeoid type; **B** Ventral view of a virgulate xiphidiocercaria with a fin fold on the tail. Both illustrations are diagrammatic and not to scale, and have the details of the excretory system omitted. (AC: apical cap of spines; AGC: aperture of gland cells; C: caecum; CB: caudal bodies; D.GL.C.: ducts of gland cells; E: eyespot; EB: excretory bladder; EP: excretory pore; F: furca; FF: finfold; GC: gland cell; GR: genital rudiment; M: mouth; OS: oral sucker; PH: pharynx; ST: stylet; T: tail; TH: tailstem setae; TS: tailstem; VO: virgula organ; VS: ventral sucker.)

AMPHISTOME CERCARIAE

Cercariae with a simple tail and a large body. The ventral sucker is large and lies at the posterior end of the body. Cercariae developing in rediae.

DISTOME

The majority of cercariae can be included in this category which is characteristic in having the ventral sucker lying some distance from the posterior end and in approximately the anterior third of the body.

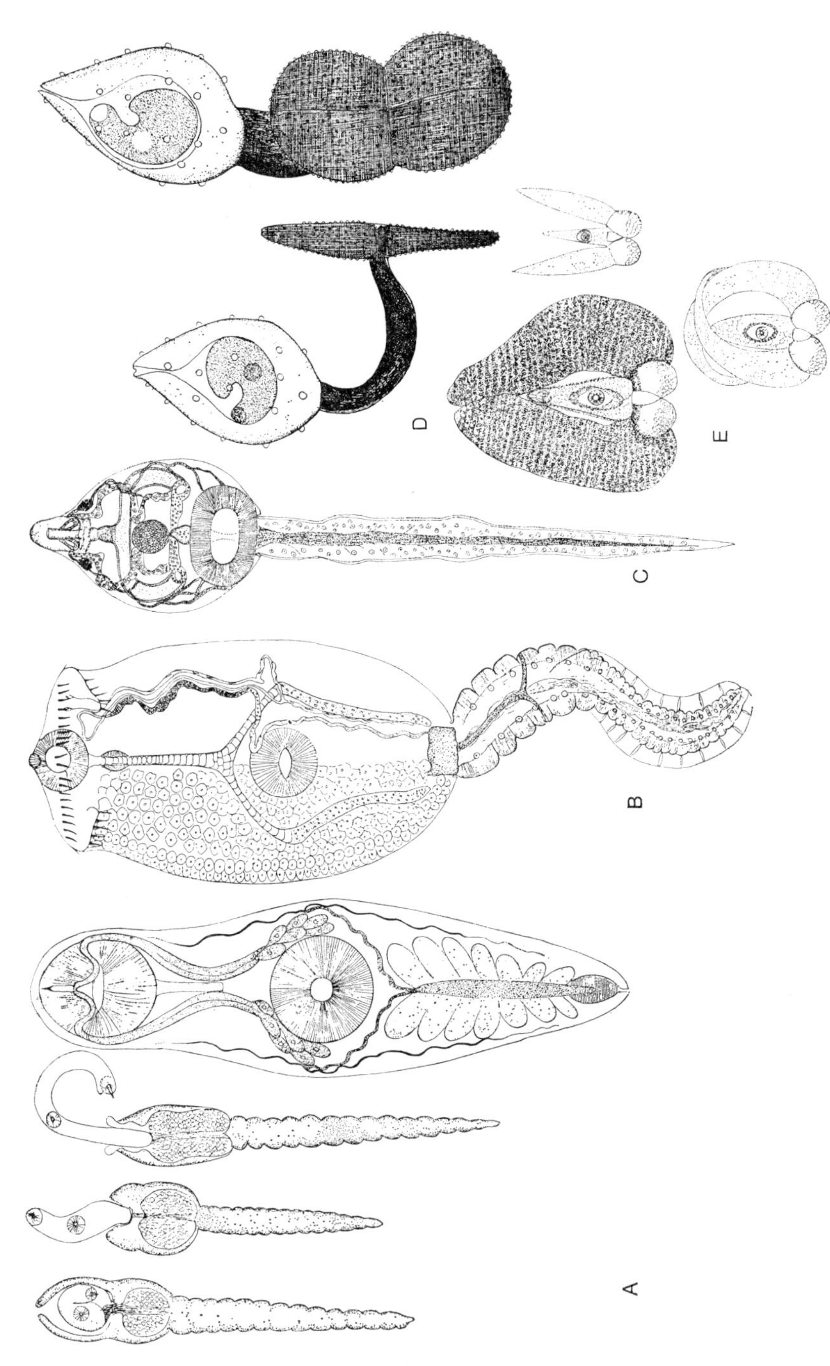

FIG. 39 Examples of cercarial types. **A** Cysticercous—*Cercaria vitellilobae*; **B** Echinostome—*Cercaria echinostomi*; **C** Amphistome—*Cercaria diplocotylea*; **D** Cystocercaria—*Cercaria splendens*; **E** Furcocercaria—*Cercaria*;

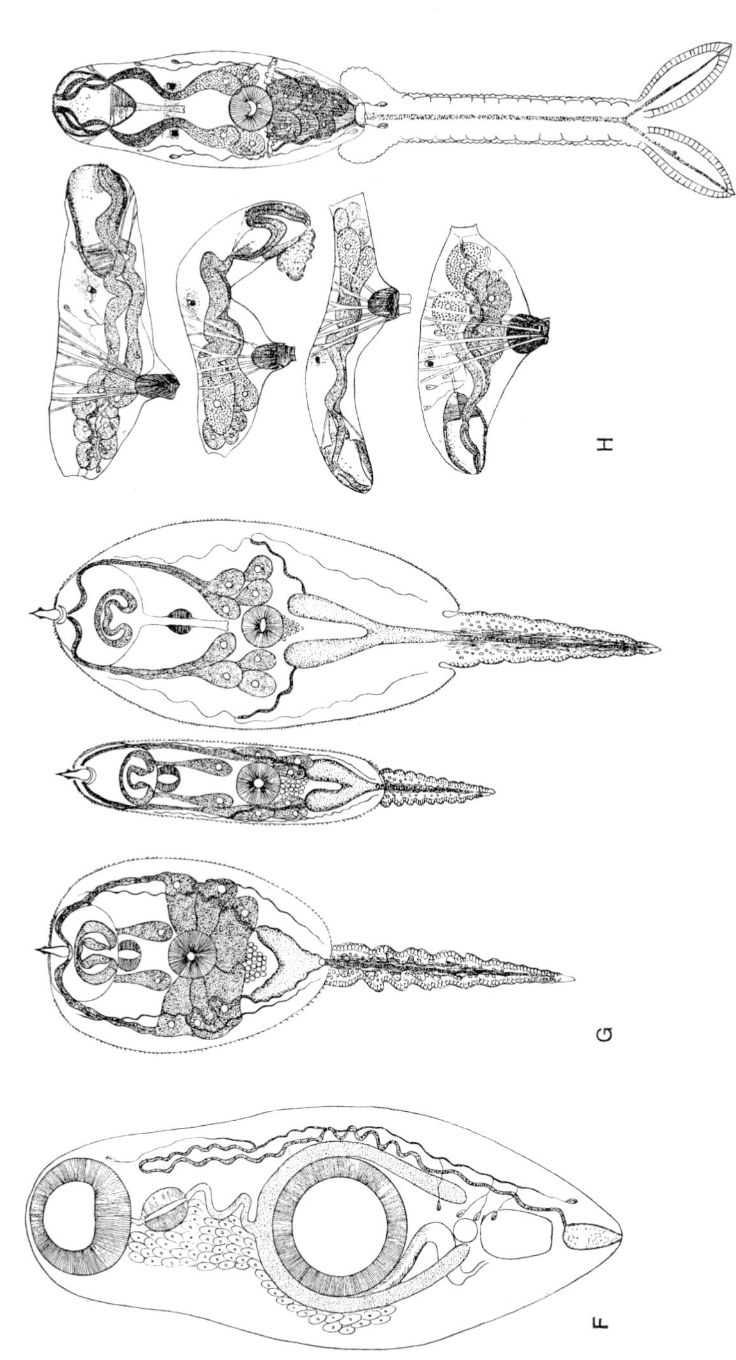

F Cercariaeum—*Cercariaeum limnaeae auriculariae*; **G** Xiphidiocercaria—*Cercaria nodulosa* and *Cercaria virgula*; **H** Furcocer caria—*Cercaria ocellata*, lateral views; *Cercaria Bilharziellae polonicae* ventral view. From WESENBURG-LUND, 1934.

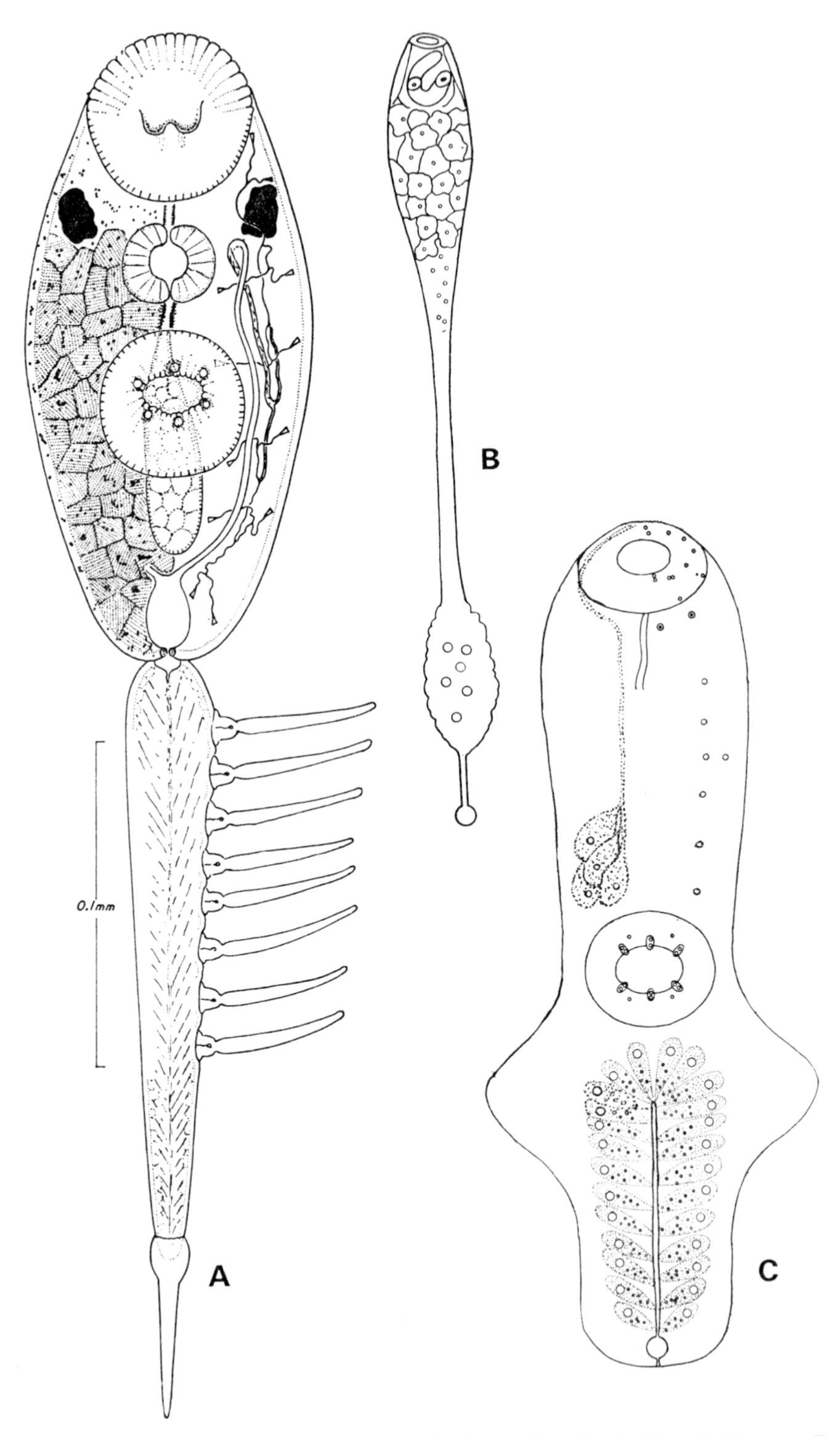

FIG. 40 Examples of cercariae. **A** The cercaria of *Haplosplanchus sp*. Note the rhabdocoele-like gut. From CABLE, 1954. **B** *Cercaria papillostoma* (Gorgoderidae) entire; **C** body of *Cercaria papillostoma*. **B** and **C** from COIL, 1960.

(*a*) *Cystocercous* Anterior end of the tail enlarged to form a bulbous portion containing a cavity into which the body may become retracted. Development generally in sporocysts.

(*b*) *Gymnocephalous* Oval body with simple tail. Oral sucker devoid of piercing structures such as spines or stylets. Cercariae developing in rediae.

(*c*) *Echinostome cercariae* Cercariae with simple but powerful tail. The body possesses two lobes forming a collar at its anterior end and, associated with this collar, is a circlet of stout spines. Development in rediae.

(*d*) *Xiphidiocercariae* The oral sucker of this type is provided with a stout stylet which is used in penetration. There may be characteristic gland cells forming a virgula organ in the sucker. Tail simple and development within sporocysts.

(*e*) *Furcocercariae* Tail complex and forked comprising a tail stem and two furcae. The oral sucker generally modified as a penetration organ. Development in sporocysts.

(*f*) *Microcercous cercariae* The cercariae in this category have short stumpy tails. In other features they are variable and may or may not possess a stylet and may develop in rediae or sporocysts.

(*g*) *Cercariaea* The tail is not developed in this type and consequently no free swimming phase occurs. The oral sucker may or may not possess a stylet and development may occur in rediae or sporocysts.

(*h*) *Trichocercous cercariae* Usually marine, planktonic forms with a long tail bearing numerous bristles and hair-like processes.

Other terms applied to cercarial stages are 'rhopalocercous'—with the tail wider than the body; 'lophocercous'—with the body bearing a finfold and 'gasterostome'—with the mouth centrally located. All the categories listed above have been subdivided further on the basis of structural variations and the reader is referred to DAWES (1946; 1953) and GRASSÉ (1961) for a more comprehensive list. Many of the structures upon which this scheme is based are highly adaptive e.g. suckers and tail. It is obvious therefore that cercarial categories such as monostome, microcercous and cercariaea for example, represent an accumulation of cercariae from different adult taxonomic groups which exhibit evolutionary convergence at the cercarial stage. It is general practice to refer to cercariae described in isolation from the rest of the life-cycle as 'species'. It is important to realize that usage in this context may, as pointed out by NASIR and ERASMUS (1964), have little systematic significance in terms of adult species and generally means that the particular cercaria can be distinguished from others. Some authors have employed letters of the alphabet or roman numerals to distinguish cercariae but in the absence of a universal scheme this procedure has no real advantage over the 'species' system, as long as the limitations of the terminology are appreciated. The basic terms used in cercarial morphology are illustrated in Fig. 38 and the range of cercarial types in Figs. 39, 40.

Morphology

1. THE CERCARIAL BODY

The variations in gross morphology exhibited by the cercarial body are less extreme than those exhibited by the tail. Most cercarial bodies are globular, cylindrical or oval in appearance but this shape is subject to almost continuous change during the various activities such as swimming, creeping and penetration. This plasticity of the body provides one of the major problems which has to be overcome by the cercarial taxonomist and generally one is referred to maximum and minimum sizes in taxonomic descriptions. Many descriptions now contain measurements based on specimens fixed in warm 10 per cent formalin and this procedure, employed by most authors, provides a more universal basis for comparison.

TEGUMENT

The body is covered by a thin tegument which is formed during morphogenesis within the sporocyst or redia. The detailed structure of this cytoplasmic covering is only now being investigated and the investigation of its origin and its relationship to that enclosing the metacercaria and adult stages provides a particularly interesting challenge. The only available evidence is the observation of ROEWER (1906) who found that the original epidermis formed by the cercaria disappeared during development and that the 'cuticle' was formed from secretions produced by gland cells within the cercarial body. In view of what we know now regarding the tegument of adult trematodes it seems unlikely that these observations will be confirmed in their entirety. Other authors (LANGERON, 1924; ROTHSCHILD, 1936b; DUBOIS, 1929) have referred to a 'primitive' or 'embryonic' epithelium covering the cercarial body at certain stages in development. In the case of the cercaria of *Fasciola hepatica*, it has been suggested (DIXON and MERCER, 1967) that the tegument of this stage is really an 'embryonic epithelium' in which the precursors of the metacercarial cyst wall accumulate. During the process of encystment this layer is lost, and the metacercarial tegument is formed from the cytoplasmic residue of the keratin-rod secreting cells. Whether or not this concept is applicable to those cercariae which do not secrete a cyst remains to be determined. REES (1967) observed that, during the process of cyst formation and accumulation of products in the cercarial tegement, the spines did not appear to be disturbed and persisted throughout the entire process. This latter observation tends to support the concept of the persistence of a cercarial tegument throughout the process of encystment and its physical and temporal continuity with that of the metacercarial tegument. The ultrastructure of the cercarial tegument has been briefly described by KRUIDENIER and VATTER (1958) but more detailed descriptions have been made by BILLS and MARTIN (1966) and REES (1967). These and unpublished observations made in the author's laboratory indicate that the tegument is cytoplasmic, bounded on the outside by a plasma membrane which is smooth and not elevated into microvilli like that of the sporocyst and redial stage. In this, and several other features, the tegument of this stage resembles that of the adult trematode. The tegument rests on a basement layer which is traversed at interval by the cytoplasmic extensions of the sub-tegumentary cells as well as by nerves running to sense organs. The cytoplasm of the tegument contains mitochondria as well as secretory bodies. Spines, when present, rest on the basement layer, do not have associated musculature and exhibit, under the electron microscope, the lattice-like structure typical of spines of the adult. The tegument of some gymnocephalous cercariae appears granular under the light microscope and this is due to the presence of keratin scrolls which have been secreted into the tegument (DIXON and MERCER, 1965; MERCER and DIXON, 1967; DIXON and MERCER, 1967). The spines may be very fine or stout and divided into a shaft and base. They are generally more frequent in the region of the oral sucker although in certain xiphidiocercariae, heavily spined caudal pockets occur on either side of the body at the point of insertion of the tail. In echinostome cercariae a distinctive collar of very stout spines is arranged around the oral sucker. Most of these spines are backwardly pointing and it is generally assumed that they function in maintaining within the host tissues the position of penetrating and migrating cercariae.

In addition to these spines, many cercariae possess long hair-like structures generally arising from a small papilla protruding from the tegument (Plate 5-5). These structures, which must represent sense organs of some kind, are more frequent in the region of the oral sucker or on the posterior regions of the body. Also included in this category are the so-called tactile hairs present on the tail stem of many furcocercariae. Although similar in their gross morphology under the light microscope, it seems likely that observations with the electron microscope will reveal the existence of a number of distinct types of sense organ. Beneath the basement layer, and attached to it, lie the outer circular and inner longitudinal layers of muscle. As in other stages of the life-cycle the muscles contain fibres of two sizes.

Cercariae are occasionally coloured and NADAKAL (1960a,b,c) has identified the pigments present in certain species. Their origin and composition resembles that described by him for the sporocyst and redial stages. The pigments, mainly B carotene and a modified chlorophyll, occur in the body and in the case of B carotene may play a part in facilitating phototropism. Other functions suggested are the inhibition of the autoxidation of lipids in the tissues and the formation of patterns which make the cercarial body more distinct and obvious to a predatory host.

ORAL SUCKER

The oral sucker surrounding the mouth is well-developed and muscular, containing radial and circular muscles. The sucker is usually fairly simple in its construction but in some cases e.g. furcocercariae and xiphidiocercariae (Fig. 38) the oral sucker plays an important part in penetration and consequently becomes more elaborate in structure. Xiphidiocercariae commonly infect aquatic larval insects or other arthropods and enter the body by penetration through the relatively soft arthrodial membrane. Attachment to the arthropod and eventual penetration is aided by two additions to the basic morphology of the oral sucker. The most frequent addition is a stylet which lies in the anterior or dorsal lip of the oral sucker. This stylet is characteristic of the particular cercarial 'species' in its size and shape and generally consists of a shaft, lateral shoulders of varying degrees of prominence, and an apical spike. Associated with the stylet is a system of muscles which permit the protrusion and retraction of the stylet and these thrusting movements result in the penetration of the host body wall by the stylet.

In addition to this stylet, some xiphidiocercariae (i.e. virgulate xiphidiocercariae) possess a reservoir containing secretions from gland cells. The reservoir opens to the exterior by short ducts opening on either side of the mouth. This reservoir is known as the 'virgula organ' and exhibits a close relationship with the subtegumentary mucoid glands in both the chemical composition of its contents and its development (KRUIDENIER, 1951a,b, 1953a,b,c,d; KRUIDENIER and MEHRA, 1957). An indication of the function of these secretions was given by BURNS (1961) and a histochemical investigation of the nature of the contents of the virgula organ and the mucoid gland cells has been made by ORTIGOZA and HALL (1963a,b). Using the Pelikan Indian Ink method devised by STIREWALT (1959a) these authors were able to collect the secretions from the virgula organ and the cephalic glands. Virgula secretions formed a thin film left behind by the moving cercaria and the cephalic gland secretions emerged as droplets which absorbed water rapidly and adhered together to form sticky strands. The mucoid glands, mucoid coat covering the cercariae and the virgula contents contained acid mucopolysaccharide, lipid material and a substance resembling elastin, whereas the cephalic glands, present in the body, contained protein and carbohydrates and exhibited staining properties common to both elastin and collagen. These authors thought that it was the secretions from the mucoid glands and virgular organ which functioned in the attachment of the free-swimming larva to the host. Many workers have observed that virgulate xiphidiocercariae become entangled in and attached to arthropods by means of strands and sheets of sticky secretion (see chapter 6).

A second type of elaboration of the basic oral sucker occurs in furcocercariae, where it becomes modified to function as a penetration organ. It no longer resembles the more typical acetabulum-like oral sucker but forms a pear-shaped sac. The muscles are not radially arranged but form a thin muscular layer surrounding the mouth, duct-apertures etc. The tegument in this region becomes heavily armed with recurved spines and the region immediately anterior or dorsal to the mouth may bear a cluster of particularly stout, forwardly directed spines. The musculature of this region is so arranged that it permits invagination and eversion so that the apical cap of spines may become thrust into and withdrawn from the host tissues in a manner similar to that of the xiphidiocercarial stylet. In addition to the spines the oral sucker (also referred to as the penetration organ or anterior organ) contains the distended ends of the ducts from the so-called penetration gland cells. These spindle-shaped dilations, which may function as reservoirs for the secretion, open to the exterior by small apertures on either side of the mouth. In addition to these structures the oral sucker of some furcocercariae contain minute gland cells e.g. *Cercaria ancyli* (JOHNSTON and BECKWITH, 1947) and Cercaria of *Alaria arisaemoides* (PEARSON, 1956).

Other elaborations of the oral sucker may take the form of ridges and papillae along the edge of the sucker and these may have a sensory function or may aid in the adhesion of the sucker to the substratum. In some cercariae, e.g. cercarial stages of the Bivesiculidae (LE ZOTTE, 1954), the oral sucker may be absent. In marked contrast to the oral sucker the ventral sucker exhibits little variation. It may be posterior in position, as in the amphistome cercariae, or may be secondarily lost, as in several diverse groups of cercariae.

GLAND CELLS

A particularly characteristic feature of the cercarial stage is the presence of gland cells in the body. The number and arrangement of these cells is generally constant for a particular cercarial 'species' and is a feature frequently referred to in the description of the cercarial stage although they represent adaptive and non-persistent structures. It is unfortunate that the terminology applied to these cells is not, in the majority of cases, based on any experimental or histochemical evidence so that most of the terms are presumptive in their implications. The present state of our knowledge has been reviewed by STIREWALT (1963b). Cercarial gland cells can be referred to three main categories on the basis of existing terminology: 1, mucoid gland cells; 2, cystogenous gland cells and 3, penetration gland cells. The characteristics of these cell types will now be considered in more detail.

1. Mucoid gland cells This category of cells has been mentioned already in relation to the virgula organ of xiphidiocercariae. The nature and possible function of these cells in xiphidiocercariae has been discussed by KRUIDENIER, 1949, 1951a,b, 1953a,b,c; KRUDENIER and MEHRA, 1957; ITO and WATANABE, 1959). In gymnocephalous cercariae they have been investigated by DIXON (1966a) and in echinostome cercariae by REES (1967). These cells lie fairly close to the tegument in the dorsal, lateral or ventral regions of the body. The secretion produced by the glands accumulates in the virgula organ of xiphidiocercariae and is largely acid mucopolysaccharide in nature. This secretion is believed to form a thin coat surrounding the outer surface of the body and is particularly evident in non-emerged cercariae within the molluscan body and is thought to afford some protection to the cercariae during the intra-molluscan migratory phase prior to emergence. The secretion in emerged cercariae seems to function in facilitating attachment to the second intermediate host. This aspect is discussed in more detail in chapter 6. The secretions from the mucoid gland cells of gymnocephalous cercariae play a particularly important part in cyst formation and this is considered in a later paragraph.

2. Cystogenous gland cells These are represented by very numerous, small subtegumentary cells with dense contents. The contents of these cells may appear granular or consist of masses of small rod-like particles. Cells with rod-like contents have been reported from a wide range of cercariae (Fasciolidae, Notocotylidae, Philophthalmidae, Echinostomatidae and Paramphistomidae) and are particularly abundant in the cystforming gymnocephalous, echinostome, monostome and xiphidiocercariae. When the appropriate external stimuli are present the cercariae settle down, attach themselves to the substratum and secrete a cyst wall. This process is discussed in chapter 5.

Recent observations by DIXON (1966a; MERCER and DIXON, 1967) and REES (1967) on the cercariae of *Fasciola hepatica* and *Parorchis acanthus* respectively have provided more information concerning the histochemistry of these cells (mucoid and cystogenous) Figs. 50, 52 and 53. DIXON (1966a) has described four categories of cystogenous cells from cercaria *F. hepatica.* (a) Tanned protein cells. These are concentrated in the ventral half of the body and are filled with large, yellow-brown secretory granules. Histochemical tests reveal that these cells are rich in protein and phenols although polyphenoloxidase was not detected. These granules are secreted into the embryonic epithelium, before the cercariae leaves the redia, forming a thin layer bounded by an external plasma membrane. These cells also contribute to Layer 1 of the metacercarial cyst wall (DIXON, 1965; DIXON and MERCER, 1964). (b) Mucopolysaccharide cells. These occur around the dorsal surface of the cercariae and also around the ventral sucker. They are PAS- and Alcian Blue positive and their contents resist digestion with diastase suggesting the presence of acid and neutral mucopolysaccharide. This material also appears in the embryonic epithelium of the cercaria after it has emerged from the redia and for this reason may also be included in the category of mucoid cells described earlier. The secretion also contributes to part of Layer II of the outer cyst wall and layers IIIb and IIIc of the inner cyst wall. (c) Mucoprotein cells. These lie central and dorsal to the ventral sucker, correspond to the position occupied by the 'penetration glands' of REES (1937) and the 'salivary glands' of ROTHSCHILD (1936a). They give a slight reaction with protein tests and a positive to PAS, are resistant to diastase and Dixon concluded that they are mucoprotein in content and probably contributed to layer IIIa of the inner cyst wall. It is obvious that the term penetration gland cells can no longer be applied to these cells. (d) Keratin cells. These occupied nearly all of the dorsal half of the body and contained the characteristic rods described by many observers in the past. The rods are approximately 5 μ long and give positive reactions with the protein tests as well as containing free —SH groups and disulphide bonds. No histochemical tests for enzyme activity were carried out by Dixon in this study although on the evidence available it seems unlikely that any would be present. The rods consist of a tightly wound scroll of keratin-like material and as described in chapter 5, unwind and contribute to the keratinized layer (IV) of the cyst (Plates 4-6 and 12-5).

A similar study by REES (1967) has revealed a slightly different arrangement of cells in the cercariae of *Parorchis acanthus*. She describes: (a) dorsal agranular cystogenous glands, lying below the dorsal tegument and containing acid and neutral mucopolysaccharide and possibly glycogen or a glycoprotein; (b) ventral agranular cystogenous cells lying at the centre of the anterior half of the ventral surface and containing acid mucopolysaccharide only; (c) ventral granular cystogenous cells which are large and occupy the most of the ventral half of the body. The cytoplasm appears to be packed with small rounded granules and exhibits positive reaction for protein and lipoprotein; (d) dorsal granular cystogenous glands occupying the whole of the dorsal half of the body below the agranular cystogenous glands and lying above the ventral granular cystogenous glands. These cells contain rod-shaped granules and probably correspond to the 'keratin' bodies of DIXON (1966a). They are PAS- and mercury-bromophenol blue positive and show a-metachromasia with toluidine blue; (e) ventral plug-forming gland cells, six in number, and lying in the preacetabular region. They contain neutral mucopolysaccharide. All these gland cells contribute to the formation of a cyst wall comprising an outer wall of three layers and an inner wall of two layers.

It is apparent that the existing terminology will have to be revised as more information regarding the histochemical characteristics of the cells is obtained. These two studies also suggest that the characteristics of gland cells in different cercariae will differ in relation to the nature of the cyst wall to be secreted. The variation in the composition of metacercarial cyst walls may reflect a response to the external habitat but may also provide a selective hatching mechanism which might result in host specificity. These variations at the metacercarial stage can be traced back to the characteristics of the cercarial gland cells and therefore it seems likely that some difference will exist in the composition of the gland cell contents.

3. Penetration gland cells Many cercariae possess, in the vicinity of the ventral sucker, pairs of gland cells whose ducts run anteriorly and discharge their contents through apertures in the oral sucker (Plate 4-1). The gland cells may lie anterior, lateral or posterior to the ventral sucker and may vary in number from two to six pairs of cells. These cells are generally referred to as penetration gland cells although, in the majority of cases, their function is far from clear. A common factor is that the contents of the cells are discharged during penetration into the host. Although the secretions may, in some cases, be lytic in their function, in other examples the secretion may serve for attachment, lubrication or protection of the migrating larva. These gland cells are a very prominent feature of furcocercariae and the strigeoid and schistosome species have received particular attention.

In schistosome species the cercarial gland cells may be differentiated into anterior and posterior groups by their optical appearance, and histological and histochemical characteristics. Our knowledge of the nature of the gland cells in these species has been advanced particularly by the work of Stirewalt and her co-workers (see reviews by STIREWALT, 1963b, 1966). In schistosomes the gland cells lie in two groups, one preacetabular and the other postacetabular. The two anterior pairs are largest and acidophilic or eosinophilic with macrogranular cytoplasm. The posterior group are smaller, basophilic and have microgranular cytoplasm. The ducts run forward through the modified oral sucker and are said to open through hollow spines (STIREWALT and KRUIDENIER, 1961) although this is not supported by electron microscope observations (Plate 5-5). The postacetabular glands are thought to produce a secretion which is mucoid in nature but, in contrast to the non-schistosome types, only a small proportion is discharged within the mollusc and the greatest part is voided during the early stages of penetration of the mammalian skin. The mucoid film produced by these gland cells covers the cercarial body and presumably provides some protection during its free swimming existence and during penetration into the final host. It is also the antigen which reacts *in vitro* with antibodies in immune serum to form the characteristic pericercarial envelope. The two sets of glands can be distinguished by a variety of histological and histochemical techniques (STIREWALT and

KRUIDENIER, loc cit.). The posterior complex is PAS-positive and seems to be associated with the mucoid secretion. The anterior glands are PAS-negative but stain strongly with the alizarin red present in Purpurin. This reaction suggests the presence of calcium in these anterior glands and the observations of LEWERT and HOPKINS (1964) support this assumption. The role of calcium in penetration seems to be particularly significant as they found that treatment with chelating agents reduced the invasiveness of the cercariae and that calcium and magnesium were necessary for optimal activity of the parasites enzymes as measured by dye release from azo-substrates. The technique of stimulation of secretion by Indian ink (STIREWALT, 1959a) has enabled secretions to be collected and analysed. Unfortunately no distinction can be made between secretions from anterior or posterior glands. The secretion (STIREWALT and EVANS, 1960) contained 17·2 per cent amino acids, 9·7 per cent amino sugars and an unidentified concentration of steroids. The conclusion was that they consisted of mucocomplexes or substances formed by the association of proteins, polysaccharides and lipoids.

In addition the cercarial extracts (entire body) exhibit activity against umbilical cord hyaluronic acid (LEVINE *et al*, 1948; LEE and LEWART, 1957), against mucopolysaccharide capsules of streptococci (STIREWALT and EVANS, 1952), against heparin (LEE and LEWART, 1957), against gelatin films LEWERT and LEE, 1954, 1956), against azo-collagen (LEWERT and LEE, 1954) and haemoglobin (LEWERT, 1958). Activity of extracts against tripalmitin (LEWERT, 1958) and has been demonstrated and also elastolytic activity by GAZZINELLI and PELLEGRINO (1964). Tests for lipase, esterase and aminopeptidase using histochemical techniques were negative (STIREWALT and WALTERS, 1964) although the preacetabular glands were positive for calcium. It is obviously very difficult to assess directly the true function of these secretions and the use of whole body extracts containing gut tissues etc. does make valid conclusions more difficult. It is obvious however that the secretions become involved in a variety of functions at different times in the life-cycle. An indication of possible function of these secretions during penetration is given by STIREWALT and FREGEAU (1966). The PAS-positive postacetabular gland secretion is used by cercariae for adhesion to the skin surface during searching and initial penetration of the horny layer. Penetration of the keratagogenous layer is aided by the alkaline secretion of the preacetabular glands. The pH of this secretion is estimated to be pH 8·5 to 9·0 and an alkaline medium is known to soften and dissolve keratin. Passage through the tissues below this layer is aided by the enzymatically active preacetabular gland secretions which alters the intercellular ground substance. These workers also demonstrated that cleavage of urea-denatured haemoglobin by cercarial extracts did not occur when schistosomules were used for the preparation of the extract. This activity was regarded as representing an invasive enzyme system because its presence or absence could be correlated with the presence of the gland cells before, and their absence after, penetration. No loss of activity occurred in free-swimming cercariae up to 24 hours old, suggesting that the secretion was discharged during penetration only.

Although the gland cells of strigeoid cercariae may form two groups, lying anterior and posterior to the ventral sucker, the cells appear to be similar in appearance. The only histochemical study available on this type of cercaria is that by BOGITSH (1963) on *Posthodiplostomum minimum*. This cercaria has three pairs of gland cells, one pair anterior to the ventral sucker and two pairs posterior to the sucker. The cells were PAS-positive and contained tyrosine, arginine, basic proteins and acid mucopolysaccharides. Bogitish has interpreted this evidence as suggesting the presence of two types of secretory substance—an acid mucopolysaccharide and a carbohydrate-protein complex. The three pairs of cells were identical and ereth was no evidence of discharge of secretion whilst the cercariae were within the sporocyst or in the snail tissues. No tests were carried out to determine the presence of enzyme activity by the cells contents. Evidence obtained by ERASMUS (1958), while observing living *Cercaria X* (Strigeoidea: Diplostomatidae) migrating in fish tissues, suggested that the secretion from the four postacetabular glands was not lytic. In this case, the droplets of secretion, emerging from the duct apertures on the oral sucker, were swept quickly backwards over the body by the plunging action into the host tissues of the apical cap of spines on the oral sucker during migration. Thus the secretion first came in contact with the host tissues at the level of the ventral sucker and it seemed that destruction of the host tissues was largely by mechanical erosion. LEWERT and LEE (1954) were unable to detect collagenase activity in living unidentified strigeoid cercariae and also a plagiorchiid cercariae. In contrast to these observations, crude tests have suggested the presence of proteolytic activity in a strigeoid (DAVIS, 1936) and *Cryptocotyle cercaria* (HUNTER and HUNTER, 1937) and a hyaluronidase-like spreading factor has been described from certain xiphidiocercariae by GINETZINSKAJA (1950). The term spreading factor refers to the ability of extract injected into the skin to produce at the site of injection a diffusion of the Evans Blue dye previously introduced into the circulation. In tissues in which the ground substance has been altered the dye will diffuse widely.

The gland cells are very significant structures associated with the cercarial stage but, except for a few observations on a restricted number of species, little can be said in precise terms concerning their nature and function. They represent a remarkable morphological adaptation at this stage but their activities, and exact function remain to be determined. A further feature, which has not received investigation, is that the cells discharge their secretion in a precise sequence and at particular moments in establishing host-parasite or substrate-parasite contact. Discharge at the wrong moment would have disastrous consequences on the completion of the life-cycle. In spite of the apparent morphological simplicity of this stage, the behavioural and nervous feed-back systems which control the correct sequence of events in the process of secretion must be well defined. It is possible that the chemosensory stimuli may play a significant role in initiating secretion as both STIREWALT (1959a), using Pelikan Ink, and WAGNER (1959), using free fatty acids, were able to produce pronounced secretory activity in cercariae exposed to these substances. Wagner found that the residue from ether extracted mouse skin also stimulated penetration. There is an interesting similarity between these findings and the observations of MACINNES (1965) which suggested a positive response by *S. mansoni* miracidia to short chain fatty acids.

PROTONEPHRIDIAL SYSTEM

One of the most distinctive features of the cercarial stage and one much employed in taxonomy is the excretory or protonephridial system. In a generalized scheme (Fig. 41) the system might be represented as consisting of a terminal bladder from which arises, on either side of the body, a primary collecting vessel. At a point opposite the ventral sucker the primary collecting vessel receives an anterior and posterior secondary vessel. Each of these received a number of tertiary vessels terminating in capillaries bearing flame cells. If flame cells are present in the tail they are borne on a branch from the posterior secondary vessel. Cilia are not confined to the flame-cell, but may also occur in the collecting vessels. The excretory pore through which the bladder opens to the exterior may be terminal at the posterior end of the body, near the ventral sucker (Paramphistomes), terminally at the tip of the tail or furcae, or paired and subterminal on the

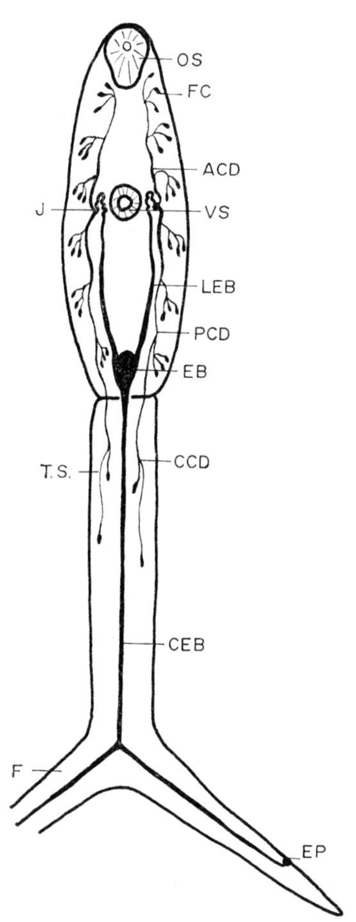

FIG. 41 Diagrammatic representation of the excretory system of a fork-tailed cercaria. The caudal flame cells are absent from the tail in non-furcate forms. (ACD: anterior collecting duct; CCD; caudal collecting duct; CEB: caudal extension of the excretory bladder; EB: excretory bladder; EP: excretory pore; F: furca; FC: flame cell; J: junction between collecting vessels and extension of the excretory bladder; LEB: lateral extension of the excretory bladder; OS: oral; sucker; PCD: posterior collecting duct; TS: tail stem; VS: ventral sucker.)

tail. A variety of commissures also exist, particularly in the Strigeoidea, linking the collecting vessels. In *Cercaria nigrita*, an amphistome cercaria described by FAIN (1953), there is an excretory plexus supplying the posterior ventral sucker.

Two main patterns have been described by DUBOIS (1929). In one 'Stenostoma' the main primary vessel opening into the bladder runs anteriorly to a position near the pharynx and then bends, continuing posteriorly for some distance before receiving the secondary branches. In the second type 'Mesostoma' the primary vessel terminates near the ventral sucker and receives the secondary branches at this point. This difference was once thought to be very significant taxonomically but it is now known that both arrangements can occur in a single subfamily. The variations in morphology and development of this system are many, and the reader is referred to LA RUE (1957) and KOMIYA (1961) for further details.

The possible significance of the cercarial excretory system to adult taxonomy was first commented on by CORT (1917) and later developed by FAUST (1918, 1919, 1924, 1932), SEWELL

(1922) and MILLER (1926b). A particularly useful contribution was the suggestion by FAUST (1924) that a flame-cell formula could be constructed which would indicate the number and arrangement of the flame-cells. Because of the bilateral symmetry of the system, only one half is considered in the formula with the expression being finally multiplied by two to represent the complete system. In constructing the formula the anterior and posterior collecting vessels are each represented by a set of brackets joined by the + sign. The number of figures within each bracket represents the number of branches arising from the anterior or posterior vessel. The value of each figure indicates the number of flame cells associated with these smaller branches. Thus a system possessing three flame cells on each of three branches arising from both anterior and posterior collecting vessels could be expressed as follows:

$$2[(3 + 3 + 3) + (3 + 3 + 3)] = 36$$

If an additional pair of flame cells were present in the tail, this number would also be enclosed in brackets as follows:

$$2[(3 + 3 + 3) + (3 + 3 + 3 + (2))] = 40$$

Although flame cells exist in miracidia and later stages the system is developed *de novo* in the cercaria. The development has been described by Hussey, Kuntz and Komiya (loc cit) and reviewed by LA RUE (1957). The globular cercarial embryo possesses a pair of collecting vessels terminating at one end, in most cases, in a pair of flame cells and opening to the exterior at the other end via a pair of primary excretory pores. As the embryo elongates and increases in size, so the system becomes elaborated by an increase in flame cell number and an extension of the collecting ducts. In *Cercaria hamburgensis* the formula changes as follows (KOMIYA, 1938):

Germ ball	$2 \times [(1) + (1)]$
Young cercaria	$2 \times [(1 + 1) + (1 + (1))]$
	$2 \times [(1 + 1) + (1 + 1 + (1))]$
Free swimming cercaria	$2 \times [(1 + 1) + (2 + 2 + (1))]$

It is obvious from this that morphological descriptions of cercariae must be based on free-swimming fully matured cercariae. At a certain stage in development the tubules at the posterior end of the body curve inwards, meet and fuse to form the precursor of the excretory bladder. This developmental pattern seems to be common to all the cercariae studied up to this stage of development. At this point, development may follow one of two lines and LA RUE (1957) has based his revised classification of the Digenea on the difference in developmental patterns exhibited at this time. In one group (Superorder Anepitheliocystidia containing the orders Strigeatoidea, Echinostomida and Renicolida) the thin-walled bladder persists in its apparently non-epithelial form although in some families i.e. Fasciolidae, the bladder may become overlain by muscle fibres and other tissues from the body wall. The second pattern is characterized (Superorder Epitheliocystidia containing the orders Plagiorchiida, Opisthorchiida) by the formation of an epithelial bladder as a result of the migration of cells to the site of the bladder precursor. The original tubule disintegrates so that the wall is cellular. Bladders of this type are usually V or Y-shaped with the cellular wall extending along the arms in some cases. In adult strigeoid trematodes the bladder becomes extended to form a reserve bladder or secondary excretory system and sometimes indications of this may be seen in the cercarial stage.

Although usually referred to as an excretory system the true function of the system has hardly been investigated. Some authors (WESTBLAD, 1922; HERFS, 1922—quoted by VON

BRAND, 1952) have observed the pulsations of the excretory bladder and detected a change in the rate of beating produced by an alteration of the osmotic tension of the external environment. Other workers have described alkaline phosphatase activity in the tubules (COIL, 1958; DUSANIC, 1959; CHENG, 1964). Morphological papers frequently refer to refractile bodies in the tubules and it is possible that these may be lipid droplets or represent inorganic substances arranged on concentric organic lamellae similar to the calcareous bodies of cestodes.

ALIMENTARY TRACT

The alimentary tract of the cercarial stage consists basically of mouth, prepharynx, muscular pharynx, oesophagus and two intestinal caeca. Various parts of the system may be absent and the caeca may be reduced to two minute sacs at the end of the oesophagus. Groups of gland cells may be present in the region of the pharynx but the function of these secretory cells is not clear. Occasionally the caeca appear to possess transverse septa. The alimentary tract persists so that there is developmental continuity through cercarial, metacercarial to adult stages. The caecal walls contain muscle fibres and their activity contributes to the contractions which may pass along the length of each caecum. The caeca are generally filled with fluid containing refractile granules and sometimes cell debris of some kind. It is possible to stain the gut of living cercaria with intravitam dyes, but extremely difficult to demonstrate the ingestion of solid particles e.g. carbon or ink particles. Nothing is known of the physiology of the cercarial gut although STIREWALT and WALTERS (1964) have demonstrated aminopeptidase activity in the caeca of *S. mansoni* cercariae. The cercarial stage is generally a very active one exhibiting considerable free-swimming activity. Such activity will involve the utilization of energy and two sources seem possible. There is good evidence to suggest that cercariae are able to utilize endogenous carbohydrate during this period. The role of the gut is not so clear and although the structure is morphologically well-defined no positive evidence of its activity during this free-swimming period has been produced.

NERVOUS SYSTEM AND SENSE ORGANS

Light receptors in the form of eyespots are frequently described from cercariae (Fig. 42). Generally there are two eyespots but in some (e.g. Notocotylid cercariae) there may be three. Electron microscope studies of three cercariae (*Macrovestibulum eversum*, *Skrjabinopsolus manteri* and *Crepidostomum* sp.) have been made by POND (1964) and POND and CABLE (1966). The paired eyespots consist of receptors possessing microvilli collectively forming the rhabdomere and separate pigment containing cells. The nature of the eye-spot pigment in a number of cercariae from *Cerithidea californica* was studied by NADAKAL (1960a). His evidence suggested that the pigment was melanin, and tyrosine, a melanin precursor, was also found. In many strigeoid cercariae 'unpigmented eyespots' have been described. They appear to be small oval vesicles apparently fluid filled and devoid of pigment. Generally there is one pair lying just anterior to the ventral sucker. Neither the exact function of these structures nor their detailed structure has been determined. Other sensory receptors are represented by the tangoreceptors within the tegument and consisting of a vesicle and sensory papilla (DIXON and MERCER, 1965) Plate 5-5. It seems very likely that the hair-like projections extending from the tegument of the tail of many cercariae, especially trichocercous forms and strigeoid furcocercariae, must also represent receptors of various kinds.

The distribution of papillae in the tegument of several species of echinostome cercariae has

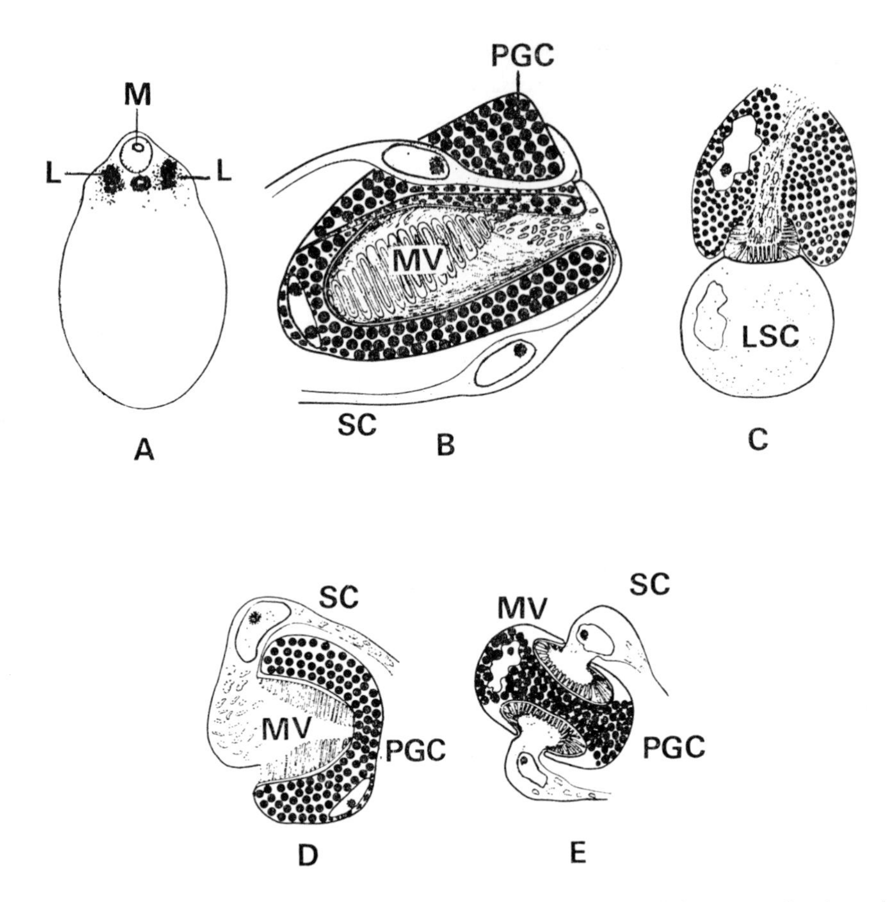

FIG. 42 Ultrastructure of cercarial eyespots. **A** Dorsal view of *Cercaria Macravestibulum eversum* showing paired lateral and single median eyespot. **B** Right eyespot of *M. eversum* from the dorsal aspect. **C** Median eyespot of *M. eversum* from the left side. **D** Dorsal view of left eyespot of *Crepidostomum* sp. **E** Lateral view of eyespot of *Skrjabinopsolus manteri*. (L: lateral eyespot; LSC: lens cell; M: median eyespot; MV: microvilli; PGC: Granular pigmented cell; SC: sensory cell). From POND and CABLE, 1966.

been described by LIE (1966b). Although the investigation was concerned with the use of papillae distribution in systematics, it seems likely that many of them represent sense organs.

The cercarial nervous system is based on the same pattern as that of the adult trematode (Fig. 52D). There is a cerebral ganglion complex in the region of the pharynx with a number of longitudinal trunks which may possess transverse commissures. Secondary branches innervate suckers, eyespots and the tail. Some of the best descriptions of the gross morphology are those of the early workers such as BETTENDORF (1897). Observations of the fine detail of the cercarial nervous system have been made by DIXON and MERCER (1965) using the cercaria of *Fasciola hepatica*. The axons resemble those of vertebrates except in that they are unmyelinated. The nerve processes contain mitochondria and vesicles and granules of various kinds. These authors describe four categories of inclusions: (a) numerous vesicles 200 to 800 Å diameter with clear or slightly electron dense contents; (b) vesicles 600 to 1000 Å with electron dense contents separated from the outer membrane by a clear space; (c) a rarely occurring dense stellate ganglion 1000 to 1700 Å diameter; (d) ovoid vesicles 2000 Å diameter also separated from the limiting membrane by a clear space. The cell bodies of the neurons are concentrated in the cerebral ganglion and the perkarya show nuclei, golgi bodies and sometimes inclusions of type (a) and (b). Also associated with the ganglion was a cell containing vesicles type (d) and endoplasmic reticulum as well as golgi bodies. It was suggested that these could be neurosecretory cells. (See comments in chapter 1 regarding the nervous system in the adult stage.) The synapses observed resemble those described from the mammalian central nervous system with dense membranes in close proximity and an asymmetric distribution of small, clear synaptic vesicles. The neuromuscular junctions exhibited closely adherent membranes of nerve and muscle cell and the presence of small vesicles in the axon terminal. Folding of the subsynaptic membrane was not described. In the cercaria of *Parorchis acanthus*, REES (1967) has noted that the ganglia are composed of a central mass of fibres with the nerve cells arranged on the outside forming a layer described as the cell rind. In this species the cell rind was present around the cerebral ganglia, transverse commissure and roots of the nerves and nerve cords.

Histochemical studies by STIREWALT and WALTERS (1964) on the cercaria of *S. mansoni* have demonstrated esterase activity and more specifically cholinesterase activity (LEWERT and HOPKINS, 1965) in the nervous system of this stage. This activity was inhibited by low concentrations of physostigmine (eserine).

Genital rudiments Many authors have described 'genital rudiments' at the posterior end of cercarial bodies. These may be represented by undifferentiated groups of cells or may be more well-defined in form. These cells can be stained with acetocarmine but their precise developmental history is far from clear.

2. THE CERCARIAL TAIL

The cercaria is basically a dispersive phase in the life-cycle and consequently its locomotive agent, the tail, is generally well developed. The morphological variation in tail structure is very large and this has formed the basis of schemes which sometimes are associated with taxonomic implications. However, CABLE (1965) has pointed out that divergence and convergence has occurred to such an extent that cercarial tails 'can be positively misleading'.

In some cases e.g. the Cercariaea, a tail is absent and these stages move about in a leech-like manner. They may accumulate on the snail tentacles (*Cercaria paludina impurae*), remain within brightly coloured sporocysts (*Leucochloridium paradoxum*) or creep from one host to another *Cercaria strigata*). This leech-like progression also occurs in the 'Microcercous' forms where the tail is extremely short. In this instance the tail may possess a terminal sucker-like structure or contain a cluster of gland cells. Secretion from these cells is said to aid in attachment and progression. In the tail of the cercaria of *P. acanthus* there is a cluster of gland cells (REES, 1967) Fig. 52C. The tip of this tail is also invaginable and forms a sucker with which the cercaria can adhere to the substratum. The gland cells open to the exterior by a duct which penetrates the tegument. The cells are PAS, mercury-bromophenol blue, Best's carmine and Sudan black B positive indicating a mixture of glyco-, muco- and lipoproteins.

In those cercariae possessing a well-developed tail it may be many times the length of the body and bear a fin-fold. Elaborations of the basic pattern are represented by the presence of many hair-like structures extending laterally (Trichocercous forms) or finger-like processes (cercariae of the Haplosplanchnidae) or by being fork-tailed. In this latter type the bifurcation may be extended to the point of attachment with the body e.g. gasterostome species (*Bucephalopsis sp.*) or there may be a tail-stem and furcae. The furcae may bear fin-folds and may be slender and tapering or they may be flat or lobe-like as in *Cercaria splendens.* In strigeoid cercariae the tailstem frequently bears long 'tactile' hair-like processes and contains in its centre vesicle-like structures referred to as 'caudal bodies'. These are often characteristic in their size, number and arrangement for a particular cercarial 'species'. In members of the Bivesiculidae the proximal end of the tailstem is dilated and hollowed to form a cavity in which the cercarial body lies. The tail is often brightly coloured and is attractive to the predatory host. A distended tail (in this case not forked) into which the cercarial body can be withdrawn, is also present in the cystocercous cercariae of the Gorgoderidae. These cercariae are not particularly active swimmers and the motion of the comma-like body serves to attract the attention of the next host required for the completion of the life-cycle. Particularly complex tails occur in the cystophorous cercariae in which a spring-like mechanism is present which projects the cercarial body through the gut wall and into the haemocoele of the appropriate crustacean host.

Although many early workers e.g. SINITSIN (1911) and HASWELL (1903) have made detailed descriptions of the anatomy of the cercarial tail, the observations of PEARSON (1956, 1959, 1961) provide a more modern interpretation. The strigeoid cercariae of *Alaria canis, A. arisaemoides, Strigea elegans* and *Neodiplostomum intermedium* possess tails consisting of a tailstem and two furcae. The tail (Fig. 43A,B) is covered by a well defined tegument variously spined and possessing tactile hair-like structures. The tegument is generally much folded presumably to allow for the flexural movements during swimming. Beneath the tegument occur the bands of circular muscle which appear to be anucleate. The longitudinal muscles extend from the base of the tailstem to the tips of the furcae and are arranged in four groups in *N. intermedium.* The arrangement of the longitudinal muscle appears to vary between different cercariae and the precise arrangement may possibly be correlated with the mode of swimming. The longitudinal fibres are bipolar with the cell body containing the nucleus bulging into the lumen of the tailstem. These cells appear to be striated whereas the circular muscle fibres are not. In forked tail forms the longitudinal and circular muscles also extend into the furcae and there are, in addition, groups of extensor and flexor muscles. The tailstem and furcae contain nerve commissures and the caudal excretory vessel which extends posteriorly from the bladder in the body. Associated with this vessel are the caudal bodies previously mentioned. They are particularly abundant and characteristic of strigeoid cercariae and may be small and numerous or large and restricted in number. Observations by ERASMUS (1959) showed that the caudal bodies of the strigeoid cercaria *Cercariae X* (Diplostomatidae) diminish in size during the free swimming period. Histochemical tests showed the bodies to be rich in glycogen and that this too diminished in quantity during the free-swimming period. Thus it seems likely that in some species they represent the site of an energy source which can be utilized during the swimming phase.

The presence of apparently striated muscle (Plate 1-3) has been commented on by a number of workers in the past (MARTIN, 1945 for strigeoid cercariae; SINITSIN, 1911 for the macrocercous species *C. equitator*). Using the electron microscope, CARDELL and PHILPOTT (1960) on *Cercaria Himasthla quissitensis* (Echinostomatidae) and KRUIDENIER (1960b) and KRUIDENIER and VATTER (1960), on the cercariae of *Schistosoma mansoni, Tetrapapillatreme concavocorpa, Schistosomatium douthitti* and an unidentified strigeoid cercaria, have made more refined observations on the structure of cercarial muscle. An apparent periodic banding of the muscle fibres was observed by these workers in *Himasthla sp., Schistosoma sp., Schistosomatium sp.* and the strigeoid cercaria. Kruidenier and Vatter (loc cit) suggested that presence of the banding might be correlated with the phylogeny of the Anepitheliocystidia. This idea has some further support by the identification of these bands in *Himasthla* by Cardell and Philpott and in an unidentified echinostome

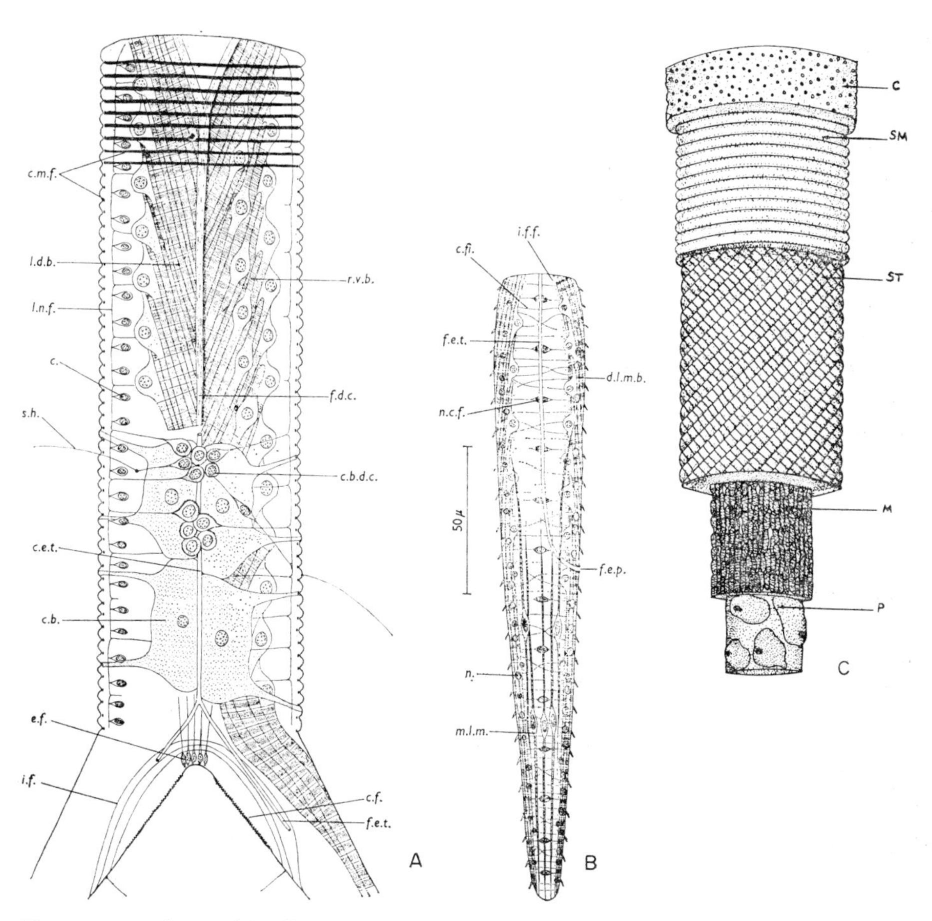

FIG. 43 The structure of cercarial tails. **A** The tailstem of the furcocercaria of *Neodiplostomum intermedium*, dorsal view. **B** The furca (interfurcal fibres and circular omitted from medial surface) of *N. intermedium*, (c.: cell on lateral nerve fibre; c.b.: caudal body; c.b.d.c.: cell bodies of dorsal commissure; c.e.t.: caudal excretory tubule: c.f.: dorso-ventral cuticular folds; c.fi.: circular fibre; c.m.f.: circular muscle fibre; d.l.m.b.: dorsal band of striated longitudinal muscle on lateral face; e.f.: extensors of furcae; f.d.c.: fibres of dorsal commissure; f.e.p.: furcal excretory pore; f.e.t.: furcal excretory tubule; i.f.: interfurcal fibres, flexors of furcae; i.f.f.: inter-facial fibres; l.d.b.: left dorsal band of striated longitudinal muscle; l.n.f.: lateral longit. nerve fibre; mlm.: striated long. muscle fibre of median face; n.: nucleus; n.c.f.: nucleus of circular fibre; r.v.b.: right ventral band of longit. muscle; s.h.: sensory hair.) Figures **A** and **B** from PEARSON, 1961. **C** The ultrastructure of the tail of the cercaria of *Himasthla quissetensis*. (C: tegument; M: mitochondrion; P: parenchymal cell; SM: smooth muscle; ST: striated muscle.) From CARDELL and PHILPOTT, 1960.

cercaria in this laboratory. Whereas PEARSON (1961) associated the banding in *N. intermedium* muscle with Z-line, J-disc and Q-discs and Cardell and Philpott with A, I and Z bands, Kruidenier and Vatter attributed the periodicity to the presence of a sarcotubular system which invades and traverses the bundles of muscle fibres at regular intervals. In *S. mansoni* the tubules were approximately 0·1 to 0·2 μ in diameter and form a complex closely associated with the cell membrane. At intervals of 0·5 to 0·75 μ bands of tubules extend completely across the muscle (Cardell and Philpott refer to a periodicity of 0·8 μ). In *T. concavocorpa* the sarcotubules are present but only occasionally penetrate the muscle and then run at various angles to the microfibrils. The tubules possess a membrane and a similarity to the elements of the endoplasmic reticulum has been commented upon. In these ultrastructural observations only one type of muscle fibre has been described. However in this laboratory two sizes of myofilament have been seen in the 'striated' muscle of the cercaria of *S. mansoni*. The filaments appear to be embedded in a less dense matrix and arranged in alternating rows. The arrangement becomes distorted in many places by the presence of sarcotubules. All these workers refer to numerous mitochondria associated with the muscles. In the simple tail of *Himasthla quissetentis* (Fig. 43C), CARDELL and PHILPOTT (1960) describe an outer layer of circular muscle lying beneath the tegument and an inner layer of 'striated' muscle which runs in a spiral fashion. Associated with the tegument and this spirally-arranged muscle are large numbers of mitochondria. These authors regard the spiral arrangement as facilitating bending of the tail in any plane and providing the basis for its whip-like motion. It seems apparent that functional specialization of the cercarial tail structure extend to the ultrastructural level as well as being evident at the level of gross morphology.

Cercarial biology

A. CERCARIAL EMERGENCE

After emergence from the sporocyst or redia and migration through the molluscan body to the point of emergence (see chapter 7 for details) from the snail, the cercaria enters a new phase in its existence and usually becomes free swimming. A variety of behaviour patterns exist at this time although, as mentioned in other chapters 6 and 7 not all cercariae become free-swimming—some remain within the sporocyst or encyst in the molluscan tissues without emergence.

The discharge of clouds of cercariae from infected molluscs is a very impressive sight and has attracted the attention of many workers, but in spite of this, our understanding of the phenomenon is still very limited. One of the most impressive observations is that of MEYERHOF and ROTHSCHILD (1940) who observed cercarial production by *Cryptocotyle lingua* from *Littorina littorea* over a period of five years. Approximately 830 cercariae per day over this period were produced by this infection.

A study of isolated infected snails shows that the emergence of cercariae is not continuous but occurs in spurts separated by irregular intervals of time. The observations of BYRD and SCOFIELD (1954) on physid snail infected with a variety of ochetosomatid cercariae showed that in those snails which survived the entire period of cercarial release, the maximum number of cercariae were shed in the early period and that this gradually diminished in relation to time. The decrease in cercarial production did not continue indefinitely but gave way to a second, although much smaller peak. After this second peak the numbers decreased until a stage was reached when no more cercariae emerged. In all these experiments the snails were infected with a single egg/miracidium. Therefore the variation could not be explained by occasional re-infection, but it is possible that the development of daughter sporocysts occurred irregularly thus influencing cercarial production. The emergence pattern over long periods of time was also studied by PEARSON (1956) using *Planorbula* species infected with *Alaria arisaemoides* and *A. canis*. In this case cercarial production was not continuous but consisted of periods of emergence which alternated with periods of non-emergence over a period of 300–480 days. In this experiment it was not stated how the snails were infected so that it is possible that the periodicity might have reflected the maturation of successive infections. The snails were kept under uniform laboratory conditions and it is again possible that the variation in maturation rate of daughter sporocysts contributed to the long term periodicity.

As well as emerging in phases, the actual numbers of cercariae produced varies from day to day. The observation of many workers suggest that emergence of large numbers of cercariae is followed by periods in which the numbers released is greatly reduced. This low level continues for a few days and then gradually increases until a high level occurs. This day to day variation in total numbers produced is again attributed to phased cercarial production within the redia or sporocyst. In some cases it is thought that the cercariae need to mature before emergence takes place and that collection of pools of cercariae occur within molluscan tissues again resulting in daily variation in numbers released (Plates 7-3 and 7-4).

The most conspicuous aspect of cercarial production is the periodicity within 24 hours. Some cercariae emerge maximally in the daytime and others at night. *Cercaria limbifera, Z* (REES, 1931), *Cercaria purpurae* (REES, 1947), *Cercaria elephantis* (CORT, 1922), *Cercaria brevifurca* and *Diplostomum flexicaudum* (GIOVANNOLA, 1936a), *Schistosomatium douthitti* (OLIVIER, 1951 *Schistosoma mansoni* FAUST and HOFFMAN, 1934 and GIOVANNOLA, 1936b), *Orientobilharzia dattai* (DUTT and SRIVASTAVA, 1962a,b), *Cercaria elvae, C. stagnicolae physellae* (CORT and TALBOT, 1936) all emerge

Table 10 Illustrating the effect of variation of illumination on the emergence of the cercariae of *F. hepatica*.

Conditions during observation period of 24 hours	Conditions during previous 24 hours	Number of snails	Number from which cercariae emerged	Total number of cercariae
Darkness	Normal day–night illumination	100	49	676
Light	Normal day–night illumination	100	40	653
Darkness	Darkness	50	25	302
Light	Light	50	17	304
Light	Darkness	50	15	102
Darkness	Light	50	22	120

(After KENDALL and MCCULLOUGH, 1951)

during the day. Others e.g. *Plagioporus vespertilionis* (MACY, 1960), emerge at night (Fig. 44) and there are some which do not exhibit any great dependency on light or darkness in their emergence pattern (e.g. *F. hepatica*, KENDALL and MCCULLOUGH, 1951; *Cercaria cambrensis*, REES, 1931) (Table 10) although it is possible that emergence is dependant on regularly alternating periods of light and darkness.

The environmental conditions during which emergence takes place has been varied in an attempt to determine the factors influencing emergence. Temperature has a particularly significant effect at the extremes over which cercarial production takes place (Fig. 45). Very few cercariae will emerge from molluscs in water below 12–15°C although KENDALL and MCCULLOUGH (1951) reported that cercarial production of *F. hepatica* ceased at 9°C. The upper limit is more variable and seems to be more dependent on the thermal death point of the mollusc. REES (1947) noted that the emergence of *Cercaria purpurae* gradually decreased as the temperature rose above 20°C with the thermal death point of the snail (*Nucella lapillus*) occurring at about 30°C. The cercariae of *Fasciola hepatica* emerge at temperatures up to 26°C but, above this, the

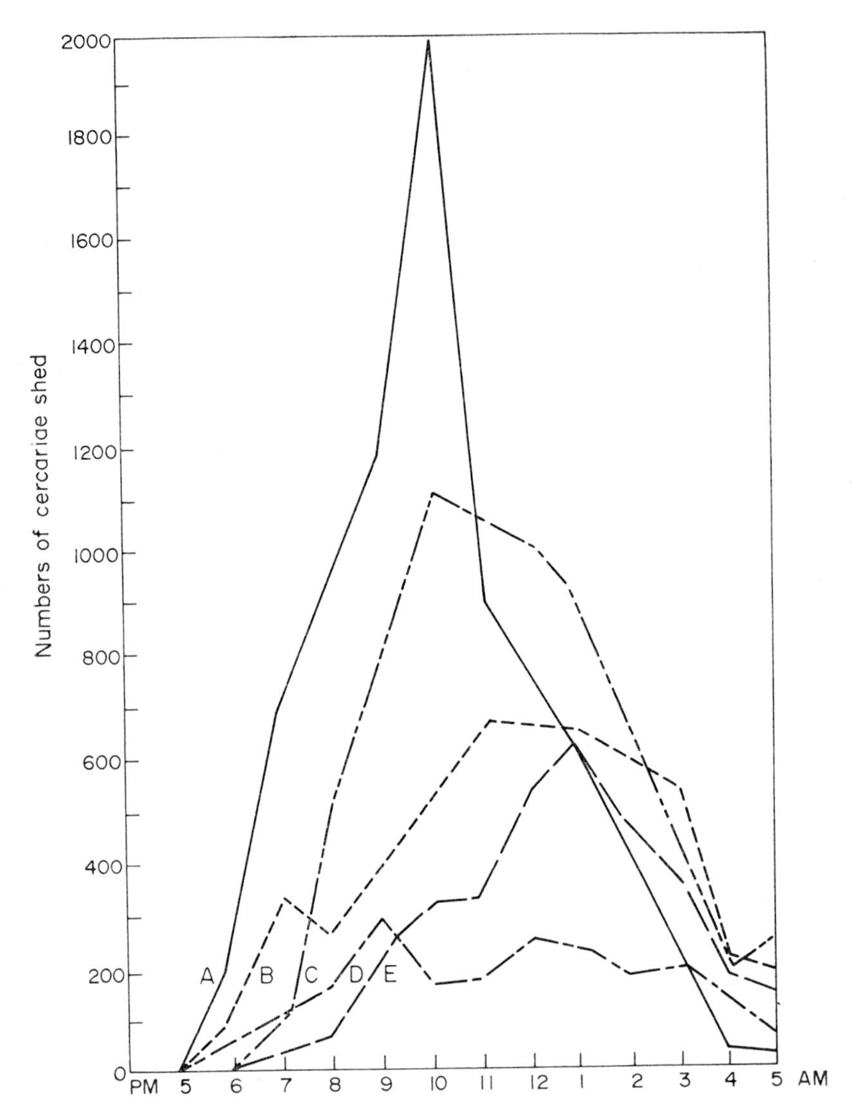

FIG. 44 The shedding period of the cercariae of *Plagiorchis vespertilionis parorchis*. Letters A, B, C, D, E represent individual snails. From MACY, 1960.

snail (*Lymnaea truncatula*) becomes moribund. In contrast, DUTT and SRIVASTAVA (1962a) recorded emergence of *Orientobilharzia dattai* from *Lymnaea luteola* at temperatures as high as 40°C after which inhibition occurred. The lower limit in this case was 12–13°C.

In those species which exhibit a distinct diurnal periodicity which is related to the light intensity, reversal of the environmental light pattern usually produces a reversal of the emergence pattern. This reversal may be very abrupt e.g. *Schistosomatium douthitti* (OLIVIER, 1951) or adjustment and change in the pattern may take a few hours to establish. Similarly the maintainance of snails under conditions of continuous light or darkness will abolish or permit continuous emergence depending on the particular species of cercaria (Fig. 46). A relationship

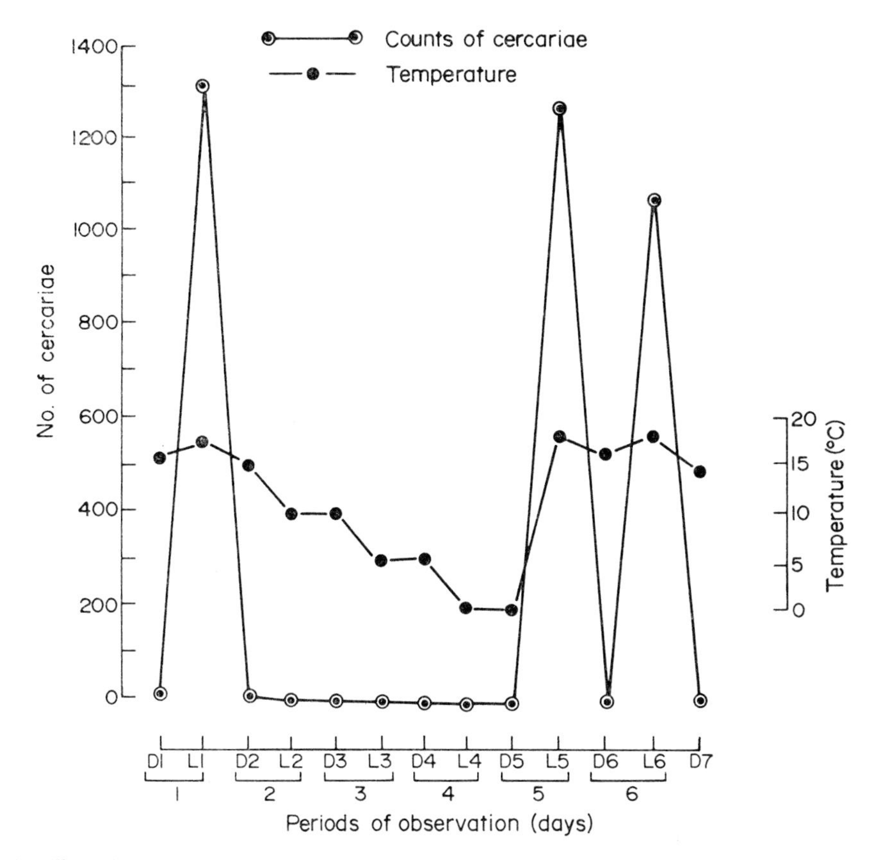

FIG. 45 The effect of temperature changes on the average number of *Cercaria purpurae* emerging from *Nucella lapillus* over a period of 6 days. The room temperature was decreased to 0°C and then returned to normal. From REES, 1947.

between emergence and light intensity seems to be characteristic of ocellate species, although this is not universal. PROBERT (1963) found that *Cercaria X* emerged at times of the day corresponding to a particular light intensity, although the snails were kept under conditions of constant temperature and the cercariae were non-ocellate forms. It is possible, as mentioned earlier, that the presence of certain pigments distributed in the body tissues may be light sensitive and play a part in this behaviour pattern. All these analyses of the emergence pattern are based on experiments carried out in the laboratory and all are open to a number of technical criticisms. In this respect the observations of ROWAN (1958) are particularly interesting. He was able to observe the daily periodicity of *S. mansoni* cercariae in the natural waters of Puerto Rico. His figures clearly indicate a daily periodicity under natural conditions (Fig. 47).

The mollusc *Nucella lapillis* is a littoral form and is exposed to variations in salinity. REES (1947) observed the effects of variation in salinity on the emergence of *Cercaria purpurae* from this snail (Fig. 48). A decrease in the salinity of the surrounding environment resulted in a reduction in cercarial emergence and none emerged when the salinity was reduced to 0·375 normal sea water. If, after exposure to this level for 24 hours, the snails were returned to normal sea water a delay of four days elapsed before cercarial emergence was resumed. With increasing salinity emergence was inhibited at 1·375 normal sea water, but in this case recovery was immediate

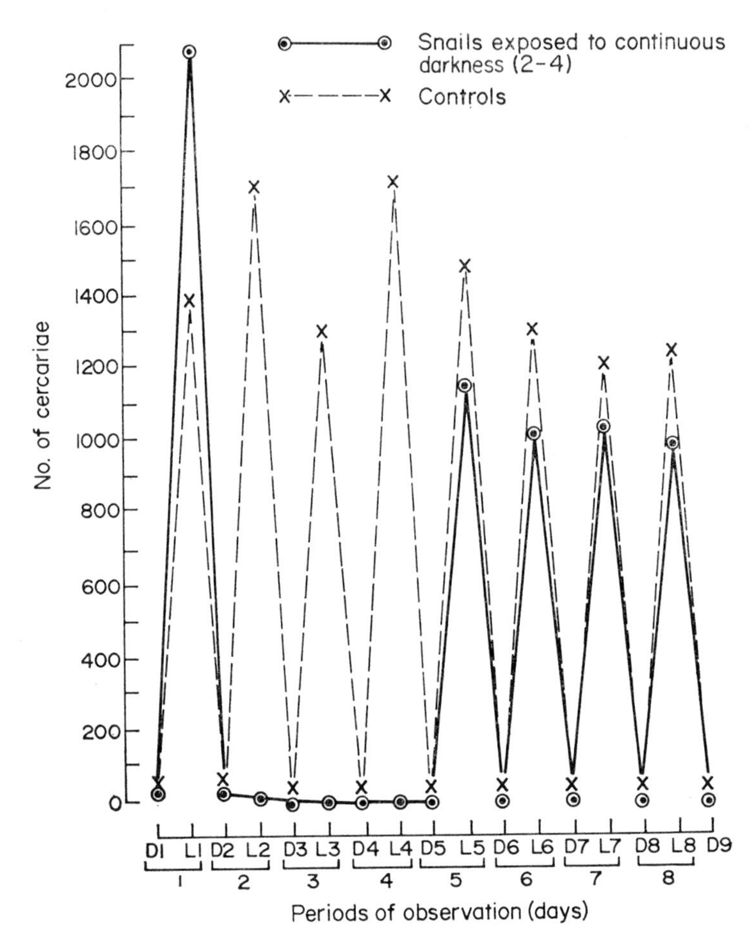

FIG. 46 The effect of changes of ambient light intensity on the emergence of *Cercaria purpurae* from *Nucella lapillis*. The snails were exposed at first to alternating periods of 12 hour darkness (D) and 12 hours light (L) and then maintained for three consecutive days in continuous darkness before returning to normal conditions. Control snails were exposed to normal diurnal variations. From REES, 1948.

when the snails were returned to normal saline. REES (1947) also observed the effects of varying salinity on emergent cercariae. In the case of decreased salinity the cercariae had a wider survival range than the snail. Increase up to 1·75N increased the rate of encystment and at 2N the cercariae became sluggish and died without encystment. In view of the varied career which many cercariae follow during the life-cycle it seems likely that the osmotic range which they can withstand would be large. STUNKARD and SHAW (1931) found that *Cercaria sensifera* (syn. *C. purpurae*) were able to survive for a few hours in 0·125N sea water, a level which was unfavourable to the host snail.

The effect of the pH of the water on cercarial emergence has been considered by BAUMAN *et al.* (1948) and KENDALL and MCCULLOUGH (1951). In the former case it was recorded that the cercariae of *Schistosoma mansoni* did not emerge from the snail at pH values below 7·2. In contrast the observations of the latter workers indicated that the cercariae of *F. hepatica* emerged over a wide range of pH from pH 5·5 to 8·5. They did note however, that a change of water

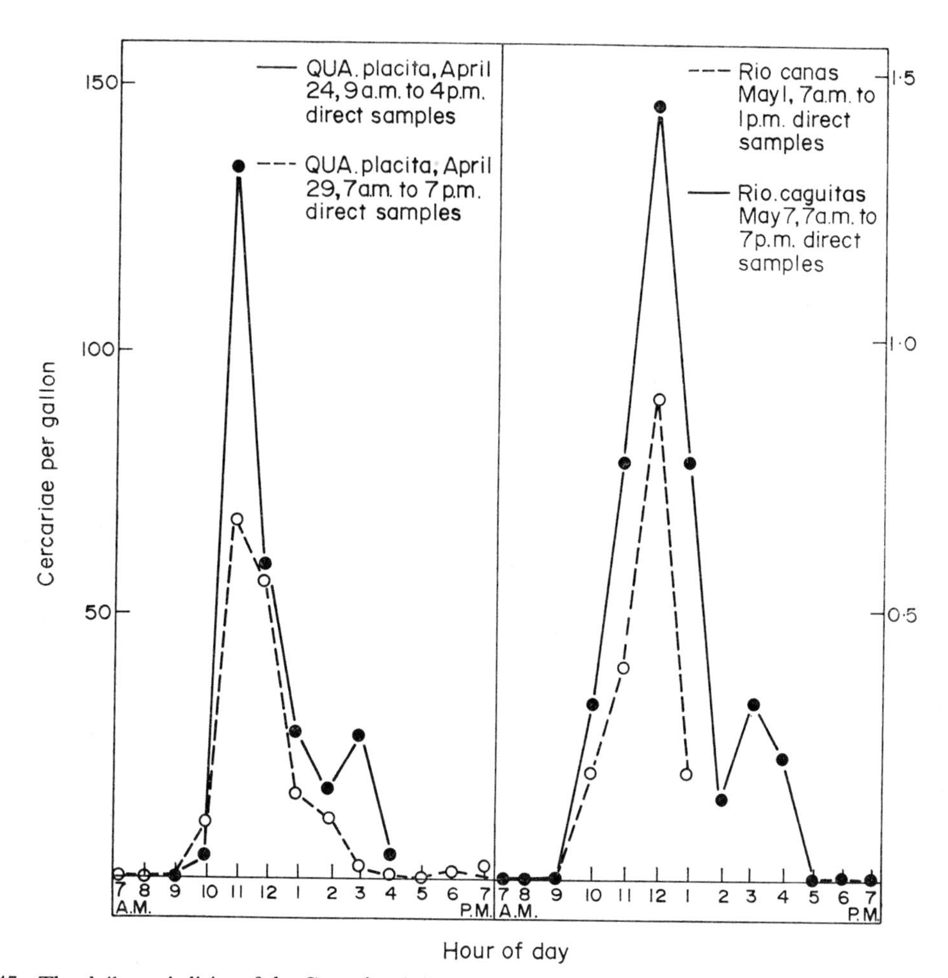

FIG. 47 The daily periodicity of the Cercaria of *Schistosoma mansoni* in Puerto Rican waters. From ROWAN, 1958.

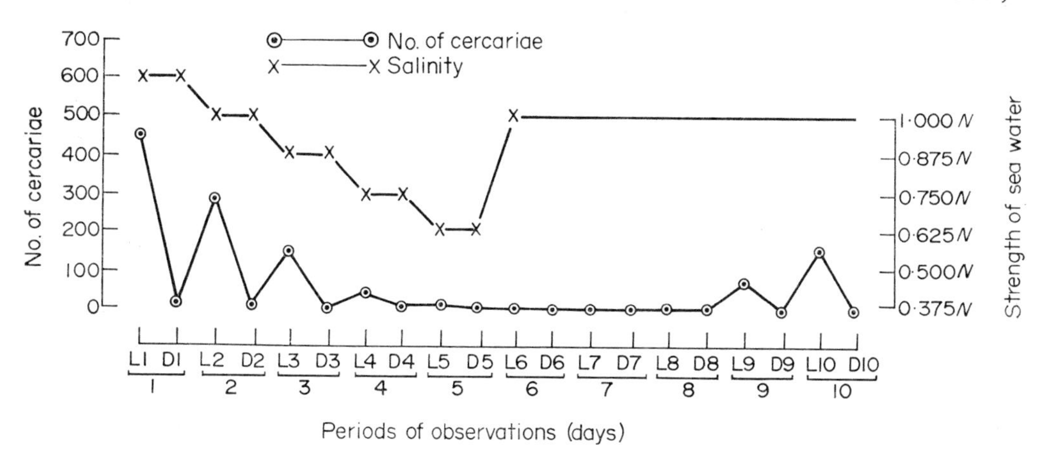

FIG. 48 The effect on changes in salinity on the emergence of *Cercaria purpurae* from *Nucella lapillus*. The snails were returned to normal sea water after five days. From REES, 1948.

Table 11 Showing that change of water induced the emergence of the cercariae of *F. hepatica*.

Group	Conditions	Number of snails	Number from which cercariae emerged
A	Snails left undisturbed	50	1
B	Snails returned to same water after mechanical disturbance	50	2
C	Snails placed in water previously occupied by other snails	50	6
D	Snails placed in fresh pond water	50	27

(After KENDALL and MCCULLOUGH, 1951).

in which the *L. trucatula* were kept did stimulate cercarial emergence (Table 11). This effect did not appear to be correlated with any degree of significance to temperature, pH or carbon dioxide tension. It is possible that in this case physical stimulation of the previously resting snail might have played a part in producing emergence. Muscular movements of the mantle could result in a physical stimulus being transmitted to the cercariae contained in the wall.

It seems as if a wide range of environmental factors affect the emergence of cercariae from the snail. The presence of food and the need for a particular temperature are features which can be explained, but the presence of a diurnal periodicity which responds to light intensity is a feature which is more obscure. The development of the sporocyst and redial generations is closely dependent on temperature and adequate nutritional supplies and it is a well known feature that cercarial emergence is more easily induced in well fed snails. SINDERMANN *et al.* (1957) found that the reduced emergence of *Austrobilharzia variglandis cercariae* from *Nassarius obsoletus* was greatly increased after the molluscs were fed. Glycogen is known to be a common carbohydrate reserve in larval trematodes and it must provide a basic energy source for these stages. Any reduction in the intake of food by the snail will affect the endogenous supplies in the trematode and will presumably alter its potentialities for active muscular movements and consequently its emergence. The variation in maturation rates in generations arising from single miracidial infections appears to be pronounced although little explanation can be offered. In most experimentation the physiology of the snail is ignored. This is unfortunate as it is becoming increasingly obvious that the presence of larval trematodes in a mollusc represents a particularly close host-parasite relationship with variations in the physiology of the one having pronounced effects upon the other. It is possible that the larval generations possess interacting cycles of activity which respond in various ways to changes in the external environment through their effect on the snail host. A study of the diurnal activities of the molluscs in the physiological sense as well as behavioural aspects might elucidate more clearly the factors affecting maturation and emergence of larval trematodes from the molluscan hosts.

B. SWIMMING

In the majority of cases the cercariae leave the molluscan host and achieve a temporary free living existence. The duration of this period is very variable and is related to a number of environmental factors. In some cases encystment may be rapid so that the actual free period is very short indeed. In other species the cercariae exhibit a fairly prolonged swimming phase.

Dubois (1929) has constructed a series of graphs showing the activity of several species of xiphidio-echinostome and furcocercaria and from these recognized several phases of activity. An initial phase, where all the cercariae were actively swimming, was followed by a second phase with the cercariae at the bottom of the container. At first they exhibited slow movements but these decreased until all movement ceased. With the xiphidiocercariae the first phase lasted 20 hours but they remained on the bottom of the contained exhibiting some slight activity for 2–3 days. The echinostome species also had a free swimming period of about 20 hours but after this they died very rapidly. Considerable variation was exhibited by the furcocercariae. Some species ceased swimming after 10–15 hours at 15–17°C, whereas others swam for 45–70 hours. He also showed that the duration of the free swimming period was a function of temperature. With *Cercaria A* Szidat the period of 18 hours at about 12°C was reduced to between 3–4 hours at 35°C. Schreiber and schubert (1949a) observed the survival of *Schistosoma mansoni* in several media at 30°C. Fifty per cent (described by these authors as the half life) of the cercariae were dead within 8–16 hours in spring water pH 7·4 and no cercaria survived beyond 18 hours. In tyrode solutions containing 0·1 per cent glucose the half life was extended to about 20 hours with considerable numbers surviving 26 hours. The observations of many workers show that the mortality within a population of freshly emerged cercariae varies considerably between individuals. Dutt and srivastava (1962a) found that at 13–16°C, 40 per cent of the cercarial population of *O. dattai* died within 10–20 hours whereas 13 per cent survived between 34–44 hours and 1 per cent survived up to 52 hours. The reason for this variation in cercariae which have emerged in a short space of time from the same mollusc is not clear. Erasmus (1958) showed that the glycogen present in the caudal bodies of a fork tailed cercaria diminished during the free swimming period and that in cercariae lying on the bottom of the container the caudal bodies had become reduced to a mass of granules free from glycogen. The observations of Schreider and Schubert suggest that the presence of glucose in the environment might help in the conservation of these endogenous reserves. Chernin (1964) however, found that in *in vitro* cultures of snail tissues containing *S. mansoni* larvae, the addition of amino acids to a balanced saline containing glucose and trehalose did not increase survival time after emergence. The ability of cercariae to feed and the type of nutrients required by this stage are aspects which have not so far been investigated.

It is obvious that for a particular cercaria there is an optimum free swimming period which will enable it to reach its goal (e.g. leaf surface or intermediate host) and then to continue with the complex behaviour patterns of entry and migration and/or encystment. If the free swimming period is excessively prolonged then the ability of the cercaria to enter into these later phases of activity will be greatly reduced. In the case of those cercariae which have large coloured tails and which lie on the substratum where they will be seen and ingested by the host it is possible that an extension of this free period may not be so deleterious. This idea has some support from Dubois' data which showed a greatly extended second phase on the substratum in the case of xiphidiocercariae. Miller and mccoy (1929, 1930) investigated the infectivity of *Cercaria floridensis* to fish in relation to the duration of the free swimming period. After exposure to cercariae which had been emergent for up to 12 hours an average of 40–50 cysts were found in the fish; after 20–24 hours only 9–10 cysts were found and after 30–36 hours only 1–3 cysts were recovered from the fish. Although the number of cysts present was very small after 36 hours the authors state that cercariae were still alive at this time and able to crawl along the bottom of the container. It is apparent that the ability to swim and so contact the host is essential in this cycle. Dutt and srivastava (1962a) indicated that the cercariae of *O. dattai* retained their infectivity to guinea pigs up to 45 hours but animals exposed to cercariae which had been in the water for 48–52 hours were negative. This reduction in infectivity during the free-swimming period may be hastened by environmental features such as flowing water. Radke *et al.* (1961) demonstrated how the worm burden in mice diminished in relation to the velocity of the water and the distance from the origin of the cercariae. In spite of this, these authors recorded infection of mice after the cercariae had travelled 5000 feet downstream.

Table 12 Comparison of the behaviour of some schistosome cercariae found in the United States.

Cercaria	Definitive host type	Behaviour
T. ocellata	Bird	Swim actively and toward light; attach to wall of container with both suckers
T. phycellae	Bird	Swim less actively and not as directly toward light; attach to wall on light side or to bottom of container with both suckers or acetabulum only
T. stagnicolae	Bird	Swim toward light; do not attach but are suspended motionless in water close to wall of container
G. gyrauli	Bird	Perhaps not affected by light; become attached to surface film soon after emergence and are not easily disturbed into activity
G. huronensis	Bird	Reaction to light unknown; become attached to surface film soon after emergence with body parallel to surface, the tail hanging variously
G. huttoni	Bird	Swim toward light; became attached to the surface film, mostly motionless, occasional slight twitching of tail with body and tail greatly contracted; not easily disturbed into activity
A. variglandis	Bird	Reaction to light unknown; swim upward, accumulate at surface film, and lie motionless with tail curved anteriad beside the body; easily disturbed
A. penneri	Bird	Not affected by light; rest at the water surface, dorsal side up, with tail held relatively straight
H. americana	Mammal	Swim toward light; do not attach but swim actively close to the water surface on the light side
S. douthitti	Mammal	Apparently not affected by light; become attached to surface film soon after emergence and are not easily disturbed into activity
C. littorinalinae	Bird	Swim toward light; cling horizontally to the underside of the surface film
C. elongata	Unknown	Swim toward light; do not attach but are suspended motionless in water close to wall of container on light side
C. tuckerensis	Unknown	Perhaps not affected by light; crawl on bottom of container

(After Lee, 1962).

In those species possessing a stumpy tail or in which the tail is absent no real swimming pattern can be demonstrated. These cercariae move along the substratum by leech-like movements employing successive attachments using the oral, ventral sucker or glandular secretions from the posterior end of the body. In species with well developed tails distinct swimming patterns are exhibited.

Some cercariae are said to swim continuously during their free period but the majority alternate short swimming periods with long periods of flotation or sinking. The heavy bodied xiphidio and echinostome cercariae swim vigorously for short periods resulting in an ascent in the water. On cessation of this activity the cercariae sink fairly rapidly. The fork-tailed cercariae possess furcae at the end of the tail stem and at the end of the swimming phase the furcae spread out so that the rate of descent is greatly reduced. In some cases the furcae are fringed with fin folds which increase their surface area and prolongs the flotation period. The presence of fin folds and hairs on the tail will obviously affect the rate of sinking and the trichocercous cercariae are admirably adapted in this way to a planktonic existence. The rate of sinking will also be affected by the flexing of the body to an anchor-like shape and also by the presence of fat droplets in the body. The duration of these phases in *Cercaria hamata* has been given by MILLER and MAHAFFY (1930) and this cercaria swims for periods of 1·0–3·8 seconds and the sinking phase may extend from 20·2 to 107 seconds. The relative value of these periods is characteristic of many furcocercariae. While swimming, this furcocercaria will change its path suddenly and erratically. It swims tail first but in its sinking phase assumes a characteristic attitude. The furcae are held apart at an angle of 180–90° and the anterior third of the body is bent in the form of a hook. The flotation attitude of most furcocercariae is constant and often distinctive, and may be used as a character in 'specific' identification. The correlation between swimming behaviour and species distinction is illustrated by a table (Table 12) assembled by LEE (1962) concerning schistosome cercariae. Swimming is accomplished by the flexing of the tail from side to side and gives the impression in furcocercariae of vibration about two nodes, one near the distal end of the tail stem and the other near the posterior end of the body. The influence of temperature on swimming was investigated by BEVELANDER (1933) working with the cercaria of *Bucephalus elegans*. Swimming is accomplished in these forms by the contraction of the furcae and an undulating movement of the body. Bevelander found that the frequency of contraction of the tail furcae varied markedly with alterations of temperature. The contraction frequency increased from 0 at 0°C to a maximum of 102 at 28°C, after which it decreased rapidly to 0 at 40°C.

The swimming pattern of many cercariae may also be modified by response to specific stimuli. Ocellate cercariae may be negatively phototactic but are generally positively phototactic, and exhibit a very marked orientation to light which may result in the accumulation of a mass of swimming cercariae near the light source or in attachment to the side of the vessel nearest the light source. In natural conditions this pattern results generally in the concentration of cercariae in the upper levels of the water and this is thought to have some significance in aiding host-parasite contact. In rat-tailed cercariae the phototactic movements of the cercariae bring them together and the sticky posterior halves of their tails become entangled so that a mass of wriggling cercariae is formed with the bodies outermost along each radius (*Cercaria W*: MILLER, 1930; RATTENKÖNIG: WARD, 1916). MILLER and MCCOY (1929–30) and MILLER and MAHAFFY (1930) both describe cercariae *C. floridensis* and *C. hamata* which react to brief shadows by swimming upwards. *Cercaria hamata* swims intermittently and a brief shadow falling on a group of cercariae results in simultaneous swimming. Repeated shadowing at short intervals, however, results in inhibition of swimming. *C. hamata* also responds by swimming to tactile stimuli in the form of mechanical contact or pressure from a stream of water. This response maintains the swimming activity of a group of cercariae after the initial response to shadow stimuli. The authors give the following figures for swimming times:

Unstimulated	1·16 sec	± 0·17
Stimulated by:		
Shadow	2·15	± 0·16
Touch	3·24	± 0·29
Stream of water	2·66	± 0·24

A response of tactile stimuli has also been referred to by FERGUSON (1943a) and ERASMUS (1959) in the case of those furcocercariae which penetrate into fish. In both instances a touch stimulus altered the swimming pattern from one in which periods of flotation occurred to one of more continuous swimming. Once contact with the fish body was achieved swimming ceased, attachment took place and a sequence of leech-like searching movements was initiated. FOLGER and ALEXANDER (1938) found that the cercaria of *Bucephalus elegans* responded to vibrations which originated from mechanical shocks administered to the cavity slide containing the cercariae transmitted through the water. The response, which occurred after an appreciable reaction time, consisted of a cessation of movement. The reaction time varied inversely with the magnitude of the stimulus and a recovery time of 30 secs was necessary before further response took place. It is apparent that the swimming behaviour and response to stimuli varies considerably and a brief review was made by MILLER (1928). A more detailed analysis of the relationship between swimming activity and changes in light intensity as well as mechanical stimuli has been made by DÖNGES (1963, 1964) for the cercaria of *Posthodiplostomum cuticola*. He found that major responses occurred with sudden decrease of illumination and mechanical agitation. Increase in intensity, periodic increase and decrease and linear increase in intensity had no prolonged affect. The mechanical stimulus was independent of light intensity so this author has postulated a direct affect on the motor response system. In the case of optical stimuli, decrease in light intensity resulted in the transmission of activating impulses to the central nervous system, increase in light intensity had the converse affect with the release of inhibitory impulses. Spontaneous activity was regarded as the result of the activity of either a spontaneous pace-maker associated with the central nervous system or to continual discharge from the ocelli.

The reaction of a cystocercous cercaria (*Gorgodera amplicava*) to the *in vitro* stimulus of molluscan serum has been described by CHENG (1963c). The response varied in relation to the molluscan species involved and the degree of dilution. The cercariae were stimulated to emerge from the cavity in the tail and the bodies would exhibit flexural and creeping movements. It was suggested that the reaction would be significant in producing evagination of the cercarial body in the appropriate molluscan host and would allow the subsequent behaviour patterns of encystment to follow.

Pronounced negative geotaxis in the cercariae of *Opistioglyphe ranae* has been described by Styczýnska—JUREWICZ (1961). For the first 4–8 hours after emergence these cercariae swim towards and near the water surface as the result of strong negative geotaxis. They then begin to fall to the bottom but on stimulation will resume their swimming. In the final phase they remain on the bottom of the tank and eventually die. Maximum survival time at 20°C was 80 hours. These figures however vary depending on the height of water above the cercariae. In a later paper (1962) it was suggested that the cercarial behaviour pattern was correlated with the behaviour of tadpoles which were confined to the surface waters because of the poor oxygen content of the deeper layers.

C. ENCYSTMENT

A considerable number of cercarial species undergo encystment at the termination of the free swimming period. Some encyst within the same mollusc without emergence and others encyst in a second intermediate host after penetration and migration. In spite of the frequency of this pattern, there have been very few detailed descriptions of the sequences involved, and the significance of the features observed could not be appreciated until the advent and application of histochemical and electron microscope techniques.

Although the formation of the metacercarial cyst is an aspect of cercarial behaviour, the cyst itself is very much a feature of the metacercarial stage and often employed in identification. Because of this, and the fact that the actual structure of the cyst is related to its formation, the process of encystment and the histochemical characteristics of the cyst wall will be considered in the chapter on the metacercaria.

D. GENERAL PHYSIOLOGY AND HISTOCHEMISTRY

The majority of observations follow those for sporocysts and rediae simply because the histochemical studies have been performed simultaneously on infected material in digestive glands and most of the literature has been reviewed by CHENG (1963b).

Glycogen has been demonstrated in a variety of cercariae (AXMANN, 1947, ERASMUS, 1958; SNYDER and CHENG, 1961; CHENG and SNYDER, 1962a; CHENG, 1963a,b; GINETZINSKAJA, 1960; GINETZINSKAJA and DOBROVALSKII, 1962; PALM, 1962a,b; JAMES and BOWERS, 1967b) and it occurs in muscular structures such as the suckers and pharynx, in the parenchymal cells in caudal bodies in tail stems, and sometimes in association with the excretory tubules. On the basis of histochemical evidence CHENG and SNYDER (1963) and CHENG and BURTON (1965) have suggested that glucose derived from host tissues becomes absorbed by the sporocysts and immediately resynthesized as glycogen inside the cercariae. The glycogen cannot be detected in the germ ball stage but appears (i.e. histochemically demonstrable) after the germ balls elongate and start differentiation into cercariae. The passage of materials from host through sporocyst or redia to cercariae is also suggested by the observations of NADAKAL (1960a,b,c) on the similarity of host and parasite pigments and the report by LEWERT and PARA (1965) which indicated that when infected *Australorbis glabratus* were fed C^{14} labelled glucose in plant materials, emergent cercariae, which were labelled, were obtained within 48 hours and that cercariae continued to exhibit activity for as long as two months after the initial exposure.

Some information is available on lipid metabolism in cercariae. Lipids have been demonstrated histochemically in the excretory system and parenchymal cells by LUTTA (1939), GINETZINSKAJA (1961), GINETZINSKAJA and DOBROVALSKII (1962), CHENG and SNYDER (1962b), CHENG (1965) and JAMES and BOWERS (1967b). The last two publications by Cheng suggest that the lipid material consists of neutral fats and fatty acids. CHENG (1965) was also able to demonstrate lipase activity on the surface of developing cercariae. Electron microscope observations in this department indicate the presence of lipid droplets within the cytoplasm lining the excretory system of certain cercariae. The lipid material appears relatively late in the development of some cercariae (GINETZINSKAJA, 1961) and fairly early in others. The quantities which appear differ also and seem to be correlated with the behaviour of the cercaria. In non-swimming forms and poor swimmers the amount of lipid is very small or may be absent whereas in active swimmers the quantity is fairly large. The exact role of lipids in cercarial metabolism is far from clear. They may be derived from host material or may be products of cercarial metabolism. It is possible that they may function as energy sources as well as representing excretory products. Lipids will also play a part in increasing the buoyancy of cercariae during flotation periods.

CHENG (1963b) has provided data on the amino acid content of the cercaria of *Gorgodera amplicava* and of a species of *Echinoparyphium*, and the list of the amino acids identified in the bound and unbound form resembles very closely that of the molluscan host.

Little information is available concerning enzyme activity in the cercarial body. The observations on phosphatase activity on the excretory system have already been mentioned. STIREWALT and WALTERS (1964) have demonstrated aminopeptidase activity in the gut caeca and esterase activity in the nervous system of *Schistosoma mansoni* cercariae. LEWERT and HOPKINS (1965) have also demonstrated cholinesterase activity at this site. The activity of whole cercarial extracts, particularly those of schistosome cercariae, have already been considered in the section dealing with gland cells. The presence of malic dehydrogenase activity in extracts of cercariae of *Schistosoma mansoni* has been demonstrated by PINO *et al.* (1966). Two fractions were found in the extracts and these differed in their electrophoretic mobilities from the mammalian (mouse) isoenzymes also investigated.

A study of the respiratory characteristics has been carried out mainly by Vernberg and co-workers (review by VERNBERG, 1963). Her studies have shown that cercariae are able to utilize oxygen and that oxygen is necessary for survival. OLIVIER *et al* (1953) showed that under anaerobic conditions 85 per cent of cercariae of *Schistosoma mansoni* were dead within four hours, and HUNTER and VERNBERG (1955b) found that although the cercaria of *Gynaecotyla adunca* were not affected after 4–5 hours under anaerobiosis, 75 per cent were dead after 12 hours. The respiratory rate of the cercaria of *Himasthla quissetensis* (VERNBERG, 1963) was strongly dependent on oxygen tension and were adversely affected by a O_2 tension of 0·5 per cent and were dead after 18 hours. In contrast the cercaria of *Zoögonus rubellus* exhibited a respiratory rate which was dependent on oxygen tension down to 3 per cent and after this became independent with the rate at 0·5 per cent resembling that at 3·0 per cent. This cercaria is a tailless form and unable to swim rapidly and also, as the above data suggests, able to resist anaerobiosis for some time. The rates of oxygen consumption of cercariae seemed to be the highest within the life-cycle (HUNTER and VERNBERG, 1955a). The relationship of oxygen uptake to temperature (VERNBERG, 1961; VERNBERG and VERNBERG, 1966) indicates that the upper thermal limits of the cercariae seem to be correlated with that of the adult host rather than that of the intermediate host and that the thermal acclimation pattern tends to be independent of that of the intermediate host. This data suggests very strongly that the cercarial stage is more attuned to existence in the next host than to that of the intermediate host from which it has originated.

5 The Mesocercaria and Metacercaria

The mesocercaria

In some strigeoid life-cycles there occurs a very distinctive stage intercalated between cercaria and metacercaria termed the mesocercaria. Previous references to this stage have been reviewed by PEARSON (1956) who also produced a revised version of BOSMA's (1934) original definition. Pearson produced a more morphological definition which is as follows: 'The mesocercaria is a definite prolonged stage in the adult generation of strigeate trematodes, which closely resembles the cercarial body, from which it develops in the second intermediate host, and does not possess metacercarial features; it develops in turn into the metacercaria in another host'. The larval generic name of *Agamodistomum* has been applied frequently to this stage.

Structurally the mesocercaria resembles an enlarged cercarial body in which the excretory system has become more elaborate (Fig 49A and B). The number of flame cells has increased and blind tubules which have developed to various degrees may be present and it is thought from these tubules that the reserve bladder system eventually arises. The spination, oral and ventral suckers and the alimentary tract resemble those of the cercarial stage. The oral sucker, in the mesocercaria of *Alaria arisaemoides*, contains the reservoirs of the ducts from the 'penetration gland cells' and also a cluster of unicellular gland cells. These are present in the cercaria but become more numerous in the mesocercaria. The cytoplasm of the unicellular glands stains with Nile Blue Sulphate and Neutral Red. In *A. arisaemoides* four 'penetration gland cells' lie anterior to the ventral sucker. The precise arrangement of these cells does seem to vary according to the particular species described.

The most recent experimental work has been carried out on *Alaria arisaemoides* and *A. canis* (PEARSON, 1956) and the following comments refer to these two species. Cercariae penetrate and develop into mesocercaria in tadpoles of *Rana pipiens*, *R. sylvatica* and *Bufo americanus* in the case of *A. arisaemoides* and into the same species with the addition of *Pseudacris nigrita* in the case of *A. canis*. Development into mesocercariae was completed in about two to three weeks. Adult amphibia were found to be refractory to infection. This feature is commented on in chapter 6. The distribution of the mesocercariae within the tadpole depended on the time after infection. In early infections, most of the development stages were in the subcutaneous lymph spaces but in the older infections they occurred in or close to the muscles of body and tail. The distribution of the mesocercariae becomes rearranged in relation to metamorphosis of the tadpole with the subsequent loss of the tail. Slight host reaction was observed in the case of mesocercariae in *Bufo americanus* in which the larvae became enclosed in a thin layer of fibrous tissue. In paratenic hosts more pronounced host reactions were observed. In the Garter snake the mesocercariae became encapsulated in masses of fibrous tissue but in avian and mammalian paratenic hosts they were enclosed singly in a capsule possessing a thin outer covering of fibroblasts and fibres, an inner layer of acellular granular material arranged in concentric layers and a central fluid filled cavity containing the mesocercaria.

The changes occurring in the so-called penetration gland cells in this developmental sequence

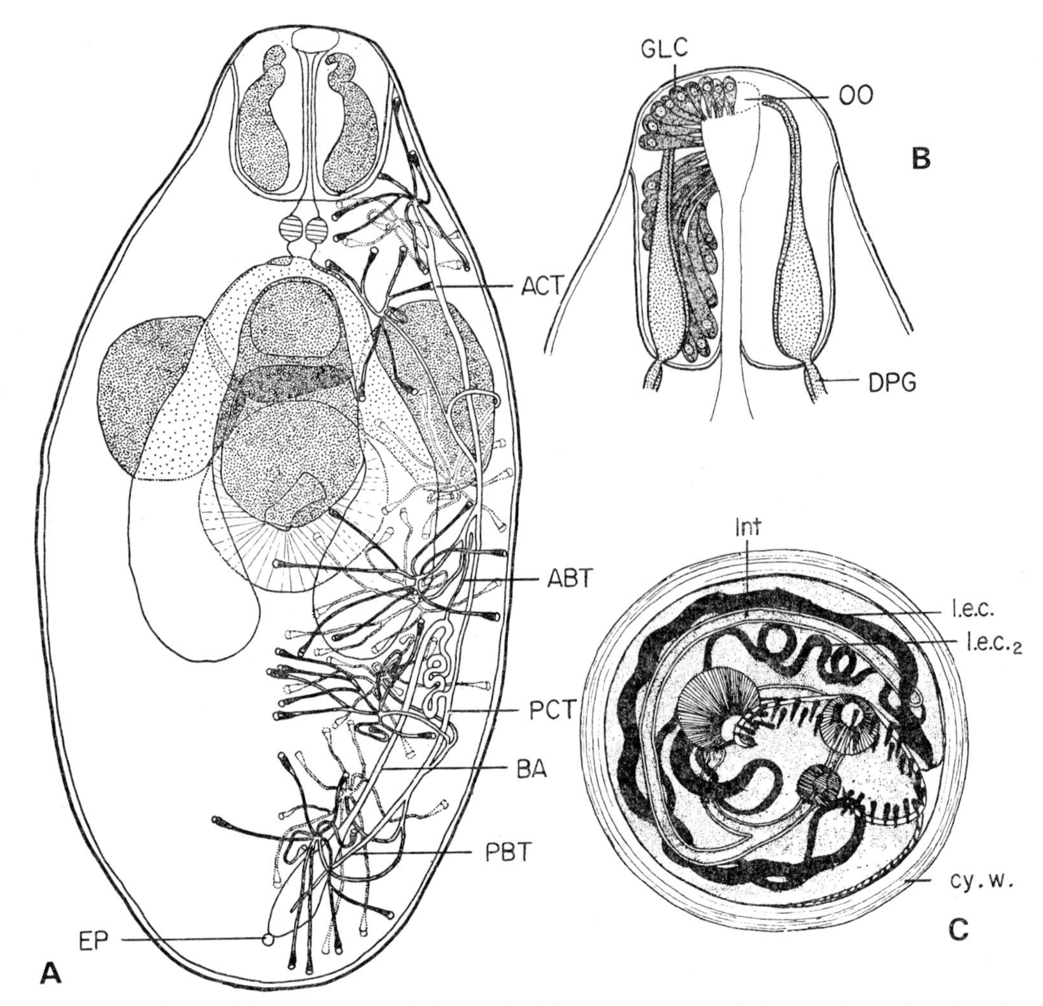

FIG. 49 **A** Dorsal view of the mesocercaria of *Alaria canis*. The excretory system is shown only on the right side and the ventral component of this system is represented by a dotted line. (ABT: anterior blind tubule, rudiment of reserve excretory system; ACT: anterior collecting tubule; BA: bladder arm; EP: excretory pore; PBT: posterior blind tubule; PCT: posterior collecting tubule.) **B** Dorsal view of the oral sucker of the mesocercaria of *Alaria arisaemoides* showing the gland cells (GLC) surrounding the mouth (OO) and the ducts of the penetration gland cells (DPG). Figures **A** and **B** from PEARSON, 1956. **C** The encysted metacercaria of *Cercaria limbifera* from the mantle cavity of *Lymnaea palustris*. (cy.w.: cyst wall; int.: intestine; l.e.c.: lateral excretory canal.) From REES, 1932.

are particularly interesting. During penetration of the tadpoles by the cercariae large amounts of material were lost from the 'penetration gland cells', but as development of the mesocercaria proceeded the cells were reconstituted. In the migratory phases through the tissues of definitive and paratenic hosts secretion of material from the gland cells of the mesocercaria took place and when migration was completed the cells were again reconstituted. Within the lungs of the host, in which metamorphosis into the diplostomulum stage occurred, the 'penetration gland cells' eventually disappeared.

Although the first intermediate and final host associated with these life-cycles appears to be fairly restricted, the mesocercaria is capable of survival and retaining its infectivity in a wide range of paratenic hosts e.g. frogs, snakes, birds and mammals. These hosts become infected after the ingestion of parasitized amphibia. It is possible that the persistence of the 'penetration gland cells', which are essentially a cercarial feature, into the mesocercarial stage may be associated with this wide host spectrum and the necessity for the persistence, for some considerable time, of the ability to migrate.

The metacercaria

The life-cycle of many digenetic trematodes contains a stage usually encysted, often described as 'resting' and termed the metacercaria. In its morphology this stage resembles the cercaria rather than the adult, although there are some conspicuous exceptions to this rule. The distribution of metacercariae within a particular environment (on vegetation, on shells or within a second intermediate host) is very closely linked to the food chain and feeding pattern of the final host. The basic disseminative stages in the digenetic life-cycle are the miracidium and cercaria both of which are free swimming (with some exceptions) and which exhibit characteristic behaviour patterns orientated towards continuing the life-cycle. These stages suffer the disadvantage of being short-lived (offset to some extent by the fecundity of the adult in the case of the miracidium, and by multiplication within the snail in the case of the cercaria) but the metacercaria represents a phase where the potentiality to continue the life-cycle and the ability to infect hosts is retained and extended over a relatively long period. The morphology and physiology of this stage is organized to this end and thus we have a period where the life-cycle is able to 'mark time' without affecting the viability of the life-cycle as a whole. In this sense the metacercaria can be regarded as 'resting', but this does not imply a phase of total physiological inactivity, a cessation of metabolic processes nor the absence of development.

The nature of the stimulus which induces encystment has not been clarified although in general terms it must be a component of the new environment in which the parasite finds itself. CHENG *et al.* (1966b), studying the effects of plasma from eight species of bivalves upon the cercarial stage of *Himasthla quissetensis*, came to the conclusion that it was the plasma which induced cercarial encystment in this species. The rapidity of the response varied with the molluscan species. Plasma heated to 60°C and 95°C did not induce encystment.

The cyst In the evolution of the digenetic life-cycle a considerable variation in morphology has occurred particularly in relation to this metacercarial stage. In various families the encysted metacercaria exhibits a different but characteristic distribution. In some (e.g. Fasciolidae, Paramphistomatidae, Notocotylidae) encystment takes place on an external substrate usually vegetation; in the Echinostomatidae, encystment may take place on an external substrate or in the primary host (same or different individual); and occasionally encystment may take place within the cercarial tail (some Azygiidae, Hemiuridae and Gorgoderidae) or within the same sporocyst (Brachylaemidae, Allocreadidae). In addition to these variations, in a large number of genera encystment takes place in a second intermediate host. In spite of this diversity the metacercaria always ends up in a position where it is likely to be eaten by the final host. In rare cases the metacercaria becomes progenetic within the first intermediate host (*Littoridina australis*) and produces eggs containing infective miracidia (e.g. SZIDAT, 1956) and consequently the life-cycle only involves the single molluscan host.

The metacercarial cyst is usually spherical (Fig. 49C) or pear-shaped in appearance and possesses a wall which, although characteristic for the species, exhibits a certain amount of variation within the Digenea (Plates 5-1 to 5-4). The wall is usually transparent, except in those cases where the presence of quinone tanned proteins give it a brown opaque appearance or where the cyst has become invested by pigmented fibrous layers, which represent the host's immunological response to the presence of the parasite. The metacercarial cyst wall has a considerable significance in the life cycle. The ability of the metacercaria to resist desiccation, mechanical abrasion and damage is largely related to the nature of the cyst wall. In those species parasitic in the tissues of second intermediate hosts, the protection of the metacercaria against the immunological defenses of the host tissues is also a function of the cyst wall. The hatching of the metacercaria in the appropriate final host may be influenced also by the biochemical composition of the wall. Excystation is a composite process involving the physical activity and possible secretion of lytic enzymes by the contained metacercaria, as well as external digestion of the cyst wall by the host digestive juices, and the resistance of these walls to digestion by different hosts will vary depending on the particular host-parasite relationship. In experimental infections metacercariae will pass through unsuitable hosts without any obvious alteration to the cyst wall. CAMPBELL and TODD (1956) working on *Fascioloides magna* in *Stagnicola reflexa*, and TAYLOR and PARFITT (1957) and KENDALL (1965), on *Fasciola hepatica* in *Lymnaea truncatula*, refer to the passage of cysts through the alimentary tract of snails leaving the metacercariae unaffected and still possessing their infectivity. After ingestion by the appropriate host the metacercaria becomes activated, and this results in movements which contribute to the dissolution of the cyst wall. The activator substances may reach the parasite by the diffusion of host substances e.g. bile components, through the wall and here again, its composition will influence the process and contribute to the host specificity of the parasite. Those metacercariae which have encysted on external substrates are also exposed to the ravages of fungi and bacteria and once more the cyst wall has a particularly significant role to play. This brief outline indicates that the cyst wall must possess a construction and composition which not only permits resistance to a wide variety of adverse agents but must also allow the diffusion of substances into and out from the cyst.

The metacercarial cyst wall may be thin, although layered, and contain a tightly coiled larva which may sometimes rotate within the rigid cyst (Plate 5-3). Many cysts are of this type e.g. echinostome, xiphidiocercariae and gymnocephalous and the nature of the contained metacercaria is revealed only by the identification of penetration stylets or collar spines through the transparent wall. The cyst wall of other species may be very thick and contain many lamellae e.g. strigeoid genera such as *Cyathocotyle* (ERASMUS, 1967a) and *Holostephanus* (ERASMUS, 1962a). These cysts are also rigid, but those of *Apatemon*, although thick and tightly investing the metacercaria, are flexible and alter their shape as the enclosed parasite moves (Plates 5-1 and 5-2). In other cases, the cyst wall is thin and loosely encloses the parasite which appears to lie in a large bag. Such cysts are seen in the genera *Posthodiplostomum* and *Asymphylodora* for example.

The nature of the cyst wall has always been difficult to determine. The intimacy of the association with the enclosed larva has proved a great hindrance in the collection of empty cyst walls for analysis and the difficulties involved in the impregnation of the cyst with paraffin wax and other supporting media has also greatly restricted investigation. The introduction of the cryostat and low temperature sectioning of frozen material, together with the application of histochemical techniques has greatly improved the prospects for detailed investigation of intact cysts with their component structures undisturbed. Using these methods it is possible to collect numbers of cysts, support them in blocks of mammalian liver which are then quenched in liquid air or nitrogen (ERASMUS, 1967a). Prepared in this way cysts can be readily sectioned and the walls subjected to a wide range of tests.

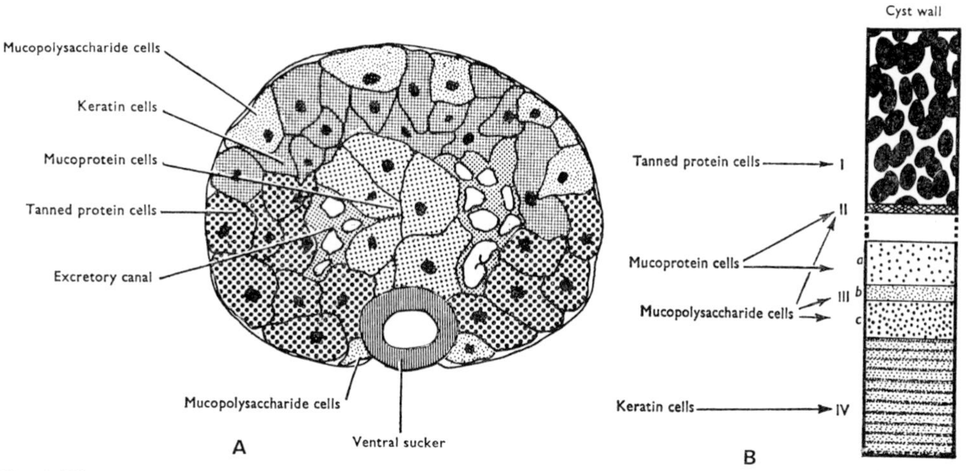

FIG. 50 **A** Diagrammatic representation of a transverse section of the cercaria of *Fasciola hepatica* showing the distribution of the cystogenic cells. **B** Diagram giving the relationship between the layers of the cyst wall of the metacercaria of *Fasciola hepatica* and the cercarial gland cells. Figures **A** and **B** from DIXON, 1966a.

HERBER (1950) described briefly the composition of the cyst wall enclosing the metacercaria of *Notocotylus urbanensis* and concluded that it was proteinaceous in nature. Later, SINGH and LEWERT (1959) examined the same material, using sections and histochemical tests, and distinguished five layers in the wall. All layers reacted in varying degrees to the PAS and Bromophenol Blue tests and the authors concluded that all layers were muco- or glyco-protein in nature. The metacercarial cyst wall of *Fasciola hepatica* has been investigated recently by DIXON and MERCER (1964), DIXON (1965) and DIXON and MERCER (1967). The wall has been shown to consist of four major layers (Figs. 50 and 51). The first and outermost layer which is incomplete ventrally, is composed of tanned protein and probably provides the major protection for the

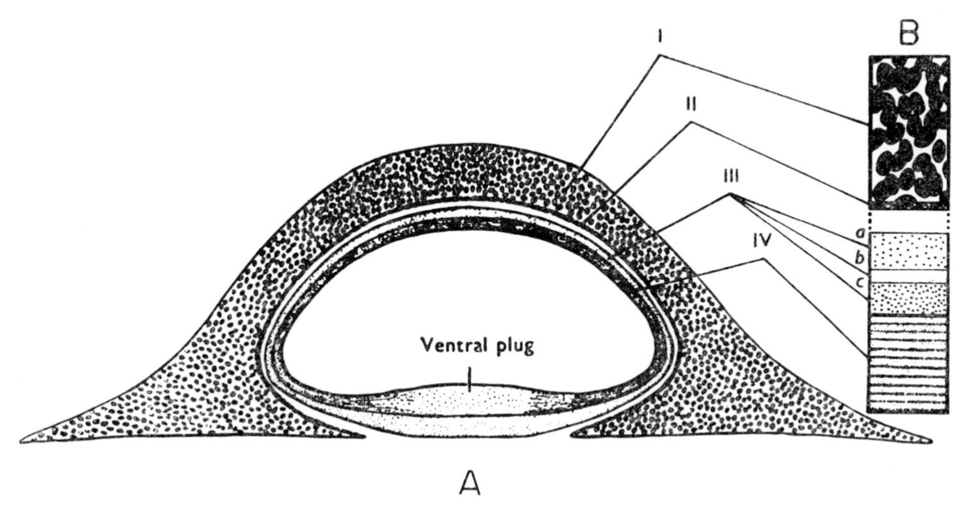

FIG. 51 The structure of the metacercarial cyst wall of *Fasciola hepatica*. **A** The cyst in vertical section; **B** details of the subdivision into layers. From DIXON, 1965.

metacercaria. This outer layer bears on its inner surface a thin fibrous layer of mucoprotein and acid mucopolysaccharide. The function of this material is somewhat obscure. The third layer is not uniform in composition but contains an outer region of muco-protein, a median acid mucopolysaccharide region and an inner zone of neutral mucopolysaccharide. The exact thickness of the components of this layer III varies between different specimens. The fourth and innermost layer consists of lamellae of protein stabilized by disulphide linkages held in a protein-lipid matrix. This layer is incomplete ventrally, where there occurs a thickened mucopolysaccharide plug. The fourth layer resembles keratin and again this type of protein could be expected to provide considerable protection to the metacercaria. The ventral mucopolysaccharide plug is the region through which the metacercaria excysts in the final host, and it is possible that this plug becomes digested by enzymes secreted by the metacercaria. An electron microscope study (DIXON and MERCER, 1964; DIXON and MERCER, 1967) of the cyst wall showed that the outer tanned layer consisted of aggregations of irregularly shaped bodies which fused when in contact suggesting that the layer might be porous in nature. The various mucopolysaccharide layers appeared as a feltwork of very fine filament (layer II) or as a granular structureless region (layer III). The inner keratinized layer was resolved at high powers into tightly compacted lamellae about 2·5 μ thick and separated by layers of less dense material. The method of formation of this layer during encystment is very interesting and will be discussed later. BILS and MARTIN (1966), in an electron microscope study of the cyst wall of *Acanthoparyphium spinulosum*, found that the wall material was laid down in concentric lamellae 0·03 μ thick. Portions of 'rods' could be seen sometimes within the cyst wall or attached to its surface. The inner surface of the cyst wall lay adjacent to the metacercarial tegument. They regarded the cyst wall as corresponding to the outer lamellated layer of *F. hepatica* cysts.

Two other species, in which the metacercaria encysts on an external substratum, have been investigated (PIKE, 1965; PIKE and ERASMUS, 1967). The histochemical characteristics of a monostome (*Notocotylus attenuatus*) and a gymnocephalous (*Psilotrema oligoon*) cyst wall were investigated. The cyst walls differed in their thickness (0·005–0·009 mm for *P. oligoon* and 0·026–0·034 mm for *N. attenuatus*) and also in the structure and histochemical characteristics of the layers. *P. oligoon* cyst wall was composed of four layers, an outer layer of acid mucopolysaccharide, a layer of protein, a zone of neutral and some acid mucopolysaccharide and finally an inner layer of protein which gave reactions attributed to SH— and SS— groups. This layer also possessed a zone strongly positive to PAS tests. In *N. attenuatus* there were three basic layers to the wall. The outermost contained acid mucopolysaccharide but also gave positive reaction to protein tests. The other two layers contained protein and gave positive reactions to the tests for SH— and SS— groups. In this cyst the inner layer also contained a zone of positive PAS material. It thus appears that the nature and arrangement of the layers in the cyst wall does differ although basic components of all seem to be mucopolysaccharides of various types, mucoprotein and a variety of proteins. The ventral mucopolysaccharide plug in all cysts probably provides a site for the emergence of the excysting metacercaria.

It is interesting to compare the pattern described above with the structure and characteristics of cysts formed in intermediate hosts as these will be exposed presumably to different environmental hazards and this might be reflected in the structure. Three species have so far been examined. The metacercarial cyst of a virgulate xiphidiocercaria from caddis fly larvae (PIKE, 1965), the cyst of *Posthodiplostomum minimum* from the fish (*Lepomis* sp.) described by BOGITSH (1962) and LYNCH and BOGITSH (1962), and that of *Cyathocotyle bushiensis* from the snail *Bithynia tentaculata* (ERASMUS, 1967a).

The cyst wall of the xiphidiocercarial cyst was 0·019–0·025 mm thick and possessed a thin outer membrane which was probably of host origin. The main cyst wall consisted of two layers—an outer one which contained a carbohydrate-protein complex which also stained with orcein, and an inner one which consisted entirely of protein. The cyst wall was rigid and the metacercaria tightly coiled within. In *Posthodiplostomum* the cyst was a thin-walled (5 μ approx.) bag in which the larva lay quite free. The wall consisted of two layers, an outer cellular and an inner non-cellular hyaline one. The cells which made up the outer layer were fibroblastic in nature and were presumably associated with an inflammatory response resulting in encapsulation. The inner layer was more complex and HUNTER and HUNTER (1940) have suggested that the layer was derived from the parasite. Histochemically it seems to be glycoprotein in its composition. In contrast the cyst wall enclosing the metacercaria of *Cyathocotyle bushiensis* is relatively thick, is rigid, and has the larva curled up tightly inside. Under the optical microscope the wall consists of a large number of concentric lamellae which vary in their optical density. Histochemically the wall consists of acid mucopolysaccharide layers with mucoprotein interleaved. Sometimes the cysts have a brown outer coat and as this gives a strong RNA reaction it is probably an investment of molluscan amoebocytes. There is very little protein in this cyst wall and it forms a marked contrast with the other cysts. The cyst wall of the closely related genus *Holostephanus*, which occurs in fish (ERASMUS, 1962a), is identical in appearance, and it seems likely that it would exhibit similar histochemical characteristics. Preliminary investigations on the cyst wall of *Apatemon* metacercariae in leeches also indicate that the cyst wall is essentially mucopolysaccharide in nature.

It is not possible therefore to make any generalizations in relation to the points made earlier until a much larger number of cysts have been investigated. It does appear however that the nature and arrangement of layers in the cyst wall will vary not only between different larval types but also between free-encysting and parasitic forms.

The ultimate arrangement of materials in the cyst wall is largely dependent on their source and on the behaviour patterns exhibited by the cercaria/metacercaria in the formation of the cyst. Secretion of the cyst wall has been observed by several workers and again it simplifies description if cyst formation is considered in two parts depending on whether the cyst is formed on an external substrate or within the tissues of an intermediate host. The process of encystment has been observed to varying extents by WUNDER (1924, 1932), BOVIEN (1931) ROTHSCHILD (1936a), REES (1937, 1967), PIKE (1965) and DIXON (1966a). The precise details of cyst formation varies between the different species but the general pattern is indicated by the description of encystment of *Cercaria purpurae* by REES (1937, 1967). In this species after the appropriate searching and testing behaviour patterns, attachment to the substratum takes place initially by the ventral sucker and eventually by both oral and ventral suckers. The ventral agranular cystoenous glands release their secretion which consolidate the attachment to the substratum along the ventral surface. The body then contracts and the oral sucker becomes attached to the substratum aided by secretion from the ventral agranular cystogenous glands. These secretions function in attachment only, and it is likely that the thin mucopolysaccharide films detected on the outside of some cysts represents this attachment material. The dorsal agranular glands then discharge their secretion which covers the dorsal and lateral surfaces of the body. Rees regards this layer as a mould beneath which is discharged the contents of the ventral granular glands. The contents of these cells flows between the cercarial body and the outer mucoid coat (Figs. 52, 53). The oral sucker is released from the substratum and the tail becomes detached with the granules flowing between the junction of tail and body. The tail may exhibit slight activity at

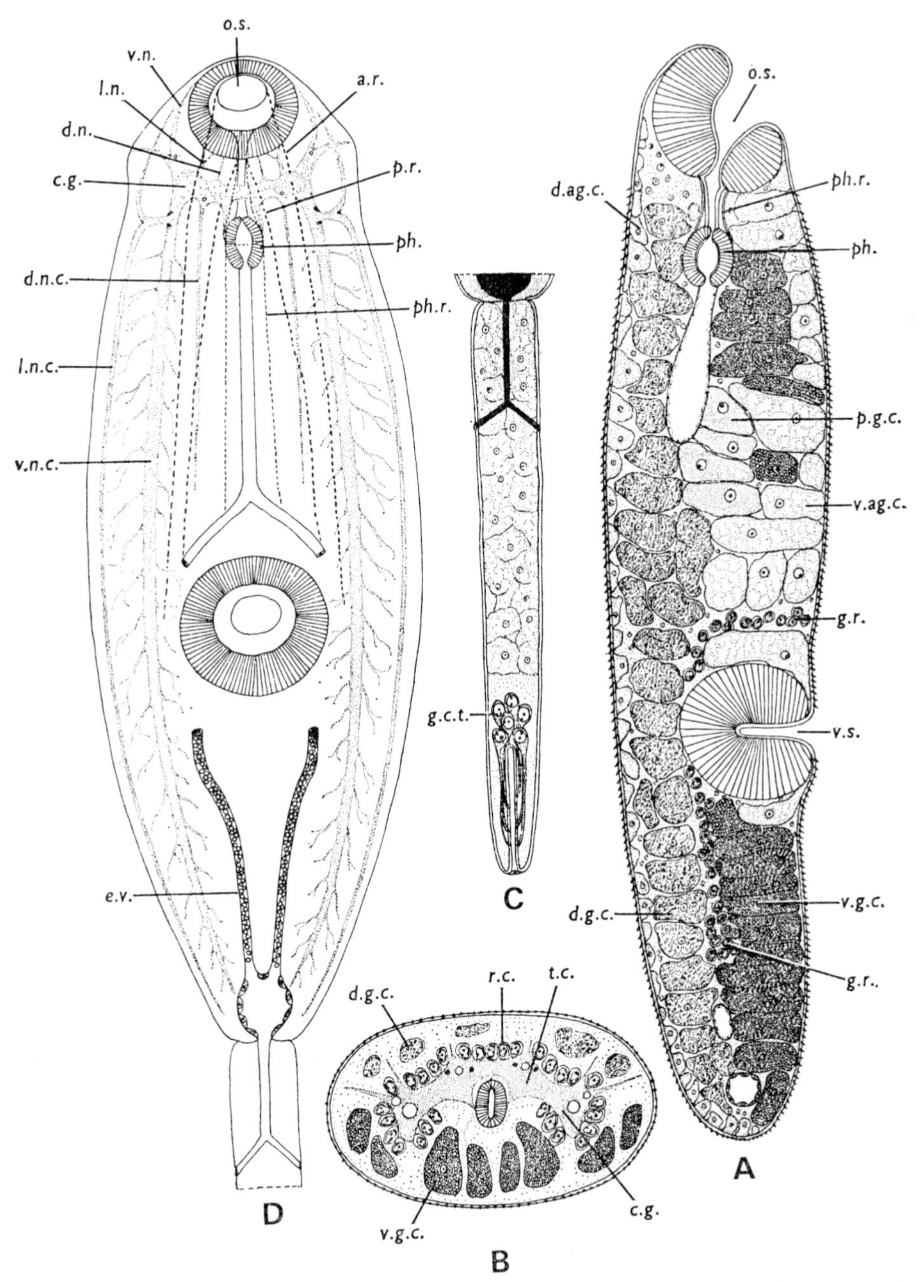

FIG. 52 The cercarial gland cells and nervous system of *Parorchis acanthus*. **A** Sagittal section of cercaria showing cystogenous gland cells; **B** Transverse section of anterior region of cercaria; **C** cercarial tail showing terminal gland cells; **D** dorsal view of cercaria showing nervous system and anterior retractor muscles. (a.r.: anterior retractor muscles; c.g.: cerebral ganglion; d.ag.c.: dorsal agranular cystogenous cells; d.g.c.: dorsal granular cystogenous cells; d.n.: dorsal nerve; d.n.c.: dorsal nerve cord; e.v.: excretory vessel; g.c.t.: gland cells of tail; g.r.: genital rudiment; l.n.: lateral nerve; l.n.c.: lateral nerve cord; o.s.: oral sucker; p.g.c.: plug forming gland cell; p.r.: posterior retractor; ph: pharynx; ph.r.: pharyngeal retractor muscle; r.c.: rind cell; t.c.: transverse commissure; v.ag.c.: ventral agranular cystogenous glands; v.g.c.: ventral granular cystogenous cells; v.n.: ventral nerve; v.n.c.: ventral nerve cord; v.s.: ventral sucker). Figures **A, B, C, D** from REES, 1967.

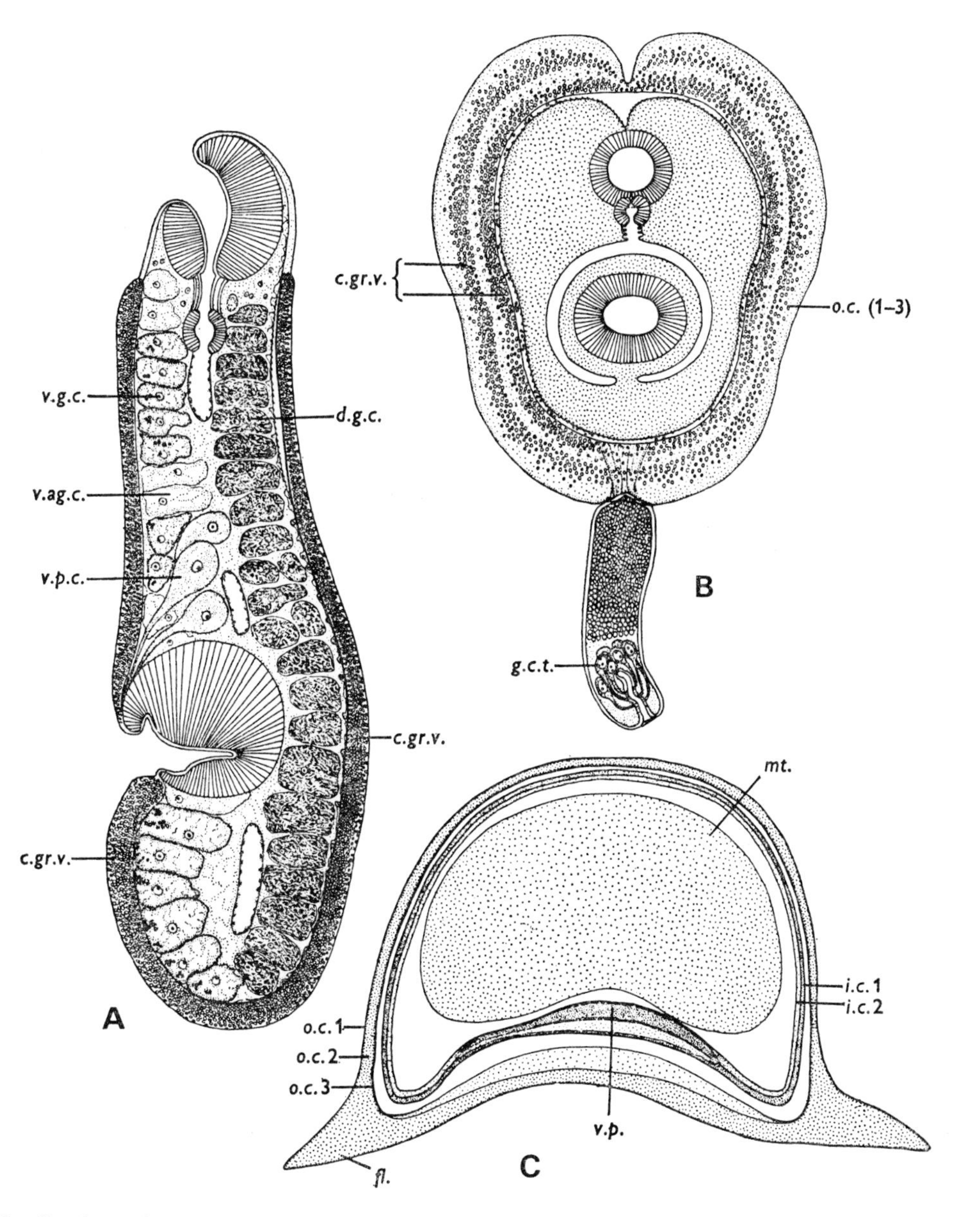

FIG. 53 Cyst formation in *Parorchis acanthus*. **A** Sagittal section through cercaria showing the gland cells and the layer of discharged granules surrounding the cercaria. **B** An encysting cercaria showing the secretion of the outer cyst wall and the loss of the tail. **C** Section through cyst showing the wall consisting of an outer wall of three layers and the inner wall of two layers. (c.gr.v.: cystogenous granules; fl.: flange of cyst wall; i.c. 1 and 2: layers of inner cyst wall; mt.: metacercaria; o.c. 1, 2, 3: layers of outer cyst wall; v.p.: ventral 'plug' region of cyst wall; v.p.c.: ventral plug-forming gland cells; other abbreviations as in Fig. 53.) Figures **A, B, C** from REES, 1967.

Table 13 A comparison of the cyst wall of *P. acanthus* and *F. hepatica*.

Parorchis acanthus		*Fasciola hepatica* (Dixon, 1965)
Histochemical reaction	Composition	Composition
Outer cyst wall—dorsal		Outer cyst wall
Outer layer (and dorsal agranular glands)		Outer layer (dorsal and ventral) tanned protein
AB + PAS + Hg BPHB-γ metachromasia	Acid mucopolysaccharide Neutral mucopolysaccharide	
Middle layer (and ventral granular glands)		
Hg BPHB + Sudan black B + α metachromasia	Protein, lipoprotein	
Inner layer (and dorsal agranular glands)		Inner layer (dorsal and ventral) mucoprotein
AB + PAS + Hg BPHB-Best's carmine + γ metachromasia	Acid mucopolysaccharide Neutral mucopolysaccharide Glycoprotein	Acid mucopolysaccharide α metachromasia
Outer cyst wall—ventral		
Outer and inner layers (and ventral agranular glands)		
AB + α metachromasia	Acid mucopolysaccharide	
Middle layer (and ventral granular glands) as on dorsal side		
Inner cyst wall		Inner cyst wall
Outer layer (more intense) Inner layer (less intense) PAS + Hg BPHB-α metachromasia	Neutral mucopolysaccharide	Outer layer (1) Mucoprotein (2) Acid mucopolysaccharide (3) Neutral mucopolysaccharide
		Inner layer Keratinized protein in protein and lipid matrix α metachromasia
'Plug' region		
PAS + Hg BPHB-β metachromasia	Neutral mucopolysaccharide	Neutral mucopolysaccharide

(After Rees, 1967).

this time which aids in its release from the cyst. The enclosed cercarial body now enters a phase of activity during which it revolves within the cyst, continually secreting material from the dorsal granular cystogenous glands. This process may continue for several hours, after which the cercaria curls up ventrally. In this way a very thin layer is secreted over the inner surface of the cyst and this completes the process.

The layers of the cyst wall, their histochemical characteristics and their relationship to the cercarial gland cells is indicated in Table 13 and Fig. 53. The cercaria of *Psilotrema oligoon* (Pike, 1965) has the outer surface of the body coated with granular material during its free swimming phase and it is this material which expands and attaches the cercaria to the substratum in the initial phases of encystment. The presence of secretory materials in the tegument of cercariae before emergence and during the free-swimming period has already been commented upon (Chapter 4). The rest of the secretory process resembles that described for *Cercaria purpurae*. The major point of variance concerns the role of the tegument in cyst formation. Wunder and Bovien refer to the accumulation of cystogenous materials in the tegument and Rothschild described the collection of cystogenous material in the outer layer of the tegument which eventually ruptured so that the outer tegument or 'primitive epithelium' became incorporated in the cyst wall. Fresh light has been shed on this aspect employing electron microscopy and histochemistry (DIXON, 1966a; DIXON and MERCER, 1967) in a study of the gland cells and cyst formation by the cercaria of *Fasciola hepatica* (Fig. 50 and 51). The four categories of cells described by Dixon i.e., tanned protein cells, mucopolysaccharide, mucoprotein and keratin cells all play a part in the formation of the cyst wall.

A number of distinct processes are involved in the formation of the cyst wall. These authors refer to the cytoplasmic covering of the cercaria as 'embryonic epithelium' and in the initial stages of cyst formation the precursors of Layer I (tanned protein) and Layer II (carbohydrate—protein) accumulate within this epithelium. The epithelium with its contents eventually ruptures and is shed to form the appropriate layers of the cyst wall. The keratinized layer is formed in a different manner. The rod-producing cells, or some of them, migrate from the deeper layers of the cercarial body to its outer surface where discharge of the keratin rods takes place. The cells are not destroyed completely in this process but remain at the cercarial surface so that their cytoplasm forms the new tegument of the metacercarial stage. Thus these authors suggest that the tegument of the metacercaria is formed from the cytoplasm of the rod-forming cells and the situation where the cell nuclei remain below the basement layer (a characteristic of most teguments) is explained by their failure to reach the surface in the early stage of the secretory process.

Particularly interesting is the observation on the keratin cells which contain the rod-like bodies described by many cercarial morphologists. The rods consist of a tightly rolled scroll which passes through the tegument, penetrates the outer plasma membrane and then becomes unrolled in the narrow space between the plasma membrane and the inner face of the cyst wall. (Plate 4-6). As the scroll unrolls the lamella becomes incorporated in the lamellated keratin-like layer of the cyst. Both REES (1937) and PIKE (1965) refer to the integrity of the cercarial tegument with its spines etc. during cyst formation and with our new concept of the cytoplasmic nature of the tegument it seems likely that the process of secretion described above represents the basic process associated with cyst formation, where temporal continuity exists between the tegument of the cercaria and metacercaria and complete destruction of the cercarial tegument may not occur. Further support for this hypothesis is provided by BILS and MARTIN (1966) who describe the intact cytoplasmic tegument lying beneath the lammellated cyst wall of *Acanthoparyphium spinulosum*, although the observations of DIXON and MERCER (1967) do not correspond with this idea.

Because of the difficulties inherent in making observations, little data is available on cyst formation within the tissues of intermediate hosts, and the situation is rendered more complex by the wide range of hosts often involved as habitats. The aquatic insect larva is often utilized in life-cycles where the final host is an insectivorous animal. Xiphidiocercariae penetrate the softer regions of the insect cuticle and encyst with the body. It has been observed that in some cases e.g. *Acanthatrium oregonense* (BURNS, 1961) the cercariae encysted on the outer surface of the gills of caddis fly larvae (*Dicosmoecus sp.*) within one minute. They then penetrated the gill tissues entering the host using the stylet and Burns mentioned that they became moribund if penetration did not occur in ten minutes. The metacercariae remained motile within the tissues of the insect until metamorphosis into the imago took place. At this time the encysted metacercariae occurred with the musculature of the insect. In a similar manner, ETGES (1960) reported that the metacercariae of *Prosthodendrium anaplocami* only encysted after the mayfly nymphs had undergone submarginal ecdysis. BROWN (1933) found that the cercaria of *Lecithodendrium chilostomum* entered the caddis fly larva, overwintered as free metacercariae and eventually encysted in the thoracic muscles of the imago. This pattern is not constant as Burns found encysted metacercariae of *Allassogonoporus vespertilionis* within the body cavity only four hours after exposure to infection. It is possible that the variations observed in this type of life cycle may reflect the hormonal and metamorphic state of the host as well as the biological requirements of the parasite. The cercariae of *Phyllodistomum simile* sometimes emerge from the sporocysts in *Sphaerium corneum* but THOMAS (1958) described occasional encystment within the sporocyst. The cercarial body emerges from the cercarial chamber within the tail, curls up and begins to revolve. Simultaneously a secretion is released from the cystogenous glands lining the excretory vesicle. The complete formation of the cyst wall takes at least 12 hours and consists of an outer and inner layer. Thomas suggests that this inner layer is the product of gland cells present in the parenchyma. The discarded stylet may become embedded in the cyst wall.

More information, together with details of cyst formation, is available on strigeoid metacercariae parasitizing intermediate hosts. Metacercariae parasitic in invertebrate hosts occur in genera such as *Cotylurus flabelliformis* (encysts in the snail *Lymnaea reflexa*) and *Apatemon gracilis*, which encysts in leeches (Plate 5-4). ULMER (1957) observed that in experimental infections, cysts were produced after 20–30 days at summer room temperature whereas CORT, BRACKETT and OLIVIER (1944) found that the rate of cyst formation was related to the severity of the infection. In light infections cysts were produced within 16–17 days, but in heavy infection many of the tetracotyles were still unencysted after 56 days. The completely encysted metacercaria of *Cotylurus brevis* was produced 23 days after infection (NASIR, 1960a). In leeches, the cyst wall enclosing the metacercariae, which had resulted from the infection of *Erpobdella sp.* by *Cercaria tetraglandis*, was formed between 51 and 91 days after infection (ILES, 1960).

The formation of the metacercarial cyst within vertebrate hosts is more complex and generally involves two processes. There is first the secretion of the true cyst, which is believed to be produced by the parasite, and then there occurs the addition of other layers and these are thought to represent the immunological reactions of the host. Most of the observations refer to cyst formation within fish. AZIM (1933) found that the formation of the cyst began 24 hours after penetration of the fish (*Gambusia sp.* and *Tilapia sp.*) by the cercaria of *Prohemistomum vivax*. In the development of *Neogogatea kentuckiensis* (HOFFMAN and DUNBAR, 1963) the cercarial bodies localized within the fish musculature had secreted a cyst wall, regarded as being of parasite origin by the authors, within two to three hours after penetration. This cyst increased in size as development continued but the connective tissue layer produced by the host did not appear until six days after infection. This outer layer increased in size and became thickened and pigmented in some cases. The formation of the cyst of *Uvulifer ambloplitis* (HUNTER and HAMILTON, 1941) and that of *Posthodiplostomum minimum* (HOFFMAN, 1958a) followed the same pattern. In the former case the parasite cyst was present within two to four days and the host cyst in ten to twenty-one days after infection. In the latter the cercarial bodies produced a cyst within 19–26 days and a connective tissue cyst layer of host origin was present at 29 days. An interesting contrast to this pattern is provided by the data of HOFFMAN (1956) on the formation of the cyst of *Crassiphiala bulboglossa*. In this case a cyst of host origin surrounded the parasite in the fish skin after six days and the cyst wall of parasite origin did not appear until twenty days after infection. Observations by the same author (HOFFMAN, 1958b) showed that the parasite cyst enclosing *Ornithodiplostomum ptychocheilus* had started at 17 days after infection and the host cyst was evident at 20 days. The cyst of *Holostephanus lühei* has a thick wall of parasite origin and an outer covering of fibrous tissue which is of host (fish) origin (ERASMUS, 1962a) and PIKE (1965) record the presence of a cyst wall after 24 hours. The well defined host reaction resulting in the enclosure of the cyst by a capsule of fibrous material is quite evident at 20 days after infection. It thus appears that the parasite cyst is formed fairly quickly in most cases, although the host component of the cyst is produced more slowly. It is likely that in the warm blooded vertebrates a host reaction would occur much more quickly. In some cases host melanophores accumulate in the host capsule and, in its vicinity and this pigmentation of the skin at the site of infection results in symptoms such as 'Black Spot' being associated with certain infections e.g. *Cryptocotyle* metacercariae in fish.

Not all metacercariae become encysted and those of the Diplostomatidae and the Brachylaeminae remain free and active. The strigeoid larvae ascribed to the larval genus *Diplostomulum* occur in the lens, humours or cranical ventricles of a wide variety of hosts. Their very precise localization is the result of a complex migration by the cercarial body after it has penetrated the second intermediate host. Various aspects of the biology of this type of parasite have been discussed in the chapters on Entry and Migration, and their morphology and development will be considered in the next section.

METACERCARIAL MORPHOLOGY

With the exception of neotenic forms and the strigeoid larvae to be discussed below, the majority of metacercaria resemble very closely the cercarial body before encystment or penetration took place. There may be some increase in size, adaptive features of the cercarial stage such as spines, stylets and gland cells may be lost, but the nervous system, alimentary system and excretory system all persist. The degree to which the reproductive system develops does vary to a considerable extent. In some cases no precursors can be detected, but in others the development may range from an undifferentiated 'anlagen' to states where testes, ovary and some ducts may be determined (Fig. 54). The extreme case occurs in those neotenic forms which are capable of producing viable eggs. Recent ultrastructural observations also suggest a persistence of the cercarial tegument; at least in those species which do not encyst. The evidence is conflicting in those species which do encyst.

The state of the excretory system at the metacercarial stage is a little more variable than the other features mentioned. This aspect has been discussed by KOMIYA (1961) and by KOMIYA and TAJIMI (1941) and they considered the relationship between the cercarial and metacercarial excretory systems under three categories.

(1) Metacercariae in which the flame cell formula does not alter from that of the cercarial stage. This group includes families such as Allocreadidae, Opecoelidae, Microphallidae, Cryptogonimidae, Fasciolidae and Echinostomatidae. The authors (loc. cit.) suggest that the absence of change from the system may be related to the smallness of the fluke or to the fact that the metabolic requirements of the metacercaria are the same as that of the cercaria. This association may be oversimplified as the exact relationship between the extent of the excretory system and metabolism of the parasite is far from understood. It also assumes that the excretory epithelium lining the ducts remains in a cytologically constant state. In fact it is possible that the system as seen in the cercaria is in excess of the requirements of that stage and complete involvement may occur later with increasing complexity of the body.

(2) In this group there is a small increase in the flame cell formula represented by the addition of only a few flame cells. The genera *Opistorchis* and *Metorchis* provide illustrations of this type of development. KOMIYA and TAJIMI (1941) gives the following data for *Metorchis*

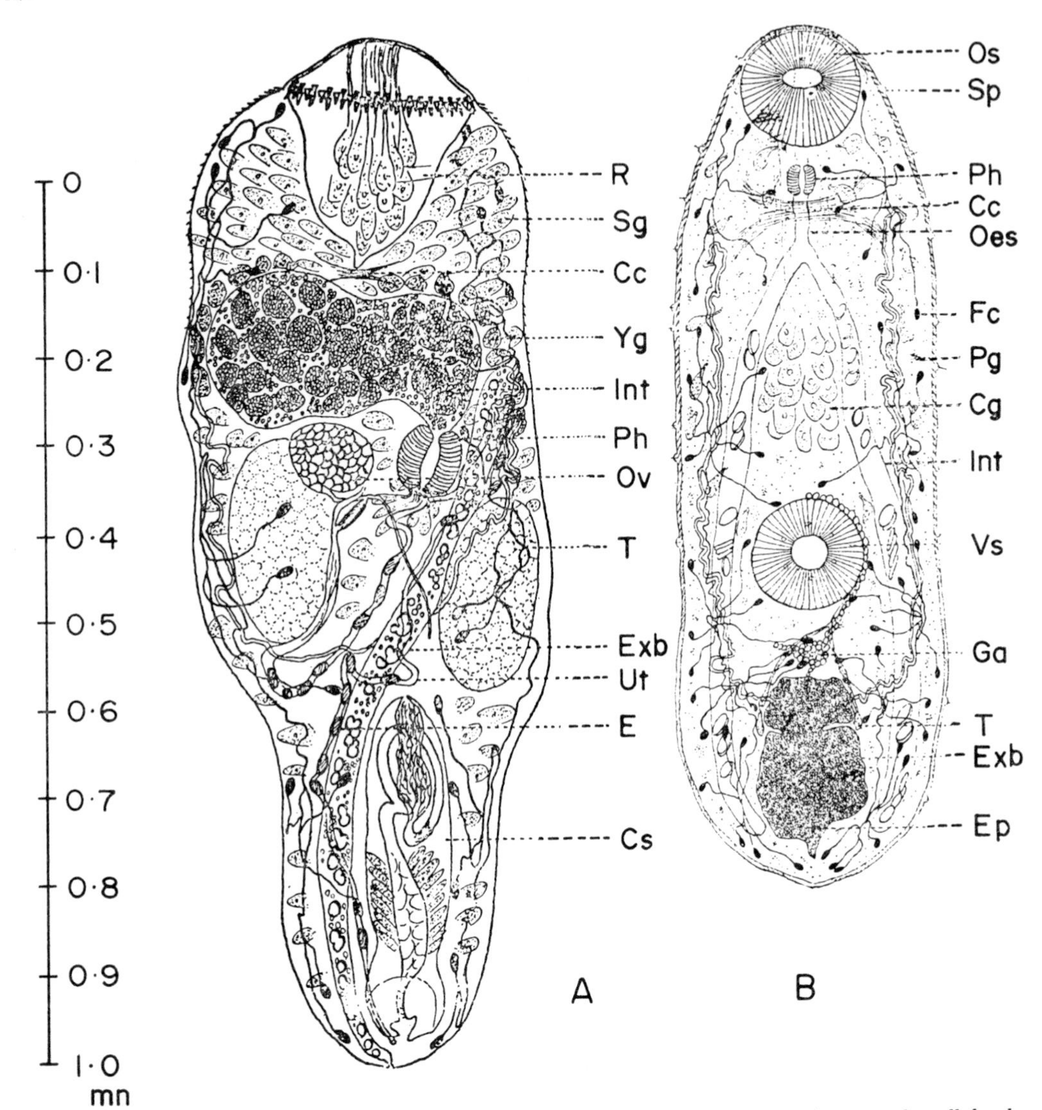

FIG. 54 **A** The excysted metacercaria of *Prosorhynchus echinatus*. The reproductive system is extremely well developed in this species. **B** The excysted metacercaria of *Metorchis taiwanensis*. (Cc: cerebral complex; Cg: cephalic glands; E: Egg; Exb: excretory bladder; Ep: excretory pore; Fc: flame cell; Ga: genital primordium; Int: intestine; Os: oral sucker; Oes: oesophagus; Ov: ovary; Ph: pharynx; Pg: pigment; R: rhyncus—glandular adhesive organ; Sg: 'skin gland'; Sp: spine; T: testis; Ut: uterus; Vs: ventral sucker; Yg: yolk gland.) From KOMIYA and TAJIMI—in KOMIYA, 1961.

orientalis

In cercaria (probably) $2 \times [(5 + 5 + 5) + (5 + 5 + 5)]$
In metacercaria, young $2 \times [(5 + 6 + 6) + (6 + 6 + 6)] \cdots$ basic form
In metacercaria, older $2 \times [(5 + 6 + 7) + (7 + 7 + 7)]$

or more generally

$$2 \times [(5 + 5^n + 5^n) + (5^n + 5^n + 5^n)]$$

where $n = 1$–4.

(3) In this group the excretory system undergoes considerable development with the appearance of new ducts and lacunae. This constitutes the appearance of the reserve bladder system typical of the strigeoid trematode and this is an addition to the primary system of portonephridia. This primary system also alters with an increase, often with mathematical regularity, in the numbers of flame cells present. Again KOMIYA (1938) provides the following data from the development of *Cotylurus cornutus*

In cercaria	$2 \times [(1 + 1 + 1) + (1 + 1 + (2))]$	
Young metacercaria	$2 \times [(1 + 1 + 1) + (1 + 1)]$	Caudal flame cells lost.
Older metacercaria	$2 \times [(3 + 3 + 3) + (3 + 3)]$	

In fully formed

Tetracotyle	$2 \times [(3^n + 3^n + 3^n) + (3^n + 3^n)]$

where $n = 2–4$.

During the development of the metacercaria, considerable changes take place and the strigeoid metacercaria is very complex in structure compared with non-strigeoid forms, and it is not surprising to see this elaboration of the excretory system.

The strigeoid metacercaria provides two conspicuous exceptions to the usual statements made about metacercarial stages. In the first place the morphology of the stage is elaborate and represents a tremendous increase in complexity on the cercarial body. The second exception is that considerable histochemical and cytological changes takes place in the development of these new systems and structures so that in these non-encysting forms, the metacercaria remains an active feeding stage until its death or ingestion by the final host.

The strigeoid metacercarial body is generally divided into two regions comprising a conspicuous fore-body and a small inconspicuous hind body which is developed to varying extents in different genera (Figs. 55, 56, 57). The fore-body may be flat and leaf-like as in genera such as *Diplostomum*, *Alaria*, *Neodiplostomum* etc. or cup-shaped and markedly concave on the ventral surface in *Apatemon*, *Cotylurus*, *Cyathocotyle*, *Holostephanus* etc. Within the fore-body lie the alimentary tract, suckers of various sorts and the major portion of the excretory system. The hind-body is generally small and conical and projects from the postero-dorsal surface of the body. It contains the excretory bladder proper with its pore to the exterior. In the final host the hind body undergoes considerable development and will contain the reproductive system and extensions of the gut caeca (see DUBOIS, 1938).

The alimentary tract of the metacercaria is well developed and consists of oral sucker surrounding the mouth, a muscular pharynx and two intestinal caeca. The gut persists through metamorphosis from the cercarial body and will survive into adult life. There is usually a ventral sucker present, although this may be secondarily lost from this stage in some species. One of the major strigeoid characteristics—the adhesive organ (organ tribocytique, holdfast organ of various authors) is absent from the cercaria but appears during the development of the metacercaria. The adhesive organ may be lobed or acetabular in form (ERASMUS and ÖHMAN, 1963) and has associated with it a cluster of gland cells. The gross features of the development of this structure have been traced by a number of workers (see below) but the possibility of its representation in the cercarial stage by a number of specialized cells has not been investigated. Thus, at a gross level, it appears to arise 'de novo' in the metacercaria. Histochemical and electron microscope studies of this organ in the adult stage (ERASMUS and ÖHMAN, 1963, 1965; ÖHMAN, 1965, 1966a,b) have demonstrated that the cells associated with the organ exhibit considerable

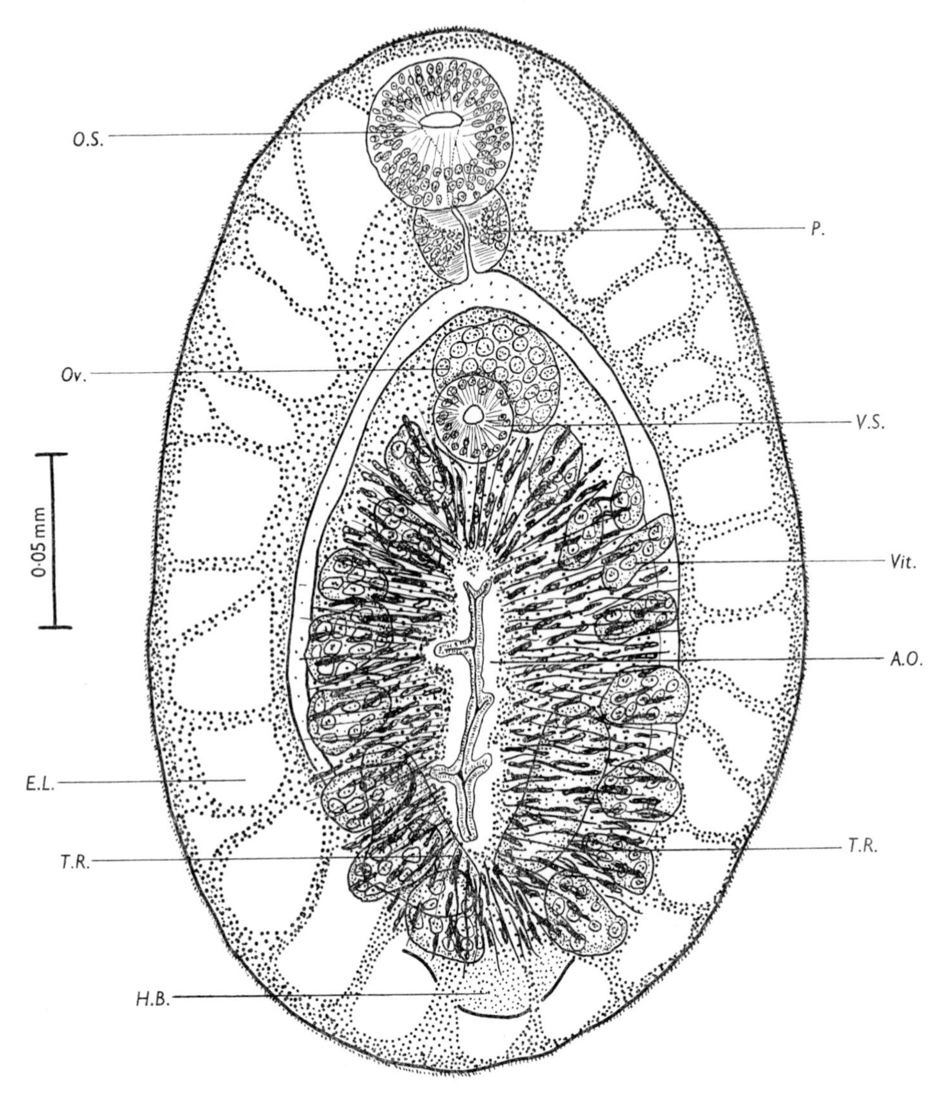

FIG. 55 The excysted metacercaria of *Holostephanus lühei* (ventral view). (AO: adhesive organ; EL: excretory lacuna; HB: hind body: OS: oral sucker; OV: rudiment of ovary; P: pharynx; T.R.: rudiment of testis; Vit.: rudiments of vitellaria; V.S.: ventral sucker.) From ERASMUS, 1962.

synthetic and secretory activity, involving the production and secretion of a number of enzymes. It is interesting to note that the cells associated with the adhesive organ of the encysted metacercaria of *Cyathocotyle bushiensis* also exhibit some of the characteristics of the adult cells. (ERASMUS and ÖHMAN, 1963). The cytoplasm of the cells in the adult was rich in RNA and exhibited acid and alkaline phosphatase activity and also esterase activity. These features could be demonstrated in sections of encysted metacercaria and also on freshly excysted metacercariae after hatching in the media devised by ERASMUS and BENNETT (1965). Thus it seems that in the case of *C. bushiensis* not only is the organ morphologically differentiated but that the cytoplasm of the cells is exhibiting some secretory activity. This may be because attachment to the gut tissues of the final host takes place within 24 hours after ingestion and that the sexually mature egg-laying stage is reached within five days. The fact that the metacercaria is well differentiated within the cyst undoubtedly facilitates this rapid maturation rate in the final host. Similar cytochemical characteristics have been noted in the gland cells associated with the adhesive organ of *Diplostomum phoxini* metacercaria (ARVY, 1954; LEE, 1962), although in this case the metacercaria is free and actively feeding in the cranial ventricles of the fish.

In addition to the adhesive organ some strigeoid trematodes possess, both in the metacercarial and adult stages, additional attachment structures known as lappets (pseudo-suckers of some authors). These consist of ear-like projections from the anterior end of the body on either side of the oral sucker. These lobes are muscular and are capable of inversion or inrolling so that host tissue may be drawn into them and attachment to the mucosa achieved. The lappets

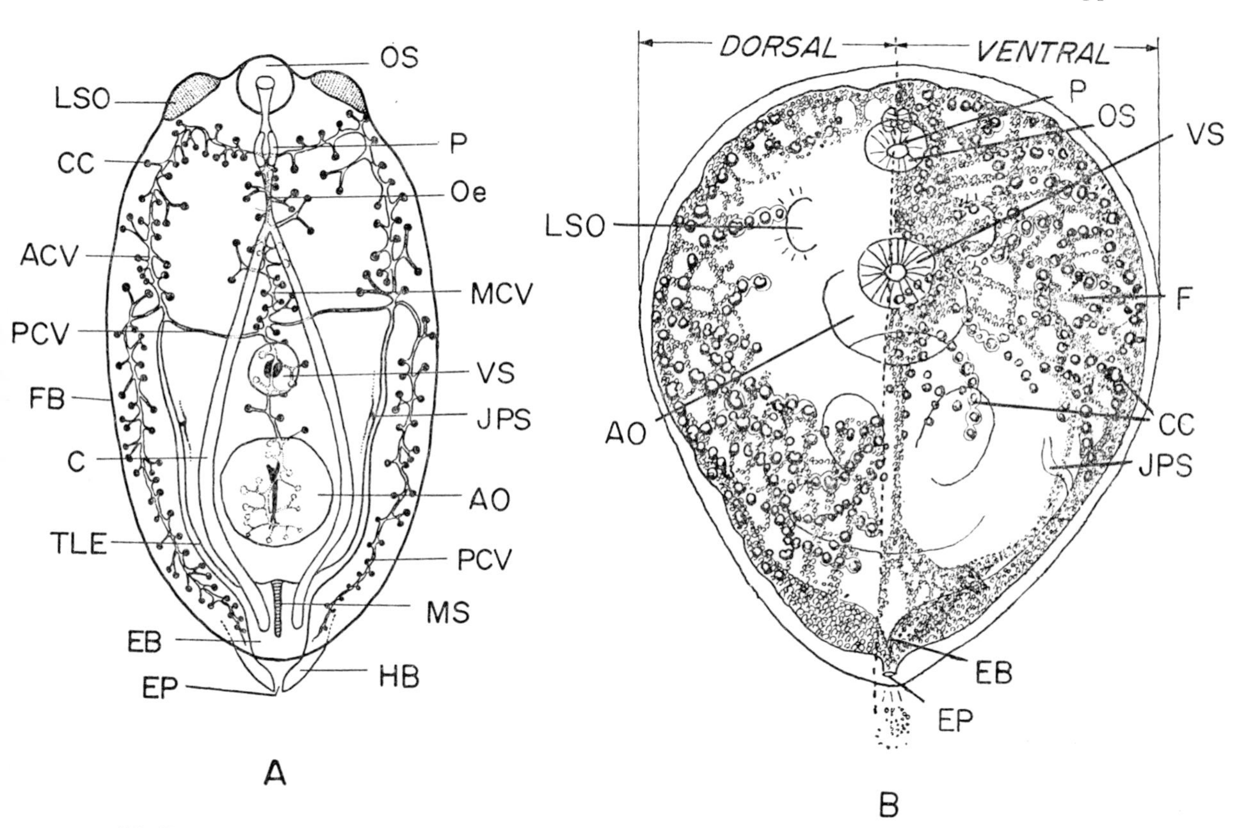

FIG. 56 Examples of strigeoid metacercariae. **A** The nonencysted diplostomulum type. The reserve bladder system is in the form of relatively narrow vessels with the calcareous corpuscles restricted to blind diverticula. From ERASMUS, 1958. **B** The encysted metacercaria '*Tetracotyle typica*'. In this form the reserve bladder system is more extensive with the individual channels less well defined. The calcareous corpuscles are not restricted to small diverticula. From KOMIYA, 1961. (AO: adhesive organ; Ac.V.: Anterior collecting vessel; C: caecum; CC: calcareous corpuscles; EB: excretory bladder; EP: excretory pore; F: lipid droplet; FB: fore-body; HB: hind body; JPS: junction with primary system; LSO: lappets; MCV: median collecting vessel; MS: median septum; OE: oesophagus; OS: oral sucker; P: pharynx; PCV: posterior collecting vessel; TLE: lateral extension of excretory bladder; VS: ventral sucker.)

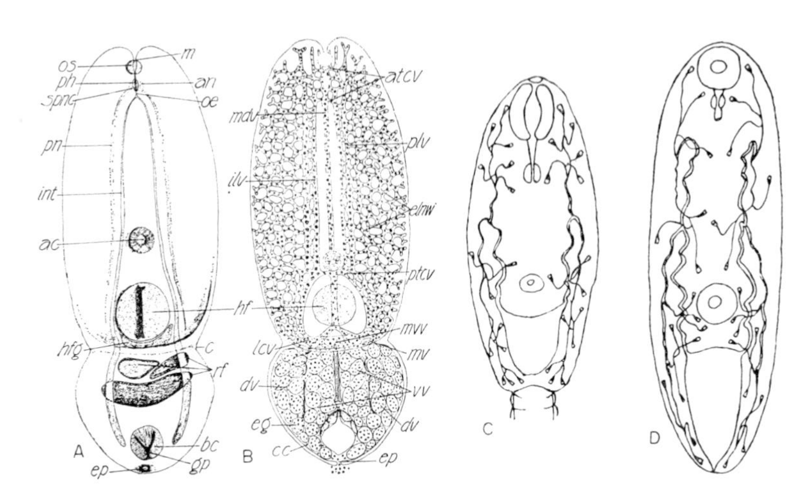

FIG. 57 Metacercarial stages. **A** Ventral view of the excysted metacercarial of *Neascus van-cleavei* showing the general anatomy. **B** The principal vessels of the reserve bladder system. (ac: ventral sucker; an: anterior nerve; atcv: anterior transverse commissural excretory vessel; bc: bursa copulatrix; c: body constriction; cc: calcareous concretion; dv: dorso-lateral excretory vessel; eg: globules in excretory fluid; elnw: extra-lateral excretory vessels; ep: excretory pore; gp: genital pore; hf: adhesive organ; hfg: adhesive organ gland cells; ilv: intra-lateral excretory vessel; int: intestine; lcv: lateral collecting vessel; m: mouth; mdv: median dorsal excretory vessel; mv: marginal vessel; mvv: median ventral vessel; oe: oesophagus; os: oral sucker; ph: pharynx; plv: primary lateral vessel; pn: posterior nerve; ptcv: posterior transverse commissural vessel; rf: reproductive rudiments; spnc: supra-pharyngeal nerve commissure; vv: ventro-lateral vessel.) From HUGHES, 1928. **C** and **D** The development of the excretory system of a non-strigeoid metacercaria. **C** the excretory system of the cercarial stage of *Clonorchis sinensis*. **D** the metacercarial stage of *C. sinensis* at an early stage of development. From KOMIYA, 1961.

also contain the ducts of the clusters of fore-body glands lying in the parenchyma of the fore-body, so that they are involved in secretion as well as attachment (ERASMUS, 1969a,c). These lappets appear in many genera such as *Diplostomum*, *Alaria* and *Apatemon*, and histochemical tests have indicated activity and secretion in the form of non-specific esterase activity in the fore-body glands and lappets of *Diplostomulum phoxini* (LEE, 1962).

An ultrastructural study of the lappets of *Diplostomum phoxini* (ERASMUS, 1969c) has indicated that the lappets are fully formed in 'mature' metacercariae from the brain of the minnow and that the two types of gland cells described in the adult lappets are present at this stage and that the cytoplasm of these cells possesses fully formed secretion bodies. This complexity further emphasizes the highly differentiated state of certain strigeoid metacercariae and its significance in facilitating rapid attachment to the host (the main function of the lappets in this species) as well as permitting the immediate assumption of the processes of digestion and absorption of host nutrients.

The other structural feature which undergoes considerable development and change in the metacercarial stage is the excretory system. The cercarial system consists of the usual protonephridial arrangement, although in a number of cases extensions of the excretory bladder are already present representing the early stages of the reserve bladder system (secondary system of some authors). Within the metacercaria the reserve bladder system is in two main parts. In genera such as *Diplostomum*, (Fig. 56A) the extension of the excretory bladder is in the form of a number of narrow channels arranged in a distinctive and regular pattern of anterior and posterior lateral ducts joined by anterior, median and posterior transverse commissures. The majority of these vessels possess small diverticula in which lie spherical bodies usually referred to as calcareous bodies, although there is little histochemical evidence to support this description. The system extends over the entire fore-body, is in continuity with the bladder and, at a specific point on each side of the body, is joined to the protonephridial system. The other arrangement seen in metacercariae of genera such as *Apatemon*, *Cotylurus*, *Cyathocotyle* and *Holostephanus* (Fig. 56B) consists of a series of large lacunae extending over the dorsal and lateral regions of the body, and rarely appearing on the ventral surface. Although appearing haphazard in arrangement, the lacunae may be organized into a W with the base at the posterior end of the body. The lacunae are generally quite large and are traversed by strands containing cells and flame cells so that the various lacunae are in continuity with one another. The lacunae are fluid filled and the refractile bodies within (referred to as fat droplets or calcareous bodies by different authors) move freely and circulate through the system with the movements of the parasite body. An intermediate type of system occurs in the larval genus *Neascus* (adult genera: *Crassiphiala*, *Posthodiplostomum*, *Uvulifer*) where the system is basically lacunar but the spaces are reduced to fairly narrow channels displaying a fairly complex pattern (Fig. 57A,B). These channels are fluid filled and possess bodies which circulate with the movements of the fluid. In all cases the system represents an extension of the excretory bladder and connects with the primary protonephridial system at specific points.

It is apparent, therefore, that considerable differences exist between the cercarial body and that of the metacercaria in the strigeoid trematodes. The pattern of development associated with these changes has been described by a number of workers. The pioneer investigations of SZIDAT (1924) and WESENBURG-LUND (1934) have been supplemented by many descriptions (e.g. BOSMA, 1934; KOMIYA, 1938; PEARSON, 1956, 61; ULMER, 1957; ERASMUS, 1958; HOFFMAN, 1956, 58a,b; NASIR, 1960a) so that the major aspects of the development are fairly clear. The pattern is complex, consisting of the persistence of certain cercarial features and the development of others *de novo*. In this way the development is not holometabolic (i.e. involving an abrupt metamorphosis) as suggested by some of the earlier workers.

Within the first seven days after infection of the second intermediate host by cercariae, the contents of the body become ill-defined, so that the only system which can be clearly distinguished is the alimentary tract. The body increases in size and at 15 days the main vessels of the reserve bladder system have appeared (Fig. 58). At 20 days (in the case of *Cercaria X*, ERASMUS, 1958) the adhesive organ was present and small diverticula were beginning to appear on the collecting ducts of the reserve system. The increase in size continues and in some species the cercarial spination persists even to this stage. The last feature to be formed are the lappets on either side of the oral sucker, and in the case of *Cercaria X* this was well after 46 days. During this period the reserve bladder system continued to expand with the appearance of the refractile bodies at about 40 days. The primary system also increases in complexity with the number of flame cells increasing rapidly. The time scale is not particularly significant as this will vary with the hosts used and the temperature at which the experimental stock is maintained, but the sequence of development seems to be fairly constant. As well as this increase in complexity with the development of new systems, the extent of growth is indicated by the following figures for *Cercaria X*. The cercarial body is approx. 0·18–0·24 mm long and 0·04 mm wide. At 45 days the body size is 0·24–0·39 mm long by 0·09–0·13 mm wide. Thus a tremendous state of cellular and metabolic activity must exist within the developing metacercaria for at least six to seven weeks. It should

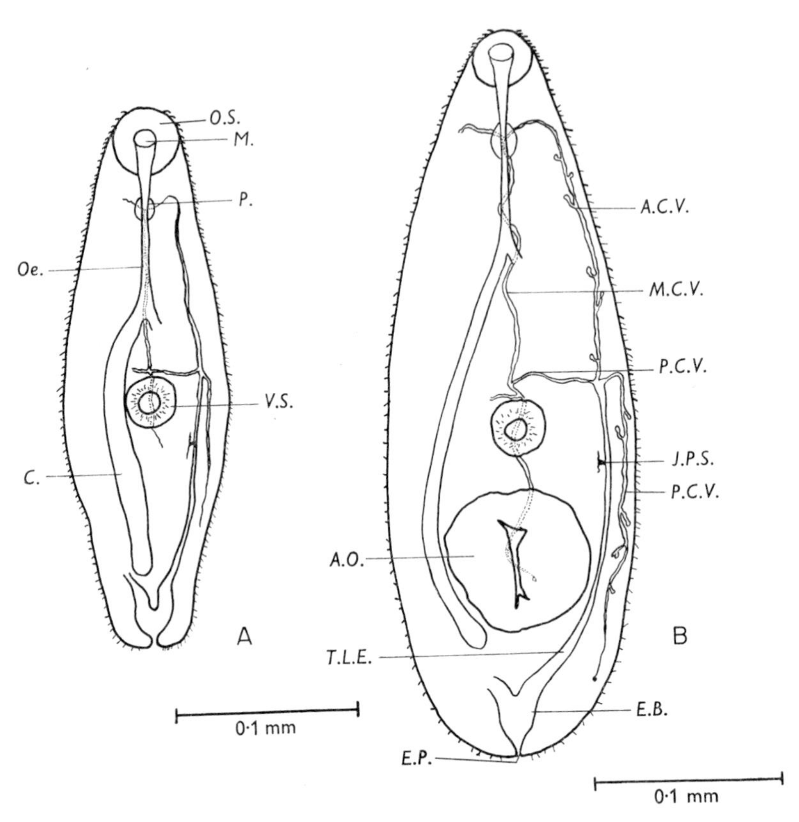

FIG. 58 The development of the reserve bladder system of the metacercaria of *Diplostomum sp.* **A** Stage 15 days after penetration of the cercaria into the second intermediate host; **B** 22 days after penetration. (ACV: anterior collecting vessel; AO: adhesive organ; C: caecum; EB: excretory bladder; EP: excretory pore; JPS: junction of primary and secondary excretory systems; M: mouth; MCV: median collecting vessel; Oe: oesophagus; OS: oral sucker; P: pharynx; PCV: primary collecting vessel; TLE: terminal extension of the excretory bladder.) Figures **A** and **B** from ERASMUS, 1958.

be emphasized that not all strigeoid metacercariae are free—some are encysted and the cyst wall is formed, to some extent, at a very early stage in development. It seems unlikely that the cercarial body contains sufficient sources of energy to sustain such a complex development and the possibility of the utilization of nutrients derived from the host by passage through the cyst wall must be considered. This suggests that the cyst wall must be permeable and allow entry of nutrients to some extent.

The distinctive morphological features exhibited by strigeoid metacercariae has encouraged the introduction of 'generic' names for these larvae. Relatively few life-cycles have been worked out and although the policy is reprehensible from a systematist's point of view, it has value in allowing the identity of larval stages to be communicated. The following larval groups and definitions are based on a synopsis given by HOFFMAN (1960).

'Tetracotyle' In fish, amphibia, leeches, snails.
Forebody oval and relatively thick, concave ventrally or cup-shaped. Hind body as a round inconspicuous prominence. Reserve bladder consisting of large lacunae covering lateral and dorsal regions of the body. Spherical refractile bodies free in the system. A pair of lappets present, usually invaginated in the encysted form. A well defined cyst wall of parasite origin. Adult genera: e.g. *Cotylurus*. *Apatemon*. (Fig. 56B).

'Diplostomulum' In fish, amphibia, snails.

Fore-body oval, slightly concave ventrally. Hindbody small and conical. Reserve bladder of well defined tubules having a regular pattern and possessing spherical refractile bodies in small diverticula arising from the main trunks. Usually with a pair of lappets. No true cyst of parasite origin. Adult genera: e.g. *Diplostomum, Tylodelphys, Hysteromorpha, Alaria, Pharyngostomum, Fibricola.* (Fig. 56A).

'Neascus' In fish.

Foliaceous forebody but with a well developed hind body. Reserve system consisting of narrow channels displayed in a complex pattern. They contain refractile bodies and diverticula are absent. Lappets absent and a cyst wall of parasite origin present. Adult genera: e.g. *Crassiphiala, Posthodiplostomum, Uvulifer.* (Fig. 57A and B).

'Prohemistomulum' In fish, snails and leeches.

Fore body generally round and deeply concave, with very small hind body. Reserve bladder consisting of lacunae arranged over the dorsal and lateral surface in an approximate *W* pattern, containing refractile bodies which circulate freely. Without lappets and usually in a well defined cyst of parasite origin. Adult genera: e.g. *Holostephanus, Cyathocotyle, Prohemistomum, Mesostephanus.*

PHYSIOLOGY AND HISTOCHEMISTRY

Relatively few studies are available on the histochemical characteristics of trematode metacercariae. Some of the data concerning enzyme activity has already been mentioned. Glycogen has been demonstrated in the parenchymal cells of a few metacercariae (ERASMUS and ÖHMAN, 1963; ŽDÁRSKÁ, 1964) and it seems probable that it is of more general occurrence than indicated by the literature. Lipids have been demonstrated in the tissues of some metacercariae (GINETZINSKIJA, 1961; ERASMUS and ÖHMAN, 1963 and ZDÁRSKÁ, 1964) and in the case of *Cyathocotyle bushiensis* metacercariae, the lipids are definitely associated with the excretory system and, in particular with the cytoplasmic lining of the reserve bladder system (ERASMUS, 1967b). The quantities of lipid and glycogen do not remain constant and ŽDÁRSKÁ (1964) has shown that over a period of six weeks the amount of glycogen in the tissues decreases and the lipid fraction increases. After this period, at 18°–28°C, the glycogen was found only near the anlagen of the genitalia and around the excretory bladder. The parasite studied in this case was *Echinostoma revolutum*. The glycogen content (VERNBERG and HUNTER, 1956) of the metacercariae of *Gynaecotyla adunca* immediately after excystment was found to be high (av. 2·774 mg . 0·21 mg glycogen/mm^3) relative to other phases of the life-cycle (Table 14). Immediately after infection of the final host the glycogen level dropped and continued to fall to an average of 0·708 $\pm$ 0·09 mg/glycogen/mm^3 96 hours after infection. This change in glycogen content resembles that which occurs in cestode plerocercoids after ingestion by the final hosts. Developing adults maintained *in vitro* in sea water plus glucose did not maintain their glycogen level, and these authors suggested that the metacercariae of the species must store the carbohydrate reserves needed to carry the adult through maturation and its very short adult life span. The respiratory rate of metacercarial stages would be particularly interesting, especially of those displaying complex developmental process as in the Strigeoidea. STUNKARD (1930) found that the survival time of *Cryptocotyle lingua* metacercaria under *in vitro* conditions was approximately 12 days under aerobic conditions but only reached a maximum of 4 days under anaerobic conditions. HUNTER and VERNBERG (1955a) found that the metacercaria of *Gynaecotyle adunca* did exhibit oxygen uptake but were able to express their results only in terms of consumption per individual without reference to mass or volume (Table 15 and 16). In this context the authors found that just prior to encystment (7–9 days after penetration of the crab) the oxygen consumption averaged 9·24 $\times$ 10^{-3} μl per hour per worm and after encystment the consumption averaged 5·62 $\times$ 10^{-3} μl per hour per worm.

Table 14 Glycogen content of *G. adunca*.

		μg glycogen/mm³*		
	Number of determinations	Range	Average	
In vitro, standard sea water solution				
immediately excysted	11	1·854–4·046	2·774	0·21
24 hour	13	1·514–2·795	1·933	0·11
48 hour	13	0·783–1·482	1·134	0·06
72 hour	10	0·625–0·833	0·722	0·02
96 hour	11	0·399–0·886	0·698	0·04
In vitro, 0·02 M glucose in standard sea water solution				
24 hour	10	1·645–3·079	2·007	0·14
48 hour	10	1·012–1·574	1·352	0·06
72 hour	10	0·875–1·427	1·134	0·05
96 hour	10	0·436–1·084	0·763	0·06
Worms recovered from experimentally fed bird hosts				
24 hour	2	1·380–1·840	1·612	
48 hour	5	1·033–1·876	1·490	0·14
72 hour	3	1·157–1·453	1·323	
96 hour	6	0·439–1·064	0·708	0·09

* One mm³ is approximately equivalent to 1 mg.
(After VERNBERG and HUNTER, 1956.)

These observations are confirmed by later experiments (VERNBERG and HUNTER, 1959) where the rate of oxygen consumption is related to nitrogen content. The oxygen consumption of the cercarial stage of *Gynaecotyla adunca* was 0·085 ± 0·007 μl O_2/hour/μg N at 30°C whereas that of the metacercaria prior to encystment had dropped to 0·056 ± −0·004 μl_2/hour/μg N. The authors also found a slow drop in oxygen consumption by the adult worm during the first 72 hours under *in vitro* conditions. Both lipid droplets and calcareous corpuscles have been described from the excretory system of the metacercariae of *Cyathocotyle bushiensis* and the majority

Table 15 Comparative rates of oxygen consumption of stages in the life cycle of *G. adunca*.

Stage of development	μgN/worm	No. of determinations	μl O_2/hr/ug N
Cercaria	0·0027	11	0·085 ± 0·007
Metacercaria prior to encystment	0·19	11	0·056 ± 0·004
Adults, 24 hours following excystment	0·28	11	0·032 ± 0·002
Adults, 72 hours following excystment	0·28	10	0·022 ± 0°002

(After VERNBERG and HUNTER, 1959).

Table 16 Oxygen consumption of *G. adunca*.

Stage of development	Number of runs	Microliters/hr/mm³ Range	Average	Standard error	Temp. °C
Adults immediately after excystment	11	0·079–0·219	0·130	0·01	23·6
Adults immediately after excystment	12	0·175–0·433	0·290	0·02	30·4
24 hours after excystment	12	0·098–0·200	0·153	0·01	30·4
48 hours after excystment	11	0·067–0·193	0·120	0·01	30·4
72 hours after excystment	10	0·058–0·163	0·104	0·01	30·4
Cercariae	19	3·17–9·50	5·35	0·4	30·4
3–4 days after penetrating crab	14	Not measurable			30·4
Immediately prior to cyst formation	15	0·105–0·260	0·159	0·01	30·4
Metacercaria	12	See Results			30·4

(After HUNTER & VERNBERG, 1955a).

of the corpuscles are discharged from the excretory aperture during the excystation of the larva. (ERASMUS, 1967b). NASIR (1960a) records that the cyst wall of *Cotylurus brevis* metacercariae from *Lymnaea stagnalis* possesses a perforation near the posterior end of the body and that refractile bodies from the excretory system are discharged through this under '*in vitro*' conditions. It is possible therefore, that even in encysted forms, discharge of material from the excretory pore may take place. It seems likely that the formation of these calcareous corpuscles in the metacercaria may be related to the metabolic adjustments necessitated by the environmental restriction imposed on the larva by the presence of the cyst wall. A few corpuscles persist in the excretory lacuna of the young adult, but after three days in the final host they have nearly all disappeared (ERASMUS, loc. cit.). MARTIN and BILS (1964) describing the calcareous corpuscles in the metacercaria of *Acanthoparyphium spinulosum* suggest that the corpuscles, which consist of concentric layers of material around a central 'nucleus', are actually deposited around modified mitochondria and possibly cytoplasmic membranes. Frequent references occur in the literature to 'refractile bodies' present in metacercarial stages and these may be either lipid droplets or calcareous corpuscles. The appearance of these substances may be related to the production of relatively insoluble excretory products in the form of lipids and calcareous body formation may represent a mechanism for carbon dioxide fixation, both very desirable processes in an animal enclosed in a cyst. It must be remembered, however, that both products exist in free swimming cercarial stages and also in metacercariae which do not encyst. It is possible therefore that a more fundamental explanation is necessary.

The cercarial body, prior to final localization or encystment, and the metacercarial body, after excystment in the final host, frequently undergoes complex migrations which often result in the very precise localization exhibited by the metacercaria or adult. The precise distribution of the Diplostomulum stage in the fish eye and brain and the migration of the excysted metacercariae of *Paragonimus* and *Fasciola* are good examples. This activity is however related to either the pre- or post-metacercarial stage and is discussed more fully in the chapters 6 and 7.

The resistance of metacercarial cysts to adverse environmental factors has been investigated in detail for parasites of economic importance such as *Fasciola hepatica* and *Paragonimus*. The literature concerning *Fasciola* has been reviewed recently by KENDALL (1965). It appears that one of the most important factors influencing the survival of metacercarial cysts is the relative humidity of the environment and this must not fall below 70 per cent for prolonged survival. Under experimental simulated field conditions *Fasciola hepatica* metacercariae remained infective for 270–340 days on herbage infected in autumn. The mortality rate of larvae was low over the winter but increased in the following spring and summer. At 25–26°C but with a relative humidity of 20 r.h. all the metacercariae died within 10 days. Under extreme conditions, such as exposure to air in direct sunlight at 98–105°F for 12 hours daily, cysts only survived for 2 days (ROSS and MCKAY, 1929). However SHAW (1932) found that metacercariae kept at 26–36°F for 11 months were still capable of infecting guinea pigs.

More exotic data is available on the survival of metacercarial cysts of *Paragonimus* from crustacea preserved in various culinary media. YOKOGAWA *et al* (1960) state that the metacercariae of *P. westermani* survived heating *in situ* (crab muscle) for 37 min at 45°C; 3 hours in a 1 per cent salt solution, and 30 min in soya sauce. In addition, encysted metacercariae dissected from crab muscle remained viable for 43 hours at 22°C in diluted millet wine containing 10 per cent alcohol. This, and other data provided by these authors indicate that many of the methods of preparing crab meat for food without cooking allow some encysted metacercariae to remain viable and infective. In this way local and specialized methods of food preparation would favour the transmission of this parasite in certain regions.

The time needed by a metacercaria before it becomes infective to the final host will vary and depend on a variety of factors both internal and external. The main feature influencing this potentiality will be the extent of the difference in morphology between the cercarial body and that of the fully differentiated metacercaria. In those instances where little morphological development has to take place the metacercaria becomes infective very soon after encystation. Examples of short developmental periods are: four days for the metacercaria of *Haematoloechus breviplexus* (Haplometridae; SCHELL, 1965); five days in the case of the cyst of *Echinostoma nudicaudatum* (Echinostomatidae; NASIR, 1960b). These short times will also be influenced by the temperature of the surrounding environment and also by overcrowding and competition for nutrients in the intermediate host. In the case of the more complex metacercariae, development is longer although this period will also be influenced by temperature and ecological factors such as competition. The metacercaria of *Telolecithus pugetensis* (Monorchidae) takes approximately one month before it is infective (DEMARTINI and PRATT, 1964). In this species the reproductive system becomes well differentiated at the metacercarial stage and presumably this feature influences the developmental period. In the strigeoid trematodes, as has been pointed out, considerable development occurs at this time. In *Mesostephanus kentuckiensis* (MYER, 1960) infectivity is attained after 22–28 days; and in *Diplostomum baeri eucaliae* after 13–23 days (HOFFMAN and HUNDLEY, 1957). The survival times of metacercariae are not very well known. ERASMUS (1958) recorded that the Diplostomulum larvae in experimental fish survived 544 days after infection. NASIR (1960b) reports that the metacercariae of *E. nudicaudatum* were still infective after 14 months, and the data given above for *F. hepatica* indicates a survival time of approximately 344 days. Senescence in metacercariae will be the result of the interaction of a number of features. There will be a depletion of endogenous carbohydrate reserves and an accumulation of excretory products as suggested by the observations of ZDÁRSKÁ (1964). Although some of the excretory products are relatively insoluble and presumably non-toxic, their accumulation within the cyst must eventually be deleterious to the fluke. The permeability of the cyst wall and the presence of apertures in it will affect this aspect. Whereas the cysts occurring in the external environment will be severely affected by changes in the relative humidity, those parasitic in intermediate hosts will be exposed to the hazards of the immunological defenses of the hosts. In invertebrate hosts, reactions seem to develop slowly in most cases and may not be a very significant factor, but in vertebrate hosts, particularly homiotherms, well defined responses are produced and these will undoubtedly affect the parasite. In addition the presence of a prior infection within the host animal may have to be considered.

Metacercariae which do not form a cyst are presumably able to resist the effects of the digestive juices of the final host after ingestion of the secondary intermediate host. HOFFMAN (1958a) studied the effects of trypsin Ringer solution at low pH and pepsin solutions on the survival of metacercariae of *Posthodiplostomum minimum in vitro*. He found that the ability of the larvae to resist pepsin solution (1 per cent pepsin in 0·5 per cent aqueous HCl) became pronounced when the larvae were 26–44 days old—40 per cent were alive at the end of two hours at 37–38°C. Pepsin solution at pH 2 was more deleterious than Ringer solution at the same pH. They were more resistant to trypsin (0·5 per cent in Ringer at pH 8 at 37°C) and 71 per cent survived a period of 18–23 hours. Antibiotics were added to this latter mixture to prevent bacterial breakdown of the trypsin. No mechanism has been suggested for this resistance to digestion at this stage of the life-cycle, but it seems likely that the hypothesis proposed for adult flukes will apply. MONNÉ (1959) suggested that the adult trematode possessed a coat of acid mucopolysaccharide external to the tegument and this served to inhibit the activity of host digestive juices at this interface. ERASMUS (1967c) has suggested that, in the tegument of adult *Cyathocotyle bushiensis*, secretion bodies which become closely associated with the external plasma membrane may release their contents over the surface of the tegument. This sequence may represent that process postulated by Monné. In view of the similarity of the tegument and its physical and temporal continuity through the life-cycle it seems likely that a similar mechanism exists at the metacercarial stage. This hypothesis is further supported by the presence of similar secretion bodies in the tegument of metacercariae. In addition some protection from digestive juices will be afforded the metacercaria by the host tissues still enclosing the metacercaria.

The encysted metacercaria has, in contrast, the problem of release from the cyst in the appropriate host and at the optimum site for migration and/or development. It seems probable that the composition of the cyst wall will play a part in affecting both these features. The need to examine more closely the morphology of metacercariae has stimulated many workers to investigate the factors affecting hatching and to devise media and processes suitable for *in vitro* hatching of metacercarial cysts.

At present one can only list the more obvious features involved as their precise role has not been very clearly ascertained. There is also the final question as to how far *in vitro* processes resemble those occurring in the living alimentary tract. (See chapter 6).

CULTURE *in vitro*

Attempts to culture adult trematodes (reviewed by HOEPPLI *et al* 1938; SMYTH, 1959; SILVERMAN, 1965; TAYLOR and BAKER, 1968) under *in vitro* conditions have sometimes originated at the metacercarial stage. This procedure has a number of advantages in that aseptic larvae may be obtained fairly easily either by excystment of encysted larvae under aseptic conditions or by the utilization of non-encysted larvae (e.g. *Diplostomulum* larvae which occur in fairly sterile environments such as the brain and eye.

FERGUSON (1940) excysted the metacercaria of *Posthodiplostomum minimum* using a pepsin mixture. Larvae were then washed in several changes of sterile Ringer and stored in half strength Tyrode's solution in the refrigerator. The most successful culture medium consisted of dilute (5:3 dilution) Tyrode solution, chicken serum and yeast extract. Sexual maturity was attained in about 10 days at 39°C and egg laying adults obtained. In the Heron this developmental stage was reached in 35–40 hours. Unfortunately the eggs obtained under *in vitro* conditions were not viable and this is probably related to the fact that the spermatozoa were inactive although apparently normal in shape.

In a later series of experiments FERGUSON (1943b) investigated the development of Diplostomulum metacercaria under *in vitro* conditions. Cercariae were allowed to penetrate the eyes of fish and localize in the lens. As a result of this migration the cercariae seemed to attain an axenic state. The axenic cercariae were recovered from the lens after it had been crushed and the debris screened off. The larvae were successfully cultured in frog Ringer plus lens tissue from a wide range of hosts or in Tyrode with fish or rat lens. The larva developed into morphologically normal diplostomuli but these, for some reason, were not infective to experimental hosts.

More elaborate and sophisticated experiments were carried out by Hopkins and his coworkers in an attempt to obtain the maturation of *Diplostomum phoxini in vitro* from the diplostomulum stage in the brain of the minnow. The larvae were obtained in this case by the removal of the brain under aseptic conditions and after sterilization of the outer surface of the fish. After washing in sterile salines the larvae were cultured under a variety of conditions (BELL and

Table 17 Criteria recommended for the recognition of developmental phases in the trematode *Diplostomum*.

Phase	Time in host *Diplostomum* (hr.)	Criterion recommended	Method of detection
(1) Cell multiplication	0–24	Mitoses counts	Aceto-orcein squashes after colchicine treatment
(2) Segmentation or body shaping	24–48		Direct observation on living material or aceto-orcein squashes
(3) Organogeny	12–24	Appearance of uterus and testes primordia	Squashes or whole mounts
(4) Early gametogeny	36–40	Appearance of 'rosette' and 'comma' stages in spermatogenesis	Squashes
(5) Late gametogeny	40–48	Appearance of mature spermatozoa	Squashes or unstained teases
(6) Egg-shell formation and vitellogenesis	55–60	Presence of egg-shell precursors in 'vitelline' cells	Histochemical tests on whole specimens. Diazo +ve, catechol +ve.
(7) Oviposition	60–72	Appearance of fully-formed egg	Direct observations on living material or catechol-treated whole mounts.

(After BELL and SMYTH, 1958).

Table 18 Evaluation of culture medium using mitosis as a criterion.

Organism	Medium	Mitoses after 24 hr. culture*	Prognosis	Maximum development achieved					
				Testis	Ovary	Cirrus	Uterus	Vitellaria	Eggs
Diplostomum	Salines	0–5	Poor	–	–	–	–	–	–
	Yolk	50–100	Fair	+	+	+	+	–	–
	Yolk and albumen	>200	Excellent	+	+	+	+	+†	+†
	Duck control	>150	Normal	+	+	+	+	+	+

* Counts made after treatment with colchicine for 4 hr. (*Diplostomum*).
† Developed abnormally.
(After Bell and Smyth, 1958).

hopkins, 1956). The success of the culture technique was assessed by a study of the mitotic activity, development of the reproductive system and appearance of quinone tanning precursors in the fluke as it developed in the duck (Table 17 and 18). This data assembled against a developmental time scale provided an excellent basis for the assessment of the suitability of *in vitro* conditions. The best results were obtained with duck serum and egg-yolk media in roller tubes and development proceeded to the stage of sperm formation after 48 hours. Vitellaria were undeveloped and no eggs were produced by the larvae *in vitro*. The experiments were continued and it was found (wyllie *et al.*, 1960) that the yolk could be replaced by a mixture of horse serum, yeast extract and amino acids, but that the yeast was essential for egg production under *in vitro* conditions. The role played by the yeast extract was further investigated by williams *et al.* (1961) who concluded that there was some substance or substances in the yeast extract other than vitamin B6 which stimulated the development of the fluke. The semi-solid nature of the culture media used in these experiments might be significant in view of the fact that these diplostomulum larvae actively ingest host tissues and that cellular material is usually found in the alimentary tract of larvae in natural tissues as well as under *in vivo* conditions (smyth, 1959). Ingested material was also observed in the caeca of metacercariae of *Philophthalmus sp.* cultured by fried (1962a) in a variety of media. Maximum longevity (5 days) was obtained in yolk-albumen-Tyrode mixture. In a further series of experiments (fried, 1962b) egg laying adults were obtained after culture of excysted metacercariae on chorioallantoic membranes of the chick for 14 to 20 days.

6 Entry into the Host

One of the most crucial periods in the trematode life-cycle occurs when the host animal is invaded. If attack and entry or attachment is successful there is a reasonable chance that development will continue and that the life-history will proceed along its cyclical path. If unsuccessful, the parasite will inevitably die within a fairly short time. The primary phase of spatial proximity between the host and infective stage is the result of the interplay of a wide range of factors (see chapter 1) involving such things as the physical state of the environment, food preferences and other behaviour patterns of the host animal, the degree of activity of the infective stage, the physicochemical environmental factors affecting its emergence and the particular behaviour pattern exhibited by each parasite species. Thus the positively phototactic behaviour of the ocellate furcocercaria *Cercaria ocellata* produces a local concentration of these stages near the surface of the water where they are most likely to contact the final host, which is in this case a duck. Many of these behaviour patterns have been discussed in earlier chapters and will not be considered further.

Even after spatial proximity has been achieved there are still many difficulties to be overcome. The infective stage is microscopic in size compared with the relatively large host body and the feeble lashing of a cercarial tail, for example, produces a form of locomotion which compares unfavourably with that of the host which may be an active, muscular teleost capable of dislodging dozens of attached cercariae by a single flick of its tail. Then there are the host's physico-chemical barriers to invasion such as a complex and thick skin, the alimentary canal, containing enzymes and juices at a pH which may be lethal, tissue components, such as macrophages, and the immunological characteristics of the host. All these difficulties of contact and invasion must be successfully countered before the cycle can continue, and so this period of entry is a vital one in the life-cycle and the infective larva is a developmental stage of paramount importance.

In the Monogenea it is the oncomiracidium which is generally the infective stage. The digenetic life-cycle however, is conspicuous in requiring the transference of developmental stages through more than one host so that entry into an animal, with all its hazards, may occur several times in the cycle. Thus the miracidium enters the first intermediate host, the cercarial stage infects the second intermediate host, the mesocercaria the third, and the metacercaria the final host in which the sexually mature adult will develop. Most of the digenetic cycles contain one or two intermediate hosts and only rarely contain three.

The oncomiracidium

Most monogeneans are ectoparasitic and consequently the problem is usually not one of entry but of contact and attachment. The eggs laid by the Monogenea are very variable in form and may be tetrahedral or fusiform in outline. They may bear one or more long filaments and sometimes hooks and are generally operculate. The fate of the eggs differs with different monogenea, in some the eggs are said to adhere to the gills of the fish host of the ovigerous fluke, but in others the eggs are carried free by the gill-ventilating current. Some eggs are thought to float in the sea, buoyancy being aided by the long filaments but others sink to the bottom of the sea. *Udonella caligorum*, parasitic on the exterior of the caligid copepod parasite, fixes its eggs directly onto the surface of the crustacean body. Large numbers of eggs are attached in this way. The eggs hatch releasing a non-ciliated stage resembling the adult parasite in all but size and sexual development. Thus in this life-history the difficulties of transference have been greatly reduced. A non-ciliated larva occurs also in the life-cycle of *Sphyranura* hatching from eggs which drop to the sea bottom. The larva is very active and moves through the water by undulating movements. After coming in contact with the appropriate perennibranchiate urodele host, it creeps over the surface of the body until the gills are reached.

In many life-cycles a ciliated larva, called the oncomiracidium, is produced. These larvae are generally cylindrical or ovoid and have differentiated internally a digestive tract, excretory and nervous systems. There are frequently present cephalic glands and also a well developed adhesive disc bearing sclerotized hooks. Many larvae possess eyespots containing a crystalline lens. The life of the larva is generally short (6–8 hours) and its duration is related to water temperature. Contact with the host is probably accidental although detailed observations on the actual infection of the host are only available for *Polystoma intergerrimum*—parasitic in the bladder of the frog. In this instance the ciliated larva creeps over the surface of the body until the spiracular opening is reached. According to LLEWELLYN (1957c) entry occurs at this point and not by inhalation through the mouth. In most cases it seems that the larva sheds its ciliary coat soon after reaching the final site on the host.

In a review of larval development of the Monogenea, LLEWELLYN (1968) quotes the observations of Combes who found that the larvae of *P. intergerrimum* migrated over the external surface of tadpoles from the gills to the cloaca. He was unable to find evidence of a gut migration and found that the surface migration occurred only at night.

In the genus *Amphibdella*, LLEWELLYN (1960) suggests that the sixteen-hooked oncomiracidium penetrates the blood vessels of the gills, instead of attaching itself to the superficial gill tissues. Within the blood stream of the torpedinid host, fertilization takes place and the amphibdellid would emerge anterior end first through the gill mucosa. Oviposition could take place through the exposed uterine aperture and the posterior end would remain embedded in the host tissues.

The development of *Polystoma intergerrimum* is remarkable in that it follows closely that of the host, and the actual entry into the frog tadpole occurs when the external gills recede and metamorphosis is about to take place. The rigid host specificity exhibited by many monogeneans suggests that many limiting factors have their influence during this period of contact, search and attachment. KEARN (1967), working with newly hatched larvae of *Entobdella soleae* observed a marked response to isolated sole scales (Fig. 59, 60). Further studies conducted at 17°C in total darkness in which the oncomiracidia were presented with two choice and three choice situations involving scales of *Solea solea* and other pleuronectid fish, revealed that the larvae were able to differentiate the scales of *Solea solea* from those of other fish in total darkness. Thus these larvae did not attach themselves to fish scales in a random manner, and Kearn suggested that the orientation was based on chemoreception of a specific substance secreted by the skin of the fish host and that orientation patterns such as this might have considerable significance in the evolution of host specificity within the Monogenea.

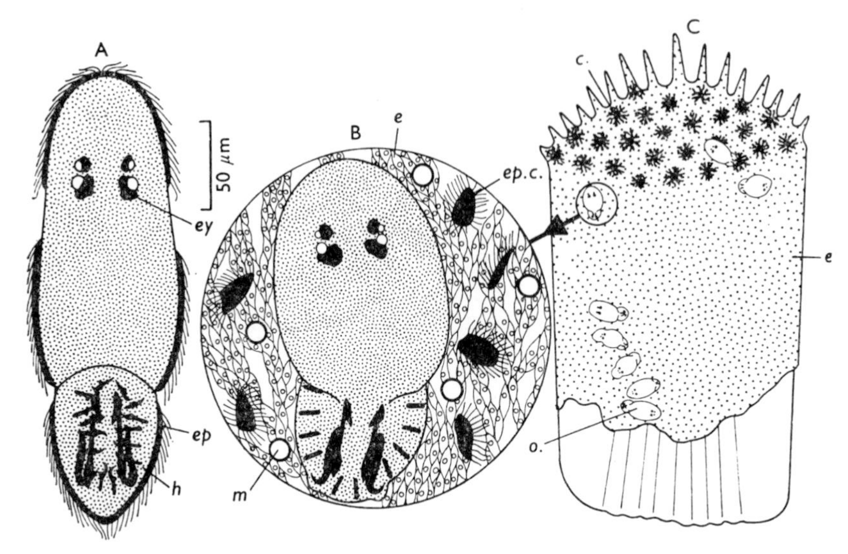

FIG. 59 **A** The oncomiracidium of *Entobdella soleae*; **B** the larva thirty seconds after attachment to the scale and after release of epidermal plates; **C** a freshly removed scale showing attached larvae. (ep: ciliated epidermal cells of larva; ep.c: larval ciliated cells which have been shed; ey: eyespot; h: haptor; o: oncomiracidium; e: fish epidermis; c: dermal chromatophore; m: epidermal mucus cell.) From KEARN, 1967.

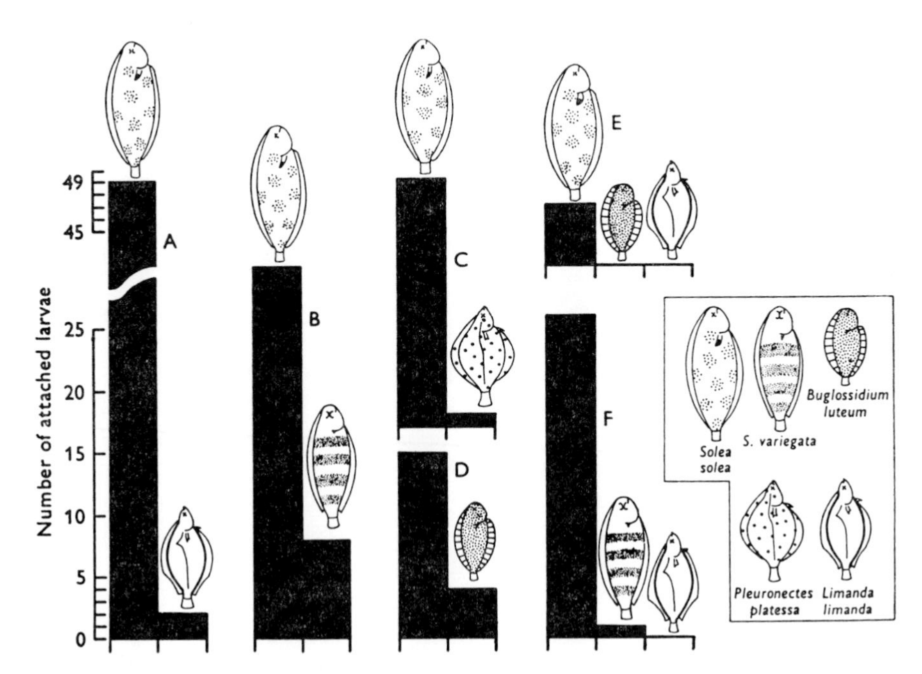

FIG. 60 The relationship between the numbers of larvae of *Entobdella soleae* attaching and the different species of fish scale offered. The larvae were offered a choice between two sorts of scales (**A–D**) or between three sorts of scales (**E, F**). Figure **A** represents four experiments; **B**—six; **C**—five; **D**—four; **E** and **F**—single experiments. From KEARN, 1967.

The miracidium

The fertilized egg deposited by the adult digenetic trematode is eliminated from the final host and reaches the surrounding environment. Here, if conditions are favourable, and development is complete, it will hatch releasing an active ciliated miracidium which has a brief free-living existence. Not all miracidia are ciliated, and those of *Halipegus excentricus* and *Otodistomum cestoides* are spined. In some cases, *Glypthelmins quieta*, *Dicrocoelium dendriticum* and *Telorchis bonnerensis* for example, the fully developed egg must be ingested by the first intermediate host and release of the miracidium occurs only in the gut of the snail. Hatching may occur in the snail's stomach as in *Haplometra intestinalis* (SCHELL, 1961) or in the intestine as in *Glypthelmins quieta* (SCHELL, 1962) and *G. pennsylvaniensis* (CHENG, 1961) and takes place within 60–70 min (SCHELL, 1961, 1962) after ingestion. Although the egg may contain a fully developed miracidium, this is quiescent and does not become active until the egg comes in contact with the digestive juices of the molluscan host (CHENG, 1961). Miracidia with this type of development are often poor swimmers with slow ciliary movement. After hatching, the miracidium penetrates the epithelium of either the stomach, the caecum or the forepart of the intestine. Penetration appears to be due to the action of apparently cytolytic secretions with the result that the epithelial cells adjacent to the penetrating miracidium become quite clear and amorphous, and the original site of penetration can be detected for some time as a minute break or slight derangement of the epithelial cells. Miracidia observed in the act of penetrating appear to undergo very little movement and examination of miracidia after penetration of the host epithelium reveals that the granular contents of the penetration glands have been lost during the passage through the host epithelium. They penetrate as far as the basement membrane and here they remain and then metamorphose to the young mother sporocyst stage.

The eggs of most Digenea hatch in the external aquatic medium and release an actively swimming miracidium. After hatching, miracidia swim rapidly and continually in straight lines or occasionally in circles with the body slowly revolving about the longitudinal axis. The attraction of miracidia to their hosts has been considered by many workers with diverse conclusions. The field has been reviewed by WRIGHT (1959) but some comments and examples are relevant here. Some miracidia e.g. *Fasciola hepatica* are said to be attracted to *Lymnaea truncatula* at a distance of 15 cm (NEUHAUS, 1953) but in many cases, and as DAWES (1960) describes for *F. hepatica*, miracidia may pass within a millimetre of the snail without exhibiting any sign of 'awareness' of its presence. Similarly YOKOGAWA *et al.* (1960) conclude that the miracidium of *Paragonimus westermanni* was not attracted to the snail host *Semisulcospira libertina*. BENNINGTON and PRATT (1960) state that the miracidia of *Nanophyetus salminicola* are not attracted to the host snail *Oxytrema silicula* and that miracidia were observed to bump into snails repeatedly and swim away without attempting to penetrate them. CRANDALL (1960) comments that although miracidia of *Heronimus chelydrae* are positively phototactic they are not attracted to the snail *Physa integra*. My own experiences with the miracidium of a *Diplostomum* species and the snail *Lymnaea stagnalis* suggest that there is no attraction exhibited by the snail and miracidia have been observed to 'test' and probe the shell and then swim away, in spite of the fact that development normally occurs in this mollusc. In the case of certain schistosome miracidia, there seems to be good evidence indicating that they are attracted to the molluscan intermediate host. FAUST and MELENY (1924) and FAUST and HOFFMAN (1934) stated that after the miracidia of *S. japonicum* and *S. mansoni* had come within a few millimetres of the snail, they head directly

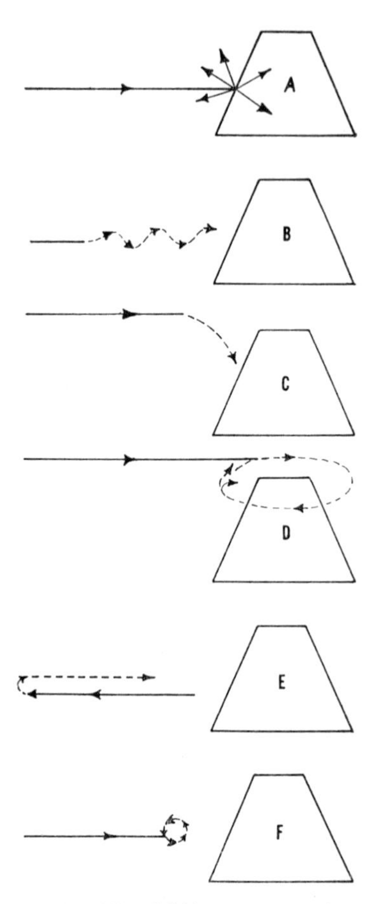

FIG. 61 The behavioural responses of the miracidia of *Schistosoma mansoni* to agar blocks. **A** normal change of direction upon contact with an obstruction i.e. contact without return; **B** increasing random turning; **C** directed turn at a distance; **D** encircle at a distance; **E** 180 degrees return; **F** circus movements. From MACINNIS, 1965.

for it. KLOETZEL (1958) produced quantitative data on this problem employing *S. mansoni* miracidia and *Australorbis glabralis* and showed that chemical attraction of miracidia to the snails took place over short distances i.e. within a 15 cm petri dish. DUTT and SRIVASTAVA (1962a) also suggest that the miracidia of *Orientobilharzia dattai* are attracted to the snail *Lymnaea luteola*. An additional quantitative study of the chemosensitivity of the miracidium of *S. mansoni* to *Australorbis glabratus* was carried out by ETGES and DECKER (1963). They were able to demonstrate moderate but definite attraction to *A. glabratus* but not to *Helisoma anceps* or *Bulinus spp*. However they were also of the opinion that both light or gravity were far more powerful stimuli in determining the orientation of *S. mansoni* miracidia than those chemical stimuli produced by the molluscan host. A more precise analysis of the behaviour of *S. mansoni* miracidia under experimental conditions was made by MACINNIS (1965) (Fig. 61). He analysed the various miracidial movements into categories following the nomenclature of FRAENKEL and GUNN (1961) and observed the reactions to pyramids of agar impregnated with various chemicals (Table 19). Distinct localization, attachment and penetration attempts were achieved with certain chemicals particularly short-chain fatty acids, some amino acids and a sialic acid. The attracting substances present

Table 19 Comparison of results obtained with different sources of the chemical stimulus.

Type of experiment	No. of experiments	Average per cent contact with return	Total no. miracidia	Contacts with return × 100 total miracidia present	Per cent attached
1. Impregnated pyramids	15	64·3	900	41·5	10·1
2. Integral pyramids	10	92·4	610	51·5	31·2
3. Live snail	8	37·9	80	21·1	3·8
4. Controls	15	2·8	600	1·6	0·0

(After MacInnis, 1965).

in *Australorbis glabratus* tissues could be removed with organic solvents, but the capacity to attract was restored by the addition of butyric or glutamic acids. There seems no doubt that swimming pattern of the miracidia of this species was strongly influenced by these chemicals (Fig. 62). Thus the evidence is mixed and host localization aided by chemotaxis seems reasonable for some species but not for others. In addition there is always the problem of interpretation of results obtained under experimental conditions. Localization in the calm of a petri dish is very different from that under the relatively turbulent conditions of an unrestricted body of water.

When miracidia do contact the snail, frenzied looping movements appear coupled with a probing of the surface. Many miracidia may fail to adhere and will swim away from the vicinity of the snail. Attachment and penetration may take place at any point on the soft parts of the snail, although Yokogawa *et al.* (1960) observed distinct differences in the number of miracidia of *P. westermanni* penetrating the various regions of *S. libertina*. After exposure to large numbers

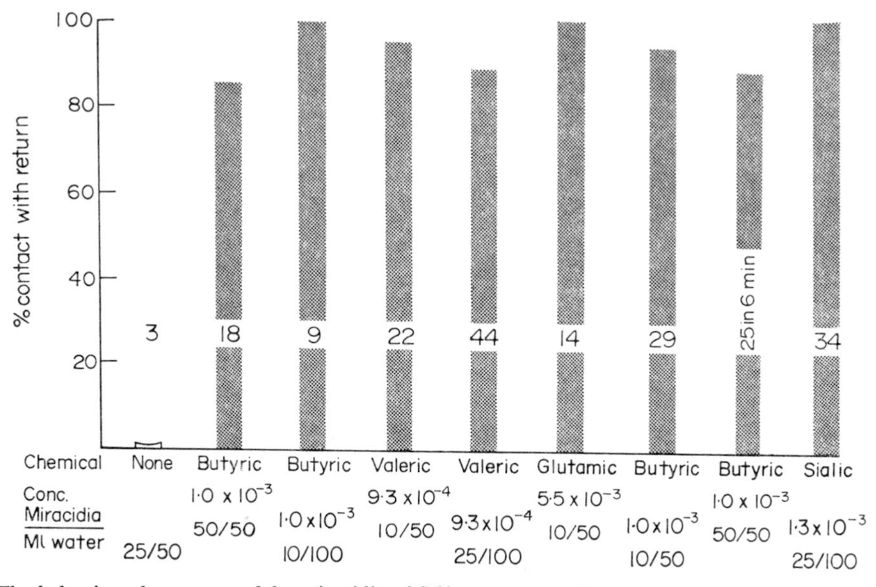

FIG. 62 The behavioural responses of the miracidia of *Schistosoma mansoni* to agar pyramids containing selected substances which were dissolved in the agar solvent during preparation of the mixture. Per cent contact with return = the number of contacts with return (figures in bars) × 100/total number of contacts. Exposed to miracidia for 15 min except where stated otherwise. Columns two and three represent agar blocks with isobutyric acid, seven and eight n-butyric acid. From MacInnis, 1965.

of miracidia for 5 hours at 30°C they recorded that 86 miracidia had penetrated the tentacles, 50 the lips, 39 the head, 83 on the dorsal side of the foot, 206 on the ventral side of the foot, 91 in the body and 181 into the anterior half of the mantle. DAWES (1960) has studied in great detail the attachment and penetration of the miracidium of *F. hepatica* into *Lymnaea truncatula* (Fig. 63). Attachment to the snail in the early stages is by suctorial contact and is not the result of a rotating, penetrating action of the anterior end. Dawes suggests that the anterior papilla becomes retracted into the anterior pit thus producing a suctorial effect over the entire area of the pit. Into this saucer-shaped space, between the anterior end of the larva and the snail's epithelium, are poured the secretions of the 'gut' and the penetration glands. These secretions appear to have a strong cytolytic effect so that the epithelial cells become loosened and disrupted, exposing the sub-epithelial layers which themselves become

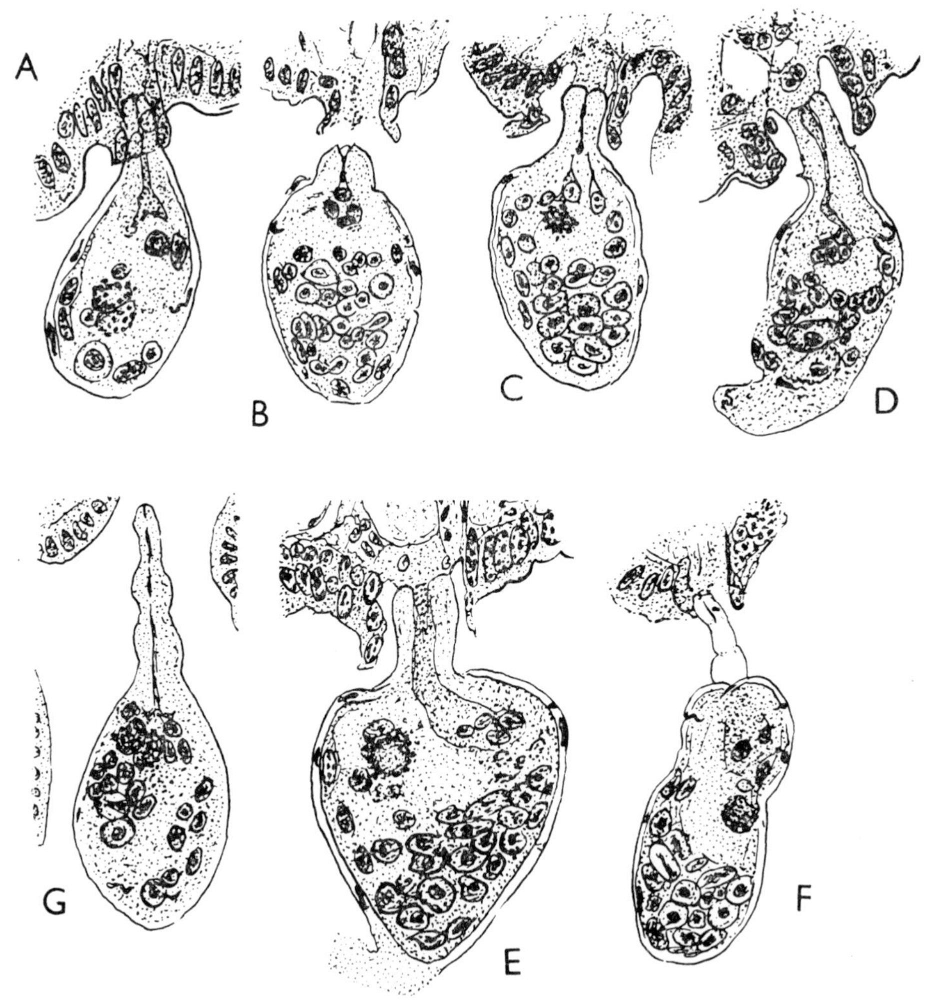

FIG. 63 A series illustrating the attachment and initial stages of the penetration of the miracidium of *Fasciola hepatica* into *Lymnaea truncatula*. The series A–F represents a period of 15–20 minutes. Note also the damage to the host tissues at the point of attachment. From DAWES, 1960.

rapidly eroded. MacInnis (1965) also described partial dissolution of agar at regions of miracidial attachment. The anterior end of the miracidium penetrates the damaged epithelium aided by the thrusting movements of the miracidial body. During this entry the ciliated cells of the miracidium become loose and are scraped off as the larva passes into the host's tissues. Loss of the ciliated epithelial cells has also been observed by YOKAGAWA (1960) during the penetration of *P. westermanni* miracidia into *S. libertina*, although CRANDALL (1960) mentions that in the miracidium of *Heronimus chelydrae* the epidermal plates are not lost immediately after penetration but may remain intact for up to 20 min after penetration. LENGY (1960) also referred to the retention of epidermal plates and described the secretion of granular material from the apical gland through two ducts during the penetration of the miracidium of *Paramphistomum microbothrium*, and suggested that the secretion has a lytic activity. LEWERT and LEE (1954) have identified PAS positive material in the penetration glands of *S. mansoni* miracidia and state that the glycoprotein basement membrane in the snail epithelium becomes altered during penetration. Thus evidence from a variety of sources supports the concept that the lateral gland cells and/or the apical gland ('gut' of Dawes) of the miracidium produces a cytolytic secretion which aids penetration. The rate of penetration varies a little; forty minutes for the miracidium of *P. microbothrium* (Lengy) 25–60 min, for *P. westermanni* (YOKAGAWA *et al.*), 30–60 min for *F. hepatica* (Dawes) and 1–5 min for the miracidium of *Spirorchis elegans* (GOODCHILD and KIRK, 1960). CRANDALL (1960) has observed that the penetration of *H. chelydrae* miracidia into *Physa integra* varied with the part attacked, it taking 2–4 min to enter the tentacle and up to 30 min to enter the foot.

The shedding of the ciliated epidermis of the miracidium during penetration leads DAWES (1960) to suggest that it is a young sporocyst which penetrates the snail epithelium and not the miracidium. The persistence of characteristically miracidial structures in stages up to 5 hours after localization (e.g. SCHELL, 1962) may constitute an objection to this idea, although Dawes is undoubtedly correct in emphasising the fact that metamorphosis into the mother sporocyst stage occurs much sooner after penetration than is generally realized.

The cercaria

After miracidial infection and a period of development in the first intermediate host a second disseminating phase of paramount importance appears. This is the cercarial stage which in some cases, e.g. Schistosomatidae, infects the final host directly and gives rise to the adult trematode. In other life-cycles the cercaria may be solely disseminating in function, ultimately producing infective metacercariae on vegetation or may be both disseminating and infective, penetrating directly into a second intermediate host and then forming a metacercaria. An important feature of the life-cycle beyond the first intermediate host phase, is that multiplication does not occur so that one cercaria is potentially capable of giving rise to only one metacercaria and eventually to one adult. Thus the prolific multiplication within the first intermediate host culminates in the production of the disseminating and infective cercarial stage which has the function of perpetuating the life-cycle and producing an ecological dispersal of the species.

The morphology of the cercarial stage is varied and may be characterized by a number of structures such as eyespots, spines and penetration or cystogenous gland cells which can be correlated with the potential behaviour and developmental pattern of this stage. The range of morphology has been indicated in a previous chapter but it is necessary to emphasize at this stage that the behavioural and developmental patterns occurring within the Digenea are also very varied and that only the more obvious features can be indicated at this stage. As this chapter is concerned with the biological problems of entry into a host, the disseminating type of cercaria culminating in encystment on vegetation or shells etc. will not be considered here. Entry into the host by the cercarial stage may be considered under two major headings. *Passive entry* in which the cercarial stage is deliberately or accidentally ingested by the host and *active entry* in which the host is directly and actively penetrated by the cercarial stage.

Passive Entry

In some cases entry into the host may be accomplished without the cercaria leaving the sporocyst in which it developed. A classical example of this was described by WESENBURG-LUND (1931) in the life-cycle of *Leucochloridium paradoxum* (see also KAGAN, 1951, 1952). The cercariae develop in branching sporocysts in the digestive gland and body of *Succinea* species (Fig. 64). Each sporocyst can be divided into three main regions—a branched central portion found in the connective tissues of the digestive gland, a narrow connecting stalk and the broodsac. In well developed infections the sporocysts extend into the snail's tentacles which consequently become greatly distended. As the sporocysts are banded with green or brown and pulsate regularly, the infected tentacles appear as very conspicuous objects. The cercariae (tailless forms) metamorphose within the sporocyst into metacercariae and the avian host is infected by becoming attracted to and ingesting the snail with its distended tentacles. Both Kagan and Wesenburg-Lund have commented on the fact that the activities of the infected mollusc become modified. Uninfected snails are somewhat negatively phototropic whereas infected snails seek the light, become more exposed and thus more easily observed by the bird and ingested. The fact that *Succinea* is a semi-terrestrial snail may indicate that the absence of the tail from the cercaria and the elimination of its free living period are really adaptations to a terrestrial existence.

In the life cycle of *Plagioporus sinitsini* (DOBROVOLNY, 1939) the sporocysts, containing the cercarial and later the metacercarial stages, accumulate in the rectum and eventually pass out to the external environment via the anus of the mollusc. The piscine final host becomes attracted to and ingests the wriggling sporocysts.

A third example in which the cercaria does not leave the first intermediate host is in the well known life cycle of *Dicrocoelium dendriticum.* Here the cercariae leave the sporocyst and encyst in the pulmonary chamber of the snail. In this case the mollusc is terrestrial (genera *Helicella*, *Torquilla* and *Zebrina*) and again this unusual cycle may represent an adaptation to terrestrial conditions.

In many cases the cercaria leaves the first intermediate host and passive entry results after the ingestion of this swimming stage by the host. In such cases the cercariae are often conspicuous either in their morphology or by their behaviour, thus increasing their chances of being discovered and ingested by the prospective host (Figs. 39 and 40). Very well known examples of these modifications are *Cercaria splendens*, SZIDAT, 1932 (Azygidae), the cercaria of *Bucephalus polymorphus* and the cercaria of *Bivesicula caribbensis* CABLE and NAHLAS (1962), all of which have highly modified, and coloured (except in Bucephalus) conspicuous tails and a behaviour pattern which attracts the attention of their future host. Also included in this group are the, not so well known, 'Trichocercous' cercariae which are mainly planctonic in their distribution and have their tails ornamented by large numbers of hair-like structures which presumably aid in flotation.

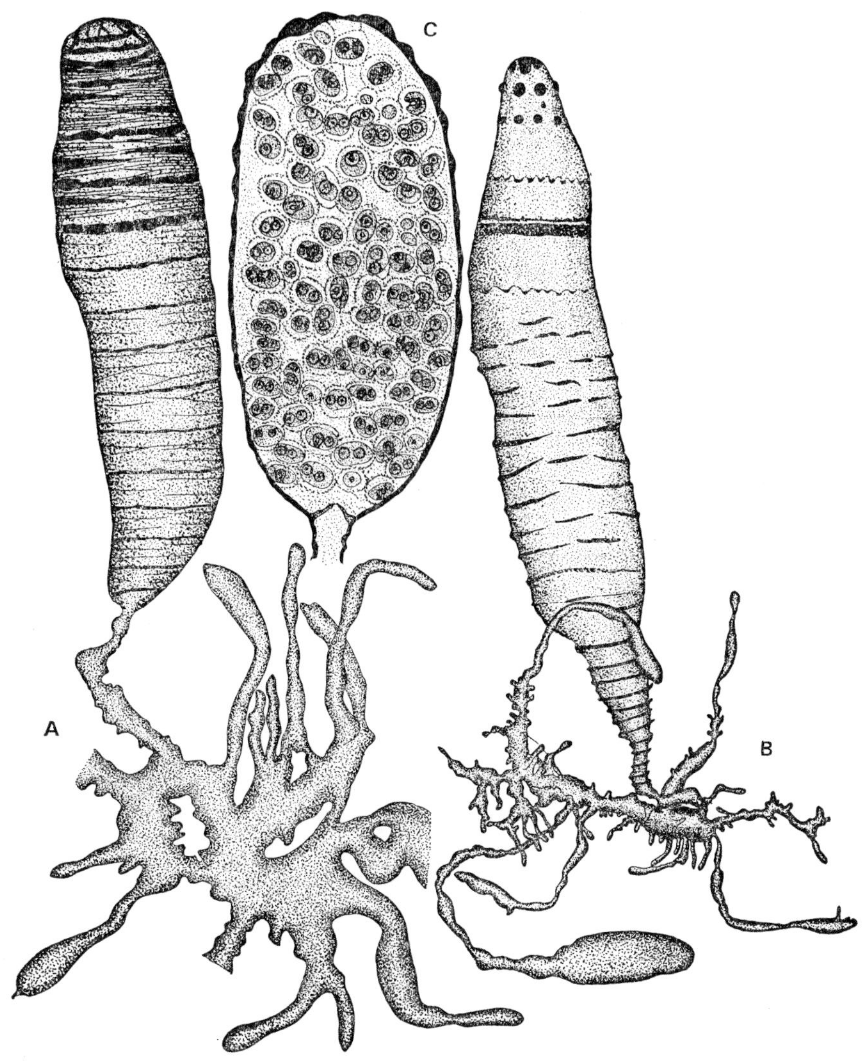

FIG. 64 The sporocysts of *Leucocloridium sp.* **A** brown and **B** green sporocysts. Note the difference in banding. **C** section through dilated sac showing contained metacercariae. From WESENBURG-LUND, 1931.

Active entry

A large proportion of emergent cercariae actively penetrate a second intermediate host or the final host, ultimately giving rise to the metacercarial and adult stages respectively. In the case of *Cercaria paludina impurae* Fil from *Bithynia tentaculata*, the cercariae enter and penetrate the host when the first intermediate host comes in close physical contact with the second intermediate host (WESENBURG-LUND, 1934). These tailless cercariae may encyst within the same snail but others emerge through the spiracle and creep over the surface of the snail ultimately accumulating on the tentacles which consequently assume a fluffy appearance. The cercariae are attached by their ventral sucker with the fore-part elevated and often being continuously extended in different directions. When infected tentacles come in contact with other snails, the cercariae 'jump' on to them, attach to and penetrate finally entering and giving rise to encysted metacercariae.

Most cercariae in this category of active penetration, have a free-swimming existence of reasonable duration (1–12 sometimes 24 hours) during which period host location and penetration must take place. Locomotion is accomplished by the action of the muscular tail and the energy necessary for this activity is derived from endogenous glycogen present in the cercarial body and tail. The characteristic caudal bodies present in the tail stem of furcocercariae are rich in glycogen and the utilization of these stores during the free-swimming period has been demonstrated by ERASMUS (1958). Cercariae without caudal bodies will presumably utilize the endogenous store of the body (AXMANN, 1947; ERASMUS, 1958; GINETZINSKAJA, 1960). See chapter 4.

Many cercariae penetrate molluscs, oligochaetes, hirudineans and polychaetes and these animals serve as intermediate hosts. Although these life-cycle patterns are well known, the behavioural and histochemical aspects of the actual method of penetration do not seem to have received much attention. NASIR (1960a) has described briefly the penetration of *Cercaria cotylurus brevis* into parasite free *Lymnaea stagnalis*. After escape from the first intermediate host, these cercariae dart in all directions and become uniformly distributed in the water. As with most furcocercariae, periods of swimming alternate with periods of rest. Nasir suggests that the cercariae are unaware of their potential host until chance contact is made with an exposed part of the snail. In most cases the cercaria takes a firm grip with its ventral sucker and explores the surrounding surface with the anterior end of the body. The tail stem with the furcae stretched apart is usually held at right angles to the body and is eventually cast off with an abrupt convulsion. The penetration of the cercariae of *Cotylurus cornutus* into *Lymnaea stagnalis* was completed within 30 min (MATHIAS, 1925 and SZIDAT, 1924). Wesenburg-Lund on the other hand, mentions that this cotylurid cercaria (*Cercaria A*, Szidat) gathers round specimens of *L. stagnalis* or *palustris*, but as later workers have often confirmed, often casts off its tail prior to penetration. Some cercariae, e.g. *Cercaria postharmostomum helicis*, enter the molluscan host via the renal aperture (ULMER, 1951) carried on the respiratory current and in this case, make their way through the ureters to the kidney and thence via the reno-pericardial aperture to the pericardium. In this site they develop into metacercariae.

Penetration into arthropods, especially aquatic larvae, occurs in many life-cycles, and entry usually takes place via the unsclerotized membrane or the thinly cuticularized gills. Recently BURNS (1961) has described the penetration and entry of several virgulate xiphidiocercariae (Lecithodendriidae) into caddis fly larvae. After emergence from the snail *Fluminicola virens*, the cercariae of *Allassogonoporus vespertilionis* remained fairly quiet unless the water was disturbed, and under these conditions they were capable of vigorous swimming movements. The cercariae swam towards the surface of the water and then slowly descended, flotation being aided by the secretion of long fine threads of mucus. BURNS (1961) noted that cercariae, with their threads, adhered to the surface of objects drawn through the water, and stonefly and caddis fly larvae, placed in the dish, rapidly become entangled with the mucus and densely covered with cercarial stages. After coming into contact with the gill of a caddis fly larva, the cercariae secreted large amounts of mucus, and when covered in mucus the tail was shed. In several cases, many cercariae formed groups adhering to one another by their mucous capsules. The cercariae became orientated with their stylets directed towards the host tissue, and, by pushing and scraping this stylet against the cuticle, eventually made an opening in the gill large enough for the cercarial body to pass through (Fig. 65). Penetration was completed in from 15 to 10 min and if not effected within that period the cercariae usually became moribund.

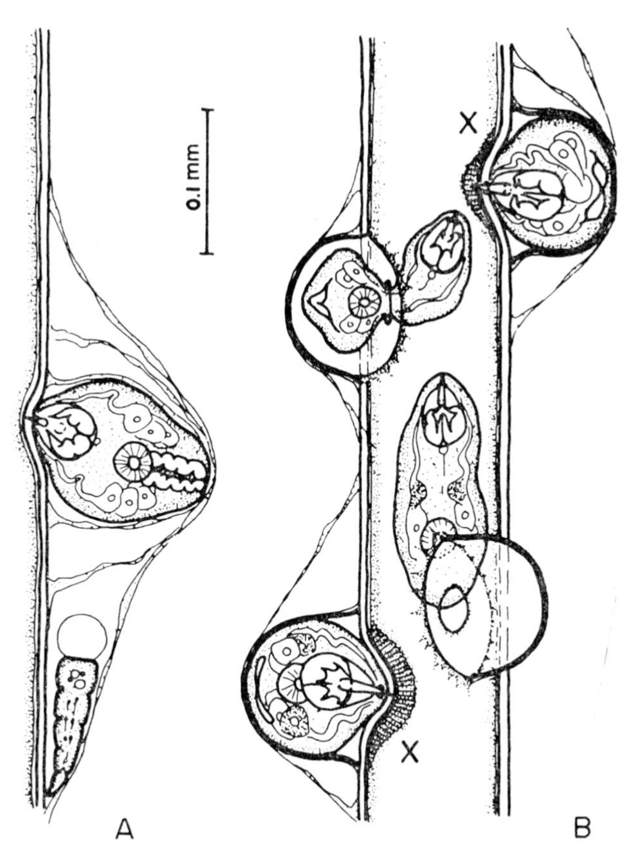

FIG. 65 The penetration of arthopod larvae by xiphidiocercariae. **A** The cercariae of *Allassogonoporus vespertilioni* penetrating the gill of *Dicosmoecus sp.* (Caddis larva). **B** The cercaria of *Acanthatrium oregonense* encysting upon and penetrating the gill of *Dicosmoecus sp.* Note the thickening of the host epidermis at points X. From BURNS, 1961.

The cercariae of *Acanthatrium oregonense* adhered to the gills by mucous threads and, after placing its ventral surface in contact with the gill, the cercariae secrete a dome-like capsule. Encystment took place within 1 min and after the tail was shed, entry into the insect larva took place in a similar manner to that described for *A. vespertilionis*. Encysted cercariae became moribund within 10 min if they were unable to penetrate the insect, and cercariae never penetrated the host without prior encapsulation. The source of the secretions comprising the cyst is unknown nor is it clear whether glandular secretions play a part in actual penetration. Many authors state that the characteristic virgula organ appears unchanged in recently penetrated cercariae and suggest that the organ contributes little if anything to the process of penetration. HALL and GROVES (1961), describing the entry of *Cercaria tremaglandis* into naiad hosts, mention that mucoid material is discharged from the virgula organ as the cercariae attach to the body surface and that the remainder is discharged soon after entry. An attempt at describing the characteristics of the gland and its contents was made by KRUIDENIER (1951b) and ORTIGOZA and HALL, (1963a,b). (Table 20.)

A number of observations are available concerning the penetration of arthropods by cercariae of the genus *Paragonimus*. KRUIDENIER (1953b) has described six pairs of mucoid glands extending from the level of the oral sucker to the base of the tail in the cercaria of *Paragonimus*

Table 20 Staining reactions of glands and secretions *Cercaria adoxovirgula*.

Test	Secretions	Mucoid glands, coat, virgula	Cephalic Glands		
			1st	2nd	3rd
PAS	++ (in part)[a]	+ to ++	+	+	++
PAS after acetone and chloroform-methanol	++ (in part)	+ to ++	+	+	++
Iodine		—	—	—	—
Bauer-Feulgen		+	+	—	+
Best's carmine		++	+ to ++	+ to ++	+ to ++
Orth's lithium carmine		+	++	++	+
Toluidine blue for Metachromasia pH 1·7	++ (in part)	++	—	—	—
pH 4·5	++ (in part)	++	—	—	—
pH 6·5	++ (in part)	++	—	—	—
Colloidal iron	++ (in part)	++	—	—	+
Millon's		+	++	+	++
Ninhydrin-Schiff	++ (in part)	++	+	+	++
Aldehyde fuchsin	+ (in part)	++	—	—	—
Acid orcein	+ (in part)	++	—	—	—
Van Gieson's	++ (in part)	—	++	++	++
Gram's		+	—	—	—
Feulgen		—	—	—	—
Sudan black B		++	+	+	+
Sudan black B after pyridine or chloroform-methanol		+ to ++	±	±	±

[a] Strongly positive reaction, ++; weak, +; weak to negative, ±. The words 'in part' signify that the secretions contain material from both mucoid and cephalic glands and the reaction cited refers only to the constituent which stains most intensely with the reagent used.

(After ORTIGOZA and HALL, 1963a).

kellicoti. They extend from the level of the oral sucker to the base of the tail, three pairs anterior and three pairs posterior to the ventral sucker. The glands *in vivo* stain a brilliant salmon-red with neutral red and their secretions are mucoid or mucopolysaccharide in nature exhibiting gamma (red) metachromasia in thionin or toluidine blue. Kruidenier suggests that these glands discharge their secretions after the cercariae have left the rediae and have entered the lymph spaces of the snail, and that the mucoids form a complete covering over the cercariae. These mucoid secretions play an important part after the emergence of the cercariae from the snail host. Groups of cercariae become entangled together by these mucoid strands which also aid attachment to the surface and appendages of the arthropod second intermediate host. Thus these cercariae, which do not swim freely in the water, are able to obtain attachment in this way. This pattern has not been described for *P. westermani* nor *P. iloktsuenensis* cercariae so an alternative method of attachment must occur in these species. The cercariae of *P. kellicotti* penetrate crayfish through the underside of the tail and at the junction of the segments. The site of infection of crabs by cercariae of *P. iloktsuenensis* does not seem to have been observed, and, in *P. westermani*, cercariae may penetrate the leg joints or may be ingested by the crab whilst still within the snail host (YOKOGAWA *et al*, 1960). These observations indicate the variations which may occur within species of the same genus and exhibiting the same life-cycle pattern.

Entry into the arthropod may occur through sites other than articular membranes and gills. Cercariae of *Prosthogonimus macrorchis* may enter the rectal respiratory chamber of various dragonfly nymphs and migrate into the muscles, finally encysting in the haemocoele. The respiratory currents drawn into the rectum presumably aid entry into these hosts.

The occurrence of strigeoid metacercariae in the eyes and brain of fish has attracted much interest and consequently a number of observations are available concerning the attachment to and entry into the fish body by the appropriate cercarial stages (BLOCHMAN, 1910; SZIDAT, 1924; DUBOIS, 1929; WESENBURG-LUNG, 1934; ERASMUS, 1958, 59). Cercariae of this type are fork-tailed, possess a well developed ventral sucker, usually armed with concentric rows of hooks, four penetration gland cells opening to the exterior via the oral sucker and an active muscular oral sucker heavily armed with a complex series of spines. The cercariae after emergence alternate periods of active swimming with periods of floating in which the body slowly sinks. During this floating period the body often assumes a characteristic position. The furcae are spread out and the body may either hang vertically downwards or may be bent at an acute angle to the tailstem. Cercariae of this type do not seem to be able to detect, from a distance, the presence of a potential host but are passively drawn to the fish on its respiratory currents or by the water movements caused by the fins and body. Once the cercaria is touched by the host it becomes very active, attaches to the stimulating surface by its ventral sucker and exhibits a series of probing actions with its oral sucker. The body anterior to the ventral sucker is actively extended and retracted in all directions and movement over the surface of the host body is leech-like involving alternate attachment by oral and ventral suckers. In *Cercaria X* (ERASMUS, 1959), the gland cells produce a secretion which is expelled in the form of droplets which have a viscous and sticky nature. This secretion can be stained with Alizarin Red S used as a vital stain. The oral sucker bears an apical cap of stout spines which lie antero-dorsal to the mouth. The cap of spines is rapidly retracted and then thrust into the tissues and must play an important role in attachment. In this species it seems that the secretion has primarily an adhesive or lubricatory function although, as there is no reason to suppose that the secretion from all four cells is the same, it is possible that part of the secretion may be lytic in its function. STIREWALT and KRUIDENIER (1961) have indicated that, in the cercaria of *Schistosoma mansoni*, the anterior and posterior sets of gland cells produce secretions which differ in their histochemical characteristics as well as in their function. The tail of this cercaria plays little part in attachment nor in penetration and usually hangs still except for an occasional flicker from side to side.

Cercariae, such as *Cercaria X*, become uniformly distributed in the water after emergence and a fish may be exposed to penetration at several different sites. In such a fish 15 minutes after exposure to infection, and examined in the form of serial sections (ERASMUS, 1959), 84·2 per cent of migrating cercariae were present in the skin and 15·8 per cent in the gills. A histogram showing the linear distribution of *Cercaria X* along the body of *Gasterosteus aculeatus* (Fig. 66) shows that although entry can occur at any point along the length of the body a local concentration is present in the posterior region of the head. This is due to a combination of features: (1) cercariae drawn along with the respiratory current first come in contact with the skin of the head and penetrate there; (2) many cercariae become trapped by the gills and penetrate at this point and into the mucous membrane of the mouth. During exposure to very large numbers of cercariae, some may creep over the surface of the body and enter the eye directly, penetrating through the conjunctiva. With normal concentrations of cercariae as are encountered in nature, the chances of direct penetration through the eye seem very remote. During entry through the skin, the cells of the epidermis clearly become displaced and each migrating cercaria lies in

Plate 1

1-1 Electronmicrograph of a transverse section through the oral sucker musculature of *Diplostomum sp.* Note that the fibres comprising the muscles are of two sizes: large and dense, surrounded by groups of twelve to fifteen smaller fibres. ×70 000. Stained with potassium permanganate and lead citrate.

1-2 Electronmicrograph of an area adjacent to Fig. 1 showing the dense sarcoplasm of the muscles, the nucleus and the dense hemi-desmosome (D) attachments of the muscle to the basement layer (BL) of the oral sucker. ×26 000. Stained with potassium permanganate and lead citrate.

1-3 Electronmicrograph of longitudinal section through a longitudinal muscle from the tailstem of the cercaria of *Schistosoma mansoni.* Note the apparently 'striated' nature of the muscle due to the interruption of the muscles fibres by completely transverse extensions of the sarcotubular system (SCT). ×18 000. Stained with potassium permanganate and lead citrate.

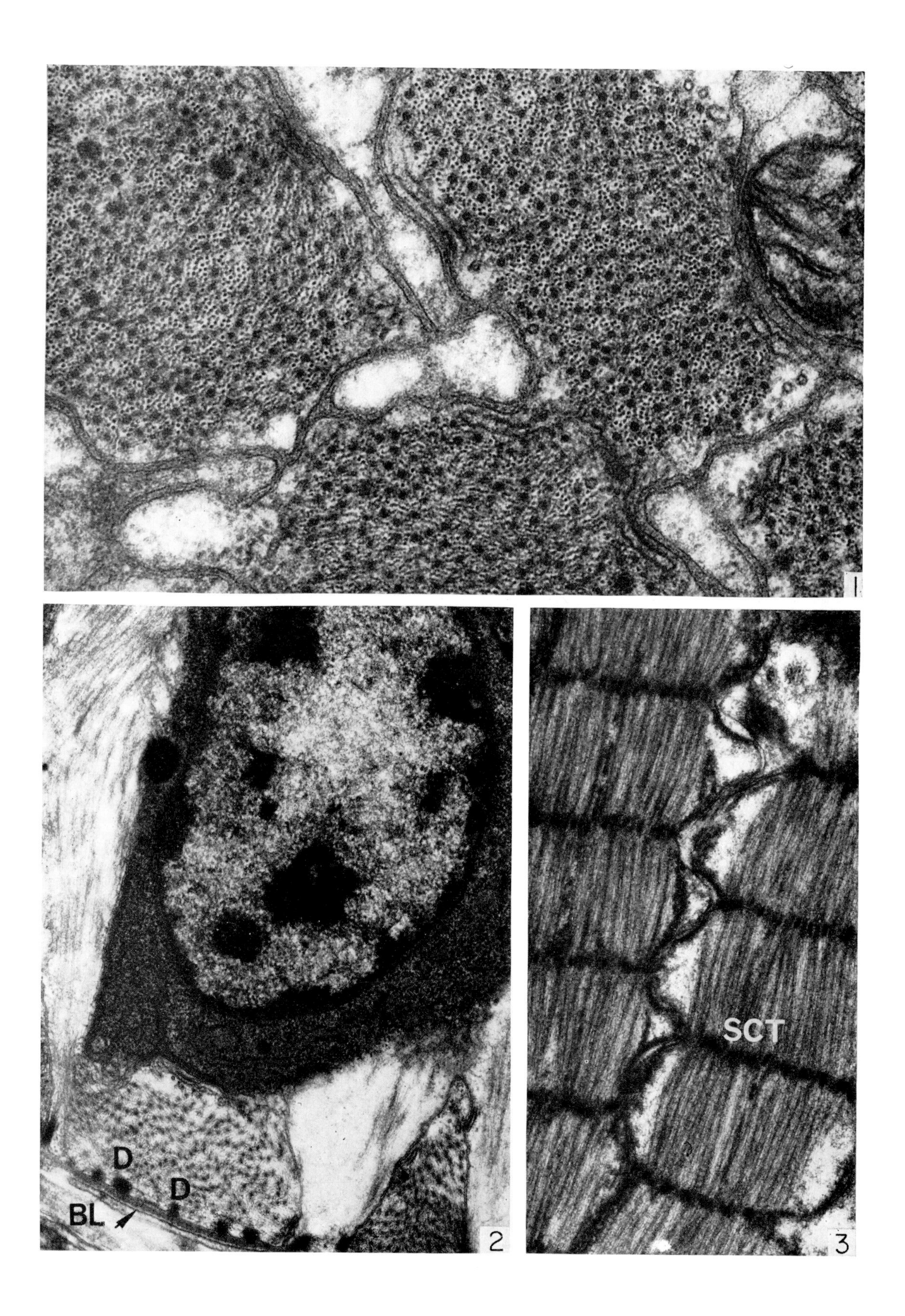
1
D
D
BL
2
SCT
3

Plate 2

2-1 Electronmicrograph of 'cerebral mass' of the miracidium of *Fasciola hepatica*. The axons (A) are unmyelinated and contain a variety of inclusions: mitochondria (M), vesicles (V) and small, dense bodies (DB). ×48 000. Stained with potassium permanganate and lead citrate.

2-2 Electronmicrograph of a transverse section through the ootype of *Apatemon sp.* Protruding into the lumen (L) of the chamber are microvilli (MV) extending from the luminal plasma membrane of the ootype epithelial cells. Discharging into the lumen are the ducts (D) of the glands from Mehlis' complex. These ducts have a supporting ring of microtubules (MT). Note the large nucleus (N) of the ootype epithelial cells. ×16 000. Stained with potassium permanganate and lead citrate.

2-3 Electronmicrograph of transverse sections of sperm tails of *Cyathocotyle bushiensis*. Each tail contains two cilium-like structures (C) comprising a nine plus one arrangement of tubules. In addition the periphery of the tail has a row of microtubules (MT). ×40 000. Stained with potassium permanganate and lead citrate.

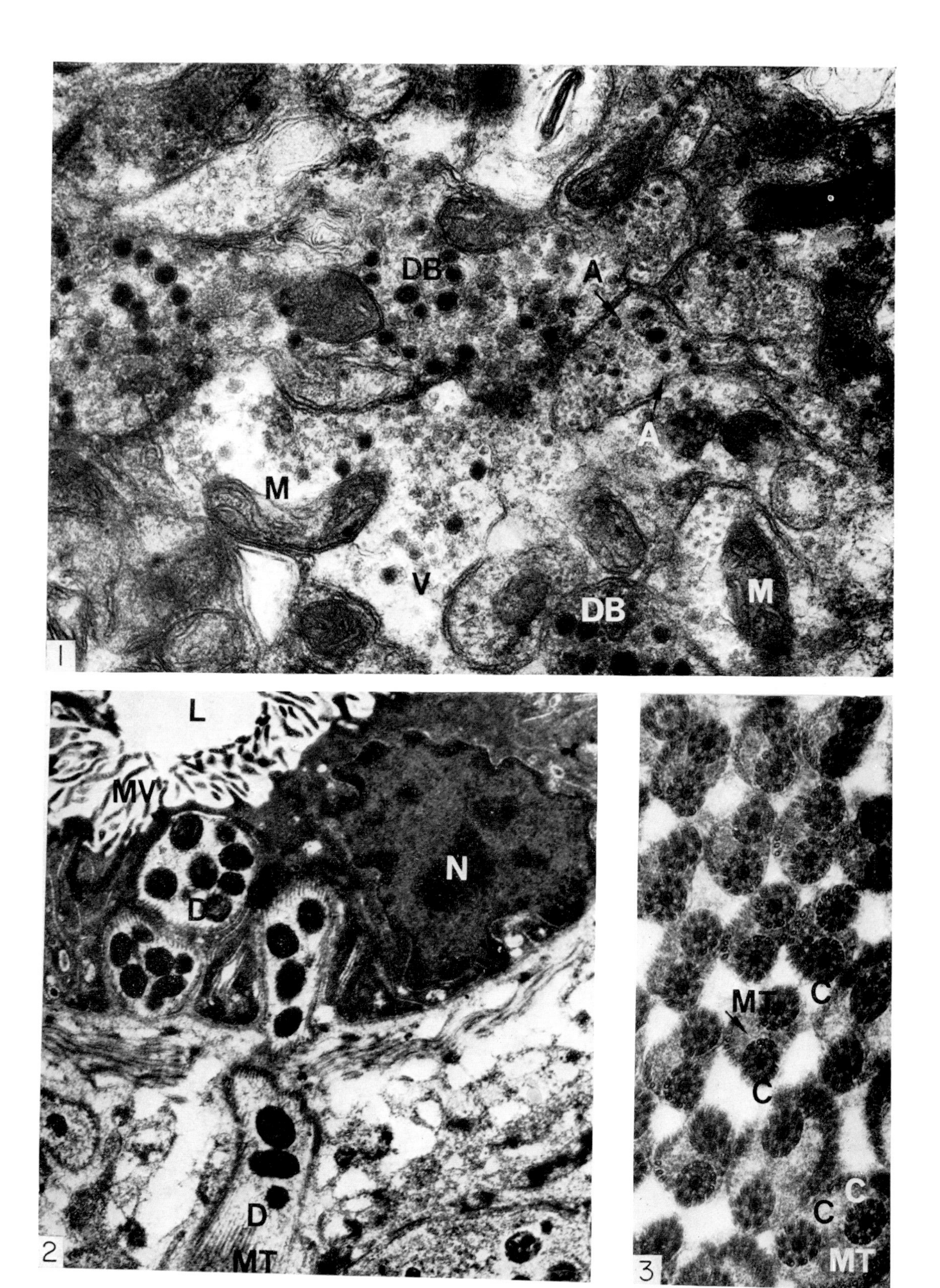
DB
A
A
M
V
DB
M
1
L
MV
D
N
D
MT
2
MT
C
C
C
C
MT
3

Plate 3

3-1 Egg of *Clonorchis sinensis*, the Chinese human liver fluke. Approx. ×15 000. Phase contrast. Note the elaborate opercular pole.

3-2 Eggs of *Diclidophora in situ* in the uterus. ×60 approx.

3-3 Egg of *Fasciolopsis buski*, phase contrast, approx. ×480.

3-4, 3-5 and **3-6** Stages in the hatching of the egg of *Fasciola hepatica*. 4: fully developed miracidium (M) within egg capsule. Note the apical cushion (AC) and the residual vitelline cells (V). The operculum (O) is still attached; 5: miracidium in the process of emerging from the egg. Note the displaced operculum (O) and the residual vitelline cells still present in the capsule; 6: the empty egg capsule with operculum (O). Approx. ×450.

3-7 and **3-8** Electronic flash photographs taken under phase contrast of the free swimming miracidium of *Fasciola hepatica*. In Fig. 7 note the extended 'terebratorium' or anterior rostrum and the paired eyespots lying above the cerebral mass. The miracidium is completely extended in Fig. 8 and is rapidly traversing the field.

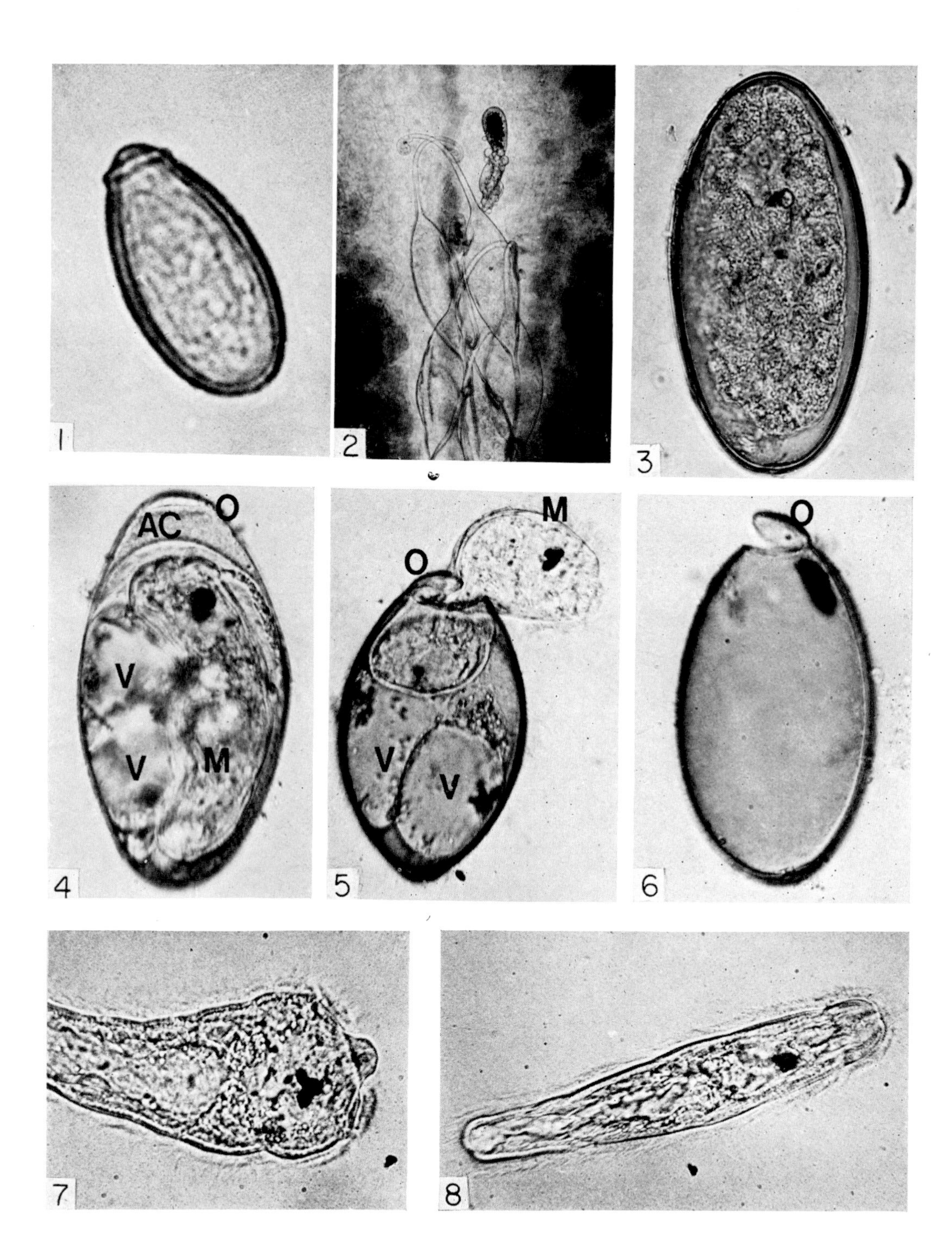
1
2
3
AC
O
V
V
M
4
M
O
V
V
5
O
6
7
8

Plate 4 Attachment.

4-1 The oral sucker of a free swimming furcocercaria (*Cercaria ocellata*) showing the discharge of the secretion from the penetration gland cells to the exterior of the larva. The strings of secretion have been stained with purpurin present in the water.

4-2 The opisthaptor of *Polystomum* from the bladder of the frog. Note the three pairs of acetabulum-like suckers and the pair of large hooks near the posterior edge. ×50 approx.

4-3 The more complex opisthaptor of *Diclidophora sp.* showing the attachment clamps at the end of short pedicels. ×22 approx.

4-4 The opisthaptor of *Acanthocotyle sp.* This has the shape of a flattened disc bearing radially arranged rows of small hooks. ×50 approx.

4-5 Detail of the attachment claps of *Discocotyle*. Note the complex arrangement of bars comprising each clamp. ×40 approx.

4-6 Electronmicrograph of the tegument of *Cercaria* Z (Echinostomatidae) six hours after penetration of the second intermediate host (*Lymnaea trunctatula*). The rod-like contents (scrolls) of the cystogenous cells have reached the tegument and are being discharged through the plasma membrane to form the third layer of the metacercarial cyst wall. (CT: cercarial tegument; L3: layer three of cyst wall; PM: plasma membrane; SC: scroll like bodies; US: the layers resulting from an unwound scroll. ×30 000.

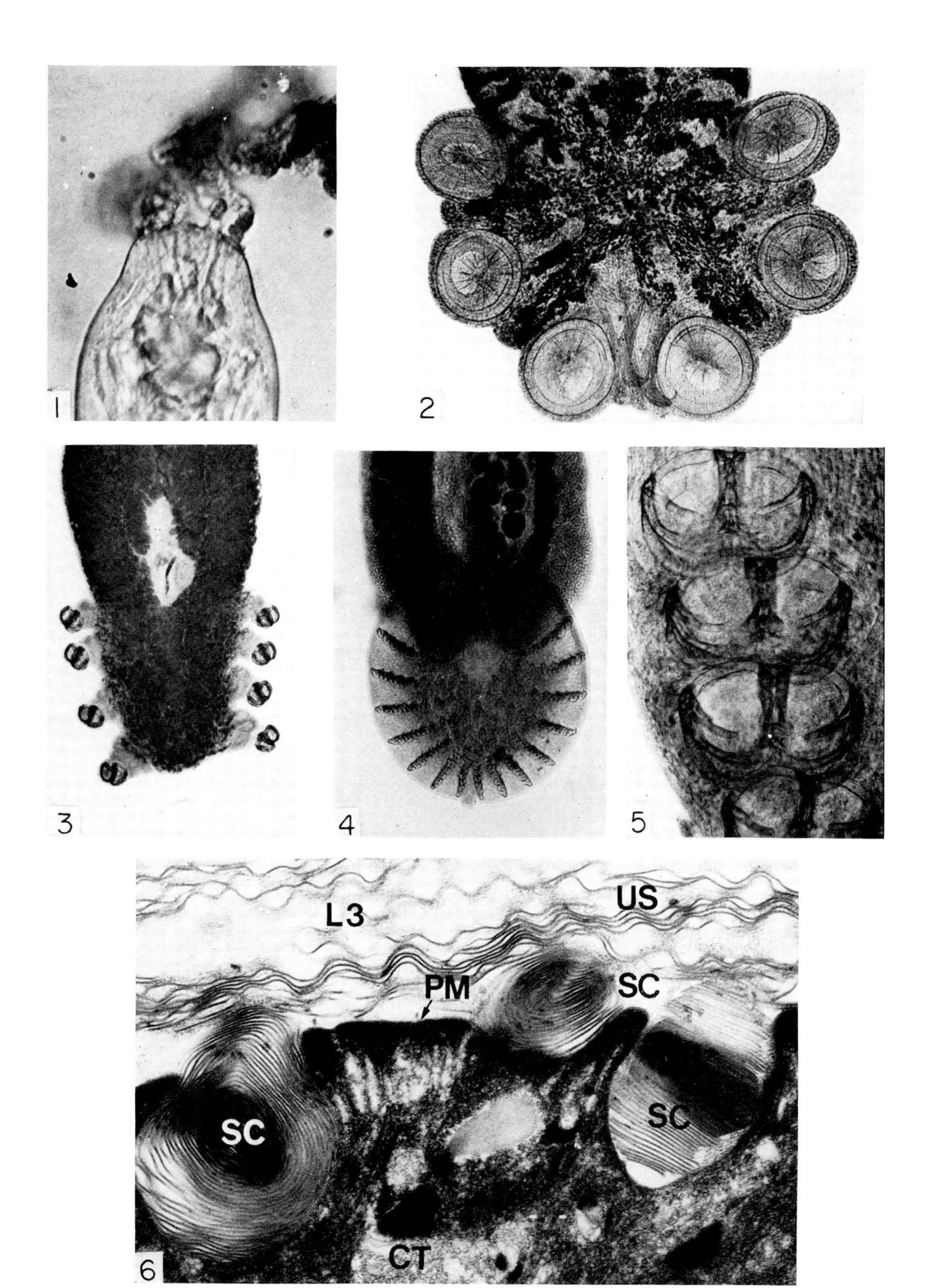
1
2
3
4
5
6
US
L3
PM
SC
SC
SC
CT

Plate 5

5-1 Phase contrast micrograph of the metacercarial cyst of *Cyathocotyle bushiensis*. Note the thick lamellate cyst wall and the contained metacercaria completely filling the cyst. Approx. ×160.

5-2 High power detail of the cyst wall of Fig. 1. Note the closely packed lamellae comprising the cyst wall. ×360 approx.

5-3 Metacercarial cyst of a xiphidiocercaria. Note that the cyst wall is thinner than that of *Cyathocotyle,* but consists of several quite distinct layers. ×120 approx.

5-4 Metacercarial cyst of the strigeoid trematode *Apatemon gracilis minor in situ* within the leech, *Erpobdella octoculata.* ×30 approx.

5-5 Electronmicrograph of a longitudinal section through the oral sucker of the cercaria of *Schistosoma mansoni*. Note the apical sense organ with its projecting cilium-like structure (C) arising from a basal vesicle (V), lying in the cercarial tegument (CT). Adjacent to the sense-organ is the aperture of one of the penetration gland cells ducts (AGL). The secretion present in this duct consists of large, membrane bound bodies containing granular material. The cercarial tegument is covered by a fine granular deposit (GD). ×20 000. Stained with uranyl acetate and lead citrate.

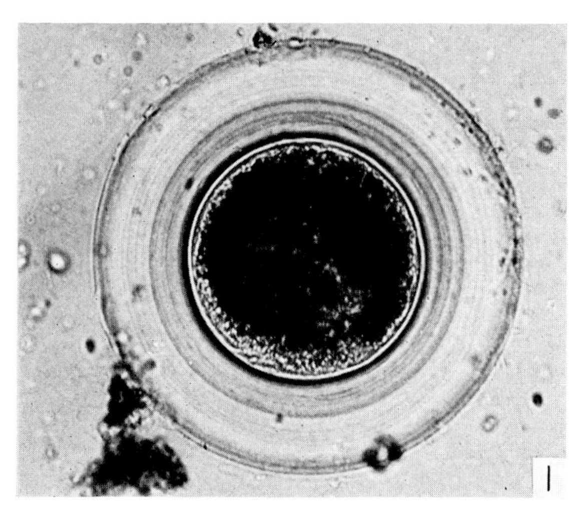
1

2

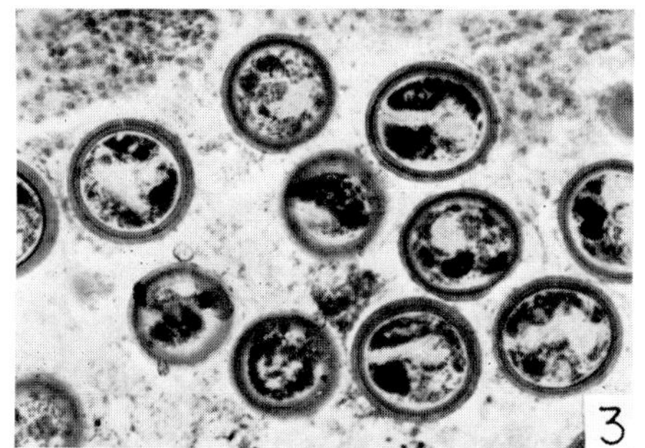
3

4

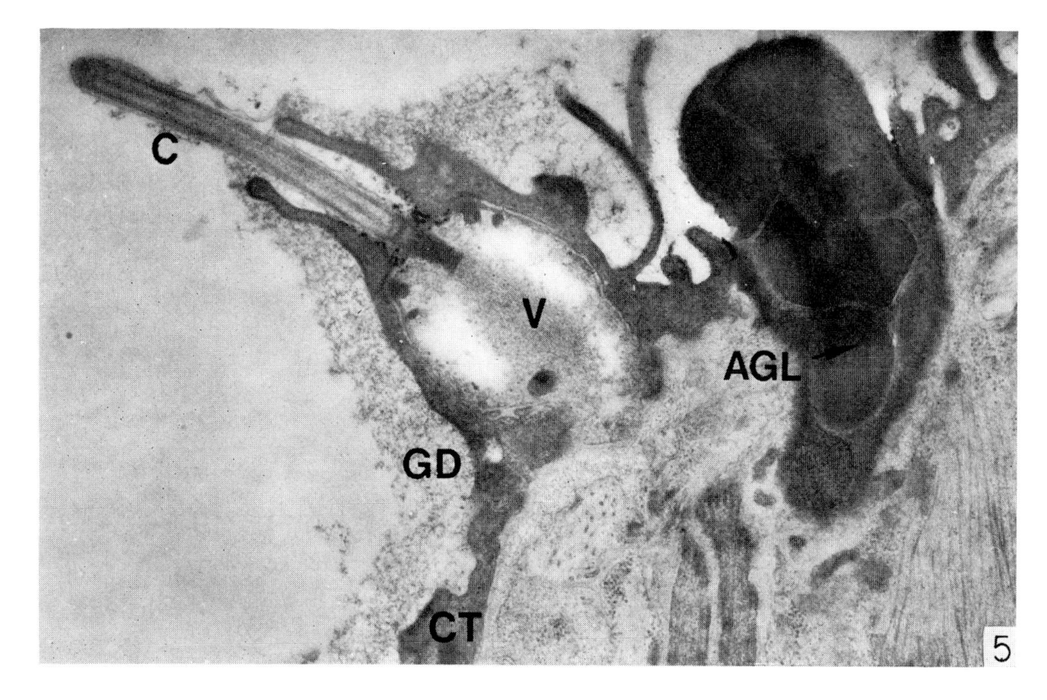
C
V
AGL
GD
CT
5

Plate 6 The excretory system (Digenea).

6-1 Electronmicrograph of a longitudinal section through a flame cell of *Cyathocotyle bushiensis*. The apical cell body contains a nucleus (N) and into the chamber of the flame cell projects a tuft of cilia (TC) and the processes from the chamber wall (P). The lumen of the flame cell is in continuity with the excretory tubule (ET). Adjacent to the flame cells are parenchymal cells (PC). ×9600. Stained with potassium permanganate and lead citrate.

6-2 Electronmicrograph of the excretory lacuna of the strigeoid trematode *Diplostomum sp.* Note the lipid droplets free (LD) in the lumen of the lacuna (L) as well as enclosed in the cytoplasmic lining (LC) of the lacuna. ×8000. Stained with potassium permanganate and lead citrate.

6-3 Electronmicrograph of cytoplasmic lining (LC) of the excretory lacuna of *Cyathocotyle bushiensis*. Note the lipid droplet (LD) still attached to the cytoplasm by thin lamellae (LM). ×18 200. Stained with potassium permanganate and lead citrate.

6-4 Electronmicrograph of excretory lacuna cytoplasm (LC) of an immature adult of *Cyathocotyle bushiensis* showing a calcareous corpuscle (CC) attached to the cytoplasm by several lamellae (LM). ×30 000. Stained with potassium permanganate and lead hydroxide.

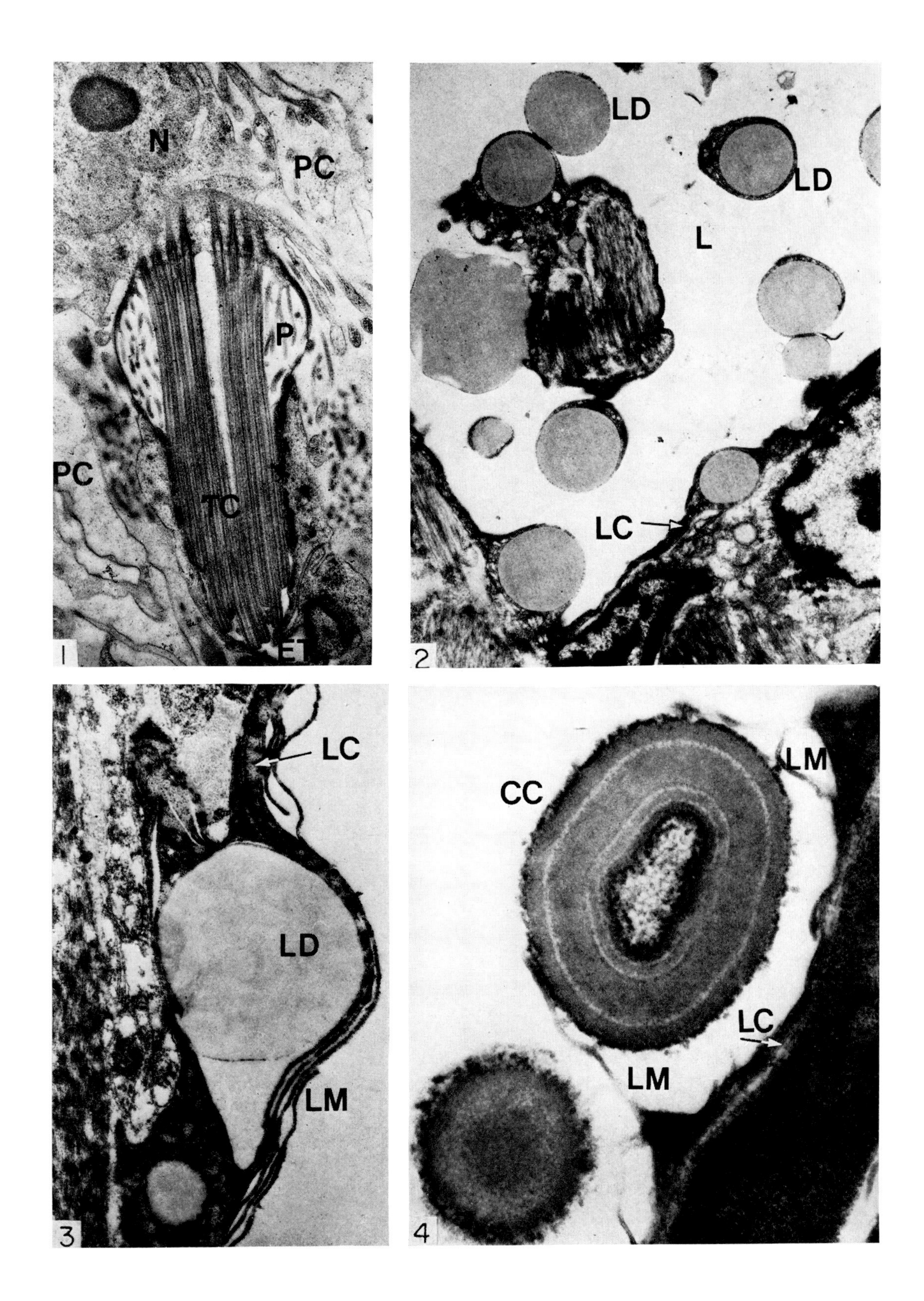
N
PC
P
PC
TC
ET
1
LD
LD
L
LC
2
LC
LD
LM
3
LM
CC
LC
LM
4

Plate 7

7-1 Photomicrograph of a section through infected digestive gland of *Lymnaea stagnalis*. The connective tissue between the remains of the molluscan gonad (G) and digestive gland (DG) is filled with a mass of daughter sporocysts (S) containing cercariae (C) in various stages of development. ×120 approx.

7-2 Section through a moribund cercaria (C) surrounded by a spherical mass of molluscan amoebocytes (A). ×580 approx.

7-3 Section through the mantle (M) of *Lymnaea stagnalis* containing a mass of cercariae (C). The mantle is distended to form a 'bleb' and it is probable that mass emergence of cercariae would occur in this region when circumstances were favourable. ×133 approx.

7-4 Section through a mantle 'bleb' (M) from which cercariae (C) are emerging to the exterior of the mollusc. Note that the cercariae emerge anterior end first and that, once perforation of the mantle epithelium has been achieved, many cercariae utilize the same route to the exterior. ×100 approx.

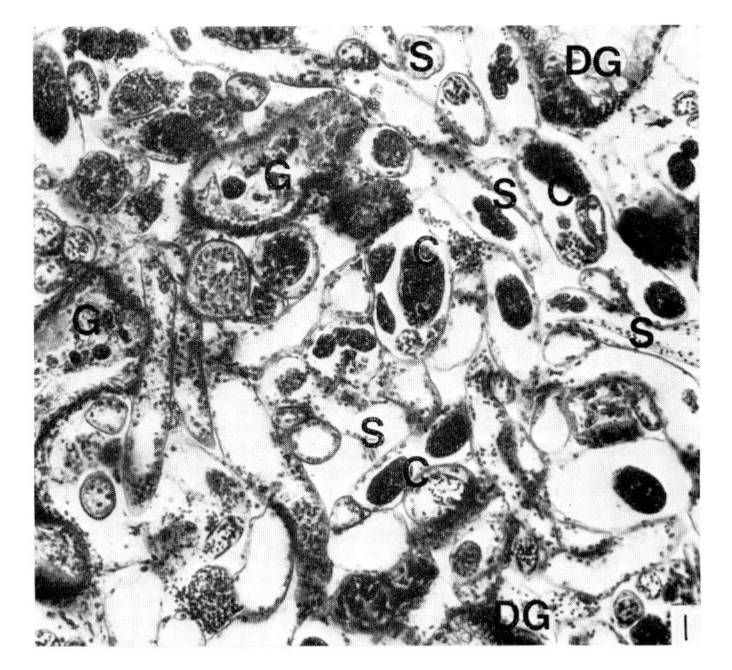
S
DG
G
S
C
C
G
S
S
C
DG
1

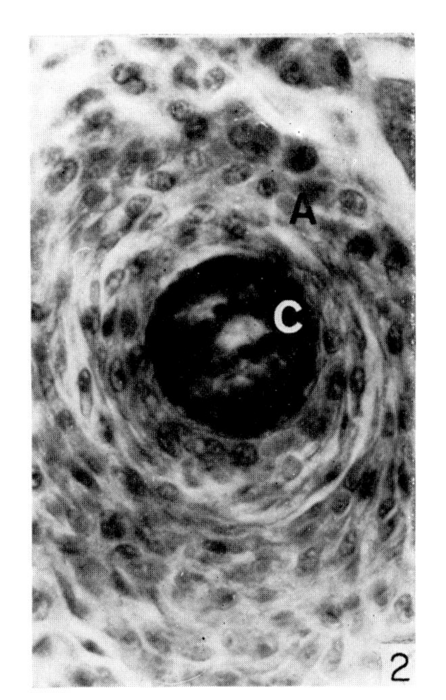
A
C
2

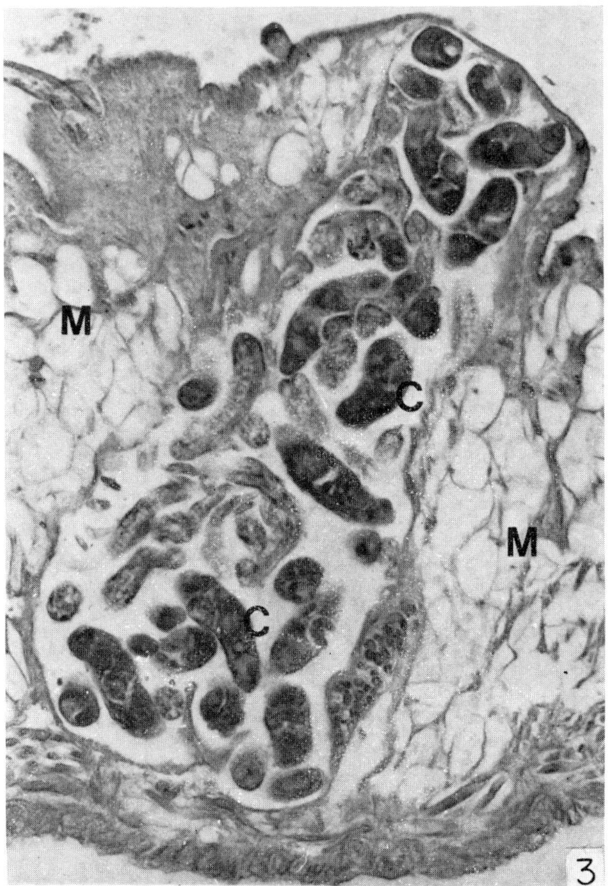
M
C
M
C
3

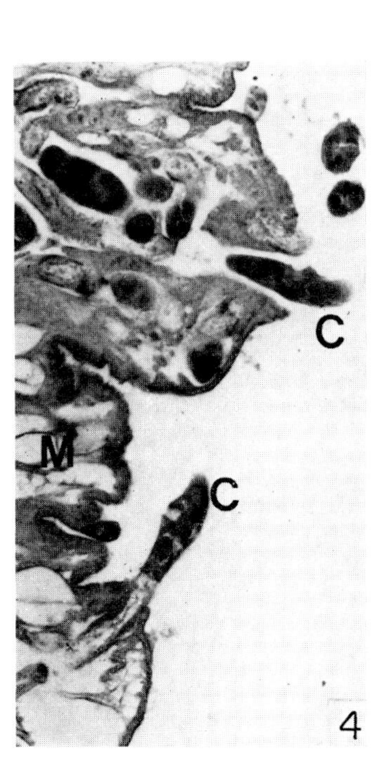
C
M
C
4

Plate 8

8-1 Section through the body wall (BW) of *Gasterosteus aculeatus* containing a cercaria (*Cercaria X*) (C) migrating through from the exterior surface of the skin (EXT). Note that the tail has been lost and that the cercaria is migrating with its oral sucker (OS) in front. The cercaria lies in a 'tunnel' (T) produced by its activity. ×400 approx.

8-2 A transverse section through an adult *Clonorchis sinensis* (CS) lying in bile duct (BD) of the liver (L). Note the thickened wall of the bile duct (W). ×24 approx.

8-3 A section showing the attachment of caudal muscles (CM) to cartilage (CT) in the tail of *Gasterosteus aculeatus*. Lying within the muscle, in a cavity (CV) produced during its migration, is a cercaria (C). Considerable damage to the musculature of the host can occur in this way. ×270 approx.

8-4 A transverse section through the ventral aorta (VA) of *Gasterosteus aculeatus* which has been perforated by a cercaria (C) migrating out through a tunnel (T) to the exterior surrounding tissues. The anterior oral sucker (OS) is clear of the vessel and the surrounding tissues contain blood (B) which has leaked from the perforation. ×150 approx.

8-5 Section of fish retina (R) showing the damage caused by a migrating cercaria (C). Note the ventral sucker (VS) of the cercaria. ×410 approx.

8-6 Surface view of the caecum of a duck which has been infected with *Cyathocotyle bushiensis*. Note that the mucosal surface has been eroded producing a lesion (L) which corresponds to the position of the adhesive organ of the attached parasite. ×16 approx.

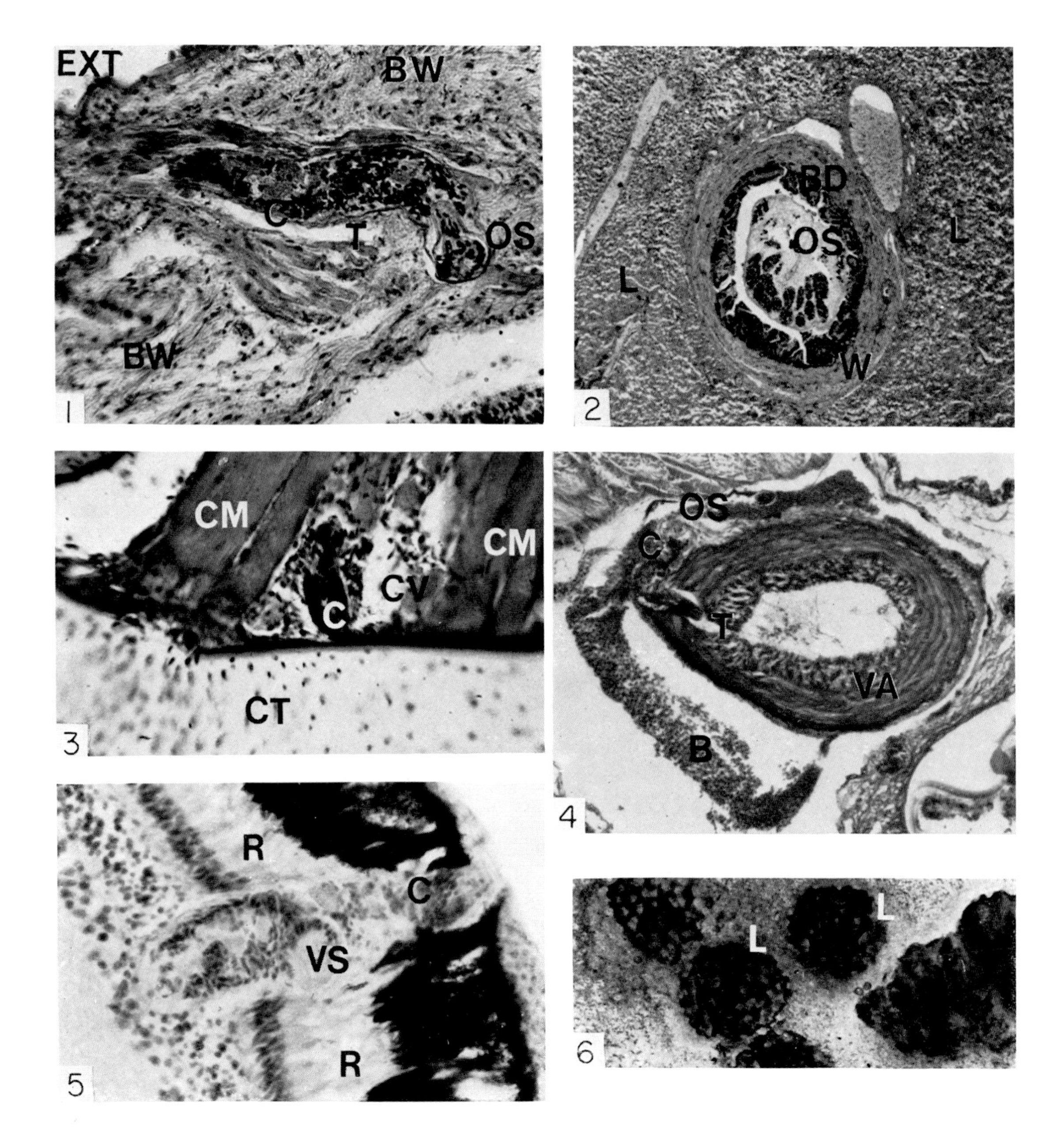
EXT
BW
C
T
OS
BW
1
L
BD
OS
L
W
2
CM
CM
CV
C
CT
3
OS
C
T
VA
B
4
R
C
VS
R
5
L
L
6

Plate 9

9-1 Electronmicrograph of a transverse section through the adhesive organ (AO) showing the transition in the form of the tegument from a normal spined tegument (ST), covering most of the body surface, to the microvillous (MV) tegument covering the surface of the adhesive organ. The lumen (L) of the adhesive organ is filled with granular debris. *Cyathocotyle bushiensis.* ×3000. Stained with potassium permanganate and lead citrate.

9-2 Electronmicrograph of a section through the general tegument of adult *Cyathocotyle bushiensis*. The tegument is bounded externally by a plasma membrane (PM) and is differentiated from the fibrous basement layer (BL) by a basal plasma membrane (BPM). Between these two membranes is the cytoplasmic matrix (TM) containing spines (SP), secretion bodies (SB) and mitochondria (not present in this photograph). Below the basement layer are the muscles (MS) of the body wall. ×32 000. Stained with potassium permanganate and lead hydroxide.

9-3 Electronmicrograph of the microvilli of the adhesive organ surface of *Cyathocotyle bushiensis* showing the sites of acid phosphatase activity (AP). Activity is represented by the deposits of dense lead phosphate crystals. Note that the activity occurs within the microvilli and is particularly evident in the distended tips (DT) of some of the microvilli. Incubation in Gomori's acid phosphatase media for 1 hour at 37°C. The section is not counterstained. ×60 000.

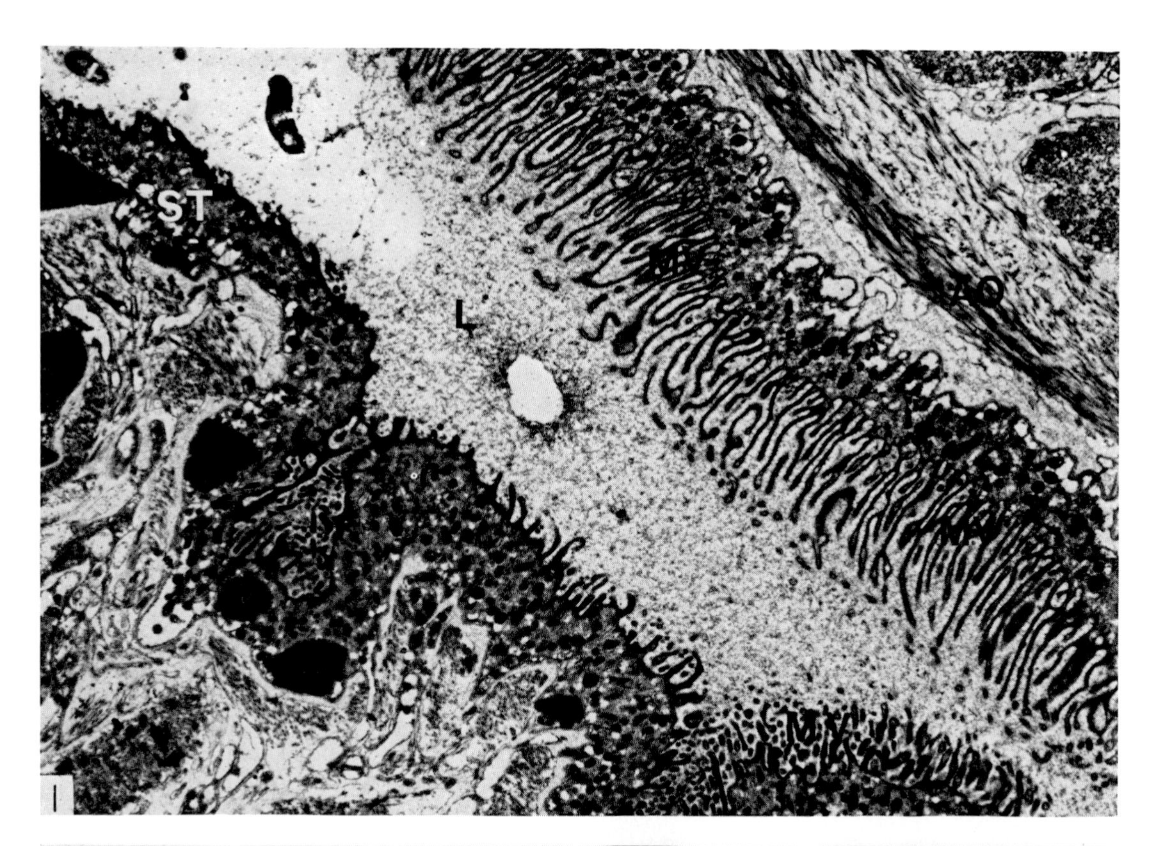
ST
L
1

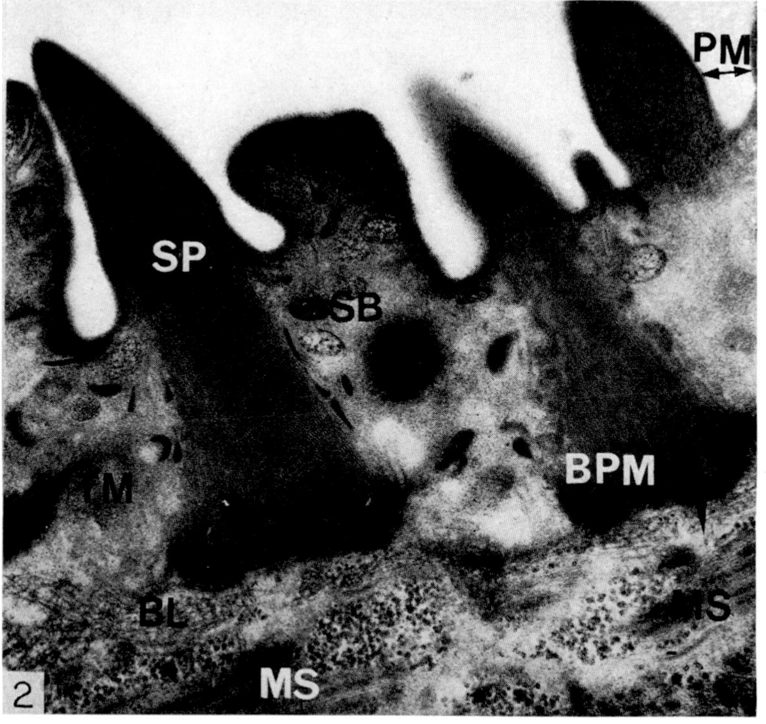
PM
SP
SB
BPM
BL
MS
2

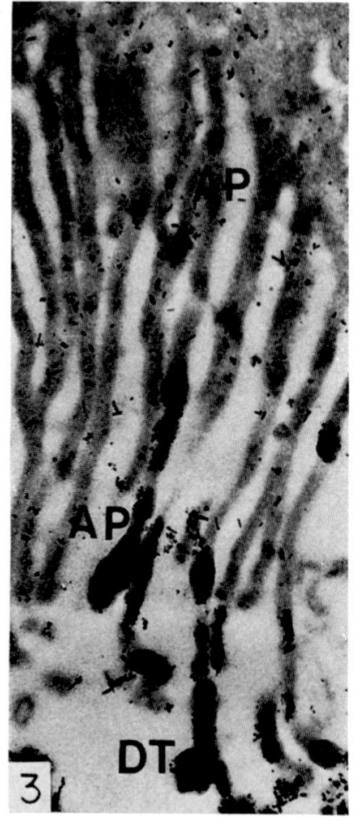
P
AP
DT
3

Plate 10

10-1 Electronmicrograph of a section through the caecal epithelium of adult *Cyathocotyle bushiensis*. The apices of the cells are shown and it is obvious that the surface of each cell is covered by microvilli (MV) which project into the lumen of the caecum (CL). The cytoplasm of the cells is rich in endoplasmic reticulum (ER) and also contains membrane bounded dense bodies (SB). Note how the microvilli fuse to form a meshwork extending between the cells. ×22 400. Stained with potassium permanganate and lead citrate.

10-2 Electronmicrograph of a section through the general tegument (GT) of *Cyathocotyle bushiensis* showing a sense organ. This consists of a cilium-like (CP) process arising from a vesicle (V) embedded within the tegument. The cilium arises from a striated rootlet system (RS) and the vesicle, attached to the tegument by desmosomes (DS) contains a mitochondrion (M) and several small vesicles (VS). ×37 000. Stained with potassium permanganate and lead citrate.

10-3 Electronmicrograph through the lappet region of *Apatemon gracilis minor*. In this specialized region of the tegument the surface possesses setae (ST) and the apertures of a mass of gland cells (GS) lying between the complex musculature (MS) associated with the lappets. The glands are unicellular and may be packed with secretion or relatively empty (EGS) after the secretion has been discharged. ×3750. Stained with uranyl acetate and lead citrate.

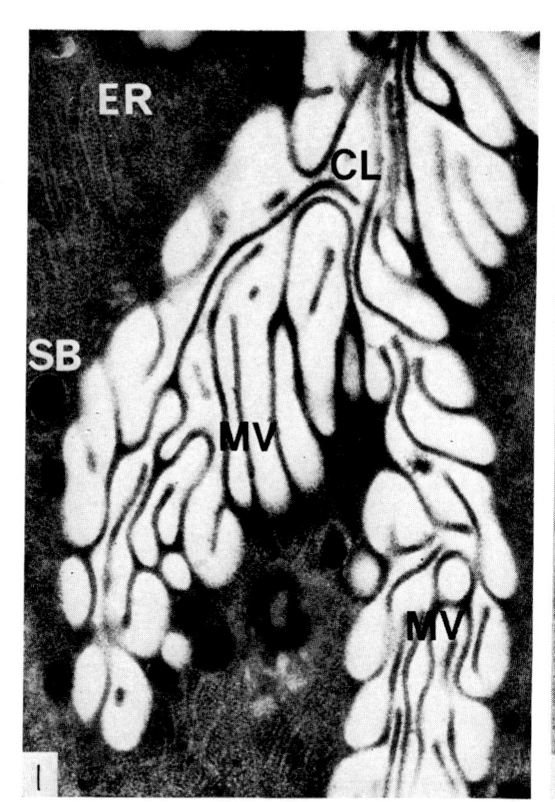
ER
CL
SB
MV
MV
1

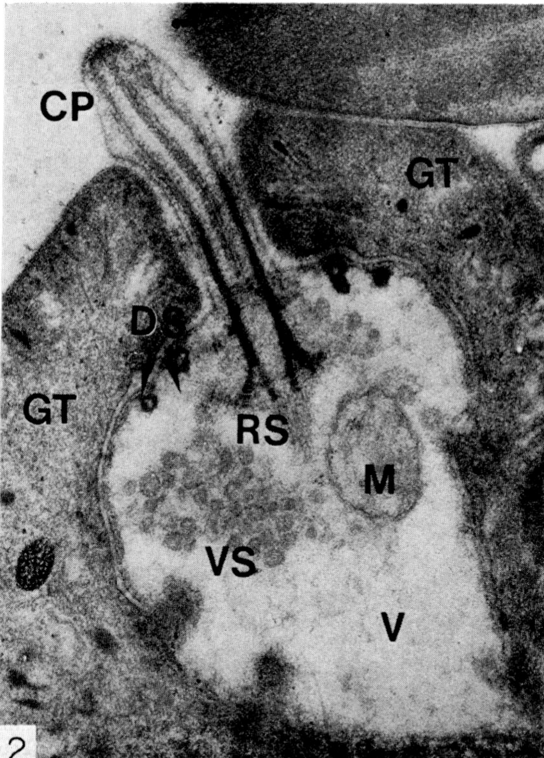
CP
GT
DS
GT
RS
M
VS
V
2

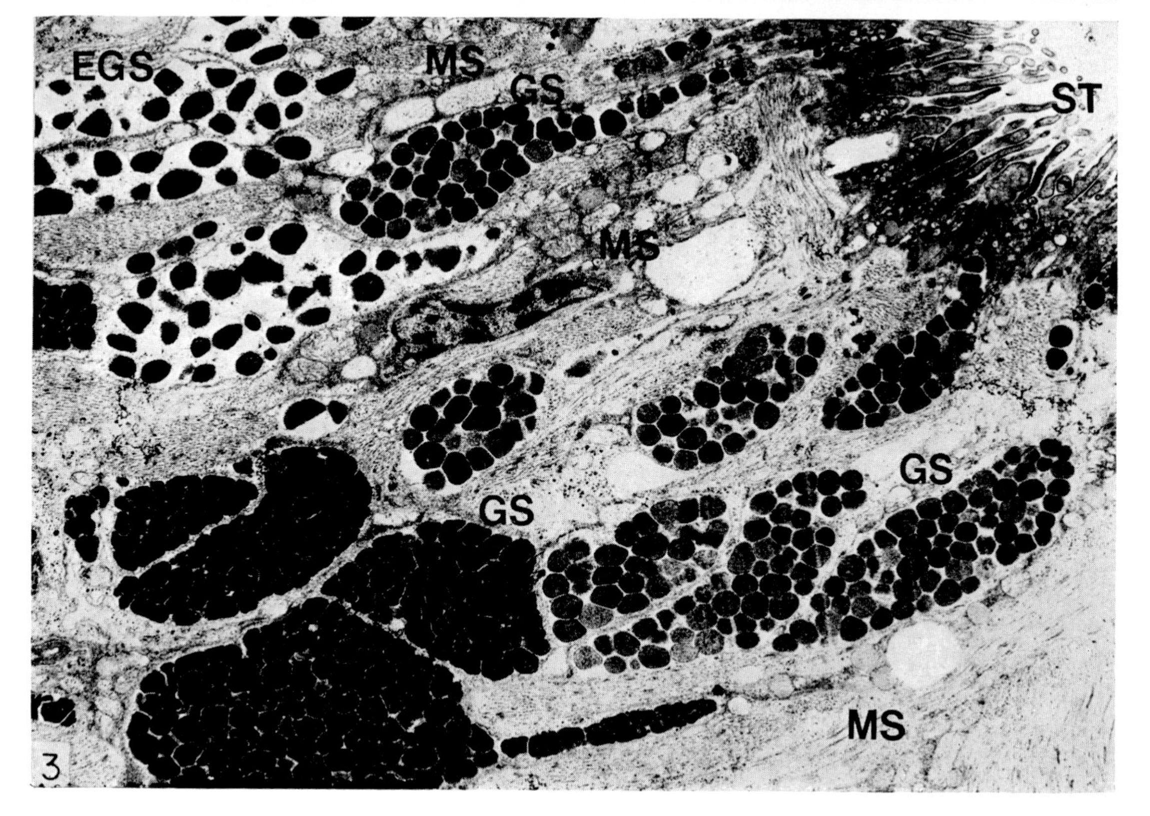
EGS
MS
GS
ST
MS
GS
GS
MS
3

Plate 11

11-1 Electronmicrograph of a transverse section through one of the adhesive organ lobes of *Apatemon gracilis minor* to show the regional differentiation exhibited by this surface which is in intimate contact with the host tissues. The general tegument (GT) is transformed into a pitted, sponge-like surface (PS) on the inner face of the adhesive organ lobe which is in contact with the host tissue plug (HP). Note the sense organ (SO) protruding into the host tissue and the large lacunae of the excretory system (EXL) in the centre of the lobe. On the other side of the host plug can be seen the other modified surface—that of the lappets (LP). ×3750. Stained with uranyl acetate and lead citrate.

11-2 Stereoscan electronmicrograph of the retracted lappet (L) of *Apatemon gracilis minor*. Note the sense organs (SO) on the surface, the setae (ST) protruding into the lumen of the lappet (L), and the apertures of the lappet gland cells (AP). Freeze dried specimen. ×6400.

11-3 Electronmicrograph of a section showing the apertures (AP) of the lappet gland cells (GS) of *Apatemon gracilis minor*. The secretion (SB) consists of discrete membrane bounded bodies, but exterior to the aperture (AP) and the general tegument (GT), is a fine granular material (GM). ×19 500. Stained with uranyl acetate and lead citrate.

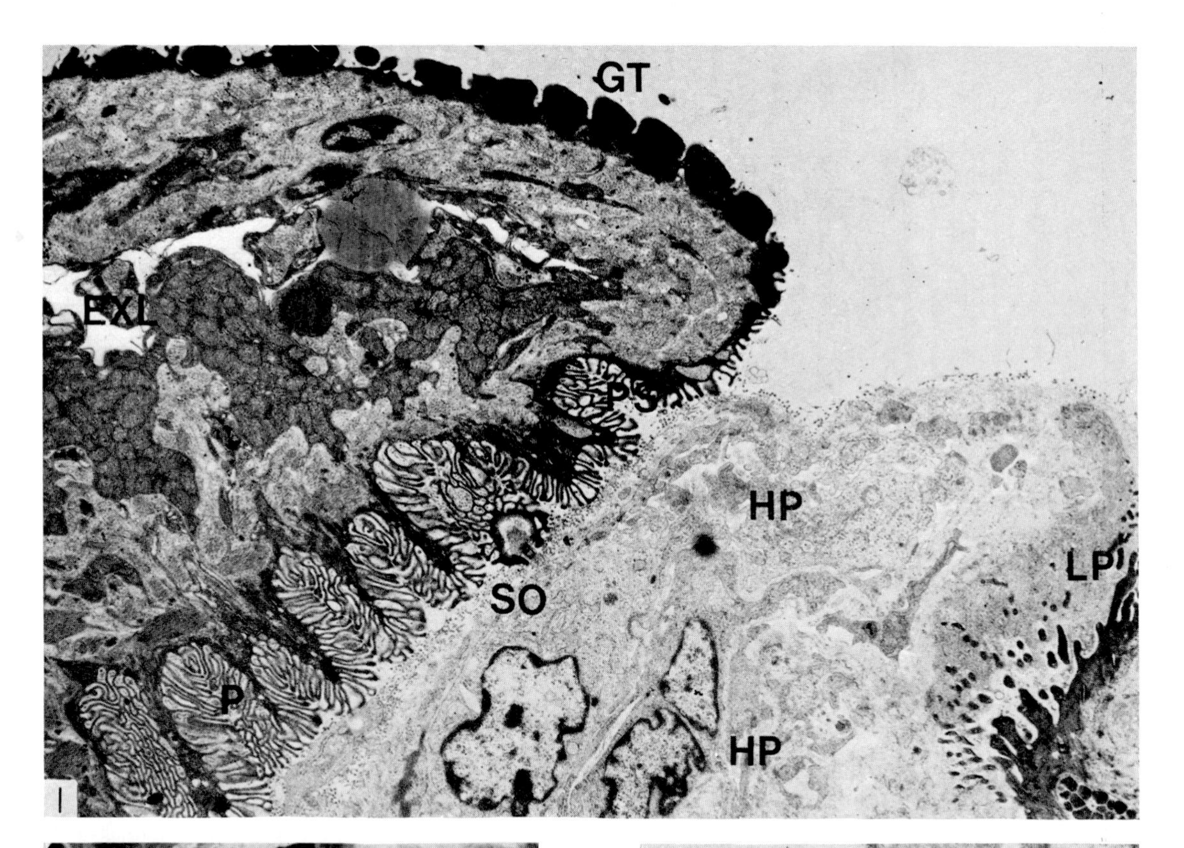
GT
EXL
PS
HP
LP
SO
P
HP
1

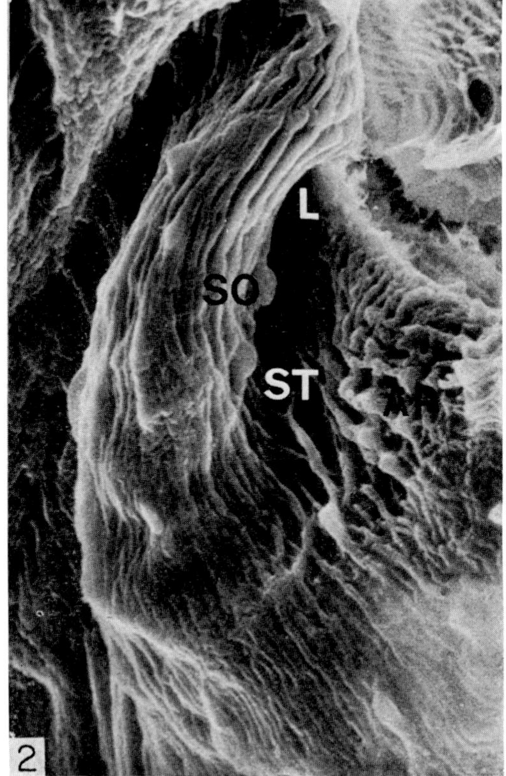
L
SO
ST
2

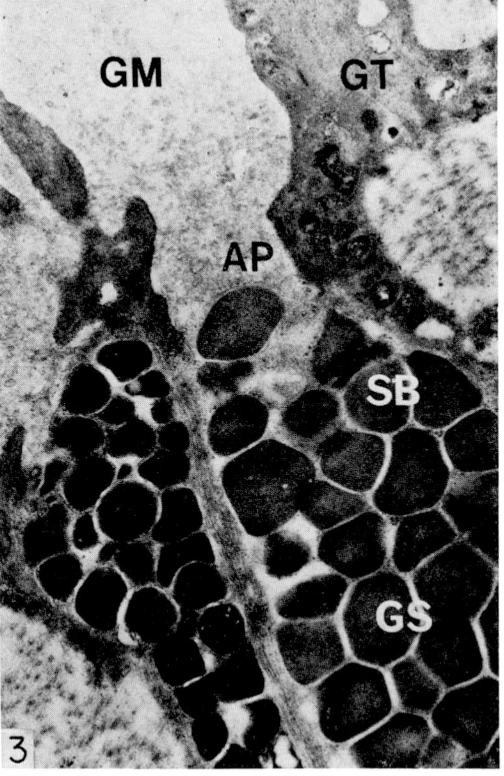
GM
GT
AP
SB
GS
3

Plate 12

12-1 Electronmicrograph of a section through *Apatemon gracilis minor* showing the intimacy of the contact between the parasite adhesive organ lobes (AOL) and a capillary (CP) within the host tissue plug (HT). The wall of the lobes is relatively narrow and the large excretory lacuna (EXL) can be seen in the left hand corner of the print. In this illustration the distance from the tegument of the parasite and the lumen of the host capillary is approximately 0·1 μ. ×30 000. Stained with potassium permanganate and lead citrate.

12-2 Stereoscan electronmicrograph of the external surface of a sense organ (SO) in the reticular (RS) surface of the adhesive organ lobe. Note the shrunken cilium like papilla (CP) protruding to the exterior. Freeze dried material. ×18 750.

12-3 Stereoscan electronmicrograph of the reticular surface of the inner face of the adhesive organ lobe of *Apatemon gracilis minor*. Freeze dried material. ×4440.

12-4 Electronmicrograph of a section through the tegument (T) of the redia of *Fasciola hepatica*. Note that the surface is elevated into a series of lamellae (LM) and beneath the tegument lies the basement layer (BL) musculature (MS) of the body wall, and a subtegumentary cell (TC). ×12 000. Stained with uranyl acetate and lead citrate.

12-5 Electronmicrograph of a 'Rod' containing cell lying below the tegument of the cercaria of *Fasciola hepatica*. The rods are sectioned in both longitudinal (RL) and transverse (RT) planes. Each rod consists of spirally wound sheet of material. ×30 000. Stained with uranyl acetate and lead citrate.

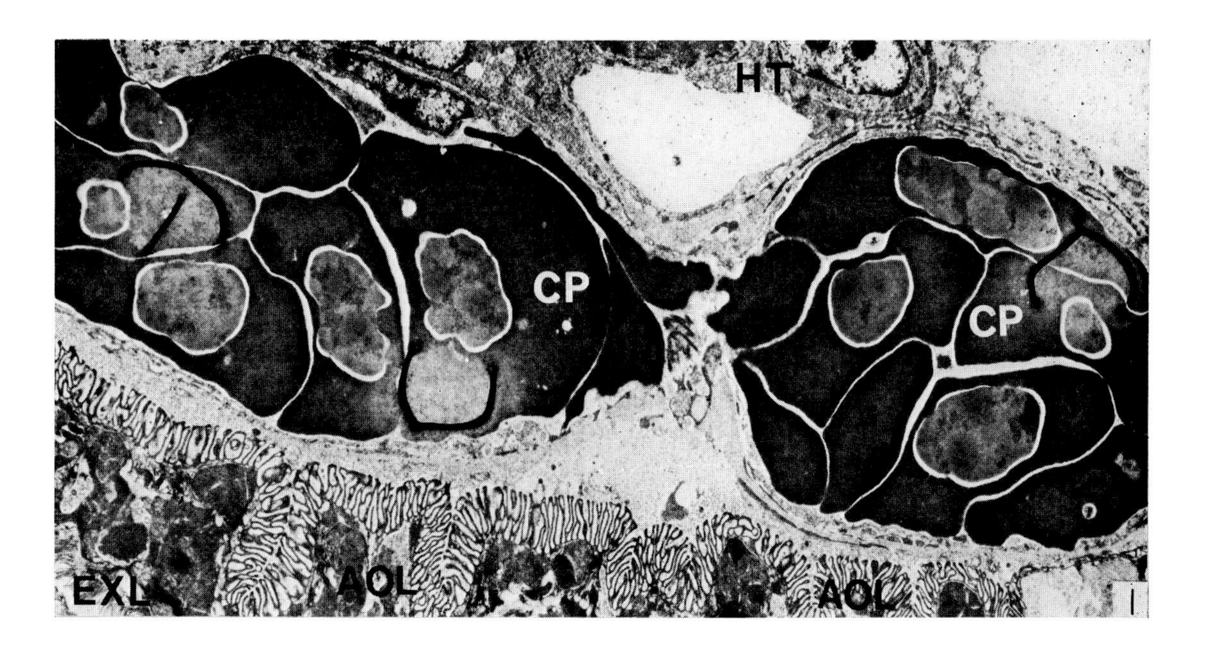
HT
CP
CP
EXL
AOL
AOL
1

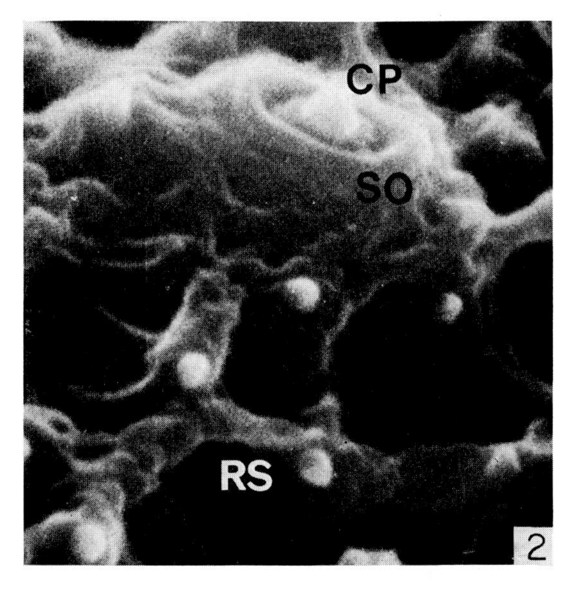
CP
SO
RS
2

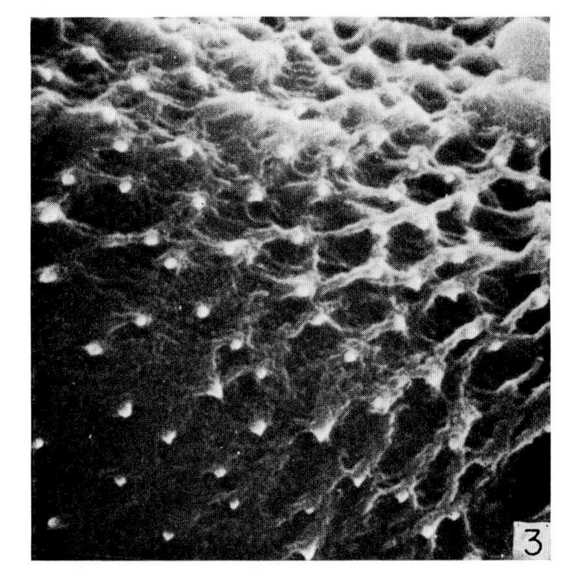
3

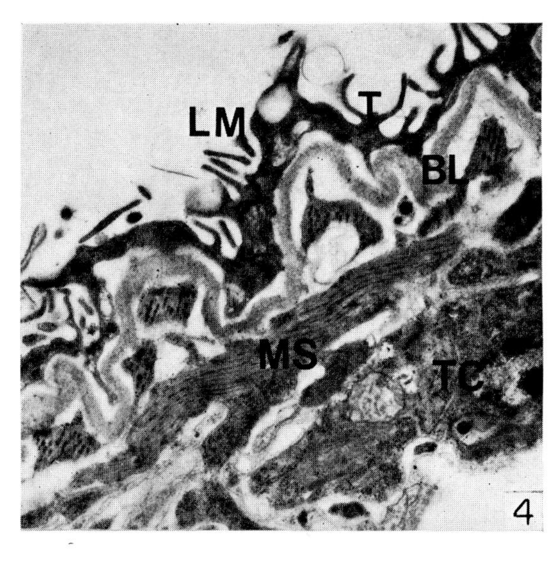
LM
T
BL
MS
TC
4

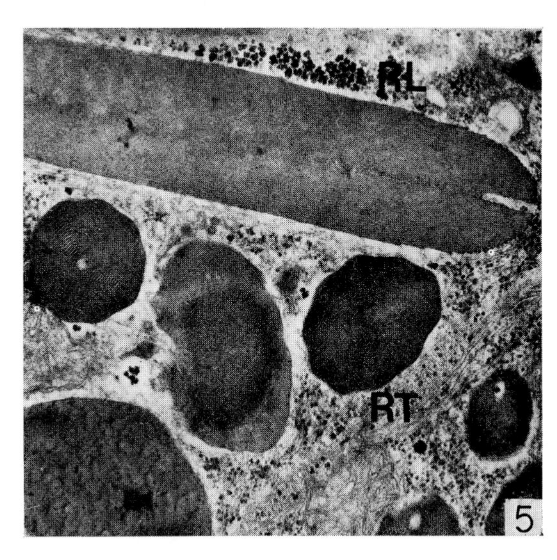
RL
RT
5

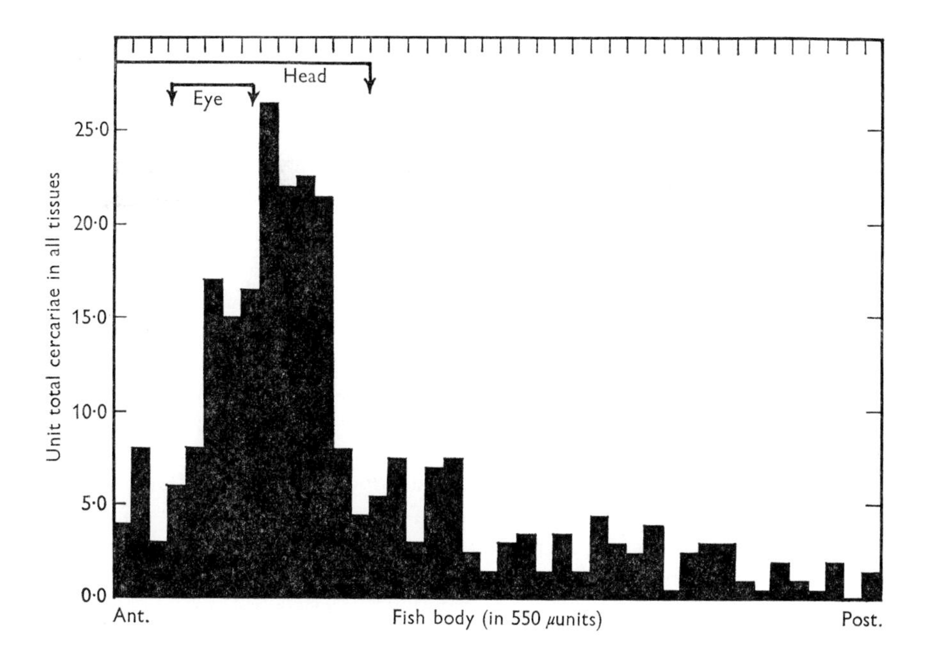

FIG. 66 The distribution of *Cercaria X* (Strigeoidea) in the tissues of the fish *Gasterosteus aculeatus*. The anterior end of the fish is to the left of the figure and the posterior end to the right. Although cercariae occur all along the fish maximum concentration occurs in the head region. Fish sectioned 57 min after exposure to cercariae. From ERASMUS, 1959.

a tunnel produced by its own activities. FERGUSON (1943a), studying the penetration and migration of *Cercaria flexicauda* in fish, has shown that after the completion of migration the gland cells are empty and that reinfection of fish using cercariae with expended gland cells does not result in migration and successful localization. So whatever the function of the gland cell secretion seems to be, there is no doubt that their function is a very essential one. Penetration into the fish skin is very rapid and is accomplished in 7–8 minutes. Speedy penetration is essential, as attachment to the outside of an actively swimming fish represents an extremely precarious period for the cercaria.

The direct penetration of fish occurs in other life-cycles. *Clonorchis sinensis* has an encysted metacercarial stage in fish which develops after the pleuroophocercous cercaria has penetrated. These cercariae hang in the water with the body down and exhibit a positive phototaxis, but they do not attack the fish directly; only after making contact with the fish body, when they attach using the oral sucker (KOMIYA and TAJIMI, 1940a). After attachment they exhibit the usual creeping and searching movements eventually penetrating the body. The tail again plays no part in penetration and is cast off before penetration is completed. The site most commonly entered was the trunk of the host. A similar pattern of events has also been described for the cercaria of *Exorchis oviformis* (KOMIYA and TAJIMI, 1940b). Entry into the fish takes about 3–5 min for *Exorchis* cercariae and 6–15 min for *Clonorchis* cercaria. Once again the rapidity of penetration is remarkable.

Members of the Sanguinicolidae, parasitic in the circulatory system of fishes, have a furcocercaria of the Cristata type. In addition to the forked tail, the body of this type bears a dorsal crest and both these structures aid in suspending the cercariae in water. When cercariae come in contact with the fish they attach immediately and penetrate mainly in the region of the gills.

In all these examples there is no evidence which suggests that the cercaria is able to detect fish at a distance, but depends on tactile stimuli produced by chance physical contact to trigger off a phase of heightened activity and attempts at penetration. It is also apparent, in the laboratory, that contact with inanimate objects such as a needle may release the behaviour pattern. Many cercariae are ornamented with long hair-like processes and these are usually particularly well developed on the tail. These processes may serve in assisting buoyancy, as has been mentioned earlier, but it is also likely that they are sensory and serve to detect the tactile stimuli so essential to penetration.

Entry into Amphibia e.g. tadpoles, follows the same general pattern described for entry into fish. No marked orientation towards the host occurs but, after chance contact, they attach immediately and begin to penetrate (PEARSON, 1956). An interesting feature of penetration into these hosts is that the constitution of the skin plays an important role, in allowing or preventing entry. The skin of the tadpole has an outer layer of cuboidal or columnar cells and is devoid of mucous glands whereas that of the adult frog has a thinner epidermis with an outer layer of squamous cells and is rich in mucous cells. Several workers e.g. HOFFMAN (1955); OLIVIER (1940); and PEARSON (1956) have commented on the ability of cercariae to penetrate amphibian tadpoles in contrast to their inability to enter through the skin of the adult frog. This behavioural difference has been attributed to the presence of gland cells and the secretion of a mucus possessing a protective function in the adult frog. This suggestion is partly supported by Pearson who mentions that cercariae of *Alaria canis* attacking *Rana sylvatica* were able to penetrate the cornea which possesses a squamous epithelium but no mucus cells.

The study of the direct penetration of cercarial stages into mammals has been stimulated by the occurrence of highly pathogenic schistosome species in man and also by the appearance of cercarial dermatitis (Swimmer's Itch) in swimmers who have been subjected to attack by species of cercariae which normally penetrate birds and mammals other than man. STIREWALT and HACKEY (1957) in the first of a long series of papers on penetration by schistosome cercariae, have shown that the behaviour of the cercaria during penetration varies with the type of mammalian skin. Differences were reflected in the ratio of successful penetrants to failures, in the choice of entry sites and in the average time spent in exploration and actual entry (Table 21). The particular site of entry, as indicated by frequency of use, of 383 penetrating cercariae varied as follows. One hundred and fifty eight cercariae penetrated the walls of wrinkles in the skin, 146 chose follicular orifices, 56 entered at the angle made by some material to the skin such as a mass of sebum, a skin scale, the curved edge of a water drop or a hair lying on the skin, 14 entered the lesion made by a previous penetrant and nine entered the tail skin of mice through the annular groove. The action of penetration follows the pattern described for penetration of the fish body. Once the site is selected, the cercaria releases its hold via the ventral sucker and attaches itself by the oral sucker so that the body assumes a vertical or oblique position in relation to the skin surface. The contents of the gland cells are extruded through the openings of ducts, and the thrusting action of the body forces the oral sucker and the droplets into the skin. The elongation of the body anterior to the ventral sucker enabled it to be thrust into narrow crevices. Droplets of secretion could be observed in the migratory tunnel during penetration. Immersion of free swimming cercariae in a suspension of Pelikan indian ink (5 parts in 95 parts saline) stimulates the cercariae to release the secretion from the pre- and postacetabular

Table 21 Exploratory and entry time for Cercariae on the skin of different hosts of *Schistosoma mansoni* ranked in order of increased entry time.

Skin	Number cercariae timed	Exploratory time in minutes		Entry time in minutes	
		Av.	Range	Av.	Range
1-2 day mouse body	56	1·8	0·3–5	2·7	0·7–11
Adult hairless mouse body	30	2·5	0·6–9	3·0	0·8–11
Adult hairless mouse ear	13	2·8	0–10	3·1	0·8–8
Adult haired (C3H) mouse tail	32	0·7	0–3	5·5	1·5–15
4–5 day mouse body	30	2·0	0–13	5·8	1·2–17
Adult monkey body	11	2·2	0·5–3	5·8	1·8–12
5 day rat body	11	4·4	1·5–14	6·9	2·0–17
Adult hairless mouse tail	19	0·8	0·2–6	6·9	2·7–12
Adult human body	21	0·4	0–1	7·2	3·1–17
Adult hamster	47	2·3	0·5–10	7·3	2·0–19

(After Stirewalt and Hackey, 1957).

Table 22 Amino acid and aminosugar composition of glandular secretions of Cercariae *Schistosoma mansoni*.

Component	% Total material	% Total amino acids	Residues	N/total material	% Total nitrogen
Nitrogen				3·73	100·0
Glycine	2·1	12·2	7	0·39	10·5
Alanine	1·1	6·4	4	0·17	4·6
Serine	0·9	5·2	3	0·12	3·2
Threonine	0·9	5·2	3	0·11	2·9
Valine	0·8	4·7	3	0·10	2·7
Leucine	0·8	4·7	3	0·09	2·4
Isoleucine	0·3	1·7	1	0·03	0·8
Phenylalanine	1·0	5·8	3	0·08	2·1
Tyrosine	0·5	2·9	2	0·04	1·1
Proline	1·6	9·3	5	0·19	5·1
Hydroxyproline	1·2	7·0	4	0·13	3·5
Aspartic acid	0·9	5·2	3	0·09	2·4
Glutamic acid	1·5	8·7	5	0·14	3·8
Lysine	0·7	4·1	2	0·13	3·5
Arginine	2·9	16·9	10	0·93	24·9
Glucosamine	4·3	—	—	0·28	7·5
Galactosamine	5·4	—	—	0·35	9·4
		Totals			
Amino acids	17·2	—	58	2·74	73·5
Amino sugars	9·7	—	—	0·63	16·9
All components	26·9	—	—	3·37	90·4

(After Stirewalt and Evans, 1960).

gland cells (Stirewalt, 1959a). Chromatographic analysis of this secretion (Stirewalt and Evans, 1960) has shown it to consist of polysaccharides in firm chemical union with proteins. The hydrolysates of these secretions contain 17·2 per cent of amino acids, 9·7 per cent amino sugars and an unidentified concentration of steroids (Table 22). Later studies (Stirewalt and Kruidenier, 1961) indicated that the pre- and postacetabular glands of *S. mansoni* cercariae could be distinguished by various histological and histochemical tests (see Table 23). These authors suggested that the postacetabular gland secretion is adhesive, lubricative and protective

Table 23 Histochemical assay of deposits of secretion from the acetabular glands of cercariae of *Schistosoma mansoni.*

Test	Reaction	Significance
PAS	+	Mucopolysaccharide, muco-or glycoprotein, glycolipid, unsat. lipid, phospholipid
after acetylation	—	1, 2-glycol groups
after KOH	+	
after saliva	+	Not glycogen
after acetone	+	Not aldehyde from glycolipids
after performic acid	—	phospholipids, or lipids with unsat. bonds
plasmal reaction	—	Not aldehydes from acetal phosphotides
bisulfite blocking	—	
Bauer	Weak	Not glycogen
Casella	—	
Feulgen's	—	Not aldehyde from desoxyribose
Iodine	—	Not glycogen
Best's carmine	+	Mucin
Mayer's mucicarmine	+	Mucus
Gormori's aldehyde fuchsin	+	Mucin, elastin, etc.
Orcein	+	Elastin
Thionin and toluidine blue metachromasia	—	Not acid mucopolysaccharide
Methyline blue binding		
>pH4	+	Muco- or glycoprotein
<pH4	—	
Ninhydrin and Millon's	—	No protein indicated*
Coupled tetrazonium	Weak	
p-Dimethylamino benzaldehyde	+	Hexosamines
Purpurine		
secretion	+	Secretion from both purpurine-positive and lithium carmine positive glands;
formed deposits	—	
Lithium carmine	+	difference in secretion as extruded and as formed deposits
Evans blue secretion	+	Change of secreted water soluble, dye-positive, carbohydrate containing protein to water-insoluble, dye-negative formed deposits
formed deposits	—	
Potassium ferrocyanide		
secretion		
formed deposits	—	
Osmium tetroxide	—	Not lipid
Sudan black B	—	

* Chromatographically, at least 14 amino acid residues have been identified in acid hydrolysates of formed deposits.

(After Stirewalt & Kruidenier, 1961).

in function while that from the preacetabular cells is primarily enzymatic (see chapter 4). The action of these secretions on host tissues has been studied in some detail by LEWERT and LEE (1954, 1956). Penetration of mammalian skin by schistosome cercariae produces profound alteration of the extra-cellular glycoprotein containing materials of the skin. The basement membrane swells in advance of the penetrating larvae, changes its staining characteristics and finally disappears at the site of penetration, water-soluble glycoprotein appears in the ground substance of the skin and an increase in free water has also been observed. *In vitro* tests have also indicated the presence of collagenase and gelatinolytic activity in the extracts of entire cercariae, and there is no doubt that this activity would be responsible for many of the histological changes described. In addition to these findings, Stirewalt (in LINCICOME, 1962, p. 219) records lysing of keratin layers, lysing and oedema of the barrier zone of the horny lamellae and dissolution of the keratogenous layers. Thus entry into the mammalian host is aided to a marked extent by the activity of these cercarial secretions (see chapter 4).

The ability of free-swimming cercariae to attach themselves and penetrate the host is very susceptible to the influence of factors in the external environment. Cercariae are relatively feeble swimmers and easily diverted and dispersed by very slight water disturbances. The effect of water velocities on the resultant worm burdens of animals exposed to *Schistosoma mansoni* cercariae has been investigated by ROWAN and GRAM (1959); and RADKE *et al* (1961). Known numbers of cercariae were released into a non-infested stream and animals were exposed to this artificially infected flowing water at various downstream stations. As the distance from the cercarial entry point increased, the number of worms maturing in the exposed animals decreased markedly, although the cercarial densities at each exposure site were similar up to a distance of 2000 ft. The authors suggested that cercarial 'fatigue', caused by the rushing waters, might have prevented these cercariae from developing in the host animal. In addition, laboratory experiments indicated that significantly fewer worms developed in animals exposed to cercariae transported at water velocities of 3·7 miles per hour. (Table 24.)

These interesting field experiments demonstrate quite clearly that water movements and velocities affect the ultimate worm burden of the exposed animals and constitute an important hazard to the parasite during this vital period of entry.

Another external feature which affects the invasive ability of the cercarial stage is the temperature of the surrounding environment. A similar relationship with respect to the miracidial

Table 24 Effect of water velocity on penetration of *S. mansoni* cercariae.

Water velocity (mph)	No. of mice	Mature worms per animal $\bar{x} \pm$ SD	% Cercariae maturing
2·0	3	(106 ± 10)	2·10
2·1	4	(84 ± 11)	1·70
2·2	5	(112 ± 27)	2·20
2·2	5	(113 ± 29)	2·30
2·3	5	(87 ± 24)	1·70
2·7	10	(43 ± 24)	0·86
3·6	5	(14 ± 4)	0·28
3·7	5	(12 ± 4)	0·24

(After RADKE *et al.*, 1961).

stage has been discussed in chapter 2. OLSON (1966) found that maximum cercarial penetration of fish occurred between 66 and 85°F and that the rate was markedly reduced between 55 and 65°C with very few penetrations occurring below 54°F. This series of experiments emphasizes further the importance of temperature to the successful completion of the digenetic cycle.

The effect of selected environmental conditions (external conditions as well as host-parasite related factors) on the penetration of *Schistosoma mansoni* cercariae into mice has been investigated by STIREWALT and FREGEAU (1965, 1968). They found that penetration into mouse tail skin did occur under extremely disadvantageous environmental conditions such as exposure in distilled water, brief contact (15 min) and temperatures as low as 7°C and as high as 45°C. Maximum penetration was however achieved under fairly defined conditions. This situation was obtained when susceptible mouse strains were exposed to 50 cercariae per mouse in natural or dechlorinated water at 27–28°C for at least one hour. The parasite-host related factors were of variable significance. Penetration into mouse tail skin was optimum under the following host-parasite conditions: (a) when the exposed snails were maintained at a constant temperature of 26–28°C; (b) when the snails had been exposed to 5 rather than 1 or 10 miracidia; (c) when the cercariae used had emerged before or after the period of low infection, which corresponded to 41 to 65 days after first cercarial emergence and (d) when the mice were exposed to cercariae within 1–2 hours after emergence from the snail. However cercarial penetration did not diminish significantly in cercariae which had been swimming after a period of up to five hours after emergence.

The hazards to which the cercariae are exposed are not over after penetration. CLEGG and SMITHERS (1968) have shown that a large proportion of the cercariae of *S. mansoni* die while penetrating the skin of the experimental host (Table 25). Cercarial mortality in this situation was greater in rats and mice than in hamsters. In mice, cercariae died within ten minutes after the start of penetration and during penetration of the Malpighian layer of the skin. The precise

Table 25 Comparison of the proportion of cercariae of *Schistosoma mansoni* which die in the skin of mice, hamsters and rats with the proportion recovered as adult worms.

Expt. no.	Animals	A Mean percentage of dead schistosomula recovered from skin	B Mean percentage of cercariae recovered as adults	C Total percentage of cercariae accounted for A + B
1	Mice	32·1 ± 9·2	41·7 ± 3·4	73·8
	Hamsters	11·5 ± 3·1	69·2 ± 11·0	80·7
2	Mice	27·5 ± 8·0	47·0 ± 12·1	74·5
	Hamsters	14·0 ± 4·6	63·5 ± 9·2	77·5
3	Mice	43·4 ± 6·6	40·0 ± 5·8	83·4
	Rats	53·9 ± 6·8	19·0 ± 3·8	72·9
4	Mice	35·7 ± 10·3	45·1 ± 10·3	80·8
	Rats	52·8 ± 3·0	21·7 ± 3·5	74·5
5	Mice (N.S.)	32·1 ± 7·7	45·3 ± 6·3	77·4
	Rats (N.S.)	42·4 ± 10·9	22·8 ± 6·2	65·2

A, Skin recoveries were made from groups of six animals each exposed to approximately 2000 cercariae. B, Adult worms were recovered from groups of six to nine animals, 7 weeks after exposure to 60–80 cercariae (mice and hamsters) or 4 weeks after exposure to 150 cercariae (rats).

N.S., Difference not statistically significant $P > 0.05$, Student's t test.

(After CLEGG & SMITHERS, 1968.)

reason for mortality at this time is not known but could be related to physical and biochemical characteristics of the ground substance and the basement membrane, and the relationship between the composition of these regions in different host species and the characteristics of the secretion from gland cells of the cercarial.

The mesocercaria

This stage, interposed between cercaria and metacercaria, occurs in the life cycle of a number of strigeoid trematodes particularly members of the genus *Alaria*. Mesocercarial stages have been described most frequently from Amphibia, and infection of the final host occurs passively following the ingestion of infected animals. The mesocercaria may be free or enclosed in a fibrous capsule of host origin. Digestion of this capsule by host enzymes results in the release of the enclosed larva into the gut of the final host.

The metacercaria

In most digenetic life-cycles, infection of the final host by the metacercarial stage involves the ingestion and digestion of the infected secondary host. Exception to this sequence occurs in mesocercaria containing strigeoid life-cycles. Thus, in the development of *Alaria arisaemoides* (PEARSON, 1956) in the fox, mesocercaria released in the digestive tract migrate to the lungs where they develop into diplostomulum type metacercariae which occur in the parenchyma of the lungs. The life-cycle becomes completed when these diplostomula migrate from the lungs into the intestine where they develop into sexually mature adults. In this way some of the hazards of transference from one host to another have been eliminated.

The majority of other metacercariae, whether encysted or not, enter the final host via its alimentary canal. Successful entry into the host in these cases depends on the ability of the larval parasite to resist the action of the host's digestive juices. Inability to survive in this new environment will immediately bring the life-cycle to a halt. Non-encysted Diplostomulum type metacercariae, if able to surmount these barriers, can continue with their localization and development. In the case of encysted metacercariae commonly found in xiphidio-, echinostome, gymnocephalous and strigeoid life-cycles, survival of the metacercariae is not enough, the release of the larva from the enclosing cystic layers is necessary. The conditions which will allow or assist excystation may be specific and in this way can play a part in producing a restricted host distribution. The factors involved in successful excystation are not clear and descriptions in the literature are very varied in their conclusions.

The earlier attempts at *in vitro* excystation of metacercarial cysts have been reviewed by MCDANIEL (1966) and DIXON (1966b). In some cases e.g. *Paragonimus westermani* metacercariae a high pH (pH 8·5–9·0) and temperature (40°C) were sufficient to induce hatching: 70 per cent hatch after three hours. It is also apparent that the metacercarial cysts of *Fasciola hepatica* will hatch without treatment with host enzymes (DAWES, 1963a; DIXON, 1964, 1966b). In other cases, excystment takes place after immersion in salines or treatment with either pepsin or trypsin containing media (e.g. STUNKARD, 1930—*Cryptocotyle lingua;* FERGUSON, 1940—*Posthodiplostomum minimum;* HUNTER and CHAIT, 1952—*Gynaecotyle adunca;* MCDANIEL, 1966—*Cryptocotyle lingua*). Some encysted forms do not respond unless treatment with trypsin solution is preceded by exposure to acid pepsin solutions e.g. *Holostephanus lühei* and *Cyathocotyle bushiensis* (ERASMUS and BENNETT, 1965) (Fig. 67A,B). Although this double treatment does not seem

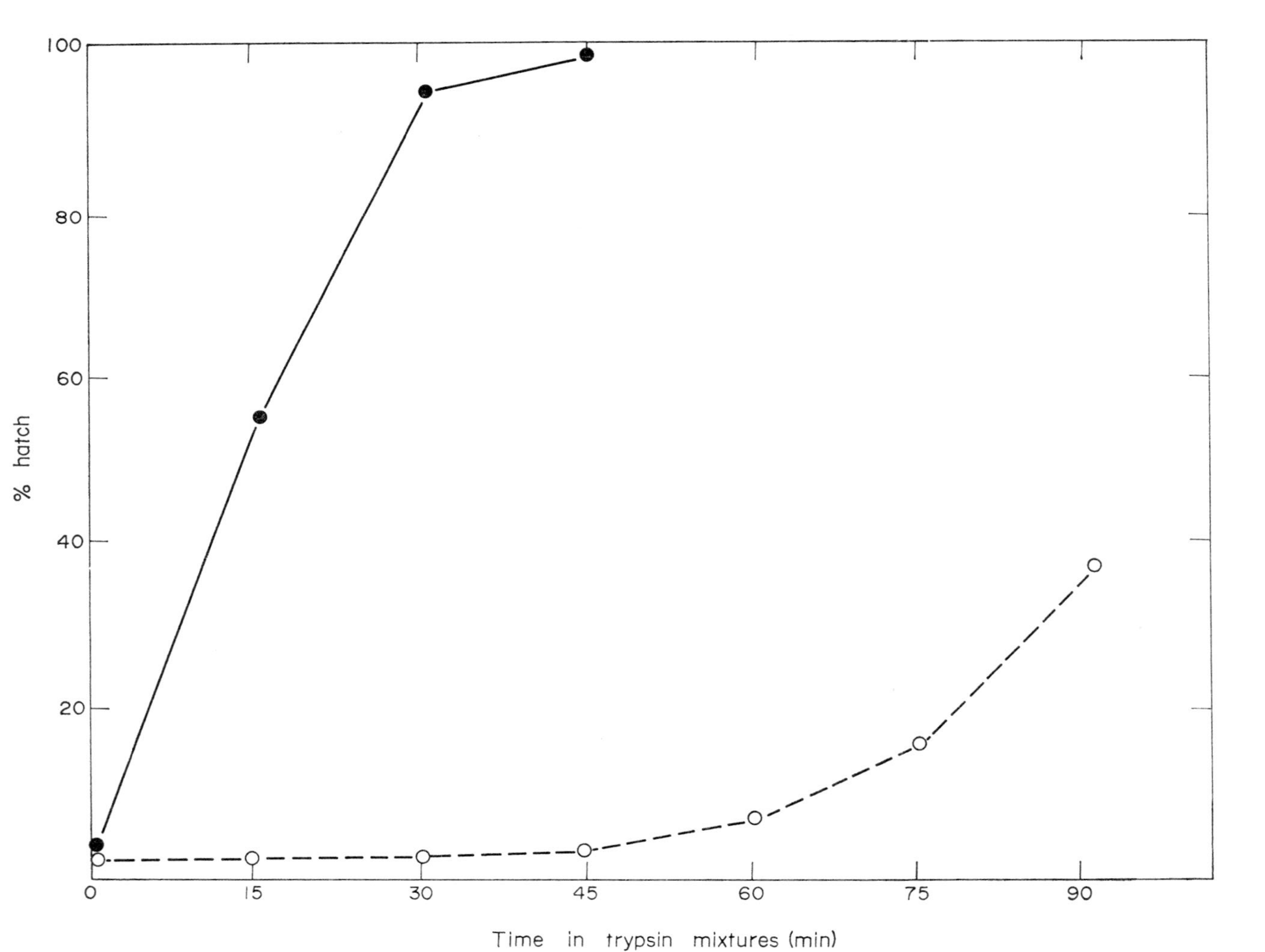

FIG. 67A The factors affecting the hatching of the metacercarial cyst of *Holostephanus lühei*. The relationship between pretreatment in pepsin-HCl mixture and hatching rate in trypsin bile salt mixture. ●——● Cysts incubated in 0·5 per cent pepsin-HCl at pH 3·1 for 60 min and then transferred to trypsin sodium tauroglycocholate mixture at 37°C. ○- - - -○ Cysts not pretreated with pepsin-HCl. Incubation in trypsin bile salt mixture as above. From ERASMUS and BENNETT, 1965.

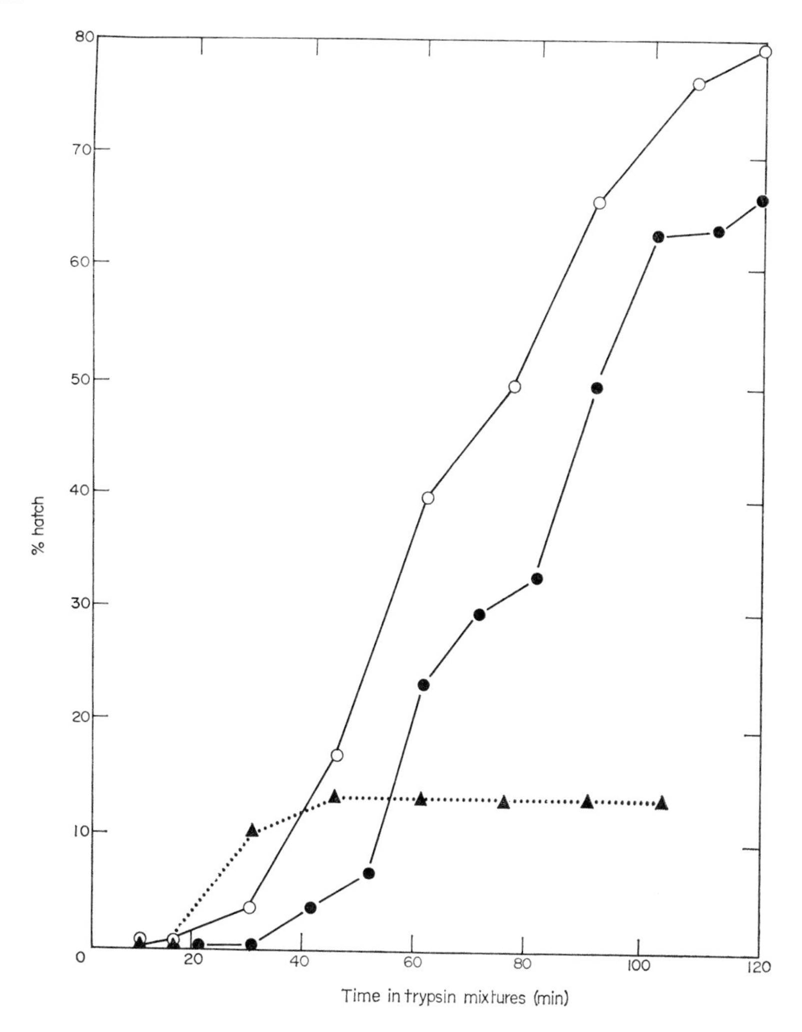

FIG. 67B The factors affecting the hatching rate of the metacercarial cysts of *Cyathocotyle bushiensis*. The relationship between excystation rate and the presence of bile salts in the incubation mixture. ○——○ Cysts pretreated for 15 min in pepsin-HCl pH 1·05 at 37°C. ●——● Trypsin + 0·5 per cent sodium tauroglycocholate pH 7·34. ▲— —▲ Trypsin only pH 7·34. From ERASMUS and BENNETT, 1965.

essential for other species it does enhance the hatching rate quite considerably (HEMENWAY, 1948; HOFFMAN, 1958a, 1959; MCDANIEL, 1966).

Other factors, which have a significant effect on the hatching process, are the presence of reducing agents and the concentration of carbon dioxide (MCDANIEL, 1966; DIXON, 1966b). In a study of the hatching of *Fasciola hepatica* cysts DIXON (1966b) considered hatching under two headings (a) activation and (b) emergence. Activation is initiated by high concentrations of carbon dioxide, reducing conditions (produced by adding sodium dithionate or cysteine to the medium) and a temperature about 39°C. (See Figs. 68, 69). Emergence, the phase following activation, is largely dependent on the presence of bile. The importance of bile in the process of excystation has been commented on by other workers (OSHIMA *et al*, 1958a,b; ERASMUS and

BENNETT, 1965; MCDANIEL, 1966), but the role of bile salts will remain uncertain until more refined preparations are used (SMYTH, 1963). DIXON (1966b) lists the following possible role of bile in hatching:

1. Increasing the effect of enzymes normally present in the gut.
2. Producing lysis of the cyst wall thus increasing the permeability of the enclosing layers.
3. By stimulating muscular movement of the enclosed metacercariae.
4. By increasing the effects of enzymes secreted by the parasite during the process of infection.

Thus after entry into the host the encysted metacercaria is exposed to (a) a sudden rise in temperature as the life-cycle usually involves a change from a poikilothermic to a homiothermic host. This abrupt change may be as vital a stimulus to the metacercaria as it is to the cestode plerocercoid (SMYTH, 1959). DAWES and HUGHES (1964) have reported a number of experiments in which *Fasciola* metacercariae have excysted and developed after direct implantation into the abdominal cavity of experimental hosts i.e. without the effects of varying pH's, digestive juices, high CO_2 content and the presence of bile. (b) exposure to the alimentary secretions from the host's stomach and duodenum; (c) the presence of bile salts with their varied

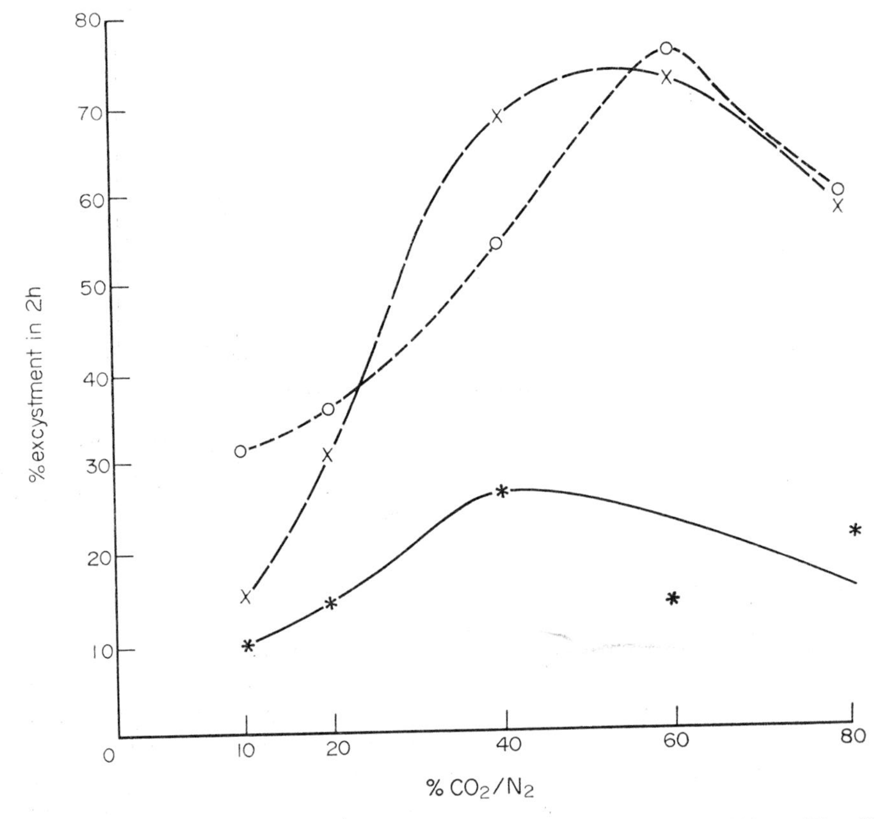

FIG. 68 The excystation of the metacercarial cysts of *Fasciola hepatica* under *in vitro* conditions. The effect of increasing carbon dioxide concentrations and of changes in the reducing agent. Unbroken line—no reductant; broken line—0·02M and 0·04M sodium dithionite; dotted line—0·02M and 0·04M cysteine. From DIXON, 1966b.

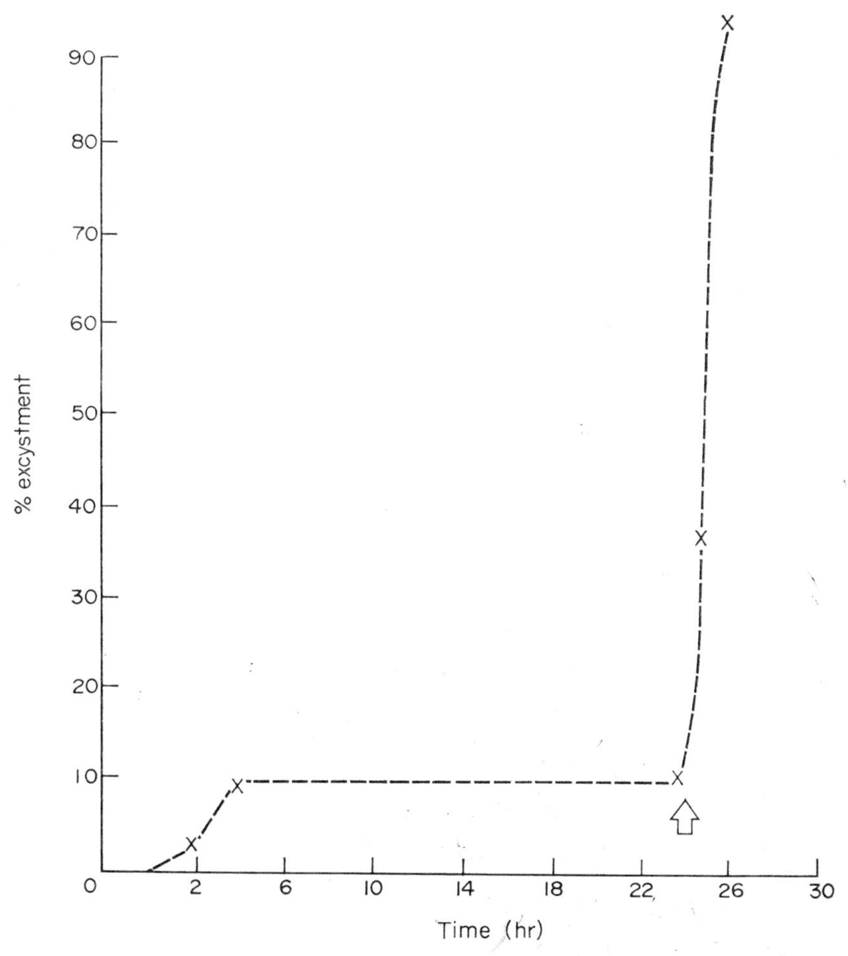

FIG. 69 The effect of bile on the hatching of *F. hepatica* metacercariae from the cyst. After treatment with the activation medium (Hanks saline plus CO_2 and reducing agent) the cysts were transferred to the emergence medium (Earle's saline slightly modified plus CO_2) and bile added after 24 hours in this medium (Arrow). From DIXON, 1966b.

biological effects which might serve to induce excystation at the appropriate place in the gut and at the most favourable moment during passage through the gut. Premature or late excystment may result in the metacercaria emerging into an unfavourable environment. (d) a change in the gaseous phase and redox potential of the environment. All these will combine and result in the hatching of the cyst with the release of the metacercaria, which might be aided also by secretion from the parasite and the mechanical activity of the suckers.

The pattern of behaviour of the metacercaria of *Fasciola hepatica* during excystation has been described by Dixon (loc cit). There is an initial phase of activity with the metacercaria rotating actively within the cyst. This is followed by a relatively quiescent stage in which the larva contracts away from the cyst wall. The final phase (initiated by sheep bile) is also an active one in which the movement is confined largely to thrusting the oral sucker against the ventral surface of the cyst. The metacercaria always emerges anterior end foremost through a small circular hole in the ventral surface. The time taken for the entire process ranged from 17 min to 1½–1¾ hours. The hole in the ventral surface of the cyst appeared to result from the digestion of the ventral plug region (neutral mucopolysaccharides). Because excystment occurs in media devoid of enzymes, Dixon suggests that this plug is digested by a substance secreted by the metacercaria.

Not all workers have been concerned with the physiological process of hatching, but simply with the development of an *in vitro* technique for hatching cysts which will permit the identification of metacercariae and the study of their morphology (see KHAW, 1935; KOMIYA and TAJIMI, 1941; FERGUSON, 1940; HOFFMAN, 1958a; ERASMUS, 1962a; ERASMUS and ÖHMAN, 1963).

7 Migration and Final Localization

Migration

One of the most remarkable phenomena associated with the parasitic mode of life is that of environmental specificity. No parasite is ubiquitous in its distribution with regard to its host animal; there is always some restriction, the extent of which varies with different parasites. This restriction or specificity may exist at two ecological levels. There is the distribution or specificity at a macro level which in effect represents the restriction of particular parasites to particular hosts. The second, is the micro-level where, within the appropriate host, the parasite occupies particular tissues, organs or fluids. It is essential in any reference to the 'host specificity' of parasites that both macro- and micro-concepts are considered. In some cases, rigid host specificity is coupled with rigid tissue specificity but in others, the two levels of specificity may vary in the extent of their determination.

In parasites with complex life histories e.g. the Digenea, the specificity may assume another dimension. Each parasitic larval stage, represents a host-parasite relationship exhibiting its own degree of specificity. Thus in a species 'A', both the host and tissue specificity may vary between the different stages of the life cycle, so that the final assessment of 'host specificity' for the species may become very complex indeed. The general assumption that in a given species host specificity refers only to the adult stage leads to gross over simplification of the concept.

The final habitat for the parasite, in terms of host and tissue specificity, represents an optimum balance established between host and parasite, after long evolutionary interaction. The site of attachment is presumably one which provides the parasite with its metabolic requirements in terms of nutrients and oxygen; is one which enables the appropriate distributive phase of the life cycle (eggs or cercariae) to emerge to the exterior and also is a site which allows the host to be exposed to a minimum of antigenic stimulation. In evolutionary terms, the establishment of such a relationship and site has probably taken considerable periods of time but in the relationship as we see it today, the attainment of the appropriate site is achieved very rapidly indeed. Once contact with the host has been obtained and entry into the body taken place, it is imperative that movement to the appropriate area of the host's body is carried out as quickly as possible. The movement through the host environment may be passive, in that the parasite is carried by respiratory currents or by the flow of the blood stream, or by the gut contents which are activated by peristalsis or ciliary movement of host structure. More active localization is present when the parasite itself migrates through the host tissues and fluids employing ciliary activity or the leech-like mode of progression involving oral and ventral suckers. The passage through host tissues is a hazardous one, in which the parasite is exposed to adverse factors such as unfavourable pH, digestive juices, peristalsis and also attack by macrophages and antibodies in a host sensitized by previous infection. Speed is also essential in that energy for migration is often derived from endogenous carbohydrate stores established in the previous host, and depletion of these stores before the completion of migration may be fatal to the parasite.

The precise location of a particular stage is often the result of the activity of the previous phase in the life cycle. For example the distribution of metacercariae in a host represents the culmination of behaviour patterns exhibited by the cercaria. Because of this temporal continuity between stages a discussion of migration and localization will be considered under five group headings.

(1) Oncomiracidium—Adult
(2) Miracidium—Sporocyst—Redia
(3) Sporocyst or Redia—Cercaria
(4) Cercaria—Metacercaria
(5) Mesocercaria—Metacercaria—Adult

ONCOMIRACIDIUM—ADULT

Adult monogenetic trematodes occur mainly on sharks, skates, holocephalans, bony fishes, amphibians and reptiles but records exist of their presence on parasitic isopods, cephalopods and aquatic mammals. Within a particular range of hosts, members of the Monogenea exhibit a strong degree of host specificity (LLEWELLYN, 1957c). HARGIS (1959) considering this host specificity has introduced two terms: *infraspecificity* representing the occurrence of a single monogeneid species on members of a single fish taxon (species, genus etc): and *supraspecificity* which is the restriction of a natural group of monogeneid species to a natural grouping of fish species. Examples of infraspecificity occurring at a species level (species specificity) are provided by the data of LLEWELLYN (1956a) and HARGIS (1959). Llewellyn, in an examination of over 2000 fishes belonging to seventeen species from both freshwater and marine habitats, found the gills to be infested with over 900 (representing 18 species) diclidophoroidean trematodes. All the parasites were found to be specific to particular host species with the exception of *Plectanocotyle gurnardi* which occurred on any one of three species of *Trigla*. (Table 26).

Hargis examined 415 fish (49 species) from the Gulf of Mexico, and collected 3335 individual parasites ranging over 75 monogeneid species. This study revealed that 68 species or 89 per cent of the collection were strictly species specific. Only eight species parasitized more than one host and of these, seven species were genus specific with regard to their hosts. BYCHOWSKY (1957) reviewing the known species, also emphasizes the high adaptation of the species of the monogenetic trematodes to the genera and species of their fish hosts. Out of 958 known species of monogenea 806 occurred on species of a single genus of fish. Within the same genus, 806 monogenean species occurred on a single fish species; 78 parasitized 2 species of the same genus, 13 on three species and only 4 species of monogenea were distributed over more than three species of fish from the same genus. (Table 27). These few selected references indicate quite clearly that specificity on the macro scale is a characteristic feature of the Monogenea.

Within their hosts, monogeneans generally occur on the fins, skin, gills, gill chamber and buccal cavity. Some are more unusual in their habitat and may be found on the eyeball of mammals (THURSTON and LAWS, 1965) in the cloacal cavity (e.g. *Calicotyle kroyeri* on *Raia clavata*), in the ureters (*Acolpenteron ureteroecetes* in *Micropterus dolomieu*), in the oviducts (*Calicotyle inermis* in *Pristiophorus cirratus*), *Enterogyrus cichlidarum* in the intestine of a fish and in the coelome (*Dictyocotyle cocliaca* in *Raia naevus*). One of the best known of all the monogeneans, *Polystoma integerrimum*, lives in the bladder of *Rana temporaria* and there is one monogenean which lives in the blood system of the host (*Amphibdella torpedinis* in *Torpedo ocellata* and *T. marmorata*). These examples appear to exhibit fairly precise microspecificity compared with those species living on the gill sand within the gill chamber. The work of LLEWELLYN (1956a, 57b,c; LLEWELLYN and OWEN, 1960) however, has shown that within the gill chamber and on the gill lamellae, particular

Table 26 Degree of infestation of hosts, and distribution of parasites on gill arches of fishes examined at Plymouth 1953–55.

Category of host sample	Host	No. of host specimens		Infestation %	Parasite	Total numbers of parasites per gill arch			
		Examined	Infected			1	2	3	4
A	*Gadus merlangus*	507	44	8·7	*Diclidophora merlangi*	53	8	1	4
	G. luscus	509	108	21·0	*D. Luscae*	7	118	91	8
	Merluccius merluccius	500	38	7·6	*Acanthocotyle merluccii*	4	28	7	5
B	*Trigla calculus*	20	19	95·0	*Plectanocotyle gurnadi*	6	22	39	25
	Trachurus trachurus	37	23	62·2	*Gastrocotyle trachuri*	41	56	46	20
			8	21·6	*Pseudaxine trachuri*	8	9	5	2
			1	2·7	*Microcotylid species*	Not known			
C	*Belone belone*	18	14	77·8	*Axine belones*	19	29	48	22
	Scomber scombrus	8	8	100·0	*Octostoma scombri*	7	15	1	0
D	*Pagellus centrodontus*	47	1	2·1	*Cyclocotyla chrysophryi*	0	0	1	0
					Microcotyle labracis	3	0	0	0
	Morone labrax	3	2	66·7	*Diclidophora pollachii*	Not known			
	Gadus pollachius	15	1	6·7	*Mazocraes alosae*	Not known			
	Alosa fallax	2	1	50·0					

A. hosts readily available, searched macroscopically.
B. hosts readily searched microscopically.
C. hosts less readily available, searched microscopically.
D. hosts less readily available, searched macroscopically.
(After LLEWELLYN, 1956a.)

species of monogenetic trematode may adopt quite specific positions (Table 26). He found that some species of trematode tended to occur on particular gill arches (e.g. *Diclidophora merlangi* occurred most frequently on the first gill arch of *Gadus merlangus* and *D. luscae* was most often located on the second and third gill arches of *Gadus luscus*), whereas others (*Gastrocotyle trachuri* or *Trachurus trachurus*) were frequently present on arches 1, 2, 3 but less frequently on the fourth arch. In addition to this fairly specific distribution between the gill arches, there was usually a characteristic method of attachment in relation to the appropriate gill arch. All the parasites were attached with their posterior ends directed towards the proximal ends of the arch and the mouth pointing towards the distal end and lying downstream relative to the gill ventilating current (Fig. 70). The precise method and position of attachment varied with the parasite species. In some (e.g. *Plectanocotyle gurnadi*, *Discocotyle sagitta*) the parasite was attached to one or the other side of a primary lamella and never to both sides. The longitudinal axis of the parasite always lies parallel to the long axis of the primary lamella, and the attachment organ of trematodes fixed in this way is usually sessile or borne on very short peduncles. A contrast in the method of attachment is shown by *Dicliophora luscae* (Plate 4-3). The opisthaptor here consists of pairs of adhesive organs which are applied to opposite sides of the same primary lamella. The opisthaptor also bears well developed peduncles which allows the parasite to span

Table 27 Occurrence of species of Monogenetic trematodes on fishes.

Number of species of Monogenoidea	On 1 species of fishes	On 2 species of fishes	On 3 species of fishes	On more than 3 species of fishes	Total
On species of 1 genus of fishes	711	78	13	4	806
On species of 2 genera of 1 family of fishes		49	15	9	73
On species of 3 genera of 1 family			8	4	12
On species of more than 3 genera of 1 family				12	12
On species of 2 genera of different families of 1 order		13	3	3	19
On species of 3 genera of different families of 1 order			8	5	13
On fishes of more than 3 genera of different families of 1 order of fishes				5	5
On species of 2 genera of different families and orders of fishes		6	1	1	8
On species of 3 genera of different families and orders of fishes			5		5
On species of more than 3 genera of different families and orders of fishes				5	5

(After Bychowsky, 1957).

the distance between the dorsal and ventral surfaces of the primary lamella so that members of a pair of adhesive organs may be attached to the opposing surfaces of the lamella. The opisthaptors are further modified in that the adhesive organs bear clamps which are used for the grasping of gill lamellae. The form of the opisthaptor with its attachment organs must be correlated with the nature of the substratum and in trematodes attached to large smooth surfaces e.g. *Polystoma* in the frog bladder, the attachment organs are acetabular in form (Plate 4-2).

It is now clear that the Monogenea exhibit specificity not only in the choice of hosts but also in their precise location on that host. The attainment of this location, the nature of which is reflected in opisthaptor structure, (Plates 4-2 to 4-5) must be the result of a complex series of behaviour patterns. Unfortunately there is hardly any experimental evidence available to enlighten us as to how localization may take place. Bychowsky (1957) refers to two contrasting patterns of behaviour exhibited by oncomiracidia after hatching. Freshly emerged larvae swim actively in straight lines and exhibit a marked positive phototaxis. This phase may last for 2–12 minutes and during this period the armature of the opisthaptor may be internal and the hooks incapable of movement. This phase is obviously important in dissemination and migration from the site of egg deposition. In the second phase the positive phototaxis is reduced and may change to a negative phototaxis. The larva frequently alters its direction while swimming, 'bumping' up against objects and 'examining' them with its anterior end. The hooks of the

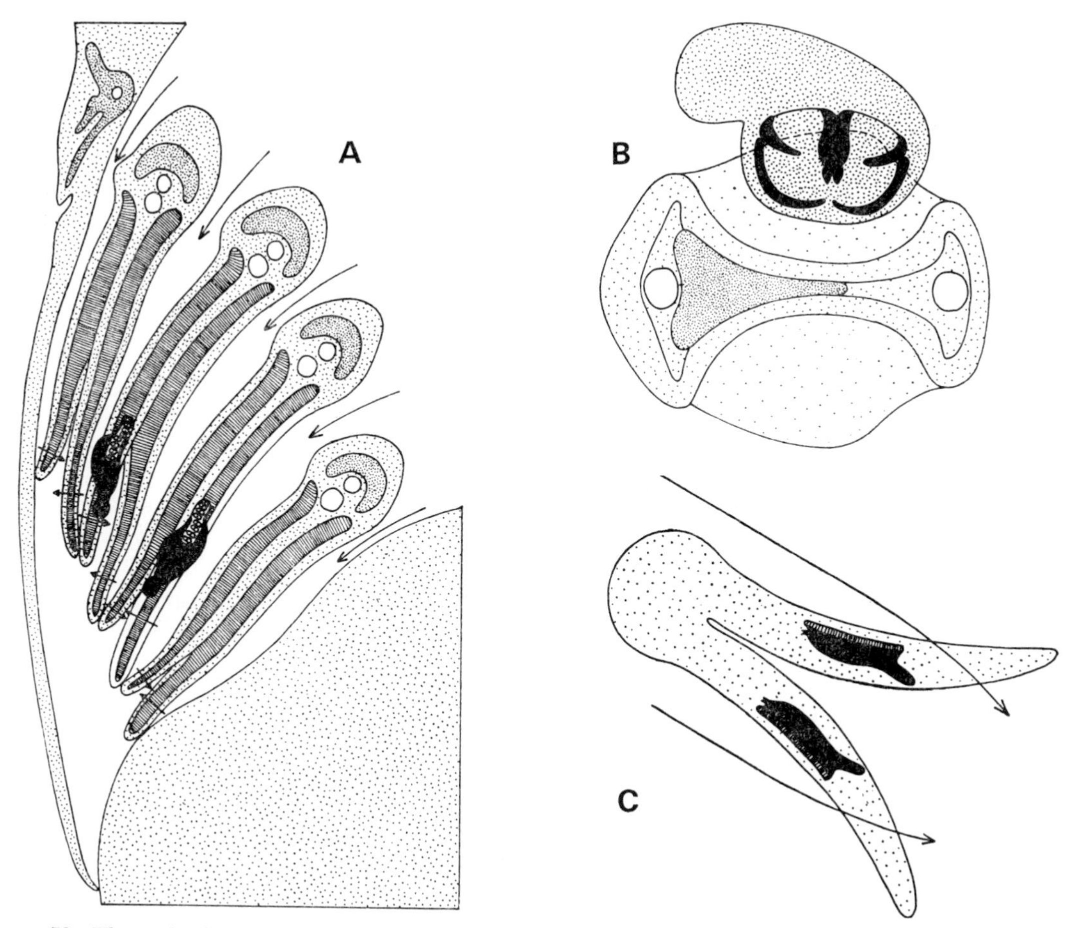

FIG. 70 The mode of attachment of some monogeneans to the gills of their hosts. **A** The typical sites of attachment of *Discocotyle sagittata* to the gills of *Salmo trutta* as indicated by a horizontal section through one side of the branchial region of the host. **B** The relative sizes of the clamp of *Discocotyle sagittata* and the secondary gill lamella of *Salmo trutta*. Figures **A** and **B** from LLEWELLYN and OWEN, 1960. **C** A diagrammatic indication of the possible origin of asymmetry in *Gastrocotyle trachuri* on *Trachurus trachurus*. The parasite on the upper hemibranch has clamps on its left side only; that on the lower hemibranch has the clamps on its right side. Figure **C** from LLEWELLYN, 1957c.

opisthaptor protrude in this phase and can be moved. This period is one of attachment and can persist for 24 hours before the death of the larva occurs.

It is generally assumed that entry into the gill chamber takes place on the respiratory current with the larva attaching itself as it passes over the gill. LLEWELLYN's (1957c) observations of the entry of the larva of *Polystoma* into the frog tadpole suggest that monogenean larvae may enter the gill chamber by entering at the operculum, swimming close to the inner surface of the operculum and then attaching itself to the nearest gill lamella. Generally cilia are shed after the attachment of the larva.

The presence of adult monogenea in the bladder cloaca, ureters and intestine (PAPERNA, 1963) necessitates a much more complex series of migrations and information is only available on the well known *Polystoma intergerrimum* where there is a remarkable correlation between the development of this larva and that of the frog. At one time it was thought that the larvae eventually migrate down the length of the gut to the cloaca and then enter the bladder. The observations of Combes (see LLEWELLYN, 1968) now indicate that the migration from gills to cloaca occurs over the external surface of tadpoles at night.

The relationship between parasite development and host life-cycle has been discussed by GORSHKOV (1964) who concluded that the former was independent of the hormonal activity of the host. In the blood inhabiting fluke *Amphibdella torpedinis* perforation of the blood vessel wall followed by entry into the blood system must take place, but no experimental observations are available.

An unusual type of life-cycle is seen in the viviparous *Gyrodactylidae.* Here there is no free swimming larva and infection of new hosts takes place by the transference from one adult to another. In *Gyrodactylus* the uterus contains a young worm which itself contains an embryo in it's uterus. Thus, instead of egg deposition, we have the release of an adult with its uterus containing a young individual. In some cases three individuals may be telescoped within the uterus of the 'mother'. In this way the uncertainties of a ciliated larval stage in the life-cycle are overcome.

Once the site, optimal for the survival of a parasite has been reached, it is essential that this position is maintained. The Monogenea possess a variety of efficient attachment structures. They may be considered under two categories (1) the anterior attachment structures (2) the posterior structures comprising the opisthaptor. Anterior attachment structures generally consist of gland cells alone, or gland cells associated with muscular cup-like depressions e.g. *Acanthocotyle, Gyrodactylus* (Fig. 71). In some, e.g. *Polystoma* a well developed oral sucker may be present. The posterior attachment disc, or opisthaptor, may bear simple acetabula-like suckers or clamps, as well as hooks of varying size (Fig. 72). The clamps represent a pincer-like arrangement of sclerites superimposed on a sucker-like base. Attachment is by pinching the tissue between two halves of the clamp and is not a suction mechanism. Movement of the clamps is produced by a complex series of muscles and a detailed description of their functioning has been given by LLEWELLYN (56a, 56b, 57a, 57b, 58, 60) and LLEWELLYN and OWEN (1960). The necessity to reduce the resistance offered by the parasite body to the ventilating current passing over the gills, may often be related to the appearance of asymmetry in the body (Fig. 70). Such asymmetry often results in the longitudinal axis of the body coming to lie parallel with the ventilating current and not across it (LLEWELLYN 56a, 62b).

Although belonging to a different order, members of the Aspidogastrea exhibit fairly simple life-cycles not involving a multiplicity of hosts and it is convenient to consider them at this point. The egg generally hatches to release a non-ciliated, fluke-like larva with suckers. (WILLIAMS, 1942—*Aspidogastrea conchicola*). The larva enters another bivalve host, via the siphon, and after entering the renal apertures reaches the pericardium via the kidney. In other cases, (e.g. *Lophotaspis*), the larva initially lives in a marine snail and reaches maturity after this is ingested by a turtle (WHARTON, 1939). Unfortunately, little is known concerning the biology of this group. The adult trematodes live in the mantle, pericardial and renal cavities of molluscs, and in the bile ducts and alimentary canal of both freshwater and marine fish. No significant comment can be made concerning their specificity nor the factors involved in their final localization.

Attachment to the host is achieved by a complex adhesive apparatus which covers most of the ventral surface. In the family Aspidogastridae the apparatus consists of a large circular or elongated sucker divided into alveoli by septa. The adhesive organ in the Stichocotylidae consists of a row of separate alveoli along the length of the body. Information, concerning attachment, comparable with that on the Monogenea, is not available.

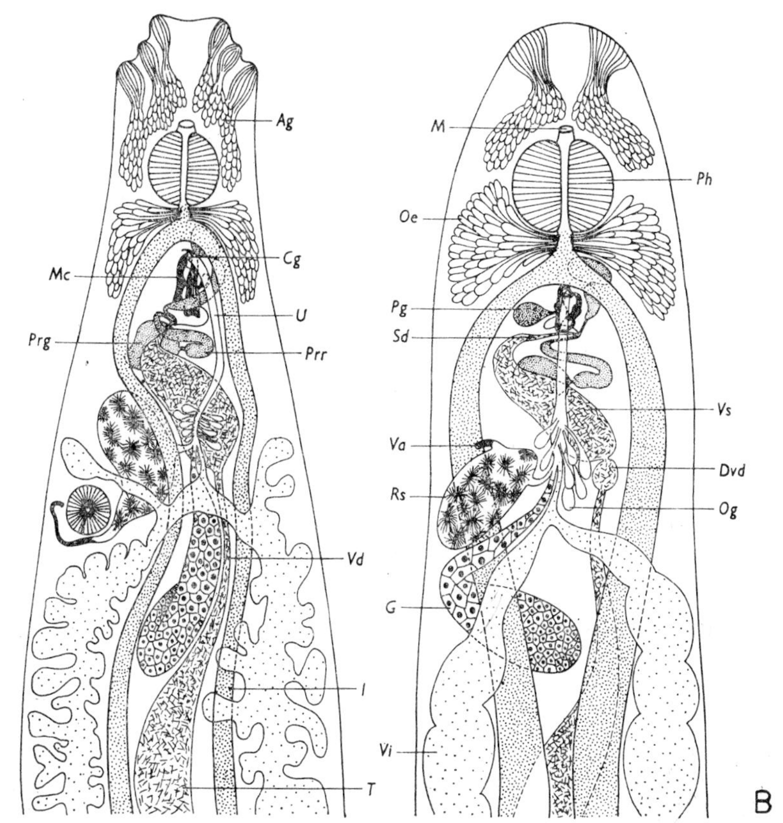

FIG. 71 The anterior end and genitalia of *Amphibdelloides maccallumi* (A) and *Amphibdella flavolineata* (B). Note the complex of glands associated with the anterior end and the pharynx. (Ag: anterior gland; Cg: common genital opening; Dvd: dilation of vas deferens; I: intestine; G: germarium; M: mouth; Mc: male copulatory apparatus; Oe: oesophageal glands; Og: ootype glands; Pg: penis gland; Ph: pharynx; Prg: 'prostate gland'; Prr: 'prostate' reservoir; Rs: receptaculum seminis; Sd: sperm duct; U: uterus; Va: vagina; Vd: vas deferens; Vi: vitellarium; Vs: vesicula seminalis; T: testis). From LLEWELLYN, 1960.

MIRACIDIUM—SPOROCYST—REDIA

The digenetic life-cycle begins with the hatching of the egg and the release of the miracidium. As mentioned in the previous chapter, emergence of the miracidium may occur in the surrounding aquatic environment or within the gut of the mollusc.

The location of the primary generation of sporocysts—the mother sporocysts—will be related to some extent to the site of entry of the miracidium. In those examples in which gut penetration takes place, the mother sporocysts are found in the immediate vicinity of the gut and may be attached to the gut wall (e.g. *Telorchis bonnerensis*, *Glypthelmins quieta*, *Haplometrana intestinalis*). The miracidium generally comes to lie between the basement membrane and the epithelium and this is where the mother sporocyst develops. In these cases the secondary or daughter sporocysts which develop from the mother sporocysts remain in the same site. This condition is not universal, and many examples exist where the mother sporocysts occur in one site and the secondary sporocysts in another (Fig. 73). A study of the development of *Cercaria X* (Strigeoidea: Diplostomatidae) in *Lymnaea stagnalis*, shows that the mother sporocyst occurs entangled around the buccal mass and protruding into the body cavity. After several days, the body cavity contains numerous small daughter sporocysts and these eventually reach the

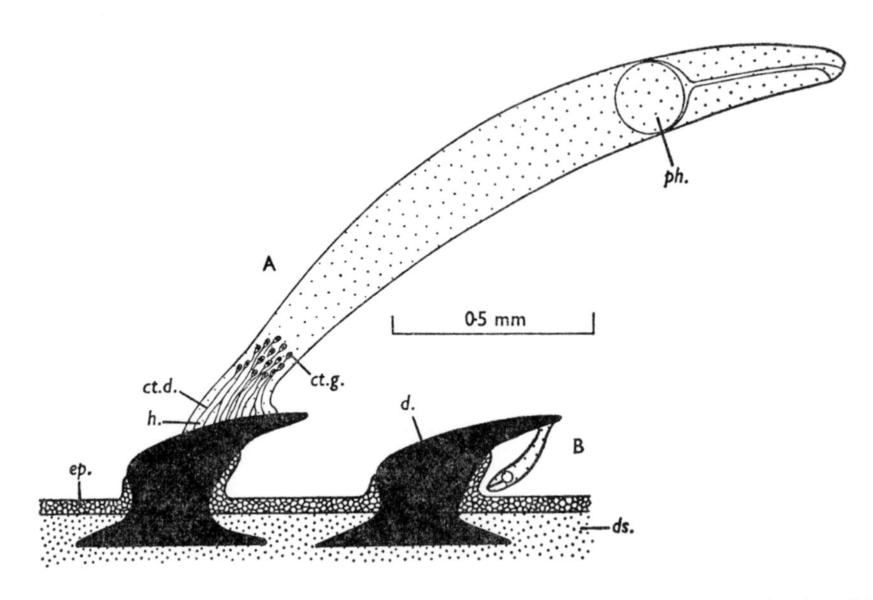

FIG. 72 The attachment of an adult (A) and a postlarval (B) specimen of *Leptocotyle minor* to the skin of the dogfish *Scyliorhinus canicula.* (d: dogfish dentile; ds: host dermis; ep: host epidermis; ct.d.: cement gland duct; ct.g.: cement glands; H: haptor; Ph.: pharynx.). From KEARN, 1965.

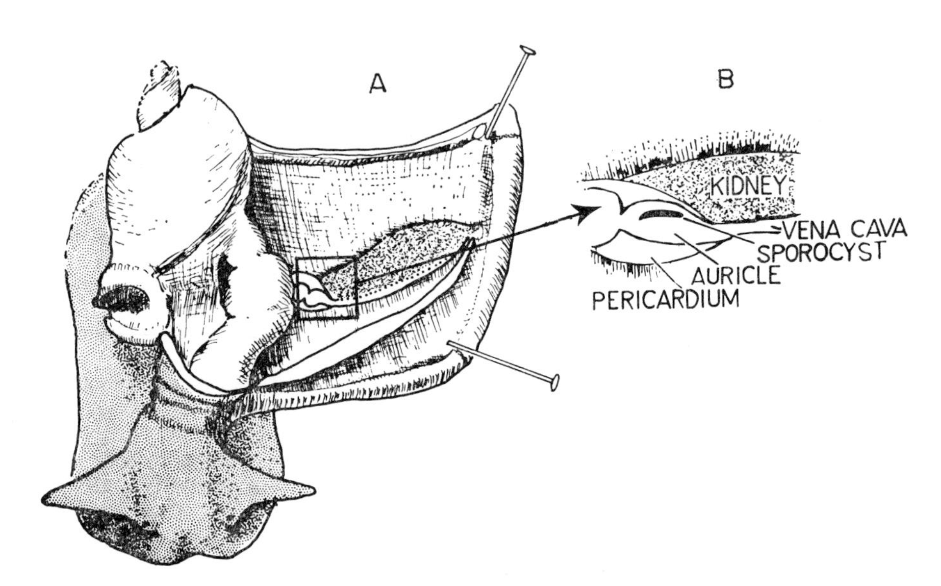

FIG. 73 The location of the mother sporocyst of *Trichobilharzia brevis* in the heart of *Lymnaea rubiginosa.* A: partial dissection of the snail to show the position of the pericardium; B: detail of pericardial region showing the location of the mother sporocyst. From BASCH, 1966.

digestive diverticulum and localize in the interlobular connective tissue. PEARSON (1956), describing the development of sporocysts of *Alaria arisaemoidis* and *A. canis*, notes the presence of mother sporocysts in the perioesophageal blood sinus in the case of *A. arisaemoides*, and in the renal and collar veins in *A. canis*. The daughter sporocysts of both species appeared in the haemocoele enclosing the digestive gland.

The time scale of the development in *A. arisaemoides* is as follows. Sixteen hours after penetration the first stage, found in the mantle, was 100 μ in length. Twenty seven hours later, a mother sporocyst 280 μ long was present in the perioesophageal blood sinus. At eighteen days after infection, daughter sporocysts were present in the haemocoele surrounding the digestive gland. In the life cycle of *Orientobilharzia dattai* (DUTT and STRIVASTAVA, 1962a) mother sporocysts were most frequently found in the mantle lobe of *Lymnaea luteola*. After seven days at 33°C, daughter sporocysts were in the digestive gland. This migration of sporocysts is often overlooked and many descriptions of cercariae contain only an account of daughter sporocysts found in the digestive gland but do not include data on the mother sporocysts.

The ability of intramolluscan stages to migrate is usually regarded as a characteristic of the redial stage. During the development of *Paragonimus kellicotti*, AMEEL (1934) and AMEEL *et al* (1951) found mother sporocysts free in the lymphatic system in almost all parts of the snail. The first and second generation of rediae were also widely distributed. A different distribution of sporocysts from that of the rediae, involving a migration of the latter stage, was described by DINNICK and DINNICK (1957) in the development of *Paramphistomum sukari* in the snail *Biomphalaria pfeifferi*. The mother sporocysts were found generally in the mantle or localized in the tissues of the pulmonary cavity. The rediae however were observed to leave the pulmonary cavity, and migrate through the snail to the digestive gland whence they were usually recovered. NAJARIAN (1953) records the sporocyst stage of *Echinoparyphium flexum* from the heart of *Lymnaea palustris*. The rediae, however, migrate from this site, apparently via the circulatory system to the digestive gland and ovotestis. These mother rediae, after reaching the digestive gland, produce large numbers of daughter rediae.

Thus within the body of the molluscan host, the larval stages may undergo migration from one region or tissue to another. The reasons for this are not clear. It may represent the movements of parasites to regions where a greater supply or more suitable nutrients are available. It is possible that the different stages such as sporocyst and redia may have differing physiological requirements which are provided by different regions of the mollusc.

SPOROCYST OR REDIA—CERCARIA

The last generation of sporocysts or rediae give rise to the cercarial stage which usually has to leave the molluscan host to continue the development of the life-cycle. The cercarial stage therefore has to undertake a migration from the sporocyst or redia, in which it develops, to the site of emergence from the mollusc. Surprisingly this aspect of this phase of the life cycle has received little attention and relatively few species have been studied. The majority of observations have been concerned with various species of *Schistosoma* and with *Fasciola hepatica*.

KENDALL and MCCULLOUGH (1951) describing the factors affecting emergence of cercaria from *L. truncatula* observed that in an actively moving snail, the cercariae were also active within the perivisceral space and that they eventually accumulated in the space surrounding the distal part of the gut. The actual emergence seemed to be a relatively passive process, and the passage through the body wall to the exterior was aided by the muscular contractions of the mantle and pneumostome of the mollusc. No details were given concerning the route of migration from the digestive gland to these anterior perivisceral spaces. DUKE (1952) gave a more detailed account of the internal migrations of cercariae from the sporocysts in the digestive gland of *S. mansoni* to the final site of emergence (Fig. 74). The daughter sporocysts lie in the connective tissue

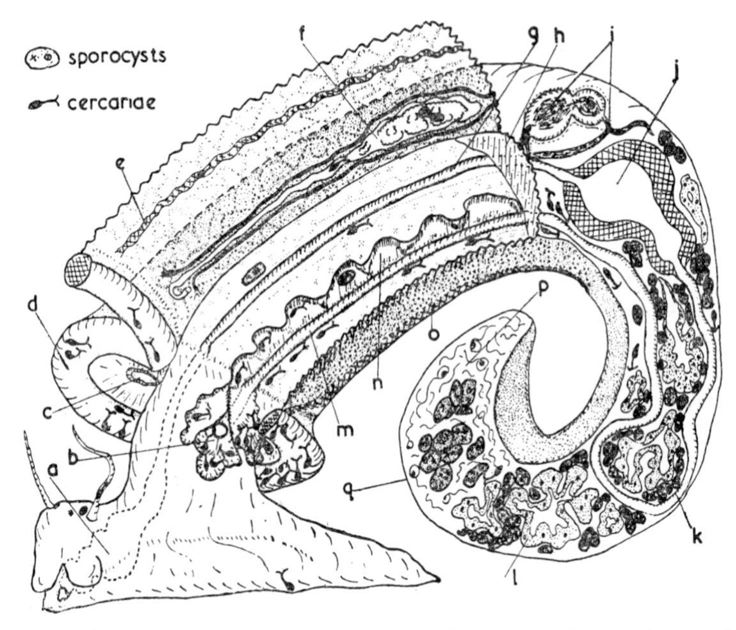

FIG. 74 A diagrammatic representation of the internal anatomy of *Australorbis glabratus* showing the sites of sporocyst development of *Schistosoma mansoni*; the migratory route of the cercariae and the points at which they emerge to the exterior of the snail. (a: buccal mass; b: pseudobranch with anus aperture; c: pneumostome; d: mantle collar; e: ridge on mantle roof; f: kidney and adjacent blood vessels; g: oesophagus; h: posterior wall of mantle cavity; i: heart with aorta; j: stomach; k: intestine; l: liver lobule; m: rectum; n: rectal ridge; o: retractor muscle; p: ovotestis; q: tunica propria). From DUKE, 1952.

between the digestive gland tubules. The cercariae emerge from the sporocysts, penetrate and emerge through the tunica so that free cercariae were present in the vascular space around the liver. Duke suggested that the main path of migration was via the venous system following the flow of blood from the visceral mass, alongside the rectum to the pseudobranch and the collar and found large numbers of cercariae in the peri-rectal spaces with the maximum concentration in the posterior and middle portions of the pseudobranch and the adjacent parts of the collar. The cercariae lay coiled up in small spaces below the collar and pseudobranch epithelium, which was eventually displaced to form a 'bleb' protruding to the exterior (Plate 7-3). Eventually the bleb ruptured releasing the cercariae to the exterior. The cercariae emerged entangled in mucus and the aperture of the 'bleb' quickly contracted sealing the wound (Fig. 75). Much of the evidence from this study suggests that the migration and emergence was an active process. These observations were confirmed later by RICHARDS (1961). A study of the non-schistosome cercaria has been made by PROBERT and ERASMUS (1965). In this case the internal migration and emergence of *Cercaria X* (Strigeoidea: Diplostomatidae) from *Lymnaea stagnalis* was followed. The snails were serially sectioned and counts made of the number of cercariae present in different tissues. Evidence from this material suggested that the migration of cercaria from the digestive

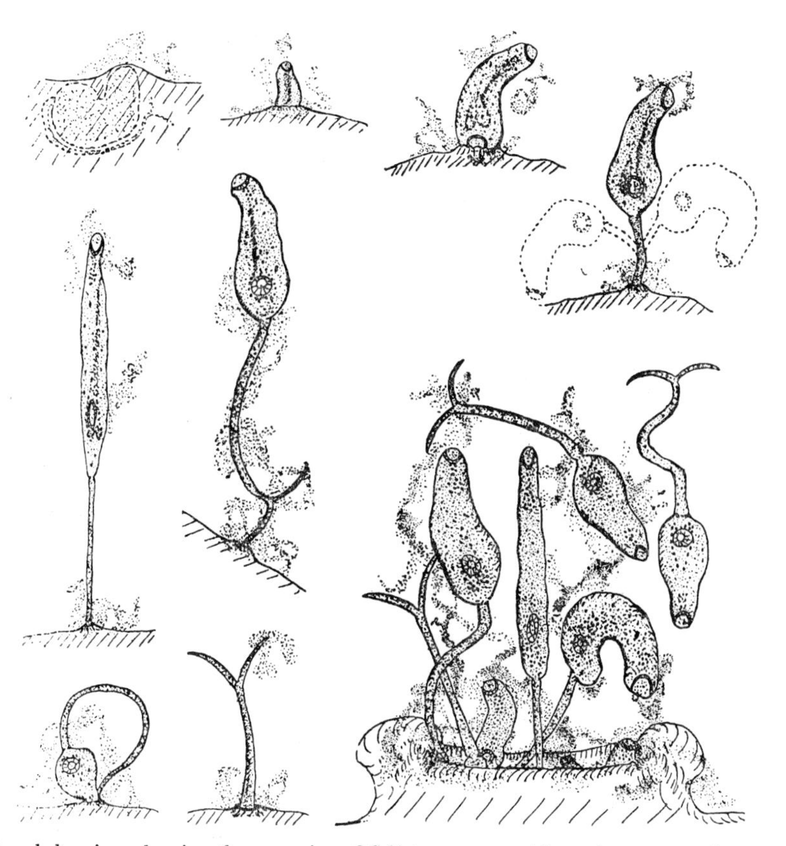

FIG. 75 Free-hand drawings showing the cercariae of *Schistosoma mansoni* in various stages of emergence from the mollusc. Note that in most cases the cercaria emerge body first and that they are covered in a thin film of mucus. From DUKE, 1952.

gland to the site of emergence is via the main venous system, particularly the rectal sinus with the final accumulation of cercariae in the blood vessels of the mantle walls (Plate 7-3). Within the mantle walls, cercariae break through the blood vessels and migrate through the tissues. In heavy infections, many cercariae move along the same route to the sub-epidermal tissues and it seems likely that once the tissues have been broken down, successive batches of cercariae would utilize the same route to the site of emergence. In this study, most observations suggested that the cercariae emerge through the outer surface of the mantle. The migration is an active one aided by the thrusting action of the spined oral sucker. The cercariae emerge, anterior end first, through the epithelium by pushing aside the epithelial cells. When large numbers of cercariae collect below the epithelium, the surface bulges in a similar manner to that described by DUKE (1952) for *S. mansoni*. The final emergence through the epithelium occurs rapidly and once the oral sucker has perforated the epithelium, the rest of the body follows in about one minute. Progress out through the punctured epithelium is aided by vigorous lateral movements of the body and also by the emerged body fixing itself firmly to the outer face of the epithelium and then pulling the tail out (Plate 7-4).

This general pattern of migration is also exhibited by the cercaria of *Neodiplostomum intermedium* in the fresh water limpet *Pettancylus assimilis* (PEARSON, 1961). Cercariae were again found in the visceral haemocoele, perirectal sinus and the mantle vessels. Within these vessels cercariae are passively carried, crawl and swim until they break through into the tissues and hence to the exterior.

Thus these more recent studies suggest very strongly that a complex migration occurs within the snail from the site of emergence from the sporocyst to the region where escape from the molluscan host takes place. In a previous chapter it has been indicated that emergence from the snail often takes place under specific environmental conditions e.g. temperature, light intensity. In some way these stimuli external to the snail become communicated to the cercariae within, resulting in their increased activity and the adoption of the behaviour patterns culminating in emergence. The effect of increasing water temperatures on cercarial emergence is fairly easy to understand, in that temperature changes may alter the metabolic state of the cercariae from a resting one to one of great activity. Similarly, the sudden presentation of food to a snail which has not fed for sometime will increase the concentration of metabolites in the circulatory system of the snail and this in turn will be associated with a corresponding increase in the endogeneous food stores of the cercarial stages. The accumulation of food resources in this way provides a source of energy necessary for cercarial activity and migration. More difficult to understand is the situation where well fed infected snails, under conditions of constant temperature, exhibit release of cercariae at times of the day corresponding to a certain light intensity. In some cases the explanation may lie in an enhanced activity of the snail at these light intensities affecting the cercarial stages within, but in other cases, this explanation is insufficient.

CERCARIA—METACERCARIA

Many metacercarial stages, particularly those which become encysted, rely on the activity of the cercarial body after penetration for their final localization. There is some difficulty in deciding the exact nature of the stage which undertakes the migration. The tail of the cercaria is cast off during penetration and consequently the active organism left embedded in the host tissues is not strictly speaking a cercarial stage. On the other hand, most of the features of the cercarial body, as it was prior to penetration, are retained, and any differences generally relate to the appearance and state of the gland cells and possibly the complexity of the excretory system. In contrast non-morphological changes detectable after penetration of the final host have been recorded from the cercariae of *Schistosoma mansoni*. These include a change in water sensitivity and in the ability to produce a cercarienhuller reaction. In the Strigeoidea, once the final site has been reached great morphological changes take place i.e. appearance of adhesive organ, anterior lappets, reserve bladder excretory system (ERASMUS, 1958) so that the fully differentiated metacercaria is clearly very different from the original migrating cercarial body. The life cycle of many xiphidiocercariae and echinostome cercariae show that once migration is complete, the cyst wall is formed and it is after this that changes take place culminating in formation of the true metacercarial stage. Thus it seems reasonable to assume that it is the cercarial body which is active in the phase from entry up to the termination of migration and the achievement of the final site. On the other hand, students of development in the Schistosomatidae use the specific term of schistosomule for this migrating cercarial body.

Not all the metacercarial stages are parasitic and some (e.g. *F. hepatica*; *F. buski*; *Paramphistomum cervi*) occur on vegetation. In these, there is no complex migration and final localization is dependent on the factors influencing the swimming behaviour of the cercarial stage. This has been discussed in chapter 4. In the case of monostome cercariae belonging to the genus

Notocotylus the cercariae may encyst on vegetation but more usually on the outer surface of shells of molluscs. These cercariae encyst very rapidly and seem to show very little specificity with respect to the nature of the substratum, although some species do show marked orientation to light. Similar superficial distribution of metacercariae is exhibited by the occurrence of cercariae of *Opechona bacillaris* on coelenterates, clinging to the manubrium of *Obelia* in both polyp and medusoid stages and also on a range of other hydroids. This marine cercaria has also been described from the Chaetognatha, e.g. *Sagitta* (LEBOUR, 1916, 1917).

Molluscs, leeches and insects are often involved in the life cycle as secondary intermediate hosts (Fig. 76). Generally the distribution within these stages is non-specific and is a result of

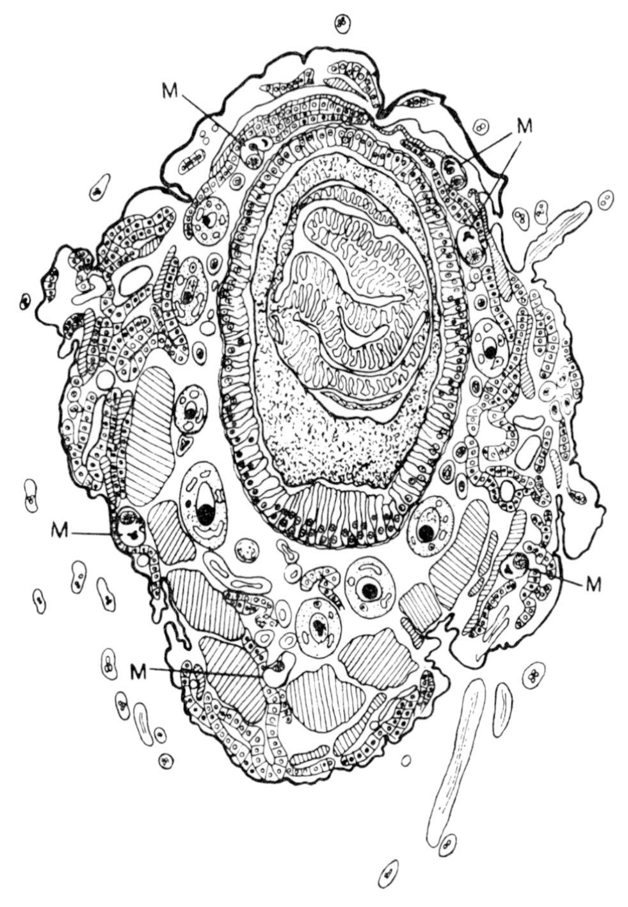

FIG. 76 A transverse section through the body of a caddis fly larva showing the distribution of the metacercarial stages of *Plagiorchis goodmani*. M: metacercaria. The distribution in this case is non-specific. From NAJARIAN, 1961a.

penetration at any point with a very brief migration after penetration, e.g. *Apatemon* metacercariae in leeches and snails. Within the leeches, encysted metacercariae may occur generally distributed in the parenchyma or in association with blood vesels and blood sinuses. In molluscs location may be general or confined to gills, digestive gland or ovotestis (ULMER, 1957). Occasionally metacercaria may be recovered from a sporocyst of another cercarial species already parasitizing the digestive gland. The low frequency with which this occurs suggests an accidental localization. In insects, metacercariae may lie free in the body cavity or occur embedded in thoracic and abdominal muscles (HALL, 1960). In others, metacercarial cysts become embedded in the fat body (NAJARIAN, 1961a). The cysts of *Paragonimus westermani* may be generally distributed in fresh water crabs although the site with the largest numbers may vary between different crabs. In *Potamon rathburi* the largest numbers were in the muscles, whereas in *Eriocheir japonica* most metacercariae were present in the gills (YOKOGAWA, *et al* 1960) Table 28.

Table 28 Number of cysts of Paragonimus metacercaria found in *Eriocheir japonicus*.

	Gills	Heart	Legs	Body	Liver	Reproductive organs	Total
Number of crabs infected	75	0	58	53	34	7	75
Total number of cysts	2995	0	1574	638	122	23	5352
Average number per crab	39.9	0	27.1	12.0	3.5	3.2	71.4

(After YOKOYAWA, 1960.)

In life cycles of trematodes parasitizing marine invertebrates, metacercariae have been described (CABLE and HUNNINEN, 1942) from the body and parapodia of *Nereis virens* and from the musculature of the lantern apparatus of various echinoderms (TIMON-DAVID, 1938). In these cases of metacercariae parasitizing molluscs, echinoderms, leeches and arthropods little specific distribution is present and consequently little migration is carried out by the cercarial stage. It must be admitted however, that no detailed study has been made of infections in these invertebrates.

Amphibia, in the form of tadpoles and young may be integrated into digenetic life cycles and carry the metacercarial stages. Generally penetration by the cercarial stage will be into the tadpole stage. It has been mentioned earlier that tadpoles, after they undergo metamorphosis, become relatively refractory to infection (HOFFMAN, 1955; PEARSON, 1956). The metamorphosis of the tadpole is also a crucial phase with respect to the distribution of metacercariae. Any cercariae which have penetrated the tail of tadpoles will have their prospects firmly curtailed after metamorphosis involving the loss of the propulsive tail. It is interesting to note that several workers on life-cycles of the strigeoid genera *Fibricola* and *Alaria* (CUCKLER, 1940a,b; HOFFMAN, 1955; PEARSON, 1956) report that prior to metamorphosis the uncysted post-cercarial stages in the tadpole undergo a redistribution thus ensuring their persistence in the adult frog. Hoffman describing the life cycle of *F. cratera*, states that in the initial stages metacercariae occur free in the body cavity of the tadpole, but just before metamorphosis, they migrate into the hind legs and become encapsulated. In the three host life-cycle of *Alaria arisaemoides* and *A. canis* the post-cercarial mesocercaria, which differs from the cercarial body primarily in the complexity of its excretory system, occurs generally distributed in the body and tail of tadpoles. In tadpoles with small hind legs a few occurred in the thighs. In the long-legged tadpole stage mesocercariae were congregated ventrally in the region of the pectoral girdle. In tadpoles with partially resorbed tails some were present in the tail, many in the thigh muscles and the majority were concentrated near the pectoral girdle. In contrast to these observations, CHANDLER (1942), studying *Fibricola texensis*, did not observe any migration of metacercaria within the tadpole body cavity.

The successful dissemination of a parasite species is often dependent on active, free moving intermediate hosts. Because of this we find fish frequently acting as secondary intermediate hosts. A great variety of conditions are present ranging from the general distribution of metacercariae on external surfaces to highly specific localization in internal organs. Some indication of the general distribution of metacercarial cysts between body and fins of 35 species of fish was given by EVANS and MACKIEWICZ (1958) (Figs. 77, 78). The greatest percentage of cysts occurred in the body and the next highest number was present on fins which are usually in motion e.g.

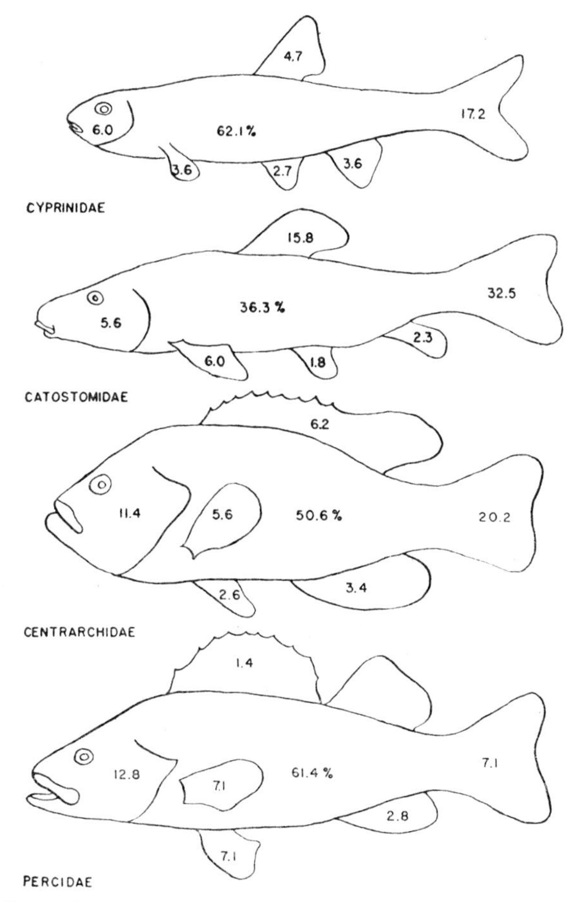

FIG. 77 The percentage distribution of metacercarial cysts in various parts of the body of certain fresh water fish. From EVANS and MACKIEWICZ, 1958.

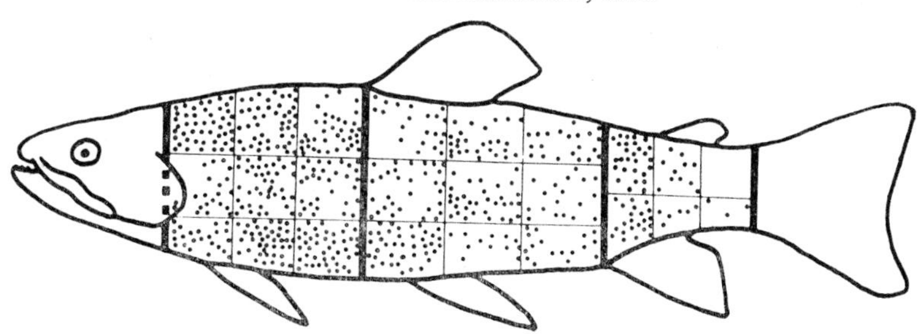

FIG. 78 The distribution of the metacercaria of *Bolbophorus sp.* in the musculature and fins of the Brown Trout from Meadow Lake, Montana. Each dot represents one metacercarial cyst. From FOX, 1962.

caudal and pectoral. Similar results were obtained by MILLER and MCCOY (1930). These observations are rather surprising in view of the difficulties involved in attachment to such continually moving surfaces. The internal distribution of the metacercarial stage of *Holostephanus lühei* has been described by ERASMUS (1962a). Although a fairly general distribution through the body of the fish was recorded, a peak was present just posterior to the gill region (Fig. 79). This suggests

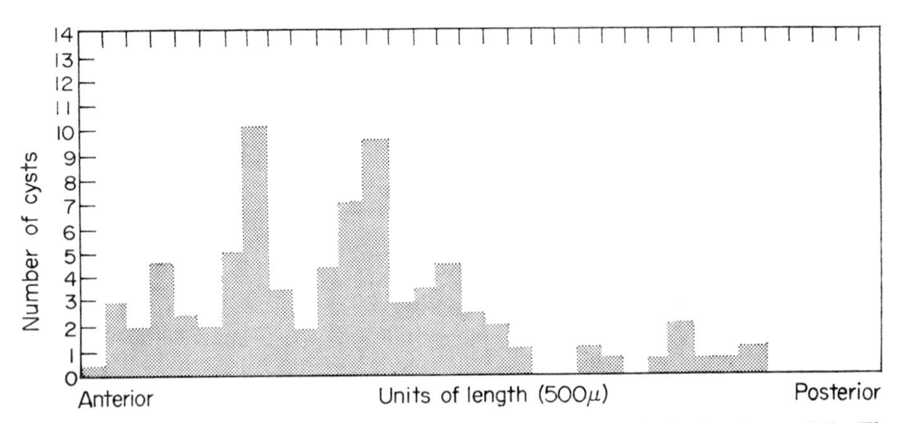

FIG. 79 The linear distribution of the metacercarial cysts of *Holostephanus lühei* in the tissues of the Three-spined stickleback. The anterior end of the fish is to the left of the diagram and the posterior end is to the right. Numbers based on counts from serial sections, so that cysts from all parts of the body are included. From ERASMUS, 1962.

very strongly that entry into the host was via the gills and once in the body, little migration took place. The species which give rise to encysted metacercariae do not become encapsuled immediately after penetration and localization. The cercarial stage of *Crassiphiala bulboglossa* comes to rest beneath the epidermis of fresh water fish. Four days after penetration the cercarial body, somewhat amoeboid in appearance, is free in the skin. In 7–14 days after infection, a host cyst has appeared, together with the characteristic pigment layer. The parasite cyst however is not formed until 20–22 days after penetration (HOFFMAN, 1956). Details of the migration performed by this cercaria are not given, although it seems likely that cercariae attach themselves to the skin and penetrate directly with little internal migration. This time scale contrasts markedly with that given for *Uvulifer ambloplitis* (HUNTER and HAMILTON, 1941) in which the parasite cyst wall is formed within 6–8 days. In *Neogogatea kentuckiensis* (HOFFMAN and DUNBAR, 1963), penetration of the cercaria through the skin of the fish host and localization in the musculature occurs in 2–3 hours and by this time the cercariae have secreted a cyst wall 7–14 μ thick. Obviously in this life-cycle little migration has taken place. The cysts of *Posthodiplostomum minimum* metacercariae occur in the liver, kidney, pericardium and spleen of fish (HOFFMAN, 1958a). Immediately after penetration, cercariae are found in the skin and superficial musculature. Although the migratory route was not studied in detail, metacercariae were found in the heart 1 hour after infection, in the kidney 1–6 hours, in the liver 18 hours later and after 48 hours metacercariae were present in kidney, liver, pericardium and spleen. Hoffman suggested that this migration might involve relatively passive movement through the renal portal system.

Very specific final localization of metacercarial stages is illustrated by members of the genus *Diplostomum*. The metacercariae live in the lens and/or humours of the eye and within the ventricles of the brain of fish and amphibia. Penetration into the host by the cercariae occurs generally over the body and through the gills so that the final localization is the conclusion of an elaborate migration. The life-cycle of *Diplostomum baeri eucaliae* has been studied by HOFFMAN and HUNDELY (1957) and HOFFMAN and HOYME (1958), and they were able to obtain information on the migration of the cercaria within the fish intermediate host. Fifteen to thirty five minutes after infection, the majority of cercariae were present in the skin and subcutaneous tissue with only a few in the brain. Five to 28 hours later, the majority of cercariae had reached the brain. Cercariae were recovered from the blood vessels and from the cranial veins and these authors suggest that although many cercariae migrate in subcutaneous tissue some move along blood vessels and cranial nerves to the brain.

More precise investigation of the features of migration to specific sites has been made by FERGUSON (1943a) and ERASMUS (1959). The entry of *Cercaria X* (Strigeoidea Diplostomatidae) into the stickleback has been described in the previous chapter. In the first few hours after exposure to cercariae, the distribution is fairly general along the length of the body with the

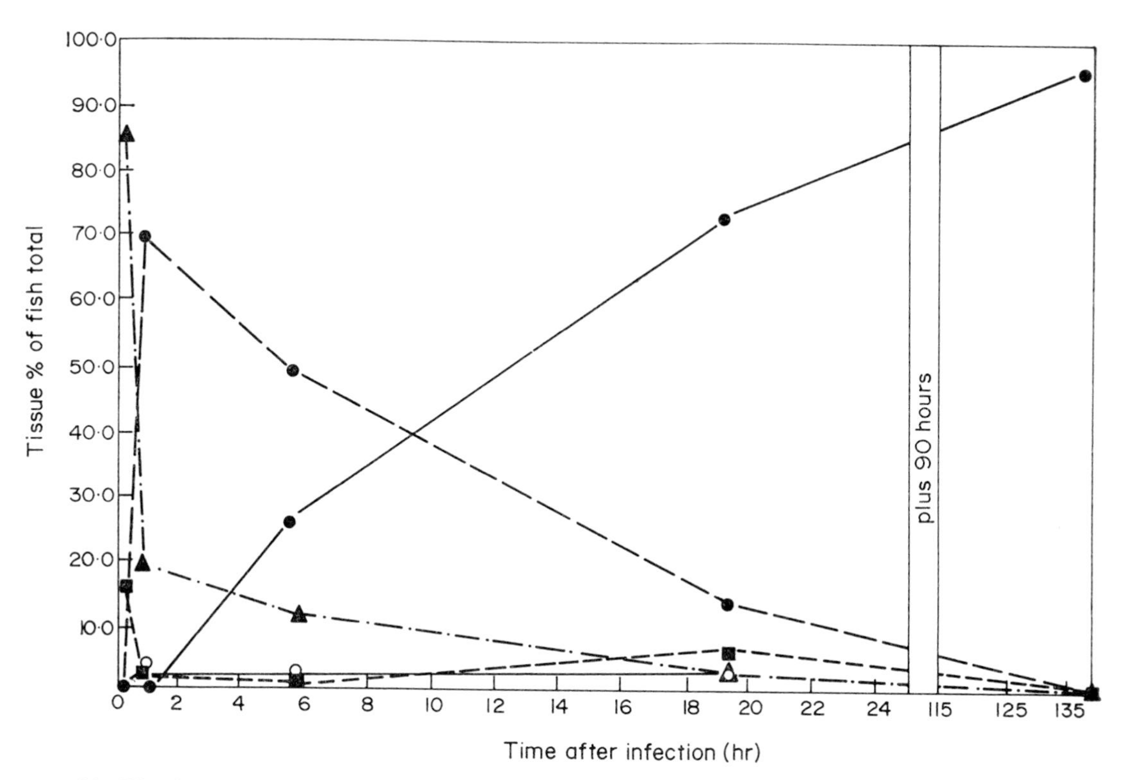

FIG. 80 The characteristics of the migration of *Cercaria X* (Strigeoidea) within the tissues of the fish intermediate host, *Gasterosteus aculeatus* as indicated by the variation in the percentage of cercariae migrating in various tissues in relation to time after infection. ●——● eye; ● – – ● connective tissue and muscle; ○——○ blood system; ▲ – · – ▲ skin; ■ – – – ■ gills. From ERASMUS, 1959.

cercariae occurring mainly in the connective tissue and muscles (Fig. 80). Over the following 140 hours a marked change takes place with respect to linear distribution and type of tissue containing cercariae. In both aspects a marked restriction develops, the cercariae migrating to the head, concentrating in the eye and moving from many tissues to the lens of the eye. Analysis of serial sections of infected fish reveals that, although some cercariae migrate via the blood stream, the majority move within the muscle and connective tissue, and examination of the eye showed that entry occurred at any point i.e. through the back of the eyeball or through the conjuctiva and was not confined to penetration at the point of entry of the optic nerve and blood vessel. This very precise localization suggests a remarkable directiveness to the migration. This is supported by the fact that the general linear distribution, exhibited by the population of migrating cercariae, seems to contract so that almost the entire population localizes in the lens, and very few remain in the trunk and tail regions of the body. If the localization was the result of a haphazard finding of the lens, then it seems very likely that the majority of cercariae present in the body far from the lens would remain there, but this is not the case. The other feature worth commenting on is the rate of migration. Nineteen hours after infection approximately 72 per cent of the cercariae migrating in the body have reached the eye. Observations on cercarial bodies migrating in tail fin preparations suggest a rate of 0·4 mm in 8 minutes. It seems that cercariae which do not localize within 24 hours are doomed to perish. This is probably related to the utilization of endogenous carbohydrate (glycogen) stored in the cercarial body. Once this has been depleted the rate of migration will diminish and ultimately the cercariae will come to a halt.

The investigation of the factors affecting and controlling such a migration is particularly difficult because of the complexity, both physiological and morphological, of the environment (fish body) in which the movements take place. The experiments of FERGUSON (1943a), who studied the migration of *Cercaria flexicaudum* in fish, do help a little in our understanding of this complex behaviour pattern. Bluegill sunfish (*Lepomis macrochirus*) 5–6·5 cm long in which only the tail fin had been exposed to cercariae showed cercariae within the lenses only two hours after infection. In view of the photoreceptive nature of the eye, the effect of presence or absence of light impinging on the fish body and eye was examined. None of the variations on this theme affected the migration and localization in any ways. However, the removal of the lens from the eye was found to affect localization. Although a few cercariae localized in the lensless eye, the number was small compared to the accumulation in the normal eye (Table 29). After the removal of a single entire eye, none of the cercariae localized in the orbit, although the intact eye on the other side of the body contained cercarial bodies. Removal of both eyes completely eliminated the localization and most cercariae were found posterior to the head in the various organs and blood stream. More precise analysis of the factors was attempted by a series of experiments in which optic nerves and blood vessels were ligated or cut in a variety of combinations before exposure to cercariae (Table 30). These results suggest very strongly that interference, particularly with the blood supply of the eye, suppresses the migration in a remarkable manner. In *Cercaria flexicauda* Ferguson suggests that this effect is related to the elimination or blocking of the main route along which entry into the eye takes place. The very closely related *Cercaria X* enters the stickleback eyeball at any point and the observations of BLOCHMAN (1910) and SZIDAT (1924) on other cercariae are similar. It is possible that some 'homing substance' is present either in the tissues or in the blood stream which cercariae are able to detect in their temporary phases of migration in the vascular system and, in addition, the presence of an intact lens in the eye is necessary for the assumption of the appropriate behaviour pattern. Until more is known about the physiology of the fish eye the possibilities of this hypothesis cannot be investigated.

There is no metacercarial stage in life-cycles of members of the Schistosomatidae, and the cercaria develops directly into the adult stage. However, as the migration to the appropriate host tissue is undertaken by the cercarial body or schistosomule it will be convenient to consider these migrations now. As has been discussed previously, entry into the mammalian skin by the cercariae is related to the type of skin, its hairiness and its degree of keratinization (Table 31).

Table 29 Summary of experiments in which minnows with one lens exterpated were exposed through the tail to cercariae.

Fish No.	Time from removal of lens to exposure Days	No. of cercariae entering tail	Time after infection Hrs.	Cercariae in lens or humours	
				− lens	+ lens
1	0	50	4	0	0
2	0	50	14	0	14
3	3	30	6	1 dead	12
4	3	30	20	4 dead	25
5	7	100	5	1 alive	0
6	7	140	18	3 dead	25
7	14	Several dozen	10	0	40
8	14	Several dozen	10	0	25
9	28	130	12	8 alive	50
				6 dead	
10	28	120	24	5 dead	
				10 alive	40
11	42	Several dozen	9	10 alive	
				1 alive	15

Note. Cercariae localizing in the lens were always alive.
(After FERGUSON, 1943a.)

Table 30 The results of experiments in which trout were exposed to cercariae after various eye operations.

Exp.	Operation	Approx. no. entering fish	Approx. number of cercariae										
				Fish no.									
			In eyes	1	2	3	4	5	6	7	8	9	10
1	Optic nerve cut, 1 eye	100s	Operated	8	20	15	15	3	5	5	8	60	125
			Normal	10	60	45	30	15	10	20	30	40	125
2	Optic nerve cut, 2 eyes	Sev. dozen	Operated	20	10	5							
			Normal	30	10	10							
4	Optic nerve & blood ves. cut, 1 eye	100s	Operated	0	0	0	0	0	0	1	0	0	0
			Normal	75	40	100	30	100	125	20	75	50	125
5	Optic nerve & blood ves. cut, 2 eyes	100s	Operated	1	1	0	0	0	0				
				0	1	0	0	0	0				
9	Blood ves. cut, 1 eye	100s	Operated	0	0	0	0	0	0	0	1	3	0
			Normal	90	75	75	100	100	5	5	10	45	15

Expt. 2. Cercariae entered through tail. In all other experiments they were injected intraperitoneally.
(After FERGUSON, 1943a.)

Once the skin is penetrated, the schistosomule migrates to and eventually enters a lymphatic vessel, proceeds through the lymphatic system until entry to the venous system is achieved and then localization in the liver takes place. Here, development to the adult stage is almost completed, and then the adults migrate to the final sites in the vessels of the gut or bladder. This account represents a generalization of the development of *S. mansoni haematobium* and *japonicum*. In unisexual infections, the male schistosomes develop normally, but the females do not attain complete sexual maturity and remain in the liver. If males are introduced at this time, the females complete their maturation and begin to pair and will then migrate to the mesenteric or vesicular veins. The rate of migration and the time scale of the major movements from one tissue to another is illustrated by the observations of STIREWALT (1959b) and LICHTENBERG and SADUN (1963) (see Table 32). One hour after infection of mice through the abdominal skin, twelve per cent of migrating cercariae have reached the dermis and migration in the subcutaneous veins and lymphatics occurs between three to four days after infection. The liver is reached in 2–3 weeks and sexual maturity attained 28–42 days after infection. STIREWALT (1959a,b) points out that many migrating schistosomules do not follow this route and return to

Table 31 Percentage of cercariae of *Schistosoma mansoni* entering skin of the hosts by the sites listed.

Skin	Cercariae observed	Wrinkle wall	Annular tail groove	Skin-scale	Sebum mass	Distal hair-skin angle	Follicular orifice	Used entry site
Adult hairless mouse ear	27	26	—	7	—	4	59	4
Adult hairless mouse body	36	56	—	19	—	—	25	—
Adult hairless mouse tail	24	—	4	—	—	—	96	—
Adult haired mouse tail	63	—	13	—	—	5	82	—
1–2 day mouse body	56	86	—	2	—	—	—	12
4–5 day mouse body	37	81	—	3	—	8	5	3
Adult hamster body	88	16	—	15	9	18	42	—
5 day rat body	11	55	—	—	—	—	27	18
Adult monkey body	15	46	—	—	—	7	27	20
Adult human body	26	100	—	—	—	—	—	

(After STIREWALT & HACKEY, 1957).

Table 32 Number of schistosomula and eggs in various organs following exposure to 200 cercariae of *S. mansoni*. Host: Swiss albino mice.

Tissue	Time in days									
	1 hr	1	2	3	5	7	14	28	40	50
Skin	++++	+++	++	++	+	*	—	—	—	*
Lymph node	—	—	—	+	—	*	*	*	*	*
Lung	*	*	—	*	*	+++	++	—	—	+E
Liver	*	*	—	*	*	—	+	+++	++++EE	+EEEE
Intestine	*	*	*	*	*	*	*	*	*	+EEEE

* No sample taken.
+ 1 or 2 schistosomula in 5 or more sections.
++ from 3 in 5 sections to 1 schistosomula per section.
+++ from 2 to 7 schistosomula per section.
++++ 8 or more schistosomula per section.
E to EEEE = relative number of eggs.
(After LICHTENBERG and SADUN, 1963).

Table 33 The distribution (per cent) of schistosomula in mouse abdominal skin after exposure to cercaria of *Schistosoma mansoni*.

	Time after exposure								
	5 min	15 min	1 hr	5 hrs	1 day	2 days	3 days	4 days	5 days
Epidermis	84	77	62	63	42	20′	2	31′	
Pilosebaceous apparatus	10	9	10	15	24	16	13′		
Sebaceous gland	3	4	10	16	28	12	1		
Pilar apparatus (below gland)			2		2	12	6	12	
Dermis	3	10	12	6	4	28	45	38	
Veins			4			12	6		
Hypodermis							27″	19″	
No. of mice	3	4	1	2	2	1	2	3	4
No. of larvae	59	150	50	54	57	55	92	16	0

20′ few walled off.
20″ few orientated surfacewards.
(After STIREWALT, 1959b).

the epidermis where they become encapsuled (Table 33). This wastage helps to explain the discrepancy between number of cercariae penetrating the skin and the number of worms recovered. For example 90–100 cercariae penetrating the tails of mice generally produce from 38–45 mature schistosomes at autopsy. It seems very likely that many other cercariae become diverted from the main migrating route. It would be very interesting to have some estimates of the efficiency of this migration. Stirewalt's figures given above suggest it may be below 50 per cent success. Mortality occurs rapidly after infection and CLEGG and SMITHERS (1968) found that large numbers of schistosomules died within 10 minutes of penetration.

MESOCERCARIA—METACERCARIA—ADULT

In those life-cycles containing an encysted metacercarial stage, completion of the cycle is achieved by the ingestion of the metacercaria by the appropriate final host. The genus *Alaria* is unusual in the intercalation of the mesocercaria between cercarial and metacercarial stages. As has been mentioned earlier, the mesocercariae occur in the musculature of frogs and these become ingested by the final host e.g. *Vulpes vulpes*. The released mesocercariae migrate to the lungs and there develop into diplostomula—the metacercarial stage. When this metacercarial stage of development is completed, it begins a further migration eventually localizing in the host's small intestine where it attaches itself and develops to sexual maturity. An experimental analysis of this sequence of migration would be very interesting indeed.

In contrast to this complex migration, developing metacercariae whose adults live in the alimentary tract exhibit little migration. The cyst wall breaks down with the release of the larva by the time the duodenum is reached. Further localization along the length of the alimentary canal involves very little migration.

An unusual site for adult trematodes is that exhibited by *Philophthalmus hegeneri* and *P. gralli* which, under experimental conditions, reaches sexual maturity under the nictitating membrane of a variety of avian hosts. Encysted metacercariae fed to chicks hatch, and the excysted metacercariae migrate up the oesophagus to the pharynx, through the median slit in the palate to the nasal cavities and thence via the nasolachrymal canal to the orbit. The entire migration takes about 7 days and sexual maturity with egg laying is attained in 20–25 days. More detailed study has been made on the migration of *Paragonimus westermani* metacercariae from the gut to the lung of the mammalian host (YOKOGAWA *et al*, 1960; 1962). Experimental studies show that in cats under 350 g the encysted metacercariae were already localized in the intestine wall within 30 mins after infection. In heavier cats (i.e. over 350 g), none were present in the intestine in 30 minutes, but many after 1 hour. The worms penetrate the mucosa of the middle and lower portions of the small intestine, and appear to migrate along the axis of longitudinal muscle fibres in the wall. After reaching the abdominal cavity of the cat the worms migrate into the abdominal wall and stay there. Five to ten days after the initial infection the young worms begin to move up through the diaphragm into the pleural cavity and thence to the lungs. *Haplometra cylindracea* and *Haematoloechus variegatus* are fairly common parasites of the frog *Rana temporaria*. They occur in the lung and it is possible that the migratory route may resemble that of *Paragonimus*. Migration from the body cavity to the lungs would be that much easier because of the absence from the frog of a muscular diaphragm. During the examination of frogs in parasitology classes, encapsuled adult *Haplometra* have been recovered embedded in or attached to the inner surface of the body wall. It is possible that they represent worms which have failed to complete their localization and have become lost en route.

Precise localization is also exhibited by liver inhabiting forms such as *Fasciola hepatica*, *Clonorchis sinensis* and *Dicrocoelium dendriticum*. Over the years, much attention has been given to the development and migration of *F. hepatica* and the most recent studies have been made by Dawes, employing mice as experimental host, and serial sectioning as the procedure for studying the infected material. These experiments (DAWES, 1961a,b; 1963a) showed that excystment occurred as quickly as 2–3 hours after infection. The young flukes leave the intestine and migrate through the wall in a manner similar to that described for *Paragonimus* i.e. along the bundles of muscle fibres. When entry into the liver takes place, (approx 3 days after infection) the flukes are about 0·2 mm long. Migration continues through the liver parenchyma and during this phase the parasites actively ingest portions of tissues. The route by which the flukes enter the bile duct has not yet been determined. In the mouse, sexual maturity of the parasites is reached 37 days after infection. The time scale is not rigid and variations occur within the mouse and also between the mouse and other experimental animals (rabbit and guinea pig) which have been employed.

Penetration through the gut wall, and transfer to the liver via the portal system, has been suggested by WYKOFF and LEPES (1957) as the route followed in the rabbit by the metacercaria of *Clonorchis sinensis*. Injection of metacercariae into the peritoneal cavity did not result in any liver infections. Larvae were obtained from the liver 72 hours after infection.

These few examples of metacercarial localization have been considered because details are available concerning the migratory paths and the time scale of the movement. Many other interesting examples of precise localization are known but unfortunately no investigation has been made concerning the nature of the behaviour patterns.

From what has been said in this chapter it becomes apparent that the sudden change in environmental conditions, which may occur several times in the digenetic life-cycle particularly, is the stimulus which releases a definite sequence of behaviour patterns culminating in the attainment of the site for development. The contrast in environments which brings about these patterns may be represented by the movement of the miracidium from a relatively simple aquatic environment (in majority of cases) to the complex tissues of a molluscan host. Similarly, the transference of a cercarial stage from aquatic environments to complex animal bodies may provide the stimulus. It is interesting that in most cases the change is from a relatively simple environment to a more complex one and it is possible that within complex bodies single features, such as increased body temperature of the new host, may provide the stimulus to initiate the new behaviour pattern. The observations of SMYTH (1959), on the development of *Schistocephalus solidus* plerocercoid in the final host, may be an appropriate example in this respect.

The studies of several workers (e.g. REES, 1940), have demonstrated that somatic and germ cell continuity exists throughout the digenetic cycle with its multiplicity of stages. Similarly, in the unisexual *Schistosoma* species, sex is already determined at the miracidial stage so that all the resultant cercariae give rise to unisexual infections. One of the most interesting things about the digenetic life-cycle is that the potentialities for the completely different migratory patterns necessary at different temporal phases may be present in the one individual. Thus a cercarial stage has first to migrate from its sporocyst or redial progenitor to the site of emergence. It then has to emerge, maintain a free-swimming existence, contact a host, adopt the appropriate behaviour for penetration and then migrate to the site favouring development into a metacercaria. The metacercaria, when ingested by the final host, has to excyst and then migrate under completely new environmental circumstances to the final site where sexual maturity is reached. In all this, the somatic part of the body exhibits complete continuity. The mechanism by which the appropriate behaviour pattern is released at the appropriate time has not been considered within the Trematoda.

8 Pathology and Host Responses

The life-cycle of a digenetic trematode involves a range of larval stages differing in their physiological requirements and feeding processes as well as in their morphology. This complexity is further heightened by the presence of intermediate hosts which themselves differ in morphology and physiological characteristics. The exact nature of the relationship existing between a parasite and its host will therefore vary as the life-cycle unfolds and progresses from host to host. It is also apparent that the relationship between a host and its appropriate parasitic stage cannot be defined precisely as this will alter with reference to the age, sex, nutritional state and general well being of the host as well as in relation to the nature of the parasite, its method of feeding and numbers present in the host. Because of this complexity it is obviously very difficult, if not impossible, to produce a comprehensive assessment of the effects of trematode parasites on their hosts and therefore it is intended to describe the main feature only, and these will be considered in relation to the particular developmental stage of the cycle.

EGG AND MIRACIDIUM

In the majority of cases, the adult trematode lives in a habitat which has direct access to the external environment, e.g. the gut and its accessory structures, the lungs, bladder and excretory ducts or, as in the monogenean infections, the external surface of the gills and the buccal cavity. The eggs are able to pass freely into the surrounding environment with the faeces, sputum or urine. A few trematodes have adopted the blood stream as the habitat and because this is a closed system, the passage of eggs to the exterior is less direct and in some cases becomes associated with damage to host tissues and the presence of local immune responses.

Some monogenean trematodes of the genus *Amphibdella* live in the bloodstream of the host. Some of the species are found on the gills but others inhabit the heart (LLEWELLYN, 1960). Llewellyn was able to recover immature forms only from the heart of electric rays, but RUSZKOWSKI (1931) found mature specimens and also free eggs in the blood system. Damage to the gills takes place when the adult emerges from the blood system and proceeds to the outer surface of the gill. The egg laying forms could be precocious egg layers, and in these cases, internal damage to host tissues must result when eggs become deposited in vessels with a low rate of blood flow or when they become trapped in vessels of narrow diameter.

A more complex picture emerges in the life-cycle of schistosomes and, in particular, those species inhabiting man. The fate of the spined egg capsules deposited by the female worm in the venules differs. Some will escape through the tissues of the appropriate organ, i.e. colon or bladder and pass to the exterior. Others, however, drift to the deeper viscera and become retained there and this retention also occurs in the submucosal tissues. In this way eggs trapped in the liver, submucosal tissues and pulmonary arterioles become the centres of areas of inflammation (EL MOFTY, 1962; LINCICOME, 1962). Their presence leads to the formation of sclerosing granular tissue which, when it heals, is followed by dense fibrosis. These changes are particularly serious to the host if the eggs are retained in narrow ducts or vessels when the granulomatous tissue will encircle the structure and the ensuing fibrosis will cause stenotic obstruction. The reaction in mice, during the period of seven to nine weeks after infection, has been described in histological terms by SAWADA *et al* (1956). Surrounding the eggs was an amorphous eosinophilic material positive to PAS and regarded as representing the specific antigen antibody reaction. In the liver tissue around the eggs, there was usually a marked increase in glycogen and fat, a decrease in the RNA of hepatic cells, as well as an excessive accumulation of acid mucopolysaccharide of the hyaluronic acid type in the connective tissue of the granulomas, and in the wall of the portal vessel. Because of this marked cellular response initiated by the living egg it has been suggested that antigenic material might be represented by secretions or excretions from the fully developed miracidium which the eggs contain, and that these substances diffuse through the egg shell which must therefore be permeable to some extent. The immunological picture is not clear as is indicated by some of the papers on the subject (e.g. LINCICOME, 1962; SMITHERS, 1962a,b; STIREWALT, 1963a,b; SADUN, 1963; KAGAN and NORMAN, 1963; DAMIAN, 1966) and different workers give varying degrees of support to this hypothesis. In a series of transfer experiments in which exposure to adult worms was achieved by transference of established adults from donor animals directly into the portal system of normal monkeys, SMITHERS and TERRY (1967) were able to induce resistance in the absence of migrating larval stages. This observation, together with the finding that eggs alone injected into the monkey did not elicit a response, strongly suggests that the main antigenic stimulus in these experiments is from antigenic material associated with the adult worms, but not from egg production and the subsequent presence of eggs in the host tissues.

Blood inhabiting flukes also occur in the family *Sanguinicolidae* and *Spirorchiidae*. Species of *Spirorchis* occur in the blood system of turtles and in one of the species studied (*S. elegans* by GOODCHILD and KIRK, 1960) eggs accumulated in the lung vessels and eventually appeared in and could be collected from the faeces. The sanguinicolids occurring in fish appear to be more troublesome to their hosts (DAVIS, HOFFMAN and SURBER, 1961), and can cause severe disease especially to fry and young cyprinid fish. *Sanguinicola inermis*, living in the bulbus arteriosus and large vessels of the gills, produces large numbers of eggs which eventually accumulate in the gill capillaries (in young fish) or kidneys (older fish). The eggs are conical and possess sharp lateral projections and when the eggs hatch in the blood vessels of the gills the miracidia perforate the walls of the vessels and gill tissue and escape into the water. The disease sanguinicolosis occurs in two forms which are related to the site of retention of the eggs (LYAIMAN, 1949). The gill form of the disease results from the presence of eggs in the gill vessels. These become obstructed, the circulation becomes affected and the tissues become anaemic and necrotic and marbled in appearance. Further damage is caused by the escaping miracidia and this results in the entry of bacteria and the establishment of secondary infections. A more chronic form of the disease occurs with the blockage of the renal blood vessels. This produces water imbalance with the accumulation of exudates and the appearance of dropsy, exophthalmus and a bristling or elevation of the scales. The disease can be particularly severe and may cause losses in fish farms and areas devoted to carp husbandry (BAUER, 1959; DOGIEL *et al*, 1958).

The first stage in the digenetic life cycle to contact an intermediate host is the miracidium. This may hatch externally in the aquatic environment or internally in the gut of the snail, releasing the miracidium which then proceeds to penetrate the host. The presence of penetration glands and their apparently histolytic secretion has been commented on earlier, and the main effect of this stage on its host is the destruction or lysis of a few host cells at the point of entry. As well as this highly localized effect, a more general loosening of the epithelial cells from the basement membrane may occur in the region around the point of entry (DAWES, 1959, 1960). A more biochemical change, in molluscs immediately after miracidial penetration, has been described by DUSANIC and LEWERT (1963) (Table 34). They observed the concentration of proteins and free amino acids of *Australorbis glabratus* haemolymph at varying times after exposure to miracidia of *Schistosoma mansoni*, and found that the number and relative concentration of the proteins and free amino acids differed, within one hour after exposure, from that of the normal snail. This difference was at a maximum eleven hours after exposure, but by 20 hours the concentration had returned to its normal values. The authors suggested that the changes could be the result of pathological processes induced in the snail by the miracidium, or they might represent a protective immunological response on the part of the snail. Cellular responses to miracidia penetrating previously infected snails or abnormal hosts have been described by BARBOSA and COELHO (1956) and NEWTON (1952) and PAN (1963) reported a tissue response against degenerating mother sporocysts of *Schistosoma mansoni* in *Australorbis glabratus* within 48 hours of infection. However the changes described by DUSANIC and LEWERT (loc cit) do seem to have occurred too rapidly to be considered as an immune response by an invertebrate.

SPOROCYST AND REDIA

The development of the sporocyst and redial generations within the tissues of the molluscan first intermediate host must throw a tremendous strain on the well-being of the host. The pronounced multiplication associated with these stages will require ample supplies of nutrients for its continuation, as well as physical space for the end products of the process—large numbers of daughter sporocysts and rediae. In this respect it may be significant that these stages generally occur in the molluscan digestive gland—a structure intimately associated with the digestion of food by the host, but not vital in the sense that a reduction in its size or efficiency would result in the immediate death of the host. The fact that the ovotestis and certain of the accessory reproductive structures lie in close proximity to the digestive gland has some significance in assessing the fecundity of infected molluscs. Many observers (see review by WRIGHT, 1966a) have commented on the external and internal changes apparently related to the presence of larval stages, but some of these descriptions do not take into account changes within the snail, which might be related to age differences between individuals under differing environmental conditions. The most accurate assessment of the effects of parasitization can only be made with experimental infections of laboratory reared molluscs of known age and environmental history.

The sporocyst and redial stages lie in the connective tissues between the tubules of the digestive gland (syn. liver, hepatopancreas of different authors), and this, in heavy infections, becomes almost entirely replaced by the parasitic stages (Plate 7-1). The number of digestive gland tubules appearing in transverse sections was counted by Pratt and co-workers and the effect of parasitization determined. In *Stagnicolae emarginata angulata* containing *Cercaria laruei*, *Cercaria Plagiorchis muris*, *Cercaria vogena* and *Cercaria diplostomum flexicaudum* (PRATT and BARTON, 1941) the apparent number of tubules was reduced and the tubules displaced to the periphery of the organ, with the parasites occupying the central regions of the digestive gland. In a later study, PRATT and LINDQUIST (1943) concerned with the same snail, but infected with *Cercaria laruei* or *Cercaria diplostomum flexicaudum*, no evidence of a decrease in digestive gland tissue was obtained. Although the number of tubules was fewer in the regions containing the parasites these authors suggested that the tubules became displaced posteriorly. It was apparent however, that the ultimate histological picture depended on the severity of the infection.

Changes in the histology of the epithelium lining the digestive gland tubules have been described by many authors (e.g. FAUST, 1920; FAUST and HOFFMAN, 1934; HURST, 1927; REES, 1934; REES, 1936; AGERSBORG, 1924; DUBOIS, 1929; LAL and PREMVATI, 1955; CHENG and JAMES, 1960; CHENG and SNYDER, 1962a; JAMES, 1965; WRIGHT, 1966a; JAMES and BOWERS, 1967a). In spite of the great variation in the type of parasite and the molluscan host studied, the observations of these and other workers fall into the same pattern. The epithelium changed from a columnar type to a squamous more flattened version and in some cases complete loss of epithelium took place (ERASMUS, 1958). Intermediate stages are represented by a loss of the brush border, vacuolization of the cytoplasm of the epithelial cells, abnormal mitoses and the formation of fibromata and granulomata. The destruction of the cells results in the release of inclusions, such as pigments, and these sometimes accumulate in other sites e.g. foot, which, in the case of *Littorina littorea* infected with *Cryptocotyle lingua*, became yellow brown in colour due to the presence of the carotenoids from the digestive gland (WILLEY and GROSS, 1957). JAMES (1965) found that the histological picture exhibited by the digestive gland epithelium of *Littorina saxitalis tenebrosa* parasitized by *Cercaria roscovita* corresponded to that observed in unparasitized but starving specimens of *L. saxitalis tenebrosa*. The sporocysts of this species are large and compress the digestive gland tubules at various points so that portions of the gland become sealed off (Fig. 81). Consequently as no food can enter this portion of the gland above the constriction, starvation autolysis occurs in the epithelium. In this case, the effect on the host tissues is produced very indirectly by the parasite. A similar 'blocking layer' of sporocysts was described by REES (1936). In other cases of parasitization of this snail, James (loc cit) attributed the damage produced to ingestion by rediae, or to the accumulation of the parasite's excreta. He also observed that the effect was related to the rate of development of the parasite, to the extent of the host's life-span and to the time of initial infection relative to the life-span.

Histochemical studies of infected digestive gland have revealed additional changes. The glycogen content (in histochemical terms) of the host cells becomes reduced (FAUST, 1920; HURST, 1927; VON BRAND and FILES, 1947; CHENG and SNYDER, 1962a; CHENG and BURTON, 1965; SNYDER and CHENG, 1961) and glucose has been detected on the outer surface of sporocysts and within the wall body wall (CHENG and SNYDER, 1963) suggesting that the host carbohydrate is hydrolysed and then absorbed as glucose. The neutral fat and fatty acid content of the digestive gland also increases (CHENG and SNYDER, 1962b). An increase in acid and alkaline phosphatase activity has been recorded (CHENG, 1964) and this could be related to a tissue response on the part of the mollusc (PAN, 1963) or might be produced by the regurgitation of redial gut contents rich in acid phosphatase activity (CHENG, 1964; PROBERT, 1966a) into the host tissues.

The stimulus, or factors initiating these changes, is not clear either. In the case of sporocysts devoid of mouth and anus, the substances, whether secretory and lytic or excretory and irritant, must pass through the tegument and current ultra-structural observations may clarify the nature of secretory and excretory processes by these stages. Opinion about the significance of secretion by these stages is divided, and some workers consider that many of the effects described might result from simple mechanical pressure of a mass of sporocysts on the gland tubules.

The close contact between the digestive gland and the ovotestis and associated ducts in many molluscs does suggest that the presence of heavy infection in the former organ might affect the functioning of the reproductive system in some way. In some cases the ovotestis becomes invaded by the larval stage, although this seems to occur more frequently in heavy infections and by redial stages rather than by sporocysts. Varied effects have been reported

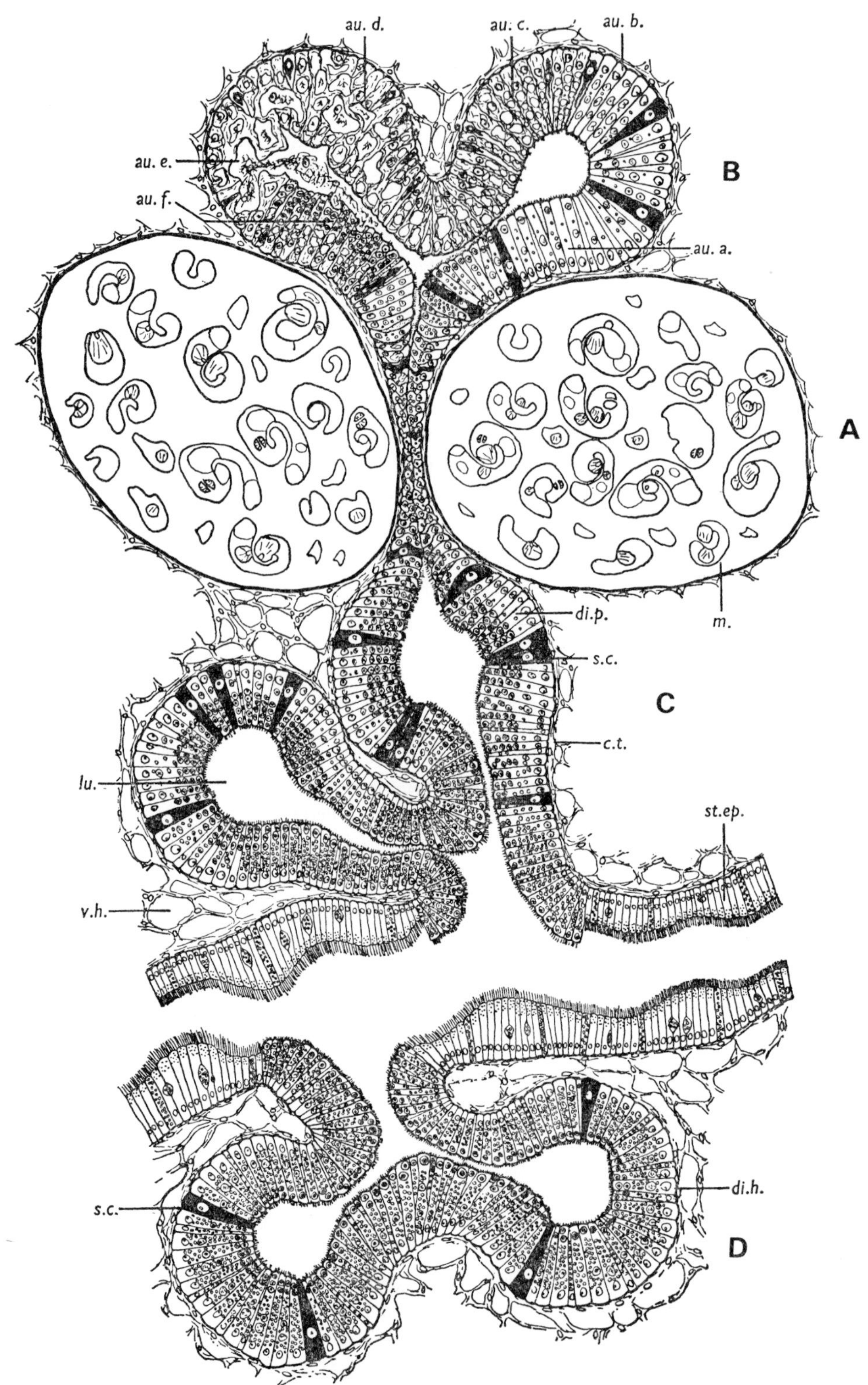

FIG. 81 The effect of sporocysts on the digestive gland tubules of *Littorina saxatilis* subsp. *tenebrosa*. **A** sporocysts compressing and blocking the tubule lumen; **B** tubules distal to the block are undergoing autolysis; **C** tubules proximal to the block with few food storage globules, many food vacuoles and secretory cells; **D** healthy tubules with many food storage globules, few food vacuoles and secretory cells. (au. a-f: various stages of starvation (*cont. opp.*)

from the inhibition of gametogenesis to castration and apparent sex reversal. The invasion of the ovotestis of *Succinea putris* by the larvae of *Leucochloridium paradoxum* was reported by WESENBERG-LUND (1931) and he found that after the ovary had been destroyed the testis continued to produce spermatozoa. Eventually in this case, regeneration took place and restored the balance. The redia of *Cercaria patellae* in *Patella vulgata* also invaded the ovotestis and either consumed the gonad or produced a reduction of the germinal epithelium (REES, 1934). REES (1936) also noted that the redia of *Cercaria himasthla secunda* actually devoured the gonadial tissue. The blood supply to the gonad was affected by the mass of sporocysts of *Cercaria milfordensis* in *Mytilus edulis* (UZMANN, 1953) so that normal gametogenesis was affected. More recently CHENG and COOPERMAN (1964) reported the invasion of the reproductive system of *Helisoma trivolvis* by the daughter sporocysts of *Glypthelmins pennsylvaniensis*. Localization was again mainly in the female system and sporocysts were found along the entire system except for the mucus gland, oothecal gland and seminal receptacle. The damage caused was mainly mechanical, although gametogenesis in both gonads was greatly affected. In very heavy and long established infections, CHENG and BURTON (1965) also noted the invasion of the gonads by sporocysts of *Bucephalus* sp. in *Cassostrea virginica*. Ova present in parasitized ovaries were reabsorbed. The time at which infection of the mollusc takes place is also significant. JAMES (1965) mentions that if juveniles of *Littorina saxatilis tenbrosa* become infected with *Cercaria pavatrema homoeotecnum* the host is prevented from reproducing at all. When infection occurs after the first reproductive cycle, as in *Cercaria littorinae rudis* and *C. lebauri*, the host is prevented from reproducing a second time. The infection of the snail after the second reproductive cycle has little significance to the species as the snails are almost at the end of their life span and die within a few months of this second cycle. In view of this information on the internal effects of parasitization, it is not surprising that some workers have reported examples of apparent sex change in samples of infected molluscs. PELSENEER (1906, 1928), REES (1936), KRULL (1935) and ROTHSCHILD (1938) have all reported molluscs which possessed an abnormally small penis and which were infected with larval trematodes. The two latter authors were of the opinion that some of the snails with a small penis were really females which had undergone sex reversal. This information is very inconclusive and, as

Table 34 Free amino acid composition of *Australorbis glabratus haemolymph**. (After DUSANIC and LEWERT, 1963).

Hours after exposure†	Amino acid					Ninhydrin reaction
	Lysine	Glycine	Proline	Threonine	Leucine	
normal	+	+	+			weak
0·75	+	+	+	+	+	strong
1·25	+	+	+	+	+	strong
2	+	+	+	+	+	weak
4	+	+	+	+	+	weak
8	+	+	+	+	+	weak
11	+	+	+	+	+	weak
14	+	+	+	+		weak
24	+	+	+			weak
2 months	+	+	+			weak

* 0·02 ml samples from pool of 10 snails. † Exposed to 10 miracidia except 2-month specimens exposed to 5–8 miracidia.

autolysis equivalent to 8, 14, 21, 31 and 40 days' starvation under laboratory conditions; c.t: connective tissue; di.h: healthy digestive gland tubule; di.p.: gland tubule showing the effects of parasitism; lu: lumen of digestive gland tubule; m: sporocysts containing metacercaria of *Cercaria littorinae rudis* Lebour; s.c.: secretory cell; st.ep.: stomach epithelium; v.h.: visceral haemocoelic space.) From JAMES, 1965.

Rothschild points out, does not take into account preferential mortality between the sexes, the sex ratio in the normal population, as well as the degree and permanence of the castration.

External changes of a different sort have been observed in infected molluscs. Shell abnormalities such as ballooning and thinning of the shell have been recorded by WESENBERG-LUND (1934) and gigantism by ROTHSCHILD (1936a,b; 1941a,b), LYSAGHT (1941) and Wesenberg-lund (loc cit) but again interpretation of this data in the absence of controlled laboratory experimentation must be very cautious. The data obtained by PAN (1963) is more significant in this respect. He studied the effect of *Schistosoma mansoni* infections on laboratory bred *Australorbis glabratus* under controlled laboratory conditions and found that the infected snails were consistently larger than the uninfected controls between the second and sixth weeks, but ultimately (after the 11th week) the growth of the infected snail became retarded. PESIGAN *et al* (1958) also reported a retardation of growth in both sexes of *Oncomelania quadrasi* infected with the Phillippine strain of *S. japonicum*. However, MOOSE (1963), using *O. nosophora* and the Japanese strain of *S. japonicum*, found that sixteen weeks after exposure to miracidia no significant difference existed between the size of male snails from control and experimental groups although the infected females were shorter than control females. He also noted a difference in length between exposed but uninfected female snails and their controls, but not between exposed but uninfected males and their controls. In spite of this additional factor his main conclusions were statistically significant.

Table 35 Comparison by analysis of variance, of egg-laying capacity of *B. truncatus* snails, infected and uninfected with *S. haematobium* cercariae.

Egg production		Snails: Infected	Non-infected	p value
Eggs per	mean	4·7	15·1	<0·001
clutch	range	0–7·2	12·9–16·0	
Clutches per	mean	0·24	0·59	<0·005
snail per day	range	0–0·58	0·39–0·70	
Eggs per	mean	1·1	8·9	<0·001
snail per day	range	0–4·2	5·1–11·2	

(After NAJARIAN, 1961b)

Controlled experiments with *Schistosoma* species in the laboratory indicate that the fecundity of the mollusc is reduced under conditions of infection. NAJARIAN (1961b) found that the egg-laying capacity of *Bulinus truncatus* infected with *S. haematobium* showed a three-fold reduction in the mean eggs laid per clutch, more than a two-fold reduction in the mean clutches laid per snail per day and an eight-fold reduction in the mean eggs laid per day compared with control snails (Table 35). This confirmed COELHO (1954) and PAN's (1963) observations with *S. mansoni* in *Australorbis glabratus*, and also PESIGAN *et al* (1958) data on egg laying by *Oncomelania quadrasi* infected with *S. japonicum*. ETGES and GRESSO (1965) however, suggest that the picture may be a little more complicated (Fig. 82). In a study of *A. glabratus* infected with *S. mansoni*, they found that the egg-laying capacity began to be inhibited in the fourth week of infection prior to cercarial emergence. After the 36th day, when cercarial emergence began to take place, egg production declined sharply and thereafter only sporadic clutches were deposited up to 90 days after exposure. At this time egg production was renewed and increased markedly.

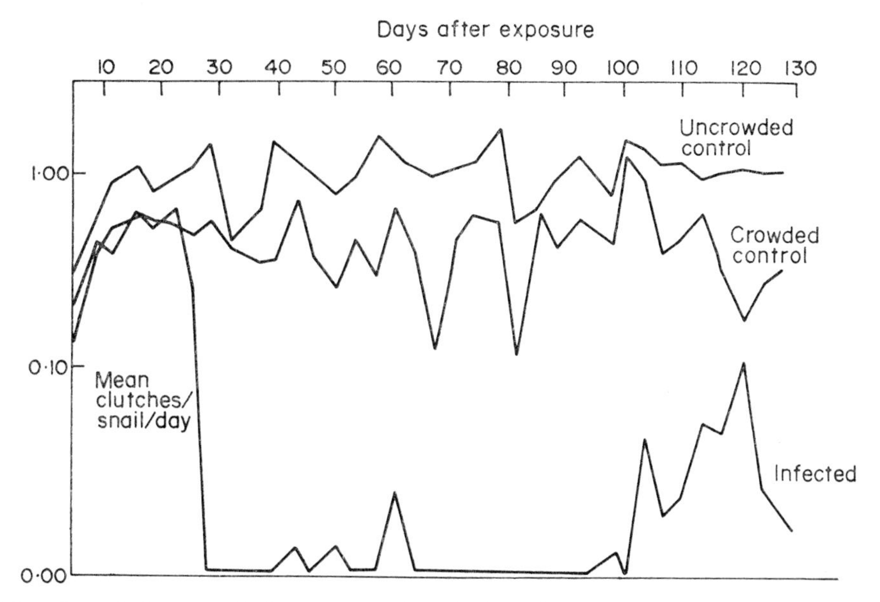

FIG. 82 The effect of infections of *Schistosoma mansoni* on the fecundity of *Australorbis glabratus*. Semilog plot of egg clutch production by *A. glabratus*. There is considerable difference between the rate of egg-laying by infected snails and uncrowded controls. From ETGES and GRESSO, 1965.

The size of egg masses from infected snails was smaller: mean eggs per clutch 12·6 for infected snail and 38·1 for the controls. Development and hatching of these egg masses was normal, although some of the eggs from infected snails contained fragments of cercaria within the perivitelline membrane.

During this experiment, described by Etges and Gresso, the air conditioning system failed and this resulted in the death of the infected snails but not the controls. This different susceptibility to temperature stress confirmed the findings of VERNBERG and VERNBERG (1963) on the differential thermal resistance of infected and uninfected mud-flat snails *Nassarius obsoleta*. As well as a reduction in fecundity, the mortality of infected snails is increased. BARBOSA (1962) stated that 39 days was the average life-span of *A. glabratus* infected with *S. mansoni* (15–17 months in uninfected snails in nature). Pan (loc cit) showed that, under laboratory conditions, the mortality rate among infected snails was only 6·7 per cent for the first six weeks but, after this period, high rates were recorded. The cumulative mortality of exposed snails was 89 per cent for the first 24 weeks compared with 35 per cent for the control snails.

Under natural conditions, many of the habitats of the Planorbids transmitting *S. mansoni* in Pernambuco are temporary and dry up. Although the majority of the snails die under these conditions, those protected by debris, leaves etc. may survive. Experimental data showed that in infected snails, degeneration of the sporocysts and cercariae occurred within 16 days of removal from the water and that after 30 days out of water, one third of them survived free of infection. However this applied only to snails harbouring mature infections, as immature infections were able to survive and remain dormant for as long as 150 days (BARBOSA, 1962). The presence of a larval infection within the snails does increase their mortality rate and reduce their chances of survival during desiccation. The molluscan host of *Fasciola hepatica* (*Lymnaea truncatula*) also lives in environments subject to periods of drought, and is capable of prolonged periods of aestivation under such conditions, but KENDALL (1965) noted that although the rate of development of the larval parasites may be retarded, loss of parasites did not seem to occur.

The mortality of infected snails is usually higher amongst very young snails than in older snails. A period of particular pathogenicity to the snail is that during which cercarial emergence takes place, and ETGES and GRESSO (1965) have shown that the egg laying capacity of the host becomes maximally affected during this period. The emergence of cercariae from the mollusc is a disruptive process and this phase of tissue damage will be particularly disadvantageous to the mollusc and might be partially responsible for some of the changes noted above.

Biochemical changes in the metabolism of infected molluscs have not been very extensively studied. HURST (1927), as well as describing histological and histochemical changes associated with digestive gland infections, also made some observations on the CO_2 elimination, O_2 consumption and respiratory quotient of infected snails. More recently, BECKER (1964) was able to detect differences between the rate of oxygen consumption of infected and parasite-free *Stagnicola sp.* The diminished oxygen consumption of the infected snails was thought to be due to the reduced activity of the snails which was related to utilization of the host's glucose by the parasite. CHENG (1963b) has analysed the serum proteins, amino acids and total blood volumes of parasite free and parasitized specimens of *Musculium partumeium*, *Physa gyrina* and *Helisoma trivolvis*. A statically significant drop in the total serum protein concentrations was observed in all cases in infected molluscs, the blood volume was also less in the parasitized specimens, and a striking reduction was observed in the number of free amino acids within the infected molluscan sera. It is obvious that the extensive multiplication occurring in infected molluscs must involve a high rate of protein metabolism and the absence of the free amino acids from the sera of these molluscs suggest that this constitutes the supply for the parasite. It is possible however that the amino acid pool may also be utilized by the mollusc for the reconstitution of damaged host tissues.

THE CERCARIA

The cercaria develops in a molluscan host, but because the cercaria infects at least one other host after leaving the mollusc the pathological effects produced by this stage differ slightly depending on the particular host-parasite association involved.

Table 36 Summary of responses to aliquots of 25 cercariae of *Trichobilharzia physellae* by human volunteers.

Case no.	Sex	Age group	Primary prickling	Appearance of macules in hours	Reactions in 24 hours			Subsequent history
					None	Macules	Papules	
1(GWH)	M	50–59	+	5		5	0	None in 14 days
2(CR)	M	30–39	–	–	–	0	10	None in 14 days
3(SAAE)	M	20–29	–	4		5	5	*
4(AWH)	F	50–59	–	14	–	0	0	None in 14 days
5 (Gay)	F	10–19	+	8		8	0	None in 14 days
6 (Jay)	M	10–19	+	4		4	0	Slight pruritus during 24 hours. No other response in 14 days

* Five macules persisted 48 hours, faded, followed by 5 papules and erythema beginning in 24 hours and persisting for 5–6 days with intermittent pruritus. (After HUNTER, 1960).

Within the molluscan first intermediate host, cercariae migrate to the site of emergence via lymphatic channels or the blood system and little damage occurs at this period, although an accumulation of migrating larvae may cause partial or complete obstruction of a blood vessel. Considerably greater damage is caused at the site of emergence and, in the process of emergence to the exterior of the mollusc, epithelia and underlying tissues become disrupted as well as blood vessels perforated. The accumulation of cercaria at particular points along the mantle produces 'blebs' which eventually rupture (see chapter 7) (Plates 7-3 and 7-4). In view of the considerable numbers of cercariae which emerge from these hosts, and the length of time over which this may occur, it is obvious that this continual perforation of mantle tissues must have a deleterious effect upon the mollusc and it is not surprising that the biological efficiency of the mollusc becomes reduced at this time. It has also been reported by different workers that distinct cellular responses by the mollusc may appear at this time.

The second phase of damage, produced by the cercaria, takes place during entry into and migration within the second intermediate or final host. The searching and testing behaviour, which is a necessary preliminary to penetration, does appear to be particularly irritating to the host animal. Molluscs and leeches exposed at this time exhibit withdrawal, contraction and wriggling movements, and chordates with limbs make attempts to wipe cercariae off their surface. The mechanical and secretory activities of cercariae result in disruption of the host epithelium and underlying connective tissues, which may be associated with histochemical changes in the ground substance of the tissue. Fish are conspicuous in their response to this attack and dash around the container in a very violent manner. The response of unusual hosts to this penetration is well exemplified by cases of schistosome dermatitis occurring in human subjects exposed to the attacks of avian and mammalian (non-human) schistosome cercariae. The condition is world wide in occurrence (CORT, 1928; BAYLIS and TAYLOR, 1930; PIRALA and WIKREN, 1957; CHU, 1958; HUNTER, 1960) and is represented by an initial pricking sensation followed by the appearance of maculae and later papules (Table 36). In man the cercaria do not penetrate deeply, are soon encapsulated and the reaction is generally over within 10 to 14 days. The degree of the response is influenced by a history of prior infection and in such cases is more intense. Much detailed information is available concerning the penetration of human schistosome cercariae into experimental hosts and the reader is referred to LINCICOME (1962) for a detailed discussion of this. The response of the skin is related to many factors such as number of active parasites, age and susceptibility of host species, previous history of host, and region of body skin invaded. Inflammation can appear within 15 min of penetration with congestion, dilation of venules, and accumulation of polymorphs occurring. This, after 18 hours to one day, develops into an acute inflammatory response and seems to be related to the act of penetration and not to subsequent migration.

The further history of the cercaria is significant in relation to its effect upon the host. In those cases where there is no precise localization of the next stage (i.e. metacercaria), little migration takes place and little tissue damage occurs. However in those cases where extensive migration to localized sites (e.g. strigeoid metacercariae restricted to the host eye or brain) takes place, extensive tissue destruction may occur. The majority of observations in this field are concerned with the migration of cercarial bodies within fish. Although in some cases (e.g. *Cercaria X*: ERASMUS, 1958, 1959) most of the migration occurs within the musculature and connective tissues (Plates 8-1 and 8-3), passage into, along and out of the blood vessels and the frequency with which this occurs depends largely on the severity of the infection within the fish. In a fish exposed to 9000 *Cercaria X* in 30 ml of water for four minutes eight complete perforations were noted in 1·2 mm of ventral aorta (Plate 8-4). Migration within the blood system occurs most frequently in the vessels anterior to the heart and in severe infections, considerable perforation and haemorrhage can occur. In addition to the blood system, cercariae often become associated with nerves and passage across and along this tissue might account for the swimming disturbances commonly exhibited by fish a few hours after infection (Plate 8-5). These symptoms will also be associated with migration through and damage to the lateral line system. It was generally found that if a fish survived a period of 24 hours after infection its behaviour would return to normal. The temporary nature of some of these disturbances is correlated with the fact that the majority of cercariae have localized in the lens within twenty four hours and that the remainder are already being rendered innocuous by the host tissues. The pathogenicity of the infection can be so severe, that small fish exposed to very large numbers of cercariae may be killed within three minutes. BLOCHMANN (1910) described mass mortality due to an infection of fish within his aquarium and SZIDAT (1924) described the mortality of fish associated with lacerations of the blood system. These situations do not normally arise and only occur under conditions of crowding and restriction to small bodies of water. Similar pathological patterns occur in non-strigeoid cercariae which undergo extensive migrations to their final sites. A review of the pathology of helminth parasites in fish has been made by WILLIAMS (1967).

THE METACERCARIA

The effect metacercariae have upon their hosts is largely dependant on whether or not they encyst. In those species where relatively short migratory distances are covered by cercariae and encystment occurs fairly rapidly, relatively little damage to host tissues takes place. The metacercariae, under such circumstances, do not feed by the ingestion of host tissues and the only effect upon the host is the stimulation of an inflammatory response (CHENG *et al*, 1966a). Some cercariae do not emerge from their molluscan hosts and encyst within the tissues of the digestive diverticulum and may result in the digestive gland becoming almost entirely replaced by a mass of metacercarial cysts and the tubules of the gland could exhibit the same histopathology as produced by sporocyst and redial stages. The inflammatory response in fish may become associated with abnormal pigmentation at the site of the metacercarial cysts and, in *Posthodiplostomum cuticola* and *Cryptocotyle lingua* where the cysts occur fairly superficially in the skin, the abnormal pigmentation produces a condition described as 'Black spot' disease. There is always the possibility that cysts may become embedded in the walls of blood vessels and in so doing produce stenosis at that point, and also reduce the efficiency of body musculature when encapsulated in that tissue (SIMMS, 1932; HUGGHINS, 1956).

A very different situation exists in the case of metacercariae which do not encyst and continue in an active state, feeding on the tissues of the host. The best studied examples of this situation are the diplostomulum larvae present in the eyes and brain of fish and amphibia. In host-parasite relationships such as these, the secondary intermediate hosts are exposed to two phases which are particularly pathogenic. With encysted metacercariae the most adverse period for the host is while cercarial bodies are migrating through the tissues prior to encystment. This hazard also occurs with infections of non-encysting metacercariae but this is followed later by a phase where the metacercariae have developed and are actively feeding on host tissues. Thus these types of life-cycles are highly pathogenic to their hosts and are often associated with outbreaks of mass mortality in the population.

The most well known species occurring in brain tissues are *Diplostomum baeri eucaliae* and *Diplostomum phoxini*. In old infections in the minnow, *Diplostomulum phoxini* accumulated in the fourth ventricle, aqueductus sylvii, optic lobes, third ventricle, lobi inferiores, occasionally the corpora striata and under the epithelium covering certain areas of the brain and spinal cord (REES, 1955, 1957). Rees mentions that the larvae do not lie free in the cavities of the brain but lie under the epithelium lining the cavities where vacuolated tissue develops. In early infections soon after exposure to cercariae, minor haemorrhages were present in the cranial tissues. SZIDAT and NANI (1951) described damage caused to various parts of the brain of *Basilichthys spp.* by the larvae of *Diplostomulum mordax* and *Tylodelphys destructor*, and HOFFMAN and HUNDLEY (1957) and HOFFMAN and HOYME (1958) have observed the appearance of a 'tumour' like structure on the posterior part of the brain of *Eucalia inconstans* infected with *D. baeri eucaliae*. In spite of the effect of these parasites on cranial tissues, no external symptoms were observed and infections may go undetected unless the brain is examined. In Great Britain the geographical distribution of *D. phoxini* is wide (ERASMUS, 1962b) and a high level of infection exists in the minnow population. As already pointed out, great mortality may occur during cercarial migration in the fish but it seems likely that the presence of the metacercarial parasites would affect also the longevity and fecundity of the fish.

The presence of diplostomuli in the eye produces more obvious effects which are easily detectable (Fig. 83A, 83B). As the metacercarial infections grow to maturity, the lenses become opaque and the visual efficiency of the eye is greatly reduced (ERASMUS, 1958, 1959; LARSON, 1965). Mass blindness in fish as a result of diplostomulum infections have been reported frequently (SZIDAT, 1927; RUSHTON, 1937, 1938; BAYLIS, 1939; PALMER, 1939; FERGUSON and HAYFORD, 1941; DAWES, 1952) and death results from secondary infection because the distended eyes often rupture and also from the reduction in feeding efficiency in those fish which are largely dependent on visual stimuli for food detection. In a given population 90 per cent of the fish may become infected and as many as 100 larvae may be recovered from each lens. LARSON (1965) describes the formation of herniations of the lens in infections of Bullheads (*Ictalurus spp*). These 'cysts' occurred dorsally and resulted from a herniation of the lens near the point of attachment of the dorsal ligament. Similar 'cysts' or 'growths' have been reported in lens infections by LA RUE, BUTLER and BERKHOUT (1926) and ERASMUS (1958). Larson points out that this herniation may be of some advantage to the fish as the parasites become removed from the main part of the lens which sometimes recovers its transparency thus allowing the continued survival of the host. However, there is the serious disadvantage that the rupture of a distended eye may result in the ejection and complete loss of the eye. In spite of this it is known that, under experimental conditions, fish with heavily infected lenses may survive for at least 544 days after infection (ERASMUS, 1958). Not all species are restricted to the lens but may occur in the humours (e.g. *D. huronense*, HUGHES and HALL, 1929; *D. trutti*, LAL, 1953). The pathology of these infections does not seem to have been described in any detail.

The responses of molluscan hosts to trematode infections

Because of the dual implication of molluscs in the digenetic life-cycle it seems appropriate at this point to consider briefly the responses of molluscan tissues to foreign agents and adverse stimuli. The mollusc becomes involved as a first intermediate host and also as a second intermediate

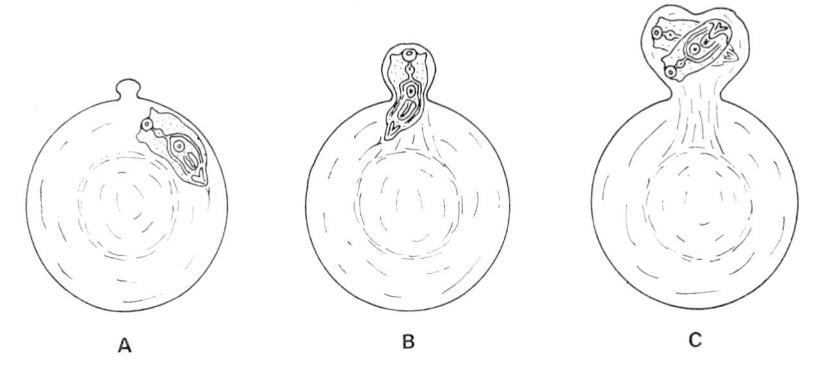

FIG. 83A The herniation of *Ictalurus sp.* lenses infected with diplostomulum (Strigeoidea) metacercariae. **A, B, C** optical sections of lenses, cysts and worms. From LARSON, 1965.

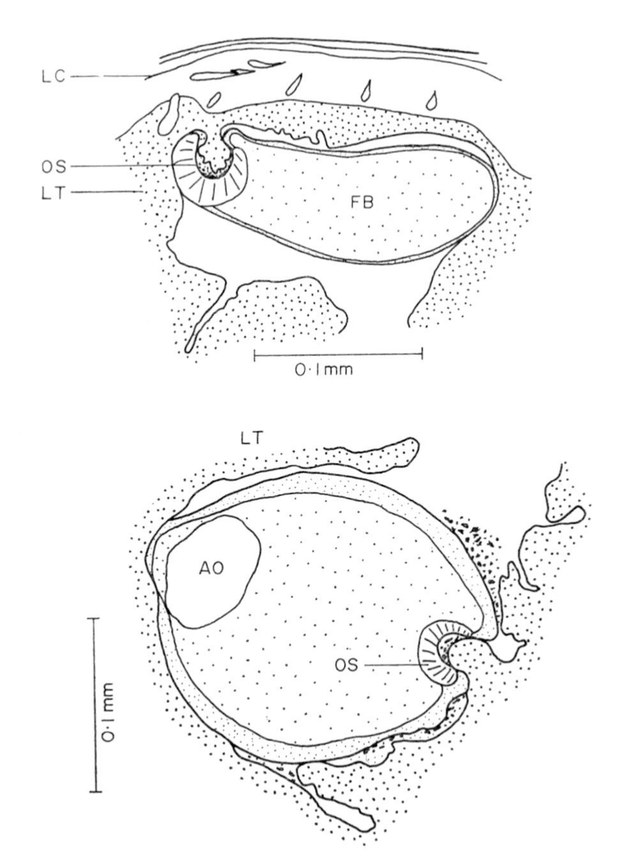

FIG. 83B Semidiagrammatic drawings of sections of fish eye containing diplostomulum larvae. Note that the larvae in both figures have drawn lens tissue into the oral sucker. (AO: adhesive organ; FB: fore-body; LC: lens capsule; LT: lens tissue; OS: oral sucker.)

host and, as has been outlined, becomes exposed to widely differing stimuli at various times depending precisely on the particular host parasite relationship prevailing at that time.

The response of molluscs associated with internal defence mechanisms has been assessed at various times by BISSET (1947), HUFF (1940), STAUBER (1961), CHENG and SAUNDERS (1962), TRIPP (1963) and WRIGHT (1966a). They resemble those of insects (SALT, 1965) and may be considered under the following categories:

(*a*) *Phagocytosis* This cellular activity occurs under experimental conditions in response to injected ink particles, starch granules, foreign erythrocytes and plant spores. It is apparent from these experiments that if the particle is digestible it becomes digested by the phagocyte, but indigestible particles are retained and are translocated through the mollusc to be eventually discharged by the migration of the phagocyte through epithelial surfaces of the alimentary tract, reproductive system and blood system.

(*b*) *Encapsulation* Particles too large to be phagocytized (12 μ and above, TRIPP, 1961) become involved in a different type of response, described as encapsulation. This is also a fairly common response by molluscs and consists of an accumulation of amoebocytes and later fibroblasts around the invading substance. The foreign object becomes completely enclosed by these cells but this does not necessarily mean that the object, if a living organism, dies immediately.

Encapsulation seems to occur within two main types of host parasite relationship. In the first instance, it may be seen in the association between larval trematodes and the molluscan species normal and usual for that particular relationship. PAN (1963) has described the tissue response of highly susceptible strains of *Australorbis glabratus* to the larval stages of *Schistosoma mansoni*. Two types of tissue response were observed. Reaction type 1 was minimal and focal in nature, occurred rarely and was usually associated with degenerating mother sporocysts within 48 hours of infection and consisted of an accumulation of fibroblasts and a few small amoebocytes. No reaction was observed against normally developing sporocysts and this may be significant with regard to the identity of the paletot associated with certain sporocysts (see chapter 3). Reaction type II reached a maximum 10 to 20 weeks after infection and then declined gradually and corresponded with the period of cercarial emergence, which also, as previously mentioned, represents the time of maximum mortality and reduced fecundity in the infected mollusc. At this time the veins of the mantle and rectal ridge become packed with migrating cercariae so that extreme venous dilation ensues with the formation of 'cercarial emboli'. In addition, multiple perforation of the mantle epithelium occurs as the cercariae emerge to the exterior of the mollusc. These stimuli seemed to provoke marked hyperplasia and hypertrophy of the local connective tissue cells as well as their transformation into amoebocytes, which circulated through the body producing granulomas and cellular accumulations around cercariae and daughter sporocysts. Well developed granulomas were described as consisting of a central core of parasitic material enclosed by hypertrophic, phagocytic amoebocytes themselves surrounded by concentric layers of hypertrophic fibroblasts (Plate 7-2). Somewhat similar responses by molluscs harbouring larval trematodes normally associated with them have been described by ROTHSCHILD (1936b) and PROBERT and ERASMUS (1965).

Encapsulation may also occur in the response of an abnormal or unusual molluscan host in infection with larval trematodes. The susceptibility and response of eight species of marine bivalves to infection with the metacercarial stage of *Himasthla quissetensis* has been demonstrated by CHENG *et al* (1966a). Infection was achieved in all but one species, and in most cases, the parasite secreted an inner cyst wall which was covered externally by a wall of host origin. The outer cyst, produced by the host, consisted of leucocytes and myofibres or connective tissue fibres.

Table 37 Susceptibility of the Bahian Strain of *A. glabratus* to infection with *S. mansoni* from Paulista, Pernambuco.

Exptl. groups	Snails				
				Infected	
	Total no.	Dead	Alive	Total	Per cent
Salvador					
1	180	97	83	0	0
2	160	58	102	3	2·9
3	100	52	48	1	2·1
Total	440	207	233	4	1·7
Paulista (control)	120	39	81	68	83·9

(After BARBOSA and BARRETO, 1960).

In a further study of the effects of the molluscan host plasma on the parasite, CHENG *et al* (1966b) suggested that it was the parasite cyst which stimulated the aggregation of leucocytes on its external surface. The apparent refractory nature of certain molluscs to infection is sometimes related to this response. This refractoriness is illustrated by data from BARBOSA and BARRETO (1960) who found that Bahian strains of *Australorbis glabratus* were not very susceptible to infection by *Schistosoma mansoni*, providing a 1·7 per cent infection rate compared to 83·9 per cent in Pernambuco snails (Table 37). The susceptibility of geographical races of *Oncomelania formosana* to infection with human strains of *S. japonicum* was described by MOOSE and WILLIAMS (1964) who found significant differences in susceptibility. Similar studies by NEWTON (1952, 1954), BROOKS (1953), COELHO (1957, 1962), and SUDDS (1960) have demonstrated cellular responses against the larvae and which correspond generally to the type I reaction described above. These observations imply that this resistance to infection, exhibited by refractory snails, may in part be due to these cellular responses. Similar resistant strains and variation in susceptibility to infection are described in the relationship between *Fasciola spp.* and its hosts (KENDALL, 1965) but the mechanisms involved are not clear.

It is possible also that the inability of larval stages to develop within the mollusc is the result of an interaction between the miracidium and the outer epithelium of the snail. Particularly valuable contributions to our understanding of these responses can be made by the study of larval trematodes transplanted into refractory and susceptible snails, and the first observations in this field have been by HEYNEMAN (1966) and CHERNIN (1966). With this technique sporocysts and rediae can be introduced into the molluscan host without involving the suspected 'resistance barrier' offered to the penetrating miracidium by the snail epidermis. Chernin found that nonvectors of *S. mansoni* such as *Helisoma anceps*, *H. caribaeum* and *Lymnaea palustris* were just as refractory to infection by transplants as by miracidial infection. In a somewhat similar manner Hyneman found that both miracidium and rediae of *Echinostoma audyi* developed in its normal host *Lymnaea rubiginosa* and rediae of *Echinoparyphium dunni* also developed in this host. The snail *Indoplanorbis exustus* also favoured the development of its 'normal' parasite *Echinostoma malavanum*. However each species cross-implanted into the wrong or abnormal host failed to establish itself in spite of several attempts. It is too early to elaborate to any great extent on the significance of these findings but they do suggest that resistance to infection may occur at several levels within the snail, i.e. at the external epidermis and also internally within the tissues of the snail. The relative importance of these has yet to be decided.

Humoral immunity The possibility of circulating antibodies is one which must be considered although their presence has not been conclusively demonstrated. The rather uncertain evidence in favour of natural antibodies in molluscan sera has been reviewed by CHENG and SAUNDERS (1962) and WRIGHT (1966a). In addition Cheng and Saunders were able to demonstrate a naturally occurring serum haemagglutinin from *Viviparus malleatus*. The possibility of a specific response to specific antigenic stimuli is suggested by MICHELSON (1963) who investigated the production of a miracidial inhibiting factor by snails (MICHELSON, 1964). *Australorbis glabratus* were inoculated with a variety of substances (acid fast bacillus; echinostome metacercariae; *Daubaylia potomaca*; 5 per cent bovine albumen; polystyrene spheres and *Schistosoma mansoni* eggs). Portions of tissues from the experimental snails were removed, homogenized and tested against *S. mansoni* miracidia. Immobilization of miracidia was observed only in extracts from snails inoculated with *S. mansoni* eggs and *Daubaylia potomaca*. In neither case, however, was the degree of immobilization as great as that by extracts of *A. glabratus* infected with larval *S. mansoni*. These results are very suggestive, but, until the nature of the miracidial immobilization factor is determined, the exact basis of the response cannot be decided.

Leucocytosis An increase in the number of white blood corpuscles does occur in some molluscs in response to certain stimuli, but its relevance to helminth infections is not immediately apparent.

It seems likely therefore that the cumulative effect of the phenomena described above will be to affect the range of hosts available to a parasite and also to influence the establishment of subsequent or secondary infections within the mollusc. Evidence presented by WINFIELD (1932) and NOLF and CORT (1933) showed that snails carrying the sporocysts of *Cotylurus flabelliformis* were resistant to penetration by the cercariae of the same species. The same phenomen occurred (CORT *et al*, 1945) with snails (*Stagnicola sp.*) carrying sporocysts of *C. flabelliformis* and exposed to the same cercariae. CHOWANIEC (1961) however, found that the presence of *Fasciola hepatica* did not prevent the reinfection of the snail in a small number of cases. The result of the infection of a single molluscan species with two differing species of larval trematode has been observed by LIE *et al* (1965). They found that antagonism between the two species resulted in all cases. Snails, harbouring xiphidio-cercariae, *Trichobilharzia sp.*, and a strigeoid resembling *Cotylurus*, were all exposed to the miracidia of *Echinostoma audyi*. In most cases it was found that the original infection gradually diminished and eventually disappeared having been replaced by the echinostome infection. The natural termination of the original infections did not occur in the control snails. It is significant that in all cases a sporocyst species was replaced by a redial species and they also observed the rediae feeding on damaged sporocysts. In cases of treble infections i.e. the echinostome superimposed on a double infection of *Trichobilharzia* and xiphidio cercaria, the *Trichobilharzia* stopped appearing 19 days after experimental exposure, and, in another case after 22 days the xiphidio cercaria stopped emerging, but for a time this latter snail did have three cercarial types emerging from it. The possibility of immunological mechanisms playing a part in this elimination was discarded by these authors and active ingestion of the sporocyst by the echinostome larvae seemed to be the most obvious explanation. In another series of experiments, LIE *et al* (1966), superimposed a second echinostome species on an original echinostome infection. Only 5 per cent of the surviving snails acquired the second infection whereas 89 per cent of control and previously uninfected snails could be infected. It was also found that the experimental snails were not resistant to the cercariae of their own species nor to those of other snails. In most instances it appeared as if the rediae of the original species ingested the sporocyst and rediae of the second, but in those examples where survival within the heart occurred, both species survived simultaneously, although cercarial production was evident from the original species only. In a further series of experiments BASCH and LIE (1966a,b) were able to show that this antagonism between parasite species (*Trichobilharzia sp.* and *Echinostoma sp.*) in the same snail, susceptible to both parasites, did not appear until the larvae were well developed.

The host-parasite relationships of various geographical races and resistant strains of molluscs (WRIGHT, 1966b) is rendered more complex by evidence which suggests that variation in susceptibility is an aspect of intraspecific genetic variation on the part of the mollusc (FILES, 1951; NEWTON, 1953; PARAENSE and CORRÊA, 1963). Furthermore, studies (PARAENSE and CORRÊA, 1963; KAGAN and GEIGER, 1965) show that by crossing susceptible and resistant strains of *Australorbis glabratus*, progeny intermediate in susceptibility may be obtained. By the selection of infective miracidia and passage through the resistant strain of the mollusc, the susceptibility can be increased. KAGAN and GEIGER (1965) found that the average rate of infectivity rose within twelve months by the Brazilian strain of *S. mansoni* from 25·4 per cent to 49·0 per cent for the adapted albino strain of the Brazilian line of *Australorbis glabratus.*

ADULT TREMATODE

The effect which adult trematodes might have upon their hosts is often difficult to predict in specific terms. Considerable variation exists between genera and species occupying similar habitats in the host and even within a single host-parasite relationship many factors influence the final picture but the precise location of the parasite within the host, and its method of feeding, are of particular significance. In some cases, gut contents or mucus secretions of the gut epithelium, gill and external surfaces provide the source of nutrients and in these cases little damage to host tissue results. Even this relatively innocuous method of obtaining food can reach pathological significance by the increase of the parasite population within the host to abnormal levels. More immediate effects on the host are produced by those parasites which feed directly on tissues, either by browsing on the adjacent epithelia, or ingesting blood, or by the extracorporeal digestion of host tissues which are then ingested. As well as by their feeding activities, parasites influence their environment by the secretion and excretion of substances into the lumen of the particular organ or system. These are the substances which might possibly act as antigens to the host and result in the production of an immunological response (KENT, 1963; THORSON, 1963; W. H. O., 1965; KAGAN, 1966) although the exact nature of the stimulatory substance is largely unknown. The extent of the response is influenced by a wide range of factors but, as suggested by CULBERTSON (1941), tissue inhabiting forms might produce stronger responses than those living in the gut lumen, and it is obvious that the method of feeding, and tissue damage involved, will also influence the result. The pathology, both histological and physiological, of adult trematode infections is best known for those species inhabiting man and certain of his domestic animals. e.g. *Fasciola spp.*, *Schistosome spp.*, *Paragonimus sp.*, *Clonorchis sp.*, *Metagonimus sp.*, *Opisthorchis sp.* In a general text such as this, it would be out of place to embark on a detailed consideration of the pathology of these infections and only the more obvious features will be mentioned. For further information the reader is referred to the appropriate textbooks of clinical and veterinary pathology.

External parasites on skin and gills

The majority of monogenean trematodes are included in this category. The general impression created in the literature is that, in small numbers, little effect is produced upon the host, but where a very large population accumulates on the host, severe disease appears. This situation is frequently found in restricted environments where overcrowding of the hosts occurs i.e. in fish hatcheries and stock ponds. The effects these trematodes have on their host is related to attachment and the nature of the opisthaptor and the method of feeding. It appears that attachment by opisthaptors in the form of discs causes damage greater than attachment by clamps. Species which seem to be notorious in producing disease are those of the genus *Dactylogyrus* (BAUER, 1959) and *Gyrodactylus*. The presence of large numbers of parasites causes numerous haemorrhages and ulceration of the host epithelium. The host often responds to this irritation by the formation of epithelial outgrowths. Other symptoms which appear are the excessive secretion of mucus (recorded in *Discocotyle* infections by SOUTHWELL and KIRSHNER, 1937) and the complete destruction of tissues (eye of the fish infected with *Benedenia melleni* described by MACCALLUM, 1927).

HALTON and JENNINGS (1965) concluded that a fundamental difference exists between the Monopisthocotylea and the Polyopisthocotylea as regards the dominant component of the diet. In those representatives of the former group which have been studied all were found to feed on the host's epidermis and epidermal secretions whereas in selected species from the second group blood was a major component of the diet although this was supplemented by gill tissue and secretions in some cases. This difference in feeding habits would obviously influence the pathogenecity of a particular species.

Extensive lesions in the gills are also caused by the digenean *Didymocystis* which occupies well a developed cyst in the gills, lamellae and operculum of the Tunny.

Parasites of the alimentary canal

Digenetic trematodes within the intestine rarely seem to be pathogenic and any effect is usually restricted to a mild inflammatory response at the site of attachment by oral and ventral suckers. But again, the size of the infection influences the result, and BAUER (1959) reports that massive infection of the gut of *Rutilus rutilus* with *Asymphylodora kubanica* had spread to the kidneys with the consequent appearance of dropsy. Most trematodes are attached by their oral and ventral suckers and feed by mechanical browsing on the gut epithelium. In the case of the strigeoid trematodes additional attachment structures are present in the form of the adhesive organ on the ventral surface and the lappets at the anterior end on either side of the oral sucker. The particular pattern of attachment varies—in genera such as *Holostephanus* and *Cyathocotyle* the adhesive organ is acetabular in form (ERASMUS, 1962a; ERASMUS and ÖHMAN, 1963) and attachment takes place largely by this structure. Evidence had now been accumulated which demonstrates the secretion of lytic substances through the surface of the adhesive organ with a resultant extracorporeal digestion of the host tissues at the point of attachment. Consequently infection by these trematodes is characterized by circular eroded patches where the mucosal epithelium covering the villi has been removed and in severe infections the gut wall can become eroded down to the muscular layer of the wall (Plate 8-6). In the case of genera such as *Apatemon*, attachment occurs via the adhesive organ and also by the invagination of the lappets. In such cases erosion of the mucosa occurs at both areas of attachment. In *Diplostomum*, attachment is mainly via the lappets so that damage to the gut is slight and restricted to this region (ÖHMAN, 1965, 1966b). In heavy infections the presence of these strigeoid trematodes can have a considerable effect on the host (VAN HAITSMA, 1930).

The tissues of the gut wall also become damaged during the migration of newly excysted immature flukes to their final location. This has been studied recently in the case of migrating young adults of *Fasciola hepatica* and *Paragonimus westermani* (see chapter 7). In the case of *Paragonimus westermani* (YOKOGAWA *et al*, 1960) the young flukes penetrated through the middle and lower portions of the intestine, and were generally found lying parallel to the longitudinal axis of the intestine. Penetration of the wall through to the abdominal cavity occurred as quickly as 60 minutes after infection of cats.

DAWES (1963a) described a similar migratory route followed by young *Fasciola hepatica* in experimentally infected mice. The young flukes entered the gut wall via the villi, or between the villi, and although a few attempted direct passage through the wall the majority migrated along the gut in the connective tissue between the mucosa and muscle. Eventually complete penetration took place. In both cases considerable erosion of the gut wall resulted in the formation of migratory tunnels. The displaced material from the gut wall is actually ingested so that the fluke is feeding as well as penetrating at this time. HOWELL (1966) was able to demonstrate considerable collaginase activity in extracts of young flukes (*F. hepatica*) and although the morphological site of this activity could not be indicated it is possible that enzyme activity of this type is related to the migratory behaviour of this trematode.

Parasites of the lung

Several genera of digenetic flukes have become adapted to this environment. The most well known genus occurring in mammals is *Paragonimus westermani* inhabiting the lung of man. Considerable lesions are produced in experimental animals during the migration of the fluke to the lung, and it is likely that these occur also in human subjects. The presence in the lung of the mature fluke and its eggs, provokes an intense reaction on the part of the host. This consists of leucocytic infiltration immediately around the parasite and the development of fibrous tissue which eventually forms a thick cyst wall. These cysts may be superficial but commonly occur deep in the tissues of the lungs or other organs. Within the cysts there are eggs, fluid, Charcot-Leydon crystals and worms which may be either dead and disintegrating or alive. Cysts may occur in tissues other than the lungs, and result from the presence of worms which have become diverted from the main migratory path. Lesions external to the lungs in the pulmonary cavity take the form of adhesions between the lungs and the pleural lining. In addition, eggs may enter the circulation and produce inflammatory responses similar to those described for schistosome eggs (YOKOGAWA *et al*, 1960; YOKOGAWA, 1965). It is obvious that a considerable immunological response exists and intradermal agglutination and complement fixation tests can be used to supplement the clinical diagnosis. When cyst formation occurs outside the lungs the symptoms will differ depending on the particular organ parasitized.

Two other well known genera, but which occur in the lungs of amphibians are *Haplometra* and *Haematoloechus*. Although these parasites sometimes infect the lungs very heavily no apparent discomfort is caused to the frog. The pulmonary epithelium may become columnar as a result of the infection. The parasites are also blood feeders and the gut becomes filled with the heavily pigmented material.

Parasites of the liver

This is a habitat frequently parasitized by trematodes and included species historically well known, as well as being notorious for their pathogenicity to man and domestic animals. Genera living in this habitat include *Fasciola*, *Dicrocoelium*, *Clonorchis* and *Opisthorchis*.

The pathology of *Fasciola* is complex (SEWELL, 1966; ROSS *et al*, 1967a) and varies in relation to the particular development phase being considered. It has been mentioned that the young fluke, after release from the metacercarial cyst, penetrates and migrates within the intestinal wall causing quite distinct lesions. This phase is followed by migration across the abdominal cavity and penetration of the liver capsule. The young animal then migrates through the parenchyma of the liver eventually localizing in the bile ducts. How this last manoeuvre is achieved is not yet understood (DAWES, 1961a,b). During this migration the fluke actively feeds, ingesting the parenchymal tissue with its oral sucker (DAWES, 1963b). A considerable host response appears at this time in the mouse and the burrows first become infiltrated with leucocytes and granulation tissue with the response finally terminating in fibrosis. DAWES (1963a) suggests that the migration of these flukes through the liver stimulates the biliary epithelium to undergo hyperplasia before the flukes have actually entered the bile duct. The inflammation continues so that the bile duct wall shows hyperplasia, granulation tissue and fibrous tissue. This immature parasite is still regarded as a tissue feeder at this time and browses on this hyperplastic epithelium which becomes eroded by the oral sucker and the tegumentary spines (DAWES, 1963c). The question of the nature of the food ingested by the fluke is still uncertain to some extent as some workers (e.g. STEPHENSON, 1947b; TODD and ROSS, 1966) have suggested that it is a blood feeder, at least in the adult stage. When adult and established in the liver, several other workers (see BORAY, 1969) have described the ingestion of blood which is consequently lost via the bile into the faeces of the host. In addition to the hyperplasia the bile ducts may become calcified (although this is largely a characteristic of bovine infections—ROSS *et al*, 1966, 1967b) and also contains a black pigment thought to represent haematin discharged from the fluke.

Although fascioliasis in sheep and cattle has many similarities (DOW *et al*, 1968; ROSS *et al*, 1967a) there are many clinical and histological differences (Table 38). In cattle, acute fascioliasis is rare and resistance mechanisms reduce the probability of accumulated worm burdens occurring and the resultant chronic anaemia as occurs in sheep. The movement of the migratory stage in the liver is less impeded in the sheep, whereas in cattle, movement is restricted and the migratory tracks become fibrosed and the bile ducts grossly fibrotic and calcified. In the sheep, tracks are numerous and distinct, giant cell formation and hepatic cell hyperplasia are common whereas in cattle, tracks are not so frequent, are more fibrotic and hepatic cell hyperplasia is not a feature. When large numbers of migrating stages accumulate in sheep liver, extensive haemorrhage within the liver may occur and this period represents a critical stage which may result in the death of the host.

As well as these specific changes in the liver more general alterations of the blood take place—serum globulins increase, blood iron, phosphorus, magnesium, sugar and potassium levels may fall but the calcium level may be only slightly affected (PANTELOURIS, 1965b; ROSS *et al*, 1967a). In its migration through the liver, the fluke may also introduce secondary infections of bacteria producing the condition known as 'black disease' of sheep. The entire effect of all this upon the host is considerable and can account for the death of much livestock as well as producing considerable economic loss by the rejection of infected livers at slaughter (TAYLOR, 1964).

The fluke is able to infect man but the symptoms are rather vague and variable, the most characteristic ones being a strong eosinophilia, an excess of γ-globulin coupled with digestive upsets, pain and fever.

The symptoms associated with opisthorchiasis, clonorchiasis and other bile duct infections follow approximately the same pattern with hyperplasia of the duct and fibrosis (Plate 8-2).

Parasites of the blood system

This habitat has become utilized by adult schistosomes, sanguinicolids and spirorchiids. Some of the pathological features associated with these infections have been discussed already.

The pathology of schistosomiasis will vary depending on the species and the site infected. In the case of *Schistosoma haematobium* it is the bladder which is mainly involved, but in *S. mansoni* and *S. japonicum*, the disease is mainly associated with the bowel. The disease in all results from (1) a general and localized reaction to the by-products from growing and mature worms, (2) trauma with haemorrhage as the eggs escape from the venules, (3) abscess and tubercle formation around eggs lodged in the perivascular tissue, and against this background the clinical symptoms may be considered in three periods. 1. Incubation period, 2. period of egg deposition and extrusion and 3. tissue proliferation and repair. All these aspects are well described in the literature (e.g. FAUST, 1949; EL MOFTY, 1962) but the studies of the pathophysiology of the infection have been hampered by the presence of nutritional defects and concurrent infections in human subjects, and by the technical difficulty of obtaining sufficiently large volumes of serum for analysis from experimental animals. Recently SADUN and WILLIAMS (1966) have reinvestigated the situation using ultramicrochemical techniques and experimental infections of *S. mansoni* in mice. They found that, even with relatively heavy worm burdens, no significant departures from normal values were obtained with glucose, alkaline phosphatase, phosphorus, bilirubin, creatinine, calcium, sodium, potassium, chloride and carbon dioxide values. Serum transaminase values increased with a relatively greater increase in serum glutamic pyruvic transaminase levels than with serum glutamic oxaloacetic transaminase. These changes are thought to indicate the degree of cellular necrosis and cellular permeability especially in the heart, liver, skeletal muscle, kidney and brain. In addition, significant increase in the degree of retention of bromosulfalein (indicative of a derangement of the excretory function of the liver) and an increase in total protein was observed in the later stages of heavy infections. The albumin/globulin ratios became decreased particularly in mice with heavy worm burdens.

This review of the pathology of helminth infections has been rather brief but it serves to emphasize the point made earlier that the infections produce results which differ considerably between different host-parasite relationships, so that the nature and extent of the pathological process often varies greatly at different stages of the life-cycle. It must be emphasized that these accounts are based on naturally occurring stock and that we are observing the interaction between the parasite and a host with its natural population of protozoa, bacteria etc. and sometimes other helminths. The exact role of the helminth in producing disease can be determined only on gnotobiotic hosts.

Table 38 The characteristics of the four types of fascioliasis.

Disease type	Clinical features	Total fluke burden	Mean length of flukes mm	Eggs in faeces e.p.g.	Packed cell volume %	Haemoglobin g/100 ml	Erythrocytes 10^6	Mean corpuscular volume Cu	Mean corpuscular haemoglobin concentration %	Serum Mg mg/100 ml	Serum Ca mg/100 ml	S.G.O.T. I.U./l.	Serum Albumin g/100 ml
Acute Type 1	Sudden death following acute haemorrhage	1,000 to 2,500	5·0 to 8·0	Nil	10	3·08	1·97	50·7	30·8	1·74	7·5	320	1·68
Acute Type 2	As Type 1	700 to 1,000	10 to 12						Probably as Type 1				
Subacute	Subacute anaemia with rapid loss in condition	800 to 1,500	12 to 14	1,300	12·5	3·5	1·43	87·4	28·0	2·48	7·5	140	1·07
Chronic	Gradual loss of condition anaemia, emaciation and bottle jaw	200 to 800	16 to 20	200 to 10,000	12·0	4·18	2·9	41·3	34·8	2·14	6·9	100	0·89
Normal controls	—	0	—	0	35	11·5	11·0	30·4	31·8	2·5	9·0	90	2·5

(After Ross, Dow and Todd, 1967).

Immunology of Adult Infections

It is obvious that many of the infections associated with adult trematodes affect the histology and physiology of the host to a considerable extent. It follows from this that pronounced immunological characteristics will be associated with disease of this nature. The situation with the Trematoda is particularly difficult because of the variation in the morphology and physiology of the infective stages. In schistosomiasis it could be egg, cercaria, schistosomule or adult stage which is responsible for initiating the immunological response, although SMITHERS and TERRY (1967) implicate the adult, whereas in fascioliasis and paragonimiasis it is the migrating metacercaria which is particularly suspect. The diversity illustrated by these two examples makes extrapolation and generalization difficult. The presence of a gut with the associated egestion of materials, excretory system with its products and a wide range of gland cells of different characteristics makes the discovery and identification of antigenic sources far from easy. The revised concept of the tegument, now regarded as cytoplasmic and thus capable of pinocytosis, secretion, enzyme activity and other cellular processes in its own right, adds a further and particularly difficult complexity to the problem. The effect of X-rays and Gamma rays on this cytoplasmic tegument has not been taken into account when irradiated parasites are used in an attempt to induce immunity (e.g. HUGHES, 1962; LICHTENBURG and SADUN, 1963; DAWES, 1964; DAWES and HUGHES, 1964; ERICKSON, 1965; ERICKSON and CALDWELL, 1965; HORAK, 1965).

DAWES (1964) has observed that x-irradiated flukes in mice become beset by dense aggregates of neutrophil leucocytes while burrowing in the liver. These leucocytes adhere to the fluke tegument and also penetrate the limiting plasma membrane and accumulate within the matrix. This evidence suggests that irradiation adversely affects the limiting plasma membrane of the fluke, and the fact that the tegument eventually fragments suggests that the secretory activity of the tegumentary cells is also inhibited so that replacement of this cytoplasmic layer

Table 39 Worm recovery from monkeys immunized against mouse tissues, 5 weeks after transplantation of worms from mice.

		Worm recovery 5 weeks after transfer of mouse worms			
Monkey	Immunized against	Males	Females	Total	% recovery
256	control	78	73	151	93
257	control	68	45	113	69
258	control	66	58	124	77
248	mouse liver and spleen in Freund's	0	0	0	0
249	mouse liver and spleen in Freund's	0	0	0	0
250	mouse liver and spleen	27	6	33	20
251	mouse liver and spleen	17	5	22	15
254	mouse red cells	3	0	3	2
255	mouse red cells	28	10	38	23
252	mouse serum	65	46	111	68
253	mouse serum	41	22	63	38

(From SMITHERS, 1968).

does not occur. The fibrous nature of the basement layer appears to be unaffected by this treatment and invasion of the fluke does not proceed beyond this point. The inability of these animals to survive might be related directly to the damaged tegument and the consequent reduction in its efficiency regarding nutrient absorption, excretion and respiratory processes associated with the tegumentary mitochondria. An extension of the survival time of the irradiated material could possibly be obtained by a more critical relating of X-ray dose to subcellular changes and thus provide a greater period in which immunity might develop.

The complexity of the immunological response associated with trematode infections is admirably illustrated by the host-parasite system exemplified by *Schistosoma mansoni*. As mentioned earlier, SMITHERS and TERRY (1967) clearly implicated the adult worm as the antigen source rather than the egg, invading cercarial stage or the migrating schistosomule. The results were obtained after transplanting (Figs. 84, 85) established worms from donor hosts into monkeys which were subsequently challenged with cercariae. The donor hosts were monkeys, hamsters and mice and the subsequent survival of adults in the recipient monkeys varied with the donor host species. Worms transferred from hamsters were not very successful and few eggs were laid with the worms dying after 14 days in the new host. Worms from monkeys adapted well, and eggs were produced soon after transference. Schistosomes from mice showed an initial check in egg production but this did not persist and eventually a high rate of egg production was attained suggesting that the worms had adapted to their new hosts.

In order to survive within a host every parasite must become adapted to that host and the possession by the parasite of antigens similar to those of the host might enable the parasite to avoid any intense immunological response on the part of the host (SPRENT, 1959). CAPRON *et al.* (1965) and DAMIAN (1967) both showed that schistosomes shared a number of antigens with their host. An experimental demonstration of the presence of mouse antigens in schistosomes reared in mice has been carried out by Smithers and Terry (quoted by SMITHERS, 1968) by

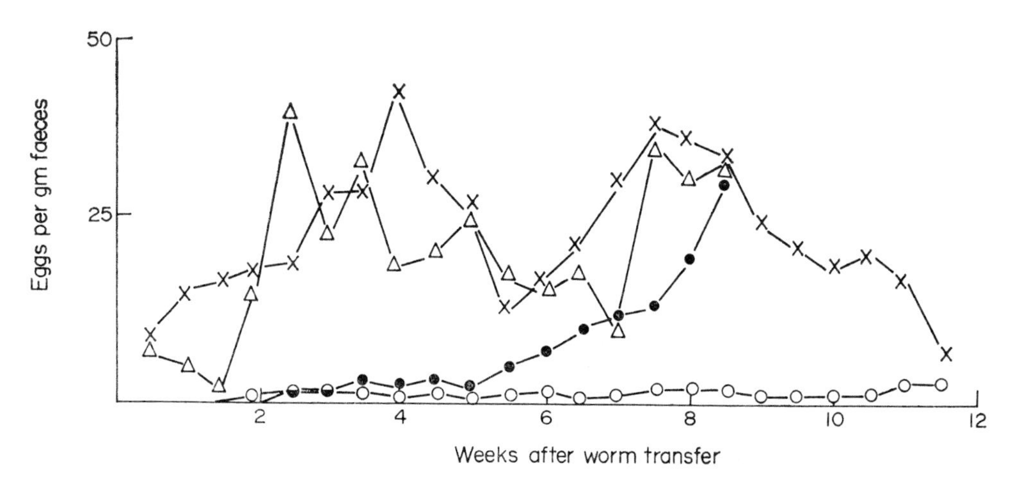

FIG. 84 The mean egg production of adult *Schistosoma mansoni* which have been transferred from various donor hosts to the recipient monkey host.

×——× } worms from monkey donor (series II)
△——△ } worms from monkey donor (series III)
●——● worms from mouse donors (series III)
○——○ worms from hamster donors (series I)

From SMITHERS and TERRY, 1967.

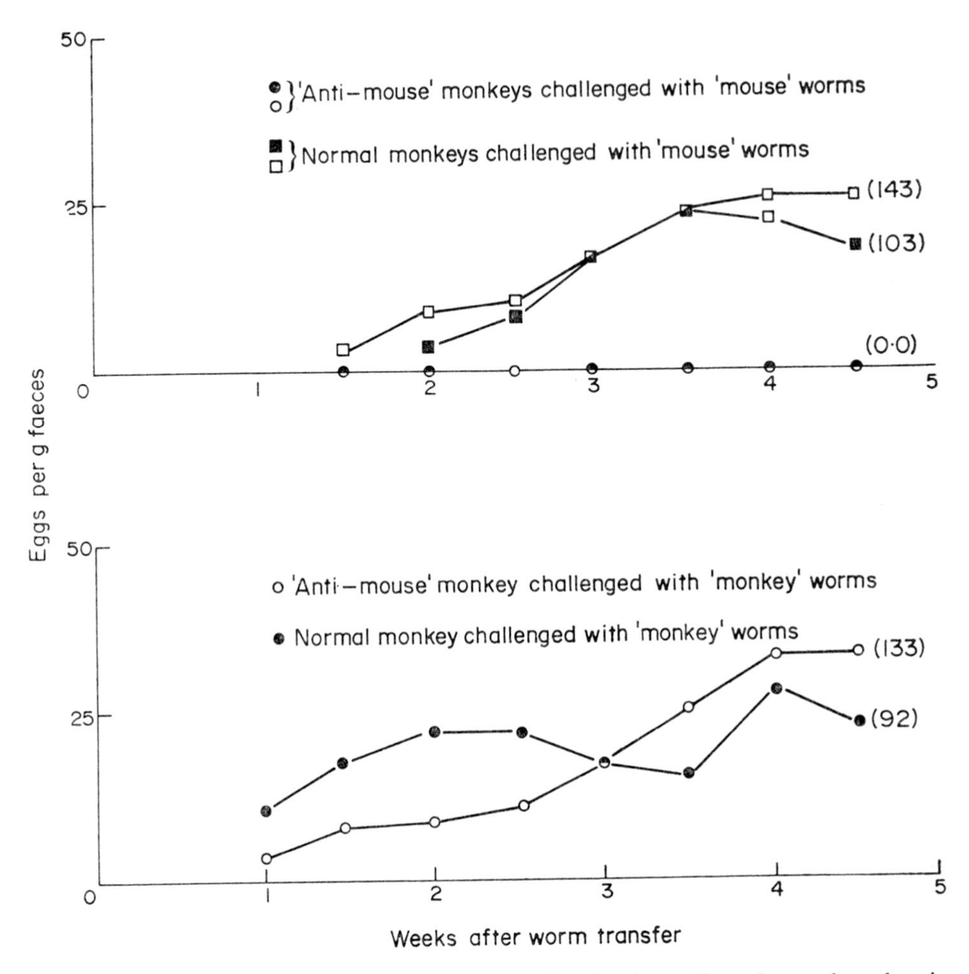

FIG. 85 The egg production of adult *Schistosoma mansoni* transferred to 'normal' monkeys and monkeys immunised 'against' mice. Figures in parentheses indicate the number of worms recovered at post mortem. From SMITHERS, 1968.

transferring schistosomes reared in mice to monkeys immunized with mouse antigens (Table 39). In these 'anti-mouse' monkeys the transferred schistosomes rapidly died whereas in normal monkeys (not previously immunized against mouse antigens) the worms began egg producing after an initial period of reduction in egg output. The authors suggested that this initial drop represented a period when 'mouse' schistosomes became readapted antigenically so that survival in monkeys was possible. Worms from hamsters did not appear to have this ability to adapt.

This 'immunological masquerade' might provide an explanation for the concomitant immunity exhibited by hosts in schistosome infections. In this host-parasite relationship, adult worms appear to survive in the blood stream whereas penetrating cercariae and migrating schistosomules become immobilized. This apparently anomalous situation is difficult to explain although the work of Smithers and Terry provides a very acceptable explanation. The nature and mode of uptake of host antigens by the worm is unknown. It may represent 'biological' contaminants such as mouse red blood corpuscles in the parasite gut or it is possible that host antigens are essential to the worms in a physiological or biochemical sense and become incorporated into their own biochemical systems.

Considerable effort is now being made to isolate, characterize and determine the origin of parasite antigens and apply this information to the development of techniques which will permit vaccination and the induction of passive immunity in hosts. The literature is extensive and the reader is referred to general papers such as those of LINCICOME, 1962; KENT, 1963; THORSON, 1963; MAGALHÃES FILHO *et al* 1965; SMITHERS and OGILVIE, 1965; YOGORE *et al*, 1965; W.H.O. REPORT 1965; KAGAN, 1966 and TERRY, 1968. A variety of humoral antibodies have been identified from trematode infections e.g. precipitins, agglutinins, CHR reaction, complement fixation as well as reagins (SMITHERS, 1967) but although specific reactions can be demonstrated *in vitro*, their function *in vivo* is not always clear. The inability to induce passive immunity by the transfer of serum rich in specific antibodies is difficult to explain. The nature and source, within the parasite, of the particular antigens responsible for a reaction has not been resolved in any precise terms. Helminth infections, including trematodes, are often characterized by marked cellular responses such as eosinophilia and, although several hypotheses exist, the actual role of eosinophils and other cells in the immunological response to helminth infections has not been determined (SOULSBY, 1967). Thus, in the immunology of trematode infections, there are still very many questions to be answered. The task, as mentioned earlier, is particularly difficult in the trematodes because of the variety of developmental stages involved, each with its characteristic physiology and host-parasite relationship, thus presenting a continually changing sequence of immunological situations.

9 Metabolism and the Host-Parasite Interface

The investigation of the physiological characteristics and biochemical processes of the Trematoda is a particularly difficult task. The majority of these parasites are small (usually well below three centimetres) and exhibit structural features which make it difficult to isolate organ systems. The parenchymal cells are closely associated with all structures so that it becomes impossible to dissect out, free of contamination by other systems, organs, such as the alimentary tract, for physiological study. The study of homogenates prepared from the entire body will contain such diverse structures as reproductive system, alimentary tract, excretory system as well as a variety of specialized cells. The implications derived from such studies must, in some cases, be considered with caution because of the heterogeneous nature of the material. It is likely that modern methods of investigation employing such techniques as differential centrifugation and histochemistry will however, resolve some of these difficulties. In spite of these technical problems, a large proportion of the basic features of trematode metabolism have been described although the details of many of the biochemical and intermediary processes have yet to be determined. As might be expected the majority of observations are confined to *Fasciola spp.* and *Schistosoma spp.* and the ubiquitous distribution of these described features within the Trematoda is largely assumed. The available information has been very adequately reviewed by several workers (BUEDING, 1949, 1962; MOULDER, 1950; GEIMAN and MCKEE, 1950; ROGERS, 1962; SMITH, 1965; VON BRAND 1950, 1952, 1960, 1966; SMYTH, 1966b). In the time which has elapsed between the two textbooks of VON BRAND (i.e. 1952 and 1966) the majority of the new information concerning helminths available in the latter text relates to groups other than the Trematoda and Acanthocephala. Because of the availability of these detailed reviews only the broad outlines of trematode physiology will be indicated below. The data available on culture *in vitro* has been discussed recently by SILVERMAN (1965) and WEINSTEIN (1966).

The physiological relationship which exists between a parasite and its host is a particularly complex one and represents an attempt to maintain a steady state between two elaborate and variable components. Because the two components of this system (host and parasite) are living organisms, and are themselves attempting to maintain equilibrium there must exist continual interchange between the two in order to allow the basic relationship to continue. All physiological and biochemical attempts to visualize and understand this interchange must, sooner or later, involve some consideration of the host-parasite interface. The interface may be regarded as that surface through which interchange of materials of physiological and immunological importance takes place. This involves the passage of substances, possibly antigenic, out from the parasite into the host in the form of excretions, secretions and egestions. The movement inwards consists of the absorption of nutrients, osmotic and ionic interchange and the eventual entry of antibodies from the host. In the trematodes (Fig. 86) three main interfaces may be postulated: 1, the surface of the alimentary tract, where ingestion, secretion, digestion and egestion may take place; 2, the general tegument, now known to be cytoplasmic, metabolically active, and capable of being involved in a wide range of cytochemical phenomena such as absorption, pinocytosis, transmembranosis and secretion; 3, the surface of attachment organs, such as suckers and

Table 40 Distribution and relative amounts of glycogen in fifteen trematode species.

Tissue	Monogenea							Digenea								
	1	2	3	4	5	6	7	8	9	10	11	12	13	14	15	16
Cuticle	0	0	0	0	0	0	++	0	0	0	0	0	0	0	0	0
Anterior parenchyma	0/+	0/+	+++	0/+	0/+	0/+	0/+	+++	++	+++	+++	+++	+++	+++	+++	++
Posterior parenchyma	0/+	0/+	+++	+	+	++	+++	++	+	++	++	+	+	+++	++	+
Body musculature	+	+	+	0	0	0	0/+	+	+	+	+	+	+	+	+	0/+
Buccal/oral sucker	0	0	+++	+	+	0/+	0/+	+++	++	+++	++	+++	+++	+++	+++	++
Clamp/ventral sucker	0	0/+	+++	0	0	0/+	0/+	+++	+	+	+	++	++	++	++	—
Pharynx	0/+	0/+	+++	+	+	++	+	++	+++	+++	++	—	—	+++	—	—
Gut caeca	0	0	0	0	0	0	0	0	0/+	0	0	0	0	0	0	0
Ovary/ova	0	0	0/+	0	0	0	0/+	0	0	0	0	0	0	+	—	0
Testes/spermatozoa	0	0	++	0	0	0	0	+	++	+	++	0	0	0	0	—
Eggs	—	—	—	—	0/+	—	++	++	++	++	++	++	++	++	—	+
Vitellaria	0/+	++	++	++	++	++	++	+	0/+	+	0/+	0/+	0/+	+	—	0/+
Excretory system	0	0	0	0	0	0	0	0	0	0	0	0	0	0	0	0

1 = *Entobdella hippoglossi*
2 = *Calicotyle kröyeri*
3 = *Polystoma integerrimum*
4 = *Diplozoon paradoxum*
5 = *Discocotyle sagittata*
6 = *Diclidophora merlangi*
7 = *Octodactylus palmata*
8 = *Opisthioglyphe ranae*
9 = *Diplodiscus subclavatus*
10 = *Haplometra cylindracea*
11 = *Haematoloechus medioplexus*
12 = *Gorgoderina vitelliloba*
13 = *Gorgodera cygnoides*
14 = *Fasciola hepatica*
15 = male *Schistosoma mansoni*
16 = female *S. mansoni*

+++, intensely stained; ++, moderately stained; +, slightly stained; 0/+, occasionally stained; —, not known.
(After HALTON, 1967d).

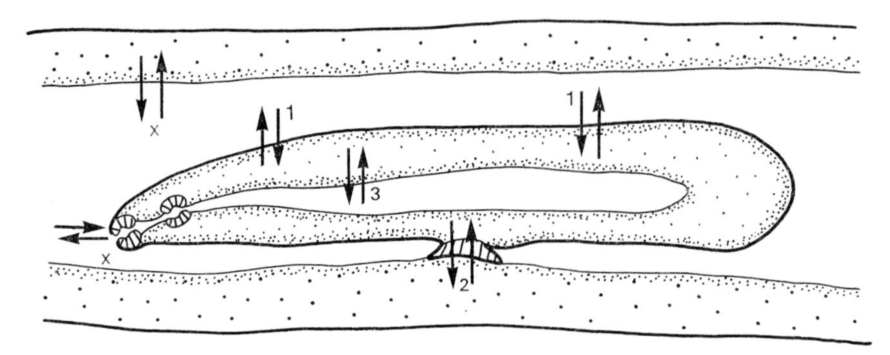

FIG. 86 The host-parasite interfaces as represented by an adult digenean parasite in a vertebrate digestive tract. Arrows represent the movement of substances into and out from the parasite and host. The various interfaces are: 1. The general tegument; 2. the tegument of attachment organs; 3. the surface epithelium of the parasite alimentary tract. Note that interchange of substances between host and parasite may also occur at those regions of the host gut indicated by X and at region 2.

adhesive organs, which may exhibit specializations associated with secretion and absorption. It will be suggested later that the physical association between parasite and host may become so intimate as to resemble contact between adjacent cells and concepts of interchange between host and parasite may have to be reassessed in these terms.

A. General metabolism

CARBOHYDRATE METABOLISM

All the available data indicates that carbohydrate metabolism is particularly well developed in the parasitic helminths, including the Trematoda. Culture *in vitro* and respiration studies show that the few trematode species observed are able to utilize simple carbohydrates such as glucose, mannose and fructose. The possibility of intracaecal digestion of complex carbohydrate has not been demonstrated and the histochemical characteristics of the caecal epithelium (described later) do not allow any conclusions to be drawn. The endogenous carbohydrate reserve in these parasites is glycogen (Table 40) and this is stored in the parenchymal cells and musculature with occasional reports of distribution in the tegument and parts of the reproductive system (vitelline gland and gonads, HALTON, 1967d). The description of the parenchymal cell in *F. hepatica* (THREADGOLD and GALLAGHER, 1966), and this cell and glycogen deposition in the cestodes (LUMSDEN, 1965, 1966), suggests that the parenchymal cell may have considerable significance in the process of glycogen deposition in both cestodes and trematodes. The glycogen content of the entire parasite is usually high (3–21 per cent dry wt., VON BRAND, 1966), but does not vary between sexes e.g. *Schistosoma* and in relation to the physiological state of the host. MÜLLER (1966) found that the glycogen content of *Haplometra cylindracea* varied with the season as did the content of the liver of the host (*Rana temporaria*) and that the lowest level in both occurred in the spring. Under conditions of starvation the glycogen is depleted, and disappears first from the peripheral tissues, particularly the parenchymal cells, and later from the musculature (VON BRAND and MERCADO, 1961). In common with most other parasitic helminths the rate of carbohydrate consumption is similar under aerobic and anaerobic conditions. The end products of these fermentations are varied and include lactic, acetic and propionic acids as well as higher fatty acids (VON BRAND, 1966). Investigations into the process of carbohydrate degradation suggest that in the trematodes, as in the other helminths, the initial steps follow those of the classical Embden-Meyerhof sequence, i.e. phosphorylative glycolysis although most of the evidence is more complete from groups other than the trematodes. Many of the enzymes involved (phosphorylase, phosphoglucomutase, hexokinase, phosphohexose isomerase, phosphohexokinase, phosphotriose isomerase, phosphotriose dehydrogenase, lactic dehydrogenase) have been identified in *Schistosoma mansoni* as well as some of the phosphorylated glycolytic intermediates (VON BRAND, 1966; BRYANT and WILLIAMS, 1962; WILLIAMS and BRYANT, 1963). The possibility of alternative pathways has not been demonstrated in the trematodes, although the major components of a pentose phosphate pathway have been demonstrated in the cestode *Echinococcus granulosus* (AGOSIN and AREVANA, 1960). In the same way some of the enzymes (succinic and malic dehydrogenase), and most of the intermediates (BRYANT and WILLIAMS, 1962) of the Krebs tricarboxylic acid cycle have been demonstrated in *F. hepatica*. The utilization of succinic and malic acid by the adults and rediae of *Himasthla quissetensis* (VERNBERG and HUNTER, 1963) also tend to confirm the existence of this cycle. There is some evidence (BRYANT and WILLIAMS, 1962) which suggests that the cycle may not be equally well developed at all stages. These authors found evidence of marked Kreb's cycle activity in the adult of *F. hepatica* but not in the miracidium. In extracts of *F. hepatica* KRVAVICA and MARTINČIČ (1964) have identified Coenzyme A.

RESPIRATION

The information available (reviewed by VERNBERG, 1963; VON BRAND, 1966) suggests that adult parasitic helminths are faculative aerobes and, although they show a marked tolerance to lack of oxygen, are able to utilize it when available. The particular relationship for any trematode will depend on the host and the tissue or organ in which it lives and, as ROGERS (1949) has shown, even mucosal inhabiting forms may encounter biologically significant oxygen tensions. In the case of miracidia, cercariae and eggs a more pronounced dependency on oxygen exists and generally development of these stages, and survival, is markedly affected by anaerobic conditions. The rate of oxygen consumption varies with the size of the parasite (larger forms having a lower oxygen consumption than smaller ones) so that in digenetic trematodes the respiratory rate varies during the life-cycle (VERNBERG and HUNTER, 1959). During the life-cycle of *Gynaecotyla adunca* the respiratory rate (as indicated by μl O_2/hour/μg N) decreases as the worm becomes larger. Before the metacercaria encysts, the body N has increased 60 times over that of the cercaria and the oxygen consumption decreased nearly 40 per cent. In the final host, 24 hours after excystment, the body N is nearly 100 times that of the cercaria and the respiratory rate nearly 64 per cent less. The small amount of evidence available indicates that the respiratory rate is dependent (i.e. directly proportional to) on the external oxygen tension, although the situation may be more complex when considering larval stages (VERNBERG, 1963). The cercariae of *Himasthla quissetensis* were 'conformers' and their respiratory rate decreased with the oxygen tension of the external environment, the redial stage, however, showed less dependency and the QO_2 in 0·5 per cent oxygen was 61 per cent of the rate in 5 per cent oxygen. The corresponding figure for the cercarial stage was 17 per cent of the rate at 5 per cent oxygen. It was also found that the cercaria of *Zoogonus rubellus* showed this usual dependency on external oxygen tension down to 3 per cent oxygen, but then seemed to regulate its respiratory rate independently of further decreases in oxygen tension. VERNBERG and HUNTER (1961) also found a correlation between the effect of increasing temperature on the respiratory rate of a number of adult trematodes and the thermal environment of the final host. The parasites normally found in poikilothermic hosts died at the normal body temperature of birds. In the case of larval trematodes (VERNBERG, 1961), the lethal thermal limits corresponded to that of the final host.

Although described as faculative anaerobes, *S. mansoni* (BUEDING, 1949, 1950) and *Gynaecotyla adunca* (HUNTER and VERNBERG, 1955b) did not show evidence of an oxygen debt following a period of anaerobiosis, but in contrast, READ and YOGORE (1955) reported that the preanaerobic rates of 0·74–0·86 mm^3 O_2/mg dry tissue/hour for *Paragonimus westermani* increased to 1·72–5·06 mm^3 after the first and to 1·58–2·74 mm after the second half-hour period of anaerobiosis. Under completely anaerobic conditions *Fasciola hepatica* has been shown to survive 20 days (ROHRBACHER, 1957) and *Schistosoma mansoni* (ROSS and BUEDING, 1950) five days.

The occurrence and physiological significance of haemoglobin has been reviewed by LEE and SMITH (1965). They report the identification of haemoglobin in nine species of trematode as well as several others where the colour of the parasite suggests that the presence of haemoglobin is probable. It is generally regarded that the parasite and host haemoglobins are quite distinct, and STEPHENSON (1947b) has shown for *Fasciola hepatica,* that although present in many tissues it is particularly concentrated in the vicinity of the uterine coils and in the vitellaria. The role of these compounds is far from clear and they could act as oxygen stores, as suggested for *Telorchis sp.* and *Allassostoma sp.* (WHARTON, 1941) or function in the transport of oxygen through the tissues. Other iron-porphyrin compounds include catalase which exhibits low activity in *Fasciola hepatica* (PENNOIT-DECOOMAN and VAN GREMBERGEN, 1942) and Cytochrome C reported from *Allassostoma magnum* (WHARTON, 1941), *F. hepatica* (VAN GREMBERGEN, 1949) and *S. mansoni* (BUEDING and CHARMS, 1951). The evidence obtained by BUEDING and CHARMS (1952) on cytochrome oxidase activity in *S. mansoni* suggests that respiration in these worms may be mediated by other respiratory catalysts and the fact that the oxygen consumption of *F. hepatica* for instance is inhibited by cyanide (capable of reversal by methylene blue in some cases) does not indicate the nature of the respiratory system present (VON BRAND, 1966). In contrast LAZARUS (1950) found that cyanide stimulated rather than depressed the respiratory rate of *Paramphistomum cervi.* It is obvious that considerably more information is needed on this aspect of metabolism in trematodes.

PROTEIN METABOLISM

The phenomenon of larval multiplication with the resultant production of large numbers of larvae, and the fecundity of the adult, egg-laying trematode, must throw considerable demands on the protein metabolism of the parasite at various stages in its life-cycle. The information obtained by CHENG (1963b) and GILBERTSON *et al* (1967) suggested that sporocyst, redial and cercarial stages could obtain amino acids from either the serum of the molluscan intermediate host or from the cells of the hepatopancreas. SENFT (1967) also records arginine depletion in snails infected with schistosomes. The possibility of proteolytic enzymes being secreted by these stages is purely speculative, although extracts of schistosome cercariae have exhibited collagenase and elastinolytic activity (STIREWALT, 1963b; GAZZINELLI and PELLEGRINO, 1964) *in vitro.* With a more purified extract, GAZZINELLI, RAMALHO-PINTO and PELLEGRINO (1966) were able to demonstrate a chemotryptic-like specificity on synthetic substrates. In view of the heterogeneous source of this material it is not possible to relate the enzyme activity to specific structures nor appreciate exactly its significance *in vivo*. In the case of the adult *S. mansoni*, a protease specific for haemoglobin was isolated from homogenates of whole worms and, in view of the evidence available demonstrating the digestion of haemoglobin by this parasite, it seems likely, though not positively proven, that this activity resides in the gut. The occurrence of large amounts of RNA (discussed later) in the caecal epithelium and the ultrastructure of the cells indicates pronounced protein synthesis by this tissue.

The uptake of a variety of amino acids under *in vitro* conditions has been demonstrated by EHRLICH *et al* (1963) PANTELOURIS (1964a,b), THORSELL and BJORKMAN (1965) and KNOX (1965a,b) for *F. hepatica* and, as discussed later some of these localized in the caecal epithelium. In schistosomes (*S. mansoni*), SENFT (1963, 1965, 1966) has described the uptake of a number of amino acids under *in vitro* conditions. He was also (1963) able to show that some of the glucose absorbed becomes shunted over into the protein cycle by a transamination procedure in which pyruvate becomes converted to alanine. Using labelled amino acids (1966) he found that ^{14}C-L-arginine was rapidly absorbed in relation to other amino acids. Homogenates of worms converted arginine to ornithine and urea and it was concluded that the arginase reaction was the most important arginolytic pathway employed by *S. mansoni*. Transaminase activity has also been identified in extracts of *F. hepatica* (DOUGHERTY, 1952). The excretion of urea has been described in *F. hepatica* (EHRLICH *et al*, 1963), and considerable arginase and ornithine transcarbamylase activity detected in extracts of *F. hepatica*, *Dicrocoelium lanceatum* and *Paramphistomum cervi* although the latter does not synthesize urea (RIJAVEC, 1966). Arginase activity was first detected in *F. hepatica* 14 days old and could not be demonstrated in eggs, miracidia nor cercariae. In contrast, ornithine transcarbamylase activity was present in all stages. Other enzymes of the ornithine cycle (carbamyl phosphate synthetase, arginino-succinic acid synthetase, arginine synthetase and arginase), have been demonstrated in extracts of *F. hepatica* (KURELEC, 1966) indicating that the urea excreted is a final product of its own nitrogen metabolism. Urease activity has not been demonstrated in *F. hepatica* (VAN GREMBERGEN and PENNOIT-DECOOMAN, 1944). Other end products of nitrogen metabolism are ammonia (*F. gigantica*, *Gastrothylax crumenifer*, *Paramphistomum explanatum*—GOIL, 1958b; *S. mansoni*—SENFT, 1963; *F. hepatica*, EHRLICH *et al*, 1963), urea (*F. hepatica* and *S. mansoni*) and uric acid *F. gigantica*, *G. crumenifer* and *P. explanatum*—GOIL, 1958b). In addition haemoglobin, coagulated proteins, albumoses and peptones have been identified in the material eliminated from the oral sucker of *F. hepatica* (FLURY and LEEB, 1926).

LIPID METABOLISM

This is probably the least understood aspect of helminth metabolism and it seems possible that lipids (i.e. fatty acids and their triglyceride esters, waxes, sterols and phospho- and glycolipids) may represent waste products resulting from anaerobic fermentations, energy reserves, or occur as structural lipids e.g. sterols or phospholipids. GINETZINSKAJA and DOBROVOLSKII (1962) suggested that in cercariae, fat may be hydrostatic in function and aid flotation or may represent an energy store. In contrast, PALM (1962a,b) found that the fat content of certain cercariae remained unchanged during the free-swimming period and thought that fat was relatively unimportant in supplying energy. The lipid content may be quite high as in *Fasciola gigantica* 12·9 per cent dry wt. or fairly low, for example 4·5 per cent for *Paramphistomum explanatum* and 1·36 per cent for *Gastrothylax crumenifer* (GOIL, 1958a). The fractions isolated are varied and for *Gastrothylax crumenifer* GOIL (1964) gives the following analysis—phospholipids, 16 per cent; unsaponifiable matter, 25 per cent; saturated acids 5 per cent; unsaturated acids 20 per cent;

glycerol 2 per cent of total lipids. The data for *F. hepatica* is similar (VON BRAND, 1928) with a slightly higher percentage (30 per cent) of phospholipids and lower percentage of unsaponifiable matter (19 per cent) and unsaturated acids (12 per cent).

Very little information is available on the ability of trematodes to absorb and digest lipids. Only faint lipase activity has been demonstrated in *F. hepatica* (PENNOIT-DECOOMAN and VAN GREMBERGEN, 1942), in the cercaria of *S. mansoni* (MANDLOWITZ *et al*, 1960) and on the surface of cercariae of *Bucephalus sp.* (CHENG, 1965). The occurrence of non-specific esterases is more general, and these have been described from the tegument of adult digeneans and from the intestinal caecum epithelium of a number of species. Non-specific esterase activity has also been found in gland cells associated with the parynx of both monogenean and digenean trematodes and in those associated with specialized structures such as the adhesive organ of strigeoid trematodes. (See later section on the tegument and alimentary tract for further details). Cholinesterase activity has also been recorded from a number of larval and adult trematodes. The significance of these enzymes in lipid metabolism is far from clear, however a biochemical characterization of *F. hepatica* esterases has been made by PANTELOURIS (1967).

The distribution of lipids within the parasites' tissues using histochemical methods has received frequent attention. Lipids occur in the tissues of sporocyst, redial and cercarial stages (see relevant chapters for details) as well as in the adult trematode. Most workers (PRENANT, 1922; VON BRAND and WEINLAND, 1924; VOGEL and VON BRAND, 1933; STEPHENSON, 1947d; PANTELOURIS and THREADGOLD, 1963; THREADGOLD and GALLAGHER, 1966; ERASMUS, 1967b,c) have identified lipid material in the parenchymal cells, caecal epithelium, vitellaria and walls of the excretory system. Of this wide distribution the occurrence of lipid droplets in the wall of the excretory system seems to be the most variable. Several workers have confirmed this association for *F. hepatica* and ERASMUS and ÖHMAN (1963), ERASMUS (1967b) and ÖHMAN (1965, 1966a,b) have clearly demonstrated this association for several strigeoid trematodes, but in *Dicrocoelium dendriticum* this does not seem to be the case. The presence of lipoproteins in Mehlis' gland and its secretion has been demonstrated in *F. hepatica* by CLEGG (1965), CLEGG and MORGAN (1966). He has suggested that the lipoprotein membrane to the outside and lining the egg shell of this trematode is derived from the secretions of Mehlis' gland and that the permeability of the egg is related to the properties of these membranes. This has been discussed in more detail in Chapter 2.

EXCRETION

It is apparent from the little information we have concerning trematode metabolism, that the excretory products may be very varied. As described above, fatty acids, ammonia, urea, uric acid, may be regarded as excretory products as well as semi-digested proteinaceous material egested through the oral sucker. In addition, Senft (loc. cit.) in his *in vitro* experiments has reported an increase in the concentration of certain amino in the culture fluid after incubation. The source and significance of these amino acids is not clear and they may represent artefacts associated with *in vitro* conditions. However SENFT (1967) has also shown that the arginine level of serum from infected mice is greatly depleted whereas the concentration of proline, alanine, ornithine and ammonia is considerably increased.

Three routes are available for the elimination of these substances. 1. The tegument. The possible functions of this layer are discussed in detail further on in this chapter, but its cytoplasmic nature does certainly implicate it in excretory processes. 2. The alimentary tract. The histochemical and ultrastructural evidence, discussed later, suggests that secretion if not excretion may take place via the caecal epithelium. Certainly, egestion of a wide range of materials must occur through the oral sucker. 3. The excretory system. This system of flame-cells and muscular bladder has often been described as the basic excretory system in the trematodes although the evidence for this assumption is somewhat scanty. This system will now be considered in more detail.

The basic pattern of the system at the cercarial, metacercarial and adult stages is fairly similar, and consists of flame-cells, capillary vessels, large collecting ducts and a bladder opening to the exterior. The number of flame cells and the complexity of their arrangement differs between genera and between developmental stages. This aspect has been considered in the chapters on the cercarial and metacercarial stages. The gross morphology of the flame cell is well known and recent ultrastructual studies have confirmed some of these features and revealed others not previously known (KÜMMEL, 1958, 60a; PANTELOURIS and THREADGOLD, 1963; ERASMUS, 1967b). The cell body in *Cyathocotyle bushiensis* (ERASMUS, 1967b) contains a nucleus with nucleoli, numbers of mitochondria, vesicles and a small quantity of organized endoplasmic reticulum (Plate 6-1). The cytoplasm does not exhibit the ultrastructural characteristics of secretory cells i.e. extensive endoplasmic reticulum and numerous golgi complexes. The flame cell chamber contains a bundle of cilia which in transverse section, exhibit the typical 9 + 2 arrangement, and which originate in a striated rootlet system in the cytoplasm of the cell. The plasma membrane lining this intracellular chamber becomes elevated to form branched processes which protrude into the lateral regions of the lumen. These appear to be longer and differ from the 'corrugations' described in the flame-cell chamber of *Fasciola hepatica* (Kummel, loc. cit.; Pantelouris and Threadgold, loc. cit.) but a function cannot be ascribed to them at present. In contrast to the flame-cell chamber, the lumen of the duct (capillary of optical microscopy) is extracellular (Fig. 87). This lumen is encircled by cells and points of contact between adjacent

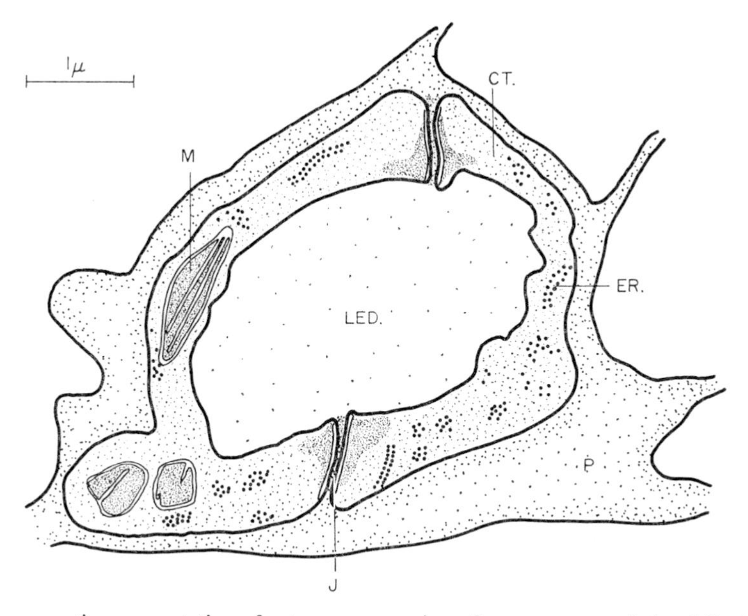

FIG. 87 Semidiagrammatic representation of a transverse section of an excretory tubule of *Cyathocotyle bushiensis*. Drawing based on an electromicrograph taken at × 30 000. (CT: lateral cells of tubules; ER: endoplasmic reticulum; J: junction between lateral cells of wall; LED: lateral cells forming wall of duct; M: mitochondrion; P: adjacent parenchymal cells).

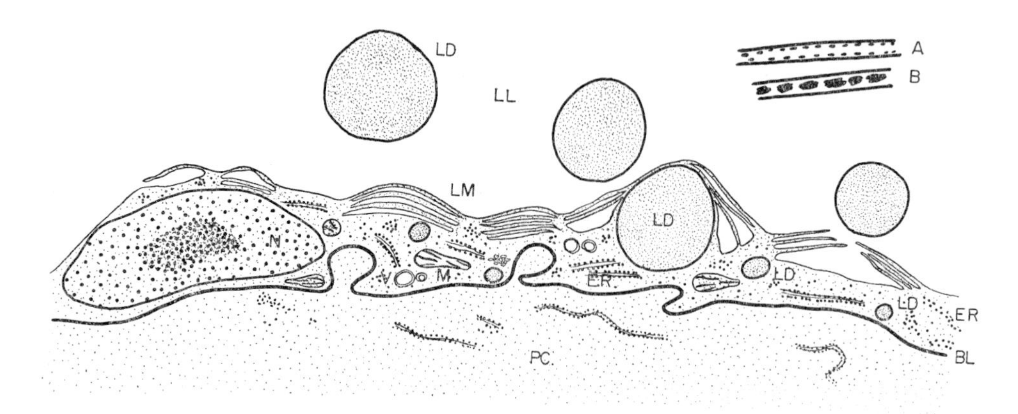

FIG. 88 Semidiagrammatic representation of the ultrastructure of the excretory lacuna of the strigeoid trematode *Cyathocotyle bushiensis*. **A** The distribution of alkaline phosphatase activity in the lamellae; **B** distribution of acid phosphatase activity. (ER: endoplasmic reticulum; BL: basement layer; LD: lipid droplet; LM: lamellae of luminal surface; LL: lumen of lacuna; M: mitochondrion; N: nucleus; PC: adjacent parenchymal cell; V: intracytoplasmic vesicle.)

plasma membranes exhibit dense terminal bars. The cytoplasm of the cells is nucleated and contains mitochondria. The nucleated nature of the tubules has also been commented on by BUGGE (1902), BYRD and MAPLES (1963) as well as by PANTELOURIS and THREADGOLD (1963). The lining of the tubule in *F. hepatica* has been described as possessing microvilli (loc. cit.) as has also that of the adult of *S. mansoni* (SENFT *et al*, 1961). These have not been observed in *C. bushiensis* nor in the excretory tubules of *S. mansoni* cercariae studied by KRUIDENIER (1959) although the tubules of a xiphidiocercaria examined did exhibit infoldings. LAUTENSCHLAGER (1961) examined the tubules of the metacercaria *Diplostomulum trituri*, and noted that the wall had filaments projecting from it.

The excretory bladder in strigeoid and other trematodes becomes enlarged during the development of the metacercaria (see chapter on metacercaria) to form a reserve bladder system which persists into the adult stage. Associated with this system, which varies in its morphology between different genera are a variety of refractile bodies which have been described as fatty or calcareous by different authors. This evidence has been interpreted as indicating the involvement of this reserve system in excretory processes. In a similar manner, the histochemical data described above refers to the presence of lipid droplets near and in the lumen of the excretory ducts, suggesting a similar association. More precise evidence has been obtained with the aid of the electron microscope (Fig. 88). The lining of the excretory lacuna in *C. bushiensis* (ERASMUS, 1967b) is cytoplasmic and contains nuclei, mitochondria, endoplasmic reticulum and lipid droplets. The layer may be very thin 0·09 μ, appears 'cuticular' under the optical microscope, and is demarcated from the underlying tissues by a plasma membrane. The surface of the layer adjacent to the lumen is covered by a plasma membrane elevated to form stacks of lamellae. The lamellae are narrow (0·02–0·03 μ) and consist of two triple-layered plasma membranes with a narrow strip of cytoplasm between. They are in contact with the general cytoplasm at either end but lie free in the centre (Plates 6-2 and 6-3). It may be significant that similar lamellate arrangements occur in highly metabolic structures such as mitochondria and chloroplasts. This cytoplasmic lining contains lipid droplets of different sizes and, although the smallest are contained within the cytoplasm, the larger ones protrude into the lumen displacing the lamellae. The lamellae hold the droplet to the cytoplasm but eventually they rupture, and the lipid is released into the lumen of the reserve bladder. Observations of living parasites in saline have shown that the droplets become discharged to the exterior via the excretory pore. There is no doubt that in *C. bushiensis* the lining of the reserve bladder is intimately connected with the accumulation and possibly synthesis as well as the excretion of lipid droplets. Histochemical tests at the ultrastructural level show that the membranes also exhibit acid and alkaline phosphatase activity. HALTON (1967c) was unable to demonstrate esterase activity in the excretory system of *F. hepatica*. The lipid droplets appear to contain neutral fat. Histochemical tests have demonstrated the presence of similar droplets associated with the reserve system of *Diplostomum spathaceum*, *Apatemon gracilis minor* and *Holostephanus lühei* (ÖHMAN, 1965, 1966a,b) and it is likely that the accumulation or formation and excretion of this material is also a function of the reserve bladder cytoplasm which these other species of parasite possess. Lipid droplets can be detected also in tegumentary cells, vitelline cells and in parenchymal cells but it appears that excretion of this material is a function of the excretory lining only. This system is a specialization characteristic of, but not restricted to, strigeoid trematodes and it may be relevant that the tegument of *C. bushiensis* contains little lipid (ERASMUS, 1967c) and does not seem to play a significant part in lipid metabolism, as far as can be determined by morphological observations.

As already described, the reserve bladder system is well developed at the metacercarial stage. Observations (Erasmus, loc. cit.) on very young adults, eighteen hours after infection, show that the reserve bladder lining is cytoplasmic but that the lamellae are poorly developed. Associated with the cytoplasm, in a similar manner as the lipid droplets of the mature adult, are excretory bodies of concentric appearance similar to those frequently described from cestodes. The layers may be arranged around a single or multiple 'nucleus' (Plate 6-4). At this stage it is difficult to detect lipid material but three days after infection, lipid formation is well evident and the excretory bodies are few in number. When the metacercaria excysts *in vitro* it discharges a quantity of refractile material, which stains with Alizarin Red S and presumably corresponds to the excretory bodies mentioned above, as well as lipid droplets, so that a considerable proportion of the excretory material becomes discharged during excystation in the final host. DIXON (1966b) also refers to a discharge of excretory materials during the excystation of *F. hepatica* metacercariae. Some of it remains, as shown by a study of 18 hour infections, but lipid excretion is not resumed until three days after infection of the final host. A similar pattern was described by VOGEL and VON BRAND (1933) who were unable to detect lipid droplets in young *Fasciola hepatica* until after twenty-four hours in the final host. The presence of 'calcareous' excretory bodies in a metacercarial stage has been described by MARTIN and BILS (1964), and it is possible that the appearance of this type of excretory body is related to the limitations imposed upon the parasite by the metacercarial cyst wall (see discussion in CABLE, 1965). The role of the excretory corpuscle is not precisely defined and it may function also in carbon dioxide fixation as has been suggested by VON BRAND *et al* (1965) for the cestode calcareous corpuscle.

The cytoplasmic lining now known to be present in the ducts and reserve bladder system may have other functions. Alkaline phosphatase activity has been described in the excretory system of a number of larval stages (COIL, 1958; DUSANIC, 1959; CHENG, 1964 and PROBERT, 1966a) as well as in that of adult trematodes (*F. hepatica*—acid and alkaline phosphatase YAMAO and SAITO, 1952, BEČEJAC and KRVAVICA, 1964; *Paragonimus westermani*—alkaline phosphatase YAMAO, 1952). In a comparative study of several species of adult monogenean and digenean trematodes, HALTON (1967a) was able to demonstrate alkaline phosphatase activity in the excretory ducts of all seven species of monogenean tested and alkaline phosphatase in the ducts of all eight species of digenean adults (see Tables 43, 44). Acid phosphatase activity was absent from the excretory system of all species.

It is possible that this enzyme activity may be associated indirectly with transport through the membranes. The possible function of phosphatases, demonstrated histochemically in membranes, will be discussed in full in the later section on tegument histochemistry. The extent of the strigeoid reserve bladder system has stimulated several workers to suggest that it may function as a hydrostatic skeleton or for the circulation or translocation of dissolved nutrients through the body.

It seems reasonable to assume, although there is no direct experimental evidence to confirm this, that the basic protonephridial system functions in osmoregulation. However, the evidence available on *F. hepatica* indicates that the adult fluke is able to survive a wide range of osmotic pressures, but does not appear to regulate its water content over a range of Δ0·40–0·81°C (KNOX and PANTELOURIS, 1966). WILSON (1967a) has shown that the flame cell activity of the free-swimming miracidium of *F. hepatica* is sensitive to changes in the temperature and light intensity of the external environment but that changes in osmotic pressure (0–1·0 per cent saline) produced no effect on the ciliary beat (Tables 41, 42). GOIL (1966) also found, in the case of *Gastrothylax crumenifer*, that the percentage increase in weight went on decreasing with

Table 41 The activity of flame cells in the free miracidium (*F. hepatica*) determined at different temperature.

Temperature	Mean frequency of flame cells + S.E. (beats/sec)	No. of animals
3	3·9 ± 0·18	15
10	8·25 ± 0·3	12
15	10·8 ± 0·23	16
20	13·2 ± 0·29	10
25	>14	10

From WILSON, 1967a.

Table 42 The activity of flame cells in the free miracidium (*F. hepatica*) determined under different conditions of external osmotic pressure (Saline conc.).

Concentration of saline (%)	Mean frequency of flame cells ± S.E. (beats/sec)	No. of animals
0·0	13·2 ± 0·29	10
0·2	12·3 ± 0·43	10
0·5	11·9 ± 0·5	9
0·8	11·8 ± 0·4	10
1·0	No free miracidia Eggs failed to hatch	10

From WILSON, 1967a.

increase in the concentration of sodium chloride solution. In a study of osmotic and ionic regulation in *Fasciola gigantica*, SIDDIQI and LUTZ (1966) suggested that in different solutions an osmotic balance is attained by the passing out and in of water and thereafter a steady weight is maintained. This parasite was more tolerant of hypertonic than of hypotonic media with a threshold concentration between 25 and 50 per cent Tyrode saline. Below this value the worms became inactive and waterlogged. They also noted that the excretory bladder became reduced in size in hypertonic media and discharged its contents at irregular intervals. In hypotonic media the bladder was distended and discharged continually. These figures may not, of course, reflect the situation *in vivo*. It has also been observed that the walls of the excretory tubules of *F. hepatica* (STEPHENSON, 1947d) and *Opisthoglyphe ranae* (SMYTH *et al*, 1945) are rich in vitamin C, and also that radioactive iron localizes in this region in *F. hepatica*. This overlap in distribution of these two substances was considered by PANTELOURIS and HALE (1962) to imply the association of vitamin C in the formation of soluble iron compounds which could be excreted.

It seems that the apparent morphological uniformity of the excretory system as seen under the optical microscope has created an over-simplified concept of function and that further investigation involving modern cytochemical methods will reveal the probable multiplicity of physiological processes associated with this system. The ultrastructure of the flame cell of the cestode *Moniezia expansa* and the trematode *Fasciola hepatica* has been described by HOWELLS (1969) and WILSON (1969) respectively. Howells has shown that pores, connecting the nephridial lumen to the intercellular space of the connective tissue, exist at the junction of the flame cell and its nephridial duct. It is suggested that the pores could be regarded as nephrostomes and this implies that the excretory system, as described in *M. expansa*, might not be protonephridial. The description by Wilson also supports this suggestion. This very significant observation will influence further our reassessement of the platyhelminth 'excretory system'.

B. The host-parasite interface

I AND II THE TEGUMENT AND ITS SPECIALIZATIONS

Until 1959 the outer surface of the Trematoda was referred to as the cuticle, but because of the implications resulting from ultrastructural studies this term is no longer suitable and the term 'tegument' will be used in this text. The nature of this outer surface has always attracted much interest but because of the difficulties inherent in its study little progress had been achieved. The early concepts associated with the tegument have been reviewed by HYMAN (1951) and she has summarized the homologies and origin of this layer as follows: (a) that the 'cuticle' is an altered and degenerated epidermis; (b) that it is the basement membrane of the former epidermis; (c) that it is the outer layer of an insunk epidermis, the cells and nuclei of which have sunk beneath the subcuticular musculature; (d) that the cells in question are not epidermal but are parenchymal cells that secrete the 'cuticle' and (e) that the 'cuticle' is secreted by ordinary mesenchyme (parenchymal) not by special cells. The relative significance of these ideas can be assessed by the reader after considering the ultrastructure of this region.

(A). ULTRASTRUCTURE

Adult tegument The first published description was based on *Schistosoma mansoni* by SENFT (1959), and SENFT *et al.* (1961). They described the tegument as a vast spongy area and did not comment on its cytoplasmic nature. THREADGOLD (1963a,b) produced a more detailed description of the tegument of *Fasciola hepatica*, described mitochondria and a variety of vesicles and recognized the basic cytoplasmic nature of the tegument. Observations by BJORKMAN and THORSELL (1964a) on *F. hepatica*, by BURTON (1964, 1966) on *Haematoloechus medioplexus* and *Gorgoderina sp.* by BILS and MARTIN (1966) on *Acanthoparyphium spinulosum*, by ERASMUS and ÖHMAN (1965), ERASMUS (1967c, 1969a,b,c, 1970) on *Cyathocotyle bushiensis*, *Apatemon gracilis minor* and *Diplostomum phoxini*, by BOGITSH and ALDRIDGE (1967) on *Posthodiplostomum minimum* and MORRIS and THREADGOLD (1968) on adult *Schistosoma mansoni* have confirmed and extended Threadgold's basic interpretation.

LEE (1966) has reviewed literature on the helminth tegument and redescribed the cytological characteristics of the tegument of adult *Schistosoma mansoni*. The unpublished observations of Lyons are also quoted in this review and she had found that the tegument of the monogeaneans *Gyrodactylus sp.* and *Tetraonchus sp.* is also cytoplasmic in nature. The description given below is based on *Cyathocotyle bushiensis* (Plate 9-2) and any differences between this and other species will be mentioned (Fig. 89). The outer surface of the tegument consists of a triple-layered plasma membrane which extends over the surface of the spines. External to this plasma membrane there is, in *C. bushiensis*, a thin layer of granular material. The tegument varies in thickness but in *C. bushiensis* is 2–6 μ wide (7 μ in *A. spinulosum*; about 15 μ in *F. hepatica*; 8–10 μ in *H. medioplexus*). The degree of irregularity exhibited by the outer surface of the tegument will depend to some extent on the state of contraction and expansion of the specimen. The body of the tegument, generally termed the matrix, consists of a finely granular cytoplasm within which occurs a number of inclusions. The most obvious, although not the most numerous, are the mitochondria. The mitochondria are typical in possessing cristae and may be irregularly distributed or arranged in rows perpendicular to the surface (*F. hepatica*). In view of the movements which must occur within the matrix as the parasite contracts it seems possible that this regular distribution may be a fixation artefact. Endoplasmic reticulum (granular form) has not been described from the matrix although THREADGOLD (1963a,b) and BJORKMAN and THORSELL (1964a) refer to smooth surfaced membranes which the latter regard as endoplasmic reticulum. None of these authors has decribed golgi complexes nor nuclei from the tegumentary matrix, although in his optical studies ALVARADO (1951) has reported nuclear fragments in the tegument of *F. hepatica*, nor has any author described the complete division of the tegument by transverse membranes. The most frequent inclusion in the matrix are small, spherical or disc-like objects referred to by most authors as secretory bodies. Rod-like or slender pear-shaped bodies also occur and these, as well as being distributed within the matrix, become closely associated with the outer plasma membrane. In *C. bushiensis* these bodies are 0·16–0·2 μ diam. and both forms are thought, (in this species), to represent differing sections of the same disc-like body. They are enclosed in a membrane and have a slightly less dense centre surrounded by a dense granular outer zone immediately beneath the membrane. In addition to these, small vesicles have also been described in *F. hepatica* tegument. It is obvious that the nature and significance of many of these inclusions is far from clear. The base of the matrix rests on the basement layer (= basement membrane of optical studies). In *C. bushiensis* this layer consists of a triple plasma membrane immediately adjacent to the matrix—below this is a narrow dense layer (0·03 μ thick) and finally a thicker lower (0·3–0·4 μ) zone of irregularly arranged fibres approximately 100 Å diameter. The plasma membrane becomes elevated into slender folds which extend for some distance into the matrix. Occasionally, the tips of the folds become distended (described also by LEE, 1966 in *S. mansoni*) giving the appearance of small vesicles containing granular material. It

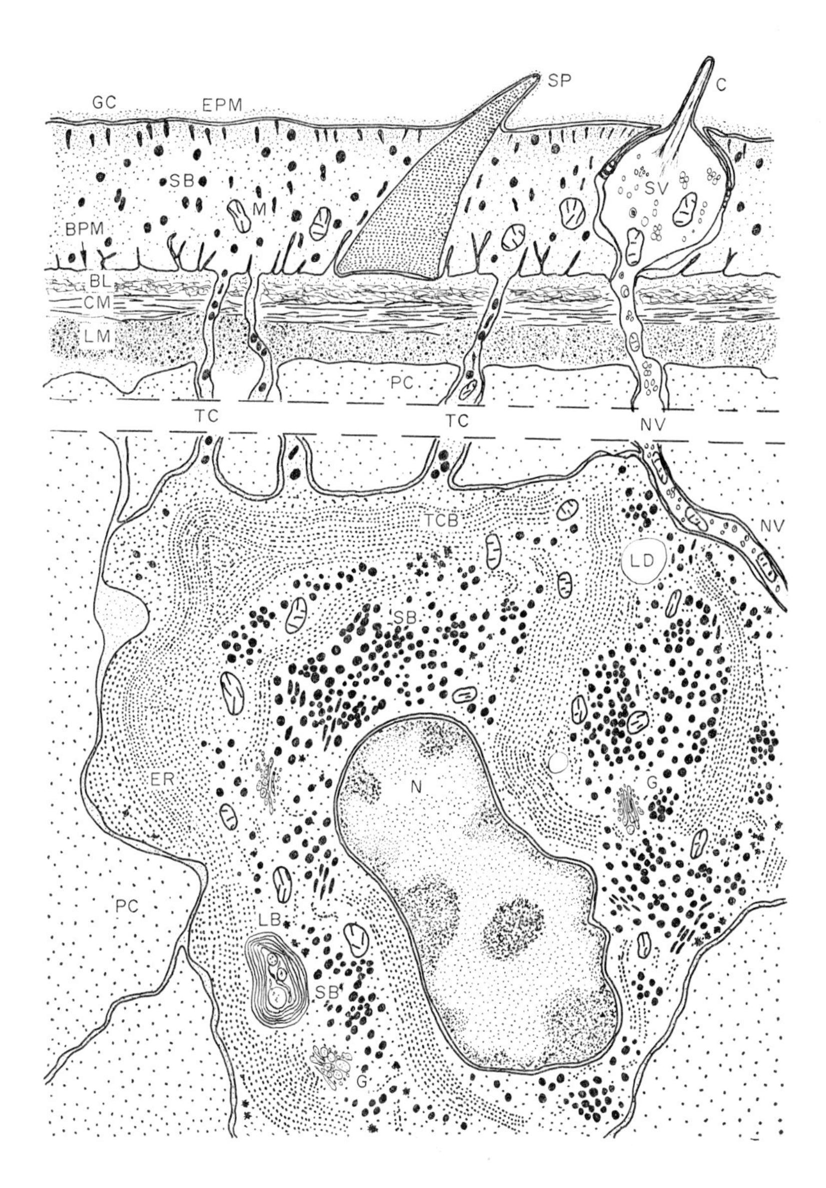

FIG. 89 Semidiagrammatic representation of the ultrastructure of the 'general' tegument, based on electronmicrographs of *Cyathocotyle bushiensis*. (BL: basement layer; BPM: basal plasma membrane; C: cilium of sense organ; CM: circular muscle; EPM: external plasma membrane; ER: endoplasmic reticulum; G: golgi complexes; GC: granular external coat; LD: lipid droplet; LM: longitudinal muscle; LB: lamellar body; M: mitochondrion; N: nucleus; PC: parenchymal cell; SB: secretory body; SP: spine; SV: vesicle of sense organ; NV: nerve process from sense organ; TC: cytoplasmic connection between external tegument and the tegumentary cell body; TCB: tegumentary cell body.)

is possible that many of the membrane bounded vesicles and smooth endoplasmic reticulum of other authors correspond to these elaborations of the basement layer. Their function may be to stabilize the matrix and permit a firm attachment to the fibrous layer below. It is also possible that the presence of this membrane system within the matrix, provides surfaces which would facilitate biochemical interchange or activities within the tegument. The fibrous layer represents the limiting, skeletal layer to the outside of the parasite, and it is to this layer that the muscles become attached. The lower limit of the fibrous zone is not clearly defined and it merges into the interstitial fibres which extend down between the cells. The structure of the basement layer corresponds closely with that described for turbellarians (PEDERSON, 1961). Spines, when present, lie with their base resting on the upper membrane of the basement layer. They do not protrude through to the fibre layer nor (in those species described) do they have muscles associated with them. The apex of the spine is covered by the outer plasma membrane with its granular coat. The body of the spine exhibits a distinct lattice pattern, which has been interpreted (BURTON, 1964) as indicating a type of crystalline protein. The basement layer becomes perforated at intervals by the nervous supply to sense organs, which lie within the body of the matrix, and by cytoplasmic tracts which plasce the tegumentary matrix in continuity with cells, termed tegumentary cells, beneath the basement layer. The tegumentary cells are large (22 μ long in *C. bushiensis*) and each possesses several connections with the matrix, and the connections may be so long that the main mass of the tegumentary cell lies associated with parenchymal cells below the peripheral musculature. These cells are quite different from the parenchymal cells and invalidate theories d and e of HYMAN (1951), but it seems likely that her third theory represents most closely, the structure as seen under the electron microscope. The cells contain a nucleus, well developed endoplasmic reticulum associated with ribosomes, phagosomes, mitochondria and golgi complexes. The cytoplasm also contains masses of secretory bodies which resemble in size and structure those present in the matrix. These bodies also occur within the cytoplasmic extensions joining the tegument, so that it has been postulated that they are synthesized in the cell cytoplasm in association with endoplasmic reticulum and golgi apparatus and, when fully differentiated, pass into the tegument via the cytoplasmic extensions perforating the basement layer.

A function for some of these bodies has been proposed by ERASMUS (1967c). In *Cyathocotyle bushiensis* the bodies become closely associated with the outer plasma membrane. Histochemical tests at the light microscope level show that this region is rich in acid mucopolysaccharide and, as this positive layer would include the peripheral layer of bodies, the plasma membrane and the external granular coat, it is possible that the external granular material represents a mucopolysaccharide or mucoprotein coat. It may be this layer which confers resistance against the digestive enzymes of the host as postulated by MONNÉ (1959). BURTON (1964) suggests that some of the bodies disintegrate within the matrix contributing to its composition. On a morphological basis only one type of secretory body can be distinguished associated with the general tegument of *C. bushiensis*. It must be stressed that morphological similarity may mask very divergent physiological and biochemical characteristics so that it is quite likely that, as techniques become more refined, different types will emerge from the general picture prevailing at present. It is also likely that the tegumentary cells themselves will become differentiated into various categories. Indeed, in *C. bushiensis* a quite different type is associated with another specialized region of the tegument to be described later. The passage of material through the tegument to the exterior of the parasite has also been suggested by DIXON and MERCER (1964) in the formation of the cyst enclosing the metacercaria of *F. hepatica*. The tegument of the monogeneans *Gyrodactylus* and *Tetraonchus* (Lyons, quoted by LEE, 1966) is cytoplasmic. Observations on the tegument of

Discocotyle indicate that the basement layer resembles that described for *C. bushiensis* in consisting of a plasma membrane, a dense zone and a thick, fibrous layer. The plasma membrane is elevated into folds containing the dense material and from the tip of the folds tonofibril-like structures run to the outer plasma membrane. This morphological feature will give considerable stability to the tegument and may be correlated with the absence of tegumentary spines seen in most digenea. The tegument is rich in mitochondria and inclusions of various types.

Little progress has been made in the understanding of trematode sense organs since the original descriptions of BETTENDORF (1897) although ROHDE (1966) has described briefly a large number of structures from the tegument of *Multicotyle purvisi.* Electron microscope studies have begun to contribute to this field and DIXON and MERCER (1965) described a sense organ from the tegument of the cercaria of *Fasciola hepatica.* This sensory structure (believed to respond to tactile stimuli) consisted of a bulb bearing at its distal end a papilla which projected to the exterior of the tegument. Basally the bulb is in continuity with a nerve fibre. This type of structure seems to be generally associated with the adult digenean tegument. Similar organs have been observed in the tegument of *Cyathocotyle bushiensis*, (ERASMUS, 1967c), *Apatemon gracilis minor* (ERASMUS, 1969b) and *Diplostomum phoxini* (ERASMUS, 1969c) (Plate 10-2; Plate 11-1). In adult *Schistosoma mansoni* (MORRIS and THREADGOLD, 1967) it consisted of a basal vesicle with a protruding cilium, but was distinctive in that the cilium was enclosed by a sheath. These authors suggested that the structure might serve to detect direction of flow of a fluid medium such as surrounds adult schistosomes *in vivo*. Observations (unpublished) by Erasmus and Robson on the cercaria of *Schistosoma mansoni* reveal that the apex of the oral sucker possesses sense organs consisting of a naked cilium arising from a vesicle (see Plate 10-2) resembling that described from *Cyathocotyle*. In *Cyathocotyle* the vesicle (1·5–3·0 μ diam) is contained in the tegument matrix and terminates in a papilla 1·0–0·8 μ long which projects to the exterior of the tegument.

The papilla contain at its periphery 9 longitudinal fibres each of which arises from a fan-like, striated rootlet system. The fibres terminate in a dense body at the tip of the papilla. In *Cyathocotyle* these tangoreceptors are particularly abundant in the tegument of the adhesive organ lobes which come into contact with the host tissues, and presumably must play a significant role in the attachment of the parasite to the host tissue. The nerves associated with these receptors have the same characteristics as described for the nervous tissue of *F. hepatica* cercaria (DIXON and MERCER, 1965). BEČEJAC, LUI *et al* (1964) have recorded acetylcholinesterase and butyrocholinesterase actively in the nerves of the oral and ventral suckers, cerebral ganglia and longitudinal nerves of *Dicrocoelium lanceatum* and UDE (1962) has suggested the presence of neurosecretory cells in the cerebral ganglia of *D. lanceatum*. The statement made frequently in general textbooks that the nervous system and sense organs of parasitic trematodes are greatly reduced is far from the truth. The general pattern of the nervous system is well described by Bettendorf (loc. cit.) although details of the innervation of suckers, pharynx, intestinal caeca and excretory system of the metacercaria of *Codonocephalus sp.* has been given recently by REISINGER and GRAAK (1962).

TEGUMENTARY SPECIALIZATION

The majority of descriptions have been of the general tegument from undefined regions of the body. In *Cyathocotyle* as in the other strigeoids, there is a distinctive adhesive organ on the ventral

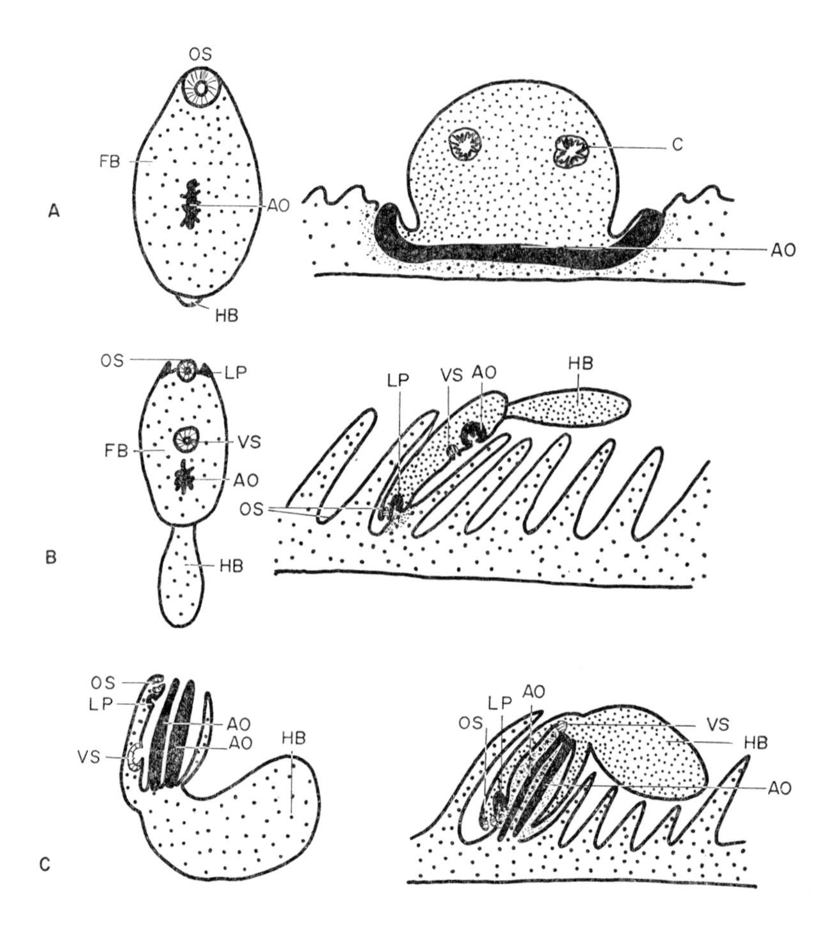

FIG. 90 The attachment of strigeoid trematodes to host tissues. **A** *Cyathocotyle bushiensis;* **B** *Diplostomum sp.;* **C** *Apatemon gracilis;* The extent of the adhesive organ surface is indicated in solid black. (AO: adhesive organ; C: caecum; FB: fore-body; HB: hind-body; LP: lappets; OS: oral sucker; VS: ventral sucker.)

surface. The ultrastructure of the tegument covering this structure differs markedly from that of the general tegument (ERASMUS and ÖHMAN, 1965, ERASMUS, 1967c) (Figs. 90, 91). The basement layer, however, is very similar but in this region becomes perforated by the cytoplasmic ducts from gland cells situated more deeply in the body. The outer plasma membrane becomes elevated into a series of what appear to be microvilli (Plate 9-1). These are generally slender and long (0·04 μ diam. and up to 2 μ long) but are rather irregular in outline and frequently distended at their tips. They are not as regular in form as the microvilli from mammalian and invertebrate brush borders nor do they appear as rigid as the cestode microtrich. Microvilli in lateral contact appear to fuse at these points. The region corresponding to the matrix of the general tegument forms an area in which accumulates the secretion from the gland cells. The entire region appears to be stabilized by fine fibres which run from the elevations of the plasma membrane of the basement layer across to the plasma membrane between the microvilli.

The gland cells associated with the adhesive organ are arranged in a corona surrounding the

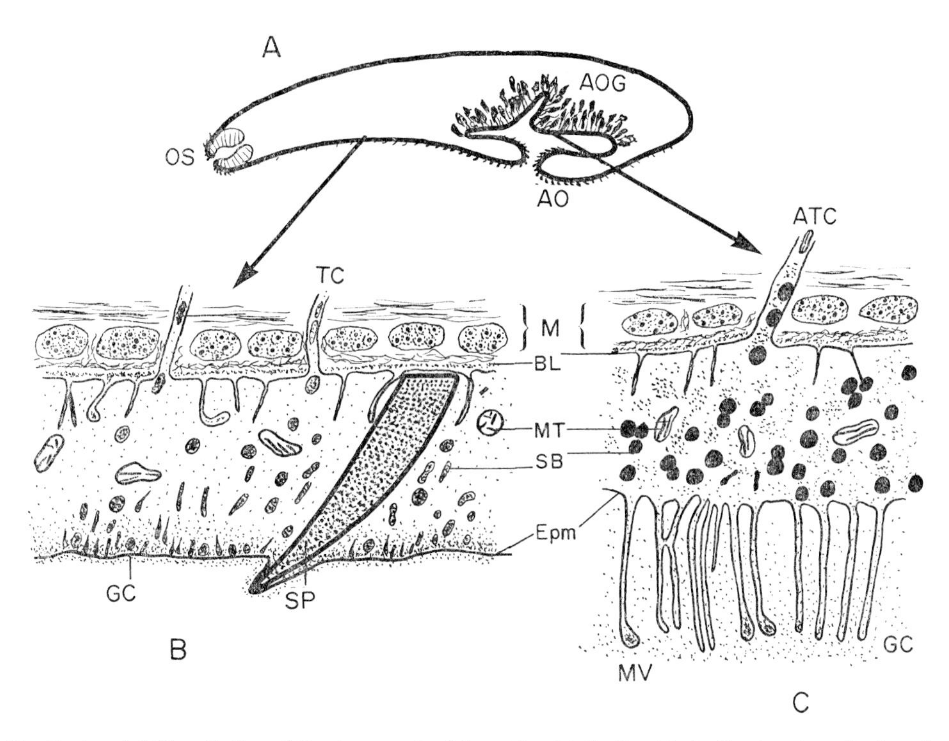

FIG. 91 Regional differentiation of the tegument of the strigeoid *Cyathocotyle bushiensis.* **A** sagittal section of the adult to show the relationship between 'general' and specialized tegument covering the adhesive organ. **B** the structure of the general tegument; **C** the structure of the specialized tegument. (AO: adhesive organ; AOG: adhesive organ gland cells; ATC: cytoplasmic connection between specialized tegument and adhesive organ gland cells; BL: basement layer; EPM: external plasma membrane; GC: external granular coat; M: muscle layers; MT: mitochondria; MV: microvilli; OS: oral sucker; SP: spine; SB: secretory bodies; TC: cytoplasmic connection between external tegument and tegumentary cell body.)

lumen and are pear-shaped, with the tapering portion extending ventrally and continuing as a slender cytoplasmic duct. Ducts from adjacent cells fuse, and the common ducts run between the muscle blocks to the basement layer. The cytoplasm of the cells contains a dense endoplasmic reticulum possessing numerous ribosomes, mitochondria and golgi bodies. The cytoplasm produces secretory bodies which have dense contents are 0·1–0·2 μ diam and bounded by a plasma membrane. The secretion, as it occurs in the ducts and adhesive organ tegument, consists of granular material, the secretory bodies and mitochondria. It appears as if the secretory process may be holocrine. This secretion is known to exhibit esterase and phosphatase activity and it has been demonstrated (ERASMUS and ÖHMAN, 1963) that it emerges to the exterior of the parasite. The route may be via the tips of the microvilli which become distended with contents and appear to bud off. It is possible that it may also pass through pores in the microvillus wall. The strigeoid adhesive organ has been suspected of playing a part in secretion and absorption and these studies have revealed a very appropriate morphological adaptation in the form of microvilli and pits which increase the surface area of the parasite in contact with the host tissues. Other features associated with this structure are discussed in the section dealing with extracorporeal digestion.

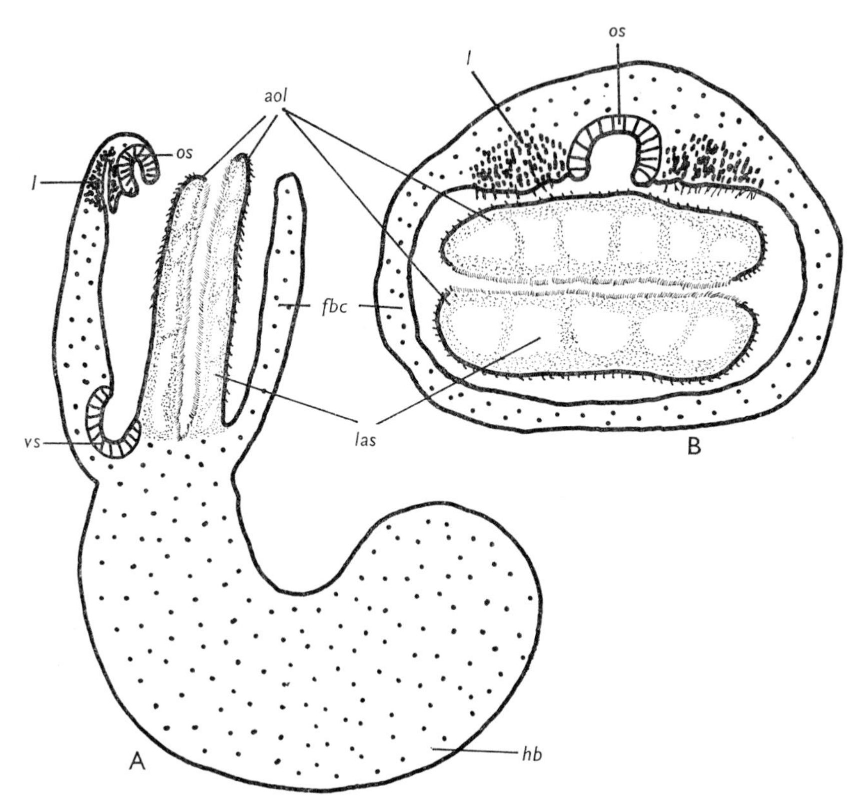

FIG. 92 Diagrammatic representations of sagittal (**A**) and transverse (**B**) sections of *Apatemon gracilis minor* to show the relative distribution of 'general' and specialized tegument. The 'general' tegument is represented by a solid line and the specialized surface by a hatched line. (aol: adhesive organ lobes; fbc: fore body cup; hb: hind body; l: lappets; las: lacuna system; os: oral sucker; vs: ventral sucker.) ERASMUS, 1969b.

This concept of tegumentary specialization has been extended to other strigeoid genera such as *Apatemon gracilis minor* (ERASMUS, 1969a,b) and *Diplostomum phoxini* (ERASMUS, 1969c; 1970). In both these genera three areas of tegumentary differentiation have been described:

(a) THE 'GENERAL' TEGUMENT This covers most of the body and corresponds in general structure to the tegument described from other adult digeneans.

(b) THE LAPPETS These are regions, one on each side of the oral sucker, which are capable of inversion and eversion and consequently play an important part in attachment to the host tissues by these two species (Plate 10-3; Plate 11-2). The tegument covering the lappets of *Apatemon* bears pointed setae and also receives the ducts from the lappet gland cells (Fig. 94). The necks of these unicellular glands are supported by a ring of microtubules (Fig. 95) and they pass through the tegument (Plate 11-3). In *D. phoxini* setae are absent although the tegument possess characteristic blunt finger-like structures. The apertures of two types of gland cells discharge to the exterior through the tegument of *Diplostomum*. Thus in these two species, the lappets are specialized glandulo-muscular regions into which host tissue is drawn during attachment to the host, and which represent areas of close host-parasite contact through which discharge of secretions from the parasite into the host can take place. Histochemical tests by

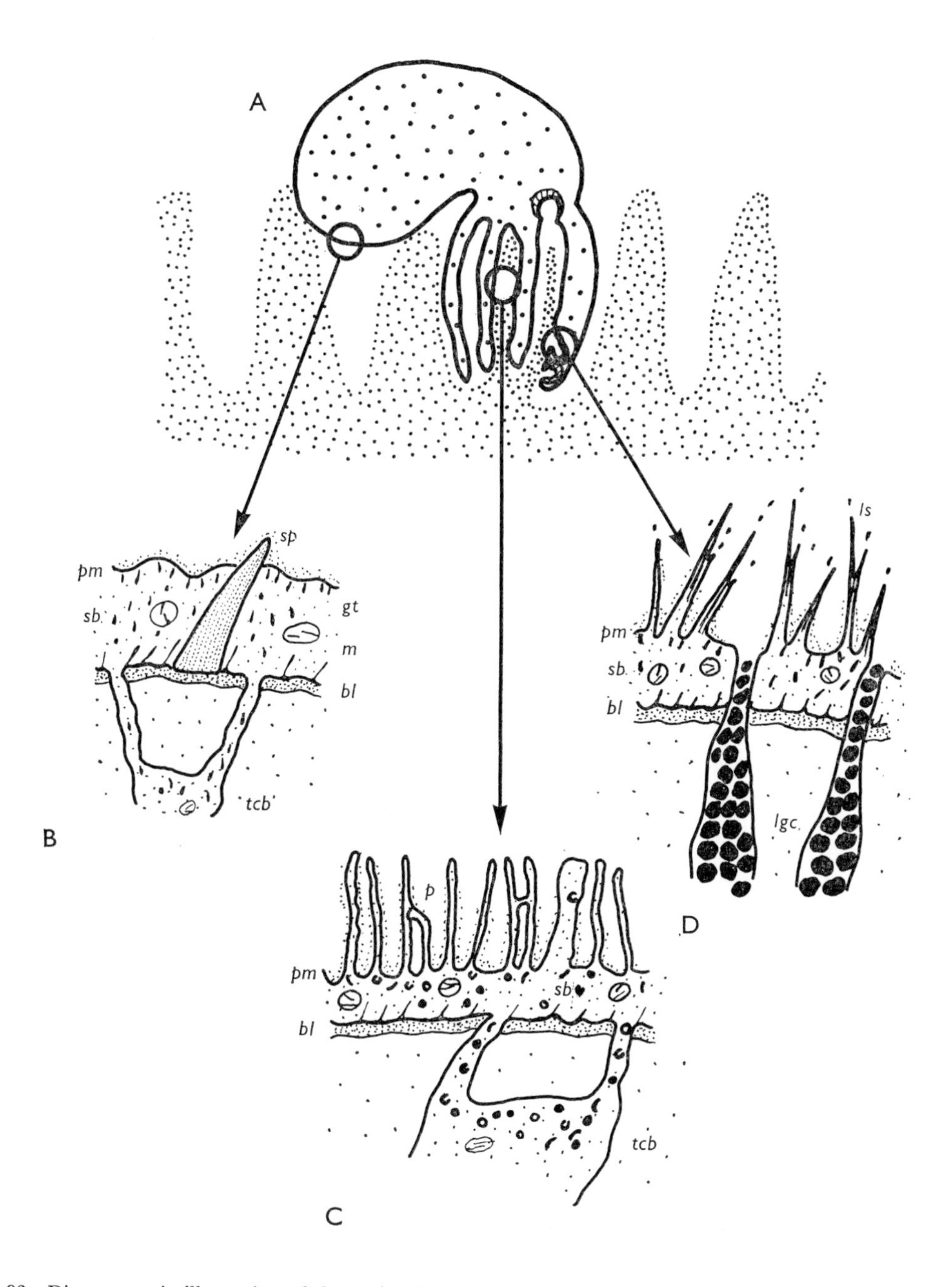

FIG. 93 Diagrammatic illustration of the regional specialization of the tegument as it occurs in *Apatemon gracilis minor*. **A** sagittal section of an adult *Apatemon* attached to the intestinal mucosa of the host (duck); **B** the 'general' tegument; **C** the specialized surface of the adhesive organ lobes; **D** The specialized surface of the lappets. (bl: basement layer; gt: general tegument; ls: lappet setae; lgc: lappet gland cells; m: mitochondria; pm: external plasma membrane; sb: secretory bodies; sp: spine; tcb: tegumentary cell bodies; p: pits of adhesive organ surface.) ERASMUS, 1969b.

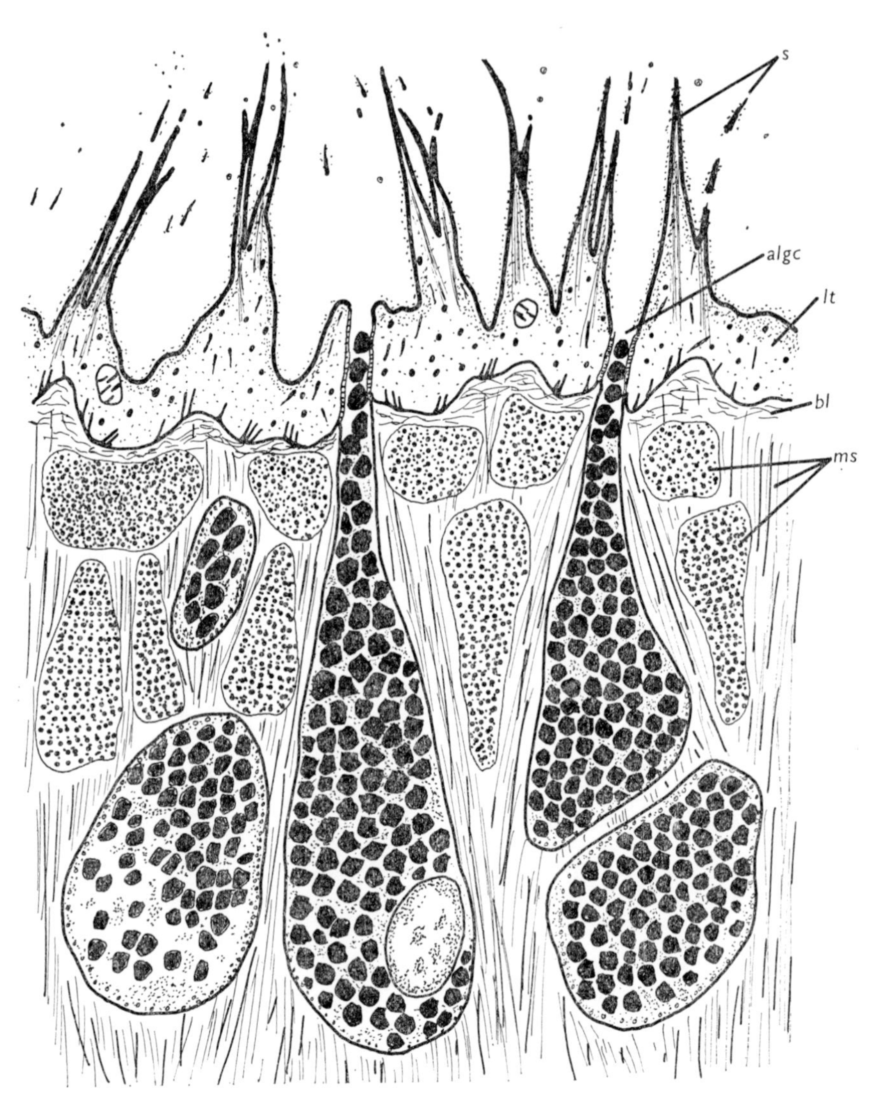

FIG. 94 Semidiagrammatic representation of the specialized surface of the lappets of *Apatemon gracilis minor*. (al gc: aperture of the lappet gland cells; bl: basement layer; lt: lappet tegument; ms: muscle blocks; s: setae.) ERASMUS, 1969a.

ÖHMAN (1965, 66a) have shown that hydrolytic enzymes are present in the lappet gland cells and it seems likely that these secretions may play a part in extracorporeal digestion of host tissues.

(c) THE ADHESIVE ORGAN As in *Cyathocotyle*, the adhesive organ of *Apatemon* and *Diplostomum* come into intimate contact with the host tissues although there are considerable differences between these two genera in the mode of contact with the host tissues and the nature of the adhesive organ surface. In *Diplostomum* the adhesive organ surface is chambered and resembles that of *Apatemon*. In *Apatemon* the adhesive organ consists of two lobes lying within the fore-body cup (Plate 11-1; Fig. 92). When attached, a host villus (devoid of mucosal epithelium) lies between the lobes so that their inner surfaces are in contact with the lamina propria of the host. This inner surface of both lobes possesses a tegument which has a pitted or sponge-like appearance (Fig. 96, 97). Studies with the Stereoscan electron-microscope show that the apical region

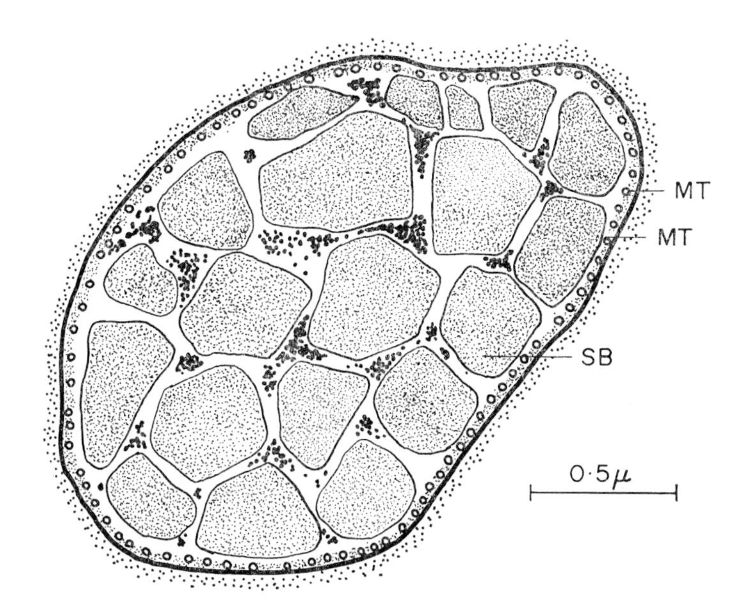

FIG. 95 Semidiagrammatic representation of a transverse section through the duct of a lappet gland cell from the lappets of *Diplostomum sp.* Note the secretion in the form of membrane bound bodies and the peripheral ring of microtubules. (mt: microtubules; sb: secretion bodies.)

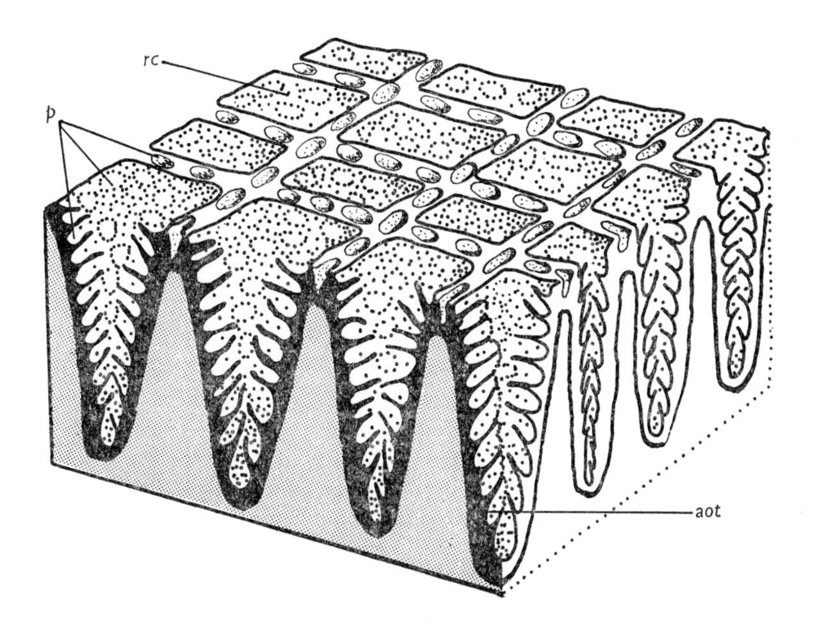

FIG. 96 A diagrammatic representation of the specialized adhesive organ lobe surface of *Apatemon gracilis minor* near the apex of the lobe. The illustration is based on both transmission and stereoscan electron micrographs. The adhesive organ tegument (aot and indicated in solid black) is finely pitted (p) and also elevated in this region to form a chambered, reticulum-like surface (rc). ERASMUS, 1969b.

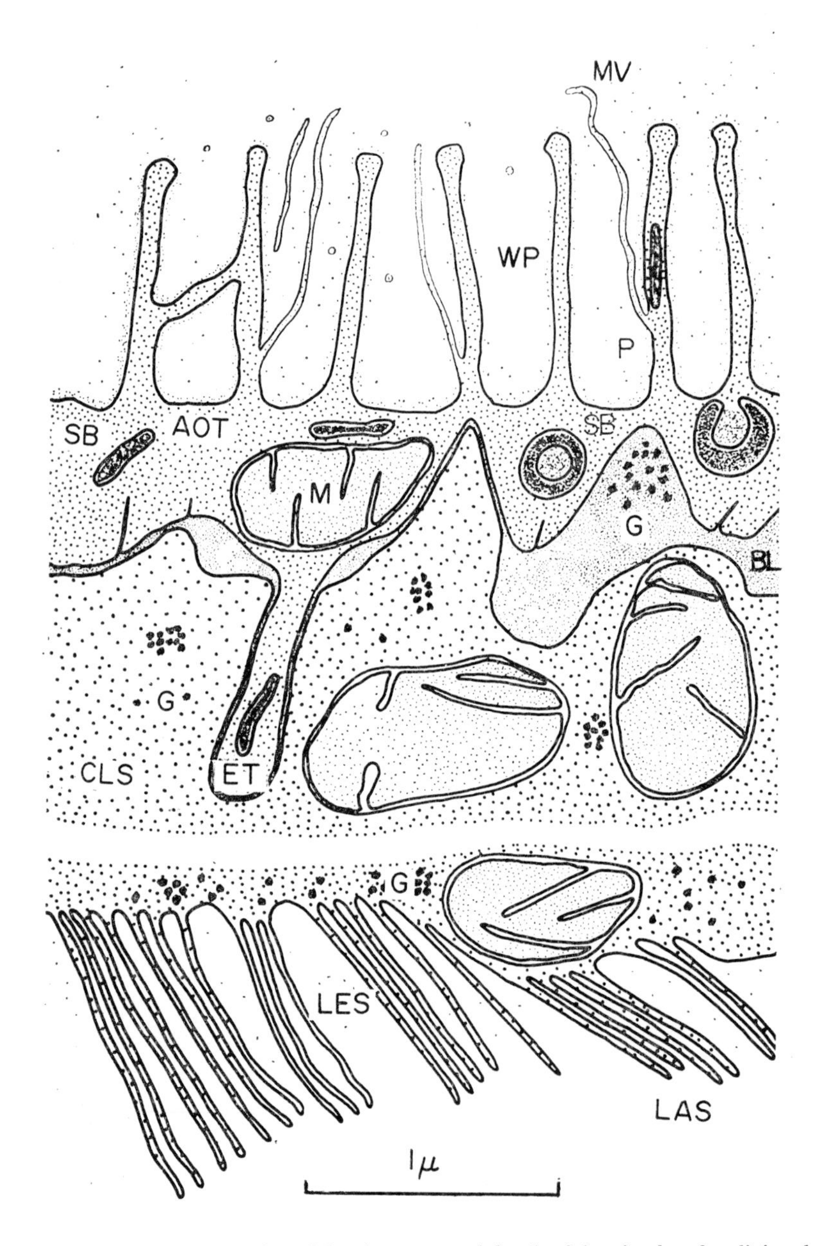

FIG. 97 Semi-diagrammatic representation of the ultrastructural details of the pitted surface lining the inner face of the adhesive organ lobes. Slender microvilli (MV) arise from the walls (WP) of the pits (P). The adhesive organ tegument (AOT) is relatively thin and contains mitochondria (M), secretion bodies (SB) and is in continuity with the more deeply placed tegumentary cell bodies by cytoplasmic extensions (ET). The tegument is bounded basally by the basement layer (BL) and adjacent to this is the cytoplasmic lining of the excretory lacuna (CLS). The innermost surface of this layer bears stacks of lamellae (LES) and is in contact with the lumen of the excretory lacuna (LAS). Note also the presence of glycogen (G). ERASMUS, 1969b.

of the lobe inner surface has a reticulum-like structure which disappears towards the middle and base of each lobe (Plates 12-2, 12-3). The entire surface, plane and reticulum, is finely pitted (1–1·25 $\mu \times 0{\cdot}3\ \mu$ diam.) from the walls of which arise single, slender microvilli. It is this sponge-like surface which comes into intimate contact with the vascular lamina propria of the mucosa and may represent a morphological basis for the 'placental' function suggested for the strigeoid adhesive organ by earlier workers (Plate 12-1). The contact between host and parasite is very close at this surface and under the light microscope it is impossible to differentiate clearly the parasite tegument from the host tissues.

The adhesive organ lobes in *Apatemon* contain the large lacunae of the excretory system. The tegument is very narrow in the lobe, so that there is considerable proximity between the lamina propria, pitted tegument and the cytoplasm and chambers of the excretory lacunae. The possible implication of the excretory lacunae in nutrient transport seems very likely under these circumstances.

Thus, morphological differentiation (of the tegumentary surface) exists in this and related forms which is only discernible under the electron microscope. Less marked differences occur in the tegumentary lining of the pharynx. Here (*C. bushiensis*) the tegument is more irregular and pinocytotic in its appearance and the basement layer very thick. BURTON (1966) also refers to regional differences in the tegument of *Gorgoderina*. These differences in the structure of the tegument allow the development of concepts which could not have been formulated previously. There is a distinct possibility that functional or physiological differentiation may be associated with these morphological differences thus making the outer surface of the trematode rather more complex than previously realised. Thus regional variation involving differences in degree of secretory activity and absorption may exist. In many ways the tegument of the adult digeneans so far studied is very similar and differences are related mainly to the form and variation of secretory bodies, the abundance of mitochondria and the presence of lipid droplets. Some of these differences are related to differences in technique, but others are undoubtedly genuine and have a biological significance. Differences in the adult tegument related to age have not been described in detail, although ERASMUS (1967c) found that the tegument of 18 hour and 3 day old adults (*C. bushiensis*) were similar in their basic features although differing in the size of the spines and concentration of secretory bodies. KUBLITSKENE (1963) has observed differences in the staining reaction with Mallory between 'cuticles' of 5 and 210 day old *Fasciola hepatica* in that the differentiation into layers was not so clear. The ultrastructural significance of this observation has yet to be determined.

Larval stages The cytoplasmic nature of the tegument has also been confirmed in sporocysts, rediae, cercariae and metacercarial stages, and there is very suggestive evidence which indicates that the tegument, once formed at the cercarial stage may persist and exhibit morphological continuity (with modifications) right through the life-cycle to the adult stage. The nature of the tegument of these stages has been described by KRUIDENIER and VATTER (1958), LAUTENSCHLAGER and CARDELL (1959, 1961), LAUTENSCHLAGER (1961), CARDELL (1962), BILS and MARTIN (1966), REES (1966), GINETZINSKAJA *et al* (1966), JAMES *et al* (1966), BELTON and HARRIS (1967) and KRUPA *et al* (1967). There is, of course, a break in the continuity of the tegument between the miracidium and sporocyst and between the sporocyst, redia and cercarial stage. It would be particularly interesting, using the electron microscope, to reassess the observations of ROEWER (1906) who described the secretion of the cercarial 'cuticle' during the differentiation of the developing germ ball.

The tegument of both sporocyst and redia is basically similar to that described from the

Table 43 General distribution and relative amount of alkaline phosphatase activity in seven species of Monogenea.

Tissue	1*	2	3	4	5	6	7
Cuticle	0**	0	0	0	0	+	++
Musculature	+	+	0	0	0	+	+
Parenchyma	+	+	+	+	+	+	++
Pharynx	—	0	+++	++	—	+	++
Ceca	+	+	++	+++	++	++	++
Testes	+	+	+	++	++	++	++
Ovary	+	+	+	++	++	++	++
Ootype	—	—	++	+++	—	+++	+++
Mehlis' gland	—	—	++	+++	—	+++	+++
Uterus	+	+	++	++	++	++	++
Eggs	—	—	—	—	+	+	+
Vitellaria	+	+	+	+	+	+	+
Exer. ducts	+	+	+++	+++	++	+++	+++

* 1 = E. *hippoglossi;* 2 = C. *kröyeri;* 3 = P. *integerimum;* 4 = D. *paradoxum;* 5 = D. *sagittata;* 6 = D. *merlangi;* 7 = O. *palmata.*

** +++, intensely stained; ++, moderately stained; +, slightly stained; 0, no stain; —, not known.

(After HALTON, 1967a.)

adult and consists of a basal plasma membrane laying on a basement layer, a cytoplasmic matrix and an outer trilaminar plasma membrane (Fig. 35). The matrix contains mitochondria and bodies and vesicles of varying characteristics and indeterminate function. The association of this outer cytoplasmic layer with 'tegumentary' cell bodies lying below the basement layer has not been universally nor conclusively demonstrated in these larval forms. The tegument of the sporocyst and redia are very similar with the external surface of the tegument extended to form structures referred to as either microvilli or ridges by different authors (Plate 12-4). An interesting contrast to these findings has been described by JAMES *et al* (1966) (Fig. 98). In the sporocyst of *Cercaria bucephalopsis haimaena* the tegument is interpreted as consisting of three regions. An outer layer immediately beneath a plasma membrane, a median nucleated zone and an inner layer of dense cytoplasm elevated into microvilli-like processes. Beneath this inner layer is the basement layer. This inner layer of the tegument of this species seems to correspond to the 'microvillous' tegument of sporocyst and redia described by other workers.

The cercarial tegument resembles more closely that of the adult. The external plasma membrane is not elevated into microvilli or ridges and the tegument contains, in addition to the structures mentioned above, spines and sense organs. There will be, in many species, continuity between the tegument of the cercaria and that of the adult or in others, between that of the cercaria, metacercaria and adult. In those life-cycles involving an encysted metacercaria (see chapter on metacercaria) there is some ambiguity concerning tegumentary continuity between these stages. The marked difference between the tegument of the intramolluscan stages and the cercaria may reflect the different host-parasite relationships exhibited by the two types of larvae. The microvilli or ridges of sporocyst and redia serve to increase the surface area and may be related to the absorption of nutrients from molluscan tissues.

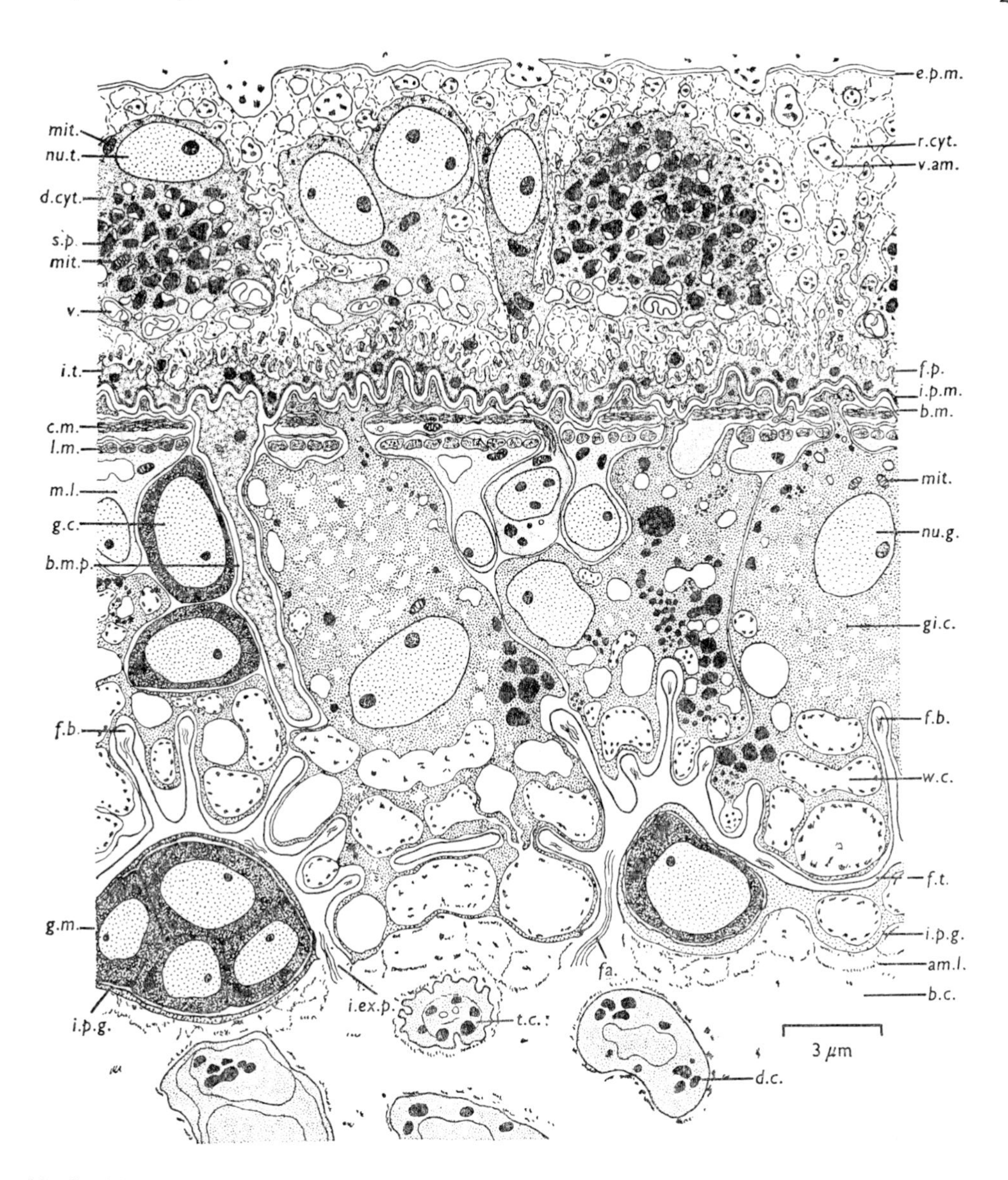

FIG. 98 Semidiagrammatic interpretation of the ultrastructure of the body wall of the daughter sporocyst of the *Cercaria bucephalopsis haimaena.* (am.l: secretion lining the body cavity of the sporocyst; b.c.: body cavity of sporocyst; b.m.: basement layer; b.m.p.: projection of basement layer into subtegument; c.m.: circular muscle; d.c.: body of developing cercaria; d.cyt.: dense cytoplasm; e.p.m.: external plasma membrane of tegument; f.b.: flame bulb; f.p.: finger-like projections of dense cytoplasm; f.t.: excretory tubule; fa: cilia at mouth of internal excretory pore; g.c.: germinal cell; g.m.: germinal mass; gi.c.: giant cell of subtegument; i.p.g.: plasma membrane of giant somatic cell lining body cavity of sporocyst; i.p.m.: internal plasma membrane of tegument; i.t.: inner region of tegument; l.m.: longitudinal muscle; m.l.: myoblasts of longitudinal muscle; mit: mitochondrion; nu.g.: nucleus of giant somatic cell of subtegument; nu.t.: nucleus of tegument of sporocyst; r.cyt.: reticulate cytoplasm of tegument.; s.p.: secretory product; t.c.: tail of developing cercaria; v: vesicular endoplasmic reticulum; v.am.: vesicle; w.c.: vacuoles.) From JAMES, BOWERS and RICHARDS, 1966.

Table 44. General distribution and relative amounts of phosphatase activity in eight species of *Digenea*.

Tissue	Acid phosphatase								Alkaline phosphatase							
	1*	2	3	4	5	6	7	8	1	2	3	4	5	6	7	8
Cuticle	0**	0	+	0	0	0	+	++	0	0	0	0	0	0	0	+++
Musculature	0	0	0	0	0	0	0	0	0	+	0	0	0	0	0	+
Parenchyma	+	0	+	0	0	0	+	0	0	0	0	+	+	+	0	+
Ceca	+++	+	++	+	+	+	+++	+++	0	0	0	0	0	0	0	+
Testes	++	0	++	0	0	0	++	0	0	0	+	+	+	+	0	0
Ovary	++	0	++	0	0	0	0	0	0	+	+	+	+	+	+	+
Uterus	++	0	++	0	0	0	0	0	0	+	0	++	+++	+++	++	+++
Eggs	+	0	+	0	0	0	0	—	0	+	+	+++	++	++	+	—
Vitellaria	0	0	0	0	0	0	0	0	+	+	+	+	+	+	+	+++
Exer. ducts	0	0	0	0	0	0	0	0	++	+++	++	++	++	++	++	++

* 1 = *H. medioplexus*; 2 = *H. cylindracea*; 3 = *O. ranae*; 4 = *D. subclavatus*; 5 = *G. vitelliloba*; 6 = *G. cygnoides*; 7 = *F. hepatica*; 8 = *S. mansoni*.

** +++, intensely stained; ++, moderately stained; +, slightly stained; 0, no stain; —, not known.

After HALTON, 1967a.

The tegument of the cercaria may be involved in cyst formation with the associated secretion and extrusion of a wide range of substances through the tegument. Furthermore, at the cercarial stage, sense organs (tango-, chemo- and photo-receptors) involved in host location are well developed, as well as the biochemical processes involved in the protection of cercariae from the tissues of molluscan and vertebrate hosts, and all these features are intimately associated with the tegument. The cercaria is also an active swimmer in most species, requiring oxygen. The tegument with its mitochondria will again be associated with vital respiratory processes.

(B) HISTOCHEMISTRY

Histochemical studies on the trematode tegument have not been particularly revealing although the evidence has more significance now that the cytoplasmic nature of this layer has been recognized. The observations to be described relate mainly to the cytoplasmic tegument external to the basement layer i.e. the 'cuticle' of optical microscopy. MONNÉ (1959) in a study of the tegument of *Fasciola hepatica*, *Alaria alata*, *Echinostoma revolutum* and *Paramphistomum* concluded that it was soft and jelly-like and did not contain tanned or keratinized protein. The major components appeared to be acid mucopolysaccharide and a diastase resistant PAS positive substance although the distribution of these substances differed between the examples studied. The basement layer contained a diastase resistant PAS positive material and also a substance resembling collagen although collagen fibres have not been identified with the electron microscope. The presence of a glyco- or mucoprotein was also suggested by BERTHIER (1954) and LAL and SHRIVASTAVA (1960) in the tegument of *F. hepatica*. In the histochemical studies (ERASMUS and ÖHMAN, 1963; ÖHMAN 1965, 66a and b) on the strigeoid trematodes *Cyathocotyle bushiensis*, *Diplostomum spathaceum*, *Apatemon gracilis minor* and *Holostephanus lühei* the general tegument did not give reactions for RNA confirming the absence of granular endoplasmic reticulum nor for fats and only very faint positives for protein as indicated by the mercury-bromophenol blue test. Tests for SS and SH were negative although LEE (1962) found that the tegument of *Diplostomulum phoxini* contained cystine except in the region of the adhesive organ and lappets. Diastase resistant, but PAS positive substances, as well as acid mucopolysaccharide material was identified in the tegument of all these strigeoid species. LAL and SHRIVASTAVA (1960) obtained positive reactions in the tegument of *Fasciola hepatica* using Biuret and Xanthoproteic tests and also concluded that the PAS diastase-resistant material was glycoprotein. Tests for phenols, —S—S bonds and unsaturated lipids were negative. A slightly different approach was made by BJORKMAN, THORSELL and LEINERT (1963) who subjected living specimens of *F. hepatica* and sections of fixed material to a variety of enzyme solutions and histochemical tests. They conclude that the 'cuticle' was proteinaceous and provided with a rim containing mucopolysaccharides or mucoproteins. Their results were unusual in that glycogen and a sudanophilic material (possibly lipid) was identified in the tegument. The presence of glycogen in the tegument of *F. hepatica* was also mentioned by PANTELOURIS (1964b) although YAMAO (1952) does not report glycogen from the tegument of *F. hepatica*, *Eurytrema spp. Paragonimus westermani* nor *Dicrocoelium lanceatum*. A negative result in the tegument was obtained by VON BRAND and MERCADO (1961) for *F. hepatica*, by MULLER (1966) for *Haplometra cylindracea* and by BURTON (1962) for *Haematoloechus medioplexus*. These variations may represent the lability of this material in the tegument whereas in other tissues (which have a more uniform record of occurrence in the literature) the deposition of glycogen may not be so temporary. Other substances described from the tegument of *F. hepatica* are ascorbic acid and iron (PANTELOURIS and HALE, 1962), and the overlap in distribution

Table 45. Distribution and effect of inhibitors on carboxylic esterase activity in *Fasciola hepatica**.

Tissue	Test	Eserine 10^{-5} M	DFP 10^{-5} M	62C47 10^{-5} M	OMPA 10^{-6} M	E600 10^{-7} M	E600 10^{-5} M	E600 10^{-2} M	PCMB 10^{-4} M	PPA 10^{-2} M	$AgNO_3$ 10^{-2} M
Tegument	+/+++**	○	○/+	+/+++	+/+++	+/+++	+	○	+/+++	+	○
Musculature	+/+++	○/+	+	+/+++	+/+++	+/+++	+/++	○	+/+++	+	○
Suckers	+	○/+	○/+	+	+	+	○/+	○	+	+	○
Pharynx	+++	○/+	○/+	+++	+++	+++	++	○	+++	+	○
Nervous system	+++	○/+	○/+	+++	+++	+++	++	○	+++	+	○
Ovary	++	+	++	++	++	++	+	○	++	+	○
Oviduct	++	++	++	++	++	++	+	○	++	+	○
Mehlis' gland	++	++	++	++	++	++	+	○	++	+	○
Laurer's canal	+++	++	++	+++	+++	++	+	○	+++	++	○
Uterus	+/++	+	+/++	+/++	+/++	+/++	+	○	+/++	+	○
Eggs	○/+	○/+	○/+	○/+	○/+	○/+	○	○	○/+	○	○
Vitellaria	○	○	○	○	○	○	○	○	○	○	○
Yolk duct	++	++	++	++	++	+	+	○	++	+	○
Testes	++	++	++	++	++	++	++	○	++	+	○
Spermatozoa	○	○	○	○	○	○	○	○	○	○	○
Cirrus sac	++	++	++	++	++	++	++	○	++	+	○
Cirrus	+++	○	○/+	+++	+++	++	++	○	+++	+	○
Seminal vesicle	+++	○	○/+	+++	+++	+++	++	○	+++	+	○
Ceca	+++	+++	+++	+++	+++	+++	+++	○	+++	+++	○
Excr. system	++	++	++	++	++	++	+	○	++	+	○

* 5-bromoindoxyl acetate as substrate.

** +++, intensely stained; ++, moderately stained; +, slightly stained; ○, no stain.

(After Halton, 1967c.)

of these two substances may reflect the process by which the iron is excreted after forming a soluble compound with the ascorbic acid.

LYONS (1966) has shown quite clearly that the sclerites of monogenean trematodes are composed of scleroproteins and not chitin. The scleroproteins are restricted to two types—sulphur containing and resembling to some extent mammalian keratin and secondly, non-sulphur containing and somewhat enigmatic in composition. The first type occurred in the haptoral hooks and the second in the clamp sclerites. In contrast to the spines of the digenea, which electron microscope studies have shown to be quite definitely restricted to the tegument, these monogenean sclerites are situated below the tegument and Lyons suggests that they are not derived from it (i.e. are not cuticularizations). The spines of those digeneans studied are proteinaceous but do not exhibit evidence of quinone tanning nor the presence of —SS or —SH groups. In very young adults of *C. bushiensis* developing spines were enveloped at their base by well defined extensions of the plasma membrane of the basement layer and also had secretory bodies aligned alongside (ERASMUS, 1967c). There seems every likelihood that in this case the spines are tegumentary in their origin.

Enzyme activity has been described within the tegument of adult trematodes by a number of workers (Tables 43, 44, 45). Alkaline phosphatase activity has been recorded by NIMMO-SMITH and STANDEN (1963) from the dorsal tegument of male and female *Schistosoma mansoni* and from the ventral tegument of female worms only. This confirms, but in more detail, the findings of DUSANIC (1959), ROBINSON (1961) and LEWERT and DUSANIC (1961). In contrast YAMAO and SAITO (1952) and BEČEJAC and KRVAVICA (1964) were not able to identify this activity in the tegument of *Fasciola hepatica* nor (YAMAO, 1952) in the tegument of *Paragonimus westermani*. Alkaline phosphatase activity has also been recorded from the tegument of *Posthodiplostomum minimum* (BOGITSH, 1966b), *Haematoloechus medioplexus* (ROTHMAN, 1968), *Cyathocotyle bushiensis, Holostephanus lühei, Apatemon gracilis minor* and *Diplostomum spathaceum* (Loc. cit.) although in *C. bushiensis* and *H. lühei* the reaction was more intense in the tegument covering the adhesive organ. This difference in activity in these two species can be correlated with ultrastructural differences in the tegument and it is possible that the difference in phosphatase distribution noted by NIMMO-SMITH and STANDEN (1963) in the male worm may also reflect a tegumentary differentiation in *Schistosoma*. It may also be related to the specialized role the ventral surface plays in the male, forming as it does the gynaecophoral canal. Acid phosphatase activity in *Schistosoma mansoni* showed a distribution similar to that described for alkaline phosphatases (NIMMO-SMITH Loc. cit.). It is particularly abundant in the tegument of *F. hepatica* and in *Paragonimus westermani* (YAMAO, 1952). MA (1964) also found intense acid phosphatase activity in the tegument of *Clonorchis sinensis*.

HALTON (1967a) was able to demonstrate alkaline phosphatase activity in the tegument of only two out of seven monogeneans tested, whereas in the digeneans, one species was positive out of the eight examined. Acid phosphatase activity was completely absent from the monogeneans and was present in the tegument of only three of the digenean species.

In *Cyathocotyle bushiensis, Holostephanus lühei* and *Diplostomum spathaceum* the activity of this enzyme was restricted to the tegument covering the adhesive organ and the lappets and in *Apatemon gracilis minor* it was even further restricted to the tegument of the lappets (Loc. cit.). Also in these strigeoid species non-specific esterase activity was generally restricted to the tegument of the lappets, and adhesive organ and was distributed in the general tegument of *Diplostomum sp.* and *Apatemon sp.* only (loc. cit.).

Non-specific esterase has also been reported from the tegument of *Schistosoma mansoni*,

S. haematobium and *S. rodhaini* (FRIPP, 1966a). In *Fasciola hepatica* the carboxylic esterase activity has been analysed in more detail by HALTON (1967c). He was able to show, with the use of specific inhibitors, that the tegument was rich in cholinesterases. FRIPP (1966b) has recorded B-glucuronidase activity from the tegument of *S. rhodaini* and found that in the male worm, activity was low in the region of the gynaecophoral canal. The tegument of *S. rhodaini* also contains leucine aminopeptidase activity (FRIPP, 1967). BOGITSH (1966a) was also able to demonstrate non-specific esterase activity in the adhesive organ gland cells and subtegumentary cells of the strigeoid trematode *Posthodiplostomum*. The enzyme activity associated with the tegument of the larval stages has already been discussed in the relevant chapters. At this point it is sufficient to recall that most of the enzyme activity associated with the adult tegument has also been recorded from the larval tegument.

These studies have revealed considerable enzyme activity in the trematode tegument, and in conjunction with the realization that it is cytoplasmic, emphasizes the fact that the tegument is a highly active metabolic region in its own right as well as being involved in subsidiary functions relating to the animal as a whole. The correlation between activity and morphology is not possible to any valuable extent at present. A possible method of analysis is the isolation of subcellular components of the tegument by ultracentrifugation. SMITHERS, ROODYN *et al* (1965) obtained a heterogeneous fraction (Fraction I: 300 g_{av} for 5 min) of *S. mansoni* males containing fragments of tegument. This fraction contained small amounts of DNA, RNA, phospholipid and exhibited acid alkaline phosphatase and succinoxidase activity. Glucose-6-phosphatase activity was not obtained in this fraction. Unfortunately, because of the heterogeneous nature of the fraction, the results have limited significance in an understanding of the biochemistry of the tegument although they do confirm the general pattern indicated by the histochemical methods.

(C) PHYSIOLOGICAL SIGNIFICANCE OF THE TEGUMENT

The role of the tegument in the physiology of the trematode is a subject which has been considered and discussed many times by many workers. In the past, concepts have been limited by the idea of a non-living cuticle and the need for the presence of pores to allow physiological continuity between the external environment and the internal tissues of the parasite. Ultrastructural studies have indicated that it is the basement layer (basement membrane) with its complex of 100 Å fibres which forms the limiting layer of the parasite defining its shape and the extent of the variation in body form. It is to this layer that muscles are attached and to which the intercellular fibres connect. It is apparent now that, between this skeletal system and the host, there is a thin layer of living cytoplasm representing the tegument. This layer is in continuity at many points with tegumentary cells and it is essential that, in considering the characteristics and functions of this tegument (probably representing a highly modified epidermis), the components lying below the basement layer must be regarded as one with the layer above. It has been pointed out earlier that the tegument represents one of the interfaces between two varying systems (host and parasite) and in the trematodes, instead of a resistant barrier as seen in the Nematoda, this consists of a layer of living cytoplasm with its associated complexity of physico-chemical and biochemical features. It is this thin layer of cytoplasm which acts as the 'buffer' between host and parasite and many of the features of the host-parasite relationship must be associated with this layer. Although the two components (tegument and cell body) can be regarded as one, there is a degree of functional differentiation between them. The distribution of RNA, as well as the ultrastructural characteristics, indicate that these tegumentary cells are regions of synthesis and that some of the products pass into the tegument. Because of the protoplasmic nature of this interface it is also possible that a cytoplasmic circulation may exist which would facilitate the transfer of materials from a region above the basement layer to the cells below. Furthermore, as THREADGOLD and GALLAGHER (1966) have shown, there is intimate contact between the tegumentary cells and parenchymal cells permitting the translocation of materials through the body of the parasite. In this way the concept of a non-living layer must be replaced by one envisaging a living, highly dynamic interface.

(D) POSSIBLE FUNCTION OF THE TEGUMENT

(a) *Uptake of materials* The osmotic relationships of helminths and particularly trematodes have not been investigated in great detail, and what evidence there is suggests that the latter helminths are able to tolerate wide ranges of osmotic pressures and carry out little water regulation under experimental conditions. STEPHENSON (1947a) did not find a great difference in the *in vitro* survival time of *F. hepatica* in solutions containing NaCl in concentrations varying from 58 to 230 nM. Similarly BUEDING (1950) found that *Schistosoma mansoni* was able to withstand wide ranges of NaCl concentrations. It is now appreciated that survival times may represent a very inadequate way of indicating the well-being of the parasite subjected to *in vitro* conditions. The observations of Stephenson were confirmed to some extent by KNOX and PANTELOURIS (1966) who found that *F. hepatica* was able to survive for at least 6 hours in Hedon-Fleig media of freezing point depressions from 0·40–0·81°C and that it did not appear to regulate its water content during this period. These authors also considered the weight changes exhibited by *F. hepatica* in media supplemented by different carbohydrates. From the data obtained they suggested that the fluke was very permeable to glycerol, less so to glucose, galactose and sorbitol and least of all to sucrose and maltose. The inability of *F. hepatica* to utilize maltose and sucrose was also suggested by Stephenson (1947a) and ROHRBACHER (1957). LAURIE (1961) found that the aspidobothrian *Macraspis cristata* absorbed glucose slowly, but not maltose nor galactose. There is obviously some variation between different species, as VERNBERG and HUNTER (1963) observed that *Himasthla quissetensis* utilized glucose and maltose and Bueding (cited in VON BRAND, 1952) reported the utilization of glucose, mannose and fructose by *Schistosoma mansoni.* The uptake of aspartic and glutamic acids (to a limited extent) and histidine, tryptophan and arginine by *S. mansoni in vitro*, as well as labelled glucose, has been demonstrated by SENFT (1963), and in a later experiment (SENFT, 1966) he was able to show that C^{14}-L-arginine was taken up more rapidly than L-histidine, D,L-tryptophan or L-methionine. The uptake of labelled amino-acids (glycine, alanine, proline, arginine, methionine and cystine) as well as glucose was studied in *F. hepatica* with ligated and non-ligated oral sucker by KNOX (1965b). She found that the tegument was permeable to all the compounds tested, but particularly so to glucose and methionine. Uptake experiments have also been carried out by NOLLEN (1968b) using *Philophthalmus megalurus* and he found that labelled glucose was taken up largely through the tegument whereas tyrosine and leucine entered mostly through the gut. Thymidine was absorbed via gut and tegument. Uptake of all compounds was rapid and occurred within 1–5 min.

These few examples indicate that particular metabolites and ions are taken into the parasite and that this intake seems to be preferential in some cases. The two routes into trematodes are the intestinal caecum via the mouth, and through the tegument. The data of STEPHENSON (1947a), ROHRBACHER (1957) and MANSOUR (1959) suggests that in *F. hepatica* uptake of glucose may take place through the tegument when the oral sucker is ligated, and a transtegumentary absorption of amino acids as well as carbohydrates has been suggested also for *S. mansoni* (ROBINSON, 1961, SENFT, 1963). The presence of a triple-layered plasma-membrane at the outside of the tegument may have considerable importance in the selective intake of metabolites. There is evidence accumulating that the uptake of glucose and glycogenesis and glycolysis in *F. hepatica* is influenced by insulin. PANTELOURIS (1964b) demonstrated a marked depletion in endogenous glycogen within 4 hours when insulin was added to the incubation media. SEKARDI (1966), on the other hand, found that insulin not only increased the rate of glucose uptake (from 4·86 mg/min/gr to 9·22 mg/min/gr but also the rate of glycogen synthesis. The techniques used by these authors differed considerably and the former employed entire worms and demonstrated the glycogen histochemically whereas the latter used bissected worms and labelled isotopes. Pantelouris found that the effect described by him was restricted to the glycogen of the tegument, peripheral tissues and muscles. The glycogen content of the vitelline glands seemed to be unaffected. Sekardi suggested that the insulin was effective at the surface membrane in enhancing the absorption of the glucose and that it would have an internal effect in influencing the active transport of glucose. It seems unlikely that the loss observed by Pantelouris was a depletion effect as VON BRAND and MERCADO (1961) found that after 12 hours starvation the glycogen level was only slightly depleted in the parenchyma and that depletion of muscular glycogen was not evident until 24 hours.

The presence of histochemically demonstrable phosphatases at surfaces involved in the translocation of substances e.g. the tegument, gut lining and the lining of excretory tubules and lacunae has always invited considerable speculation regarding their function. The older and more direct association with transport has been critised by READ (1966) and others, and the suggestion that phosphorylation of hexose sugars is an essential part of the transport mechanism seems no longer acceptable. Furthermore the use of aldehyde fixed material precludes the demonstration of specific transport ATPase in histological material. LUMSDEN *et al.* (1968) and THREADGOLD (1968) suggest that hydrolysis of phosphate esters at the membrane surface might facilitate the subsequent passage through the membrane of the organic part of the dephosphorylated molecule. In addition, the free phosphate ions might also be taken in through the membrane during this process and utilized in a variety of metabolic processes.

The function of phosphatases on internal membranes has been interpreted by THREADGOLD (1968) as follows: 1. they might provide free phosphate which could be coupled to previously dephosphorylated compounds and so reduce the possibility of them recrossing the plasma membrane; 2. they could provide free phosphate which could be utilized to phosphorylate plasma membrane lipids and so play a part in invoking membrane mobility and transport. Whatever the final role of these enzymes, their presence at a surface does seem to be associated with active transport in some way or other.

The uptake of substances of high molecular weight has not been explored to any great extent. RIJAVEC *et al.* (1962) were unable to demonstrate any absorption of radioiodinated albumen through the tegument of *F. hepatica* and KNOX (1965a) was also unsuccessful with labelled albumen, insulin and labelled algal protein. BJORKMAN and THORSELL (1964a) were more successful with ferritin. After 30 min incubation in saline containing ferritin the particles could be demonstrated within the tegument. The partices were not enclosed by membranes and these authors suggested that the ferritin entered the tegument by a process resembling transmembranosis rather than pinocytosis. Although the particles appeared to accumulate at the basement layer some were identified in tissues below the integument.

The uptake of ferritin and thorium dioxide by *Haematoloechus medioplexus* was observed by

ROTHMAN (1968). Both electron dense markers were found in the tegument of the worm after 120 min at 23–26°C. In *Apatemon gracilis minor* it has been possible to demonstrate the uptake of ferritin through the plasma membrane of the adhesive organ and its subsequent accumulation between the fibres of the basement layer (Erasmus—unpublished observations). Passage through the outer surface of the parasite does not involve pinocytosis, at least not in its gross form, and is best described as transmembranosis confirming the earlier observations of Bjorkman and Thorsell and Rothman with other species. Ferritin is an iron-protein complex approximately 110 Å in diameter with a molecular weight of approximately 460 000. These observations raise a number of interesting possibilities regarding the uptake of large molecules through the parasite tegument. The mechanism seems to involve some specificity as ROTHMAN (1968) was unable to demonstrate uptake of horse-radish peroxidase, a much smaller molecule, via the same route.

(b) *Exit of materials from the parasite* Although the egestion and excretion of substances through the mouth and excretory pore can be observed easily, the possibility that a significant movement of materials may take place out through the tegument must not be ignored. Indeed, now that the cytoplasmic nature of this layer is evident, the possibility of its involvement in this process seems more likely. The two processes related to such an outward movement of substances are secretion and excretion.

Secretion from the surface of the tegument will be involved in the production of the anti-enzyme barrier of mucopolysaccharides and its release at the surface of the parasite and also in the production of lytic secretions associated with penetration and migration and with extracellular digestion. A possible morphological basis for the production of the anti-enzyme coat has already been discussed in this chapter. The evidence suggests the possibility of a movement of secretory bodies from their site of origin in the tegumentary cell to the limiting plasma membrane of the tegument as well as the extrusion of their contents to the exterior. The route by which lytic secretion emerges has also been considered in the section on the ultrastructure of the strigeoid adhesive organ and extracellular digestion. The observations suggests (1) that local specialization of the tegument occur in regions where the extrusion of secretion takes place. These regions may be revealed by the local occurrence of enzyme activity as indicated by the histochemical tests i.e. tegument of lappets and adhesive organ of strigeoid trematodes; (2) the secretory process may involve the direct extrusion of substances to the exterior through ducts with distinct apertures, as in cercarial gland cells and in the strigeoid lappets, or it may be a more indirect process such as vesicle formation i.e. the breaking off of the bulbous ends of microvilli (Plate 9-3) or the fusion of membrane bounded secretory bodies with the limiting plasma membrane of the tegument. In this way large molecules or masses of secretion may emerge to the exterior via the parasite's tegument. Culture *in vitro* experiments have also demonstrated that leakage of substances can take place from the parasite. For example egestion of the partly digested blood from the oral sucker could occur. SENFT (1963, 1965, 1966) has shown that when *S. mansoni* is cultured in a defined medium an increase in the concentration of certain amino acids in the culture fluid takes place. The principal increase occurred with alanine, proline and ornithine although smaller amounts of other acidic and neutral amino acids were noted (SENFT, 1963). The production of alanine is particularly impressive in that three times the original mean value was produced. Experiments with uniformly labelled C^{14} glucose have demonstrated that the newly produced alanine is tagged suggesting that the carbon source of alanine is carbohydrate. However the origin of these amino acids is still open to question—they may represent a leakage or excretory product or the accumulation may represent disintegration products of the worm. Other studies (RIJAVEC, 1966) have demonstrated urea excretion in *Fasciola hepatica* and *Dicrocoelium dendriticum* although *Paramphistomum cervi* does not. Thus these investigations have demonstrated the release of considerable quantities of nitrogenous material and once again the tegument must be considered as a possible route. In this context it is interesting that KURELEC and EHRLICH (1963) found a significant loss of amino acids in *F. hepatica*, with ligated oral suckers. Unfortunately this type of experiment does not eliminate the possibility of the loss of these substances through the excretory pore.

(c) *Respiration* Biochemical studies with tissue extracts have demonstrated the presence of carbonic anhydrase (*F. hepatica*; CASTELLÁ, 1952), succinate dehydrogenase (*F. hepatica*, PENNOIT-DECOOMAN and VAN GREMBERGEN, 1942; *Schistosoma japonicum*, HUANG and CHU, 1962) and malic dehydrogenase (*F. hepatica*, PENNOIT-DECOOMAN and VAN GREMBERGEN, 1942) in various trematodes. ROTHMAN and LEE (1963), and WAITZ and SCHARDEIN (1964) have demonstrated intense dehydrogenase and cytochrome oxidase activity in the tegument of a number of cestodes. In view of the presence of mitochondria in the tegument of digenetic trematodes it seems very likely that these enzymes will occur here thus involving the tegument in oxidative and other intermediary aspects of metabolism. It is well established that the parasitic helminths studies are facultative aerobes and are able to utilize oxygen when available (VON BRAND, 1966) and it is again particularly significant that it is the cytoplasmic tegument which first comes in contact with the gaseous phase of the parasite's environment. Whereas a non-living cuticle has little significance in this context the presence of a cytoplasmic mitochondrial layer on the outside of the parasite has considerable import.

(d) *Contact* The extent and the nature of the contact between host and parasite is influenced by the nature of the tegument. The soft tegument with its outer plasma membrane is capable of extremely intimate contact with host tissues. In sections of strigeoids such as *Cyathocotyle bushiensis* and *Apatemon gracilis minor* with host tissue attached, it becomes very difficult to distinguish host from parasite tissues even at magnifications of over × 16 000. Under these circumstances the passage of materials from host to parasite or vice versa can be considered almost in the same light as the passage of materials from one cell to an adjacent one, in that the distance which has to be traversed by the molecule becomes small. Thus it is possible that the degree of intimacy at points of contact between host and parasite may be cellular in its extent and will be enhanced by local specialization of the tegument and the dynamic nature of the plasma membrane.

It should now be apparent that our concept of the host-parasite interface as represented by the tegument is being subjected to extensive reappraisal and that the results should influence considerably our ideas concerning uptake of nutrients and chemotherapeutic compounds by the parasite, as well as the nature of the source and method of transference of antigenic materials from parasite to host. The terminology of this layer has been discussed by LEE (1966) who suggests that the tegument may be regarded as a modified epidermis. This interpretation brings the outer covering of trematodes and cestodes into line with that of the Turbellaria. It must be remembered, however, that the connotation of ectodermal origin may not be applicable to the parasitic helminths.

III THE ALIMENTARY TRACT

The host-parasite interface has been defined earlier (p. 214) as that surface through which interchange of materials takes place between host and parasite. Three surfaces come into this category in trematode parasites and two—the general tegument and the tegument of attachment structures have already been described. The third surface which can be considered from this point of view is that of the alimentary tract.

The presence of the alimentary canal in members of the Trematoda is a feature which has considerable significance in our interpretation of the evolution, physiology and host-parasite relationships of these parasites. The alimentary tract consists basically of mouth, pharynx, oesophagus and intestinal caeca, although the presence and extent to which the individual components are represented varies to a considerable degree. In the Monogenea the oral sucker, as we see it in the Digenea may not be present, so that the mouth opens into a buccal funnel. In *Polystoma* there is a well developed oral sucker but in some cases, e.g. *Microcotyle*, suckers occur on the inner surface of the funnel. The funnel leads into the pharynx which in certain of the Monogenea may be protrusible and possess gland cells (KEARN, 1963) and bears a considerable resemblance to the pharynx of the Turbellaria. The intestine may consist of a simple sac or may be bifurcate with two caeca which are unbranched. Further elaboration on this pattern consists of lateral branching of the intestinal caeca and the formation of transverse connections between the caeca. The branching may eventually result in an anatomosis being formed.

In the Digenea the mouth is enclosed by an oral sucker although this is secondarily lost in some cases. The pharynx is a highly muscular, non-eversible organ which may be separated from the oral sucker by a short prepharynx. The oesophagus extends posteriorly from the pharynx and may also recieve the ducts from clusters of gland cells sometimes described as salivary glands. In some species both pharynx and oesophagus are absent. The oesophagus, when present, continues into the intestinal caeca which basically consist of two blind caeca. The variations exhibited by the caeca consist of: lateral branching sometimes with anastomoses; fusion of the caeca which then continues as a single caecum; branches may extend anteriorly on either side of the pharynx forming an H-shaped system and rarely there is a reduction to a single caecum. In some trematodes the alimentary system is said to open into the excretory system and in others, anal openings exist (STUNKARD, 1931). It is apparent from this brief account that the gut, which is usually well developed, exhibits some variation.

HISTOCHEMICAL STUDIES

MONOGENEA

In chapter 8 it was indicated that monogenean parasites may feed on the mucus, secretions of the host and some of its tissues, but in other cases the parasites feed on blood. Many workers in the past had suggested that the dark pigment in the intestinal caeca of *Polystoma* represented a degradation product of haemoglobin but the identification of haematin in the gut by Llewellyn did not take place until 1954. This observation was later confirmed by JENNINGS (1956) who also (1959) was able to demonstrate both haematin and haematoidin crystals in the gastrodermis of the parasite and concluded that digestion was both intraluminar and extracellular. It must be mentioned however that the black pigment in the gut of some Monogenea e.g. *Capsala* sp. does not seem to be haematin (KEARN, 1963).

A more extensive histochemical study was conducted by HALTON and JENNINGS (1965) on nine monogeneans. The monogenea selected represented the Mono- and Polyopisthocotylea and included tissue and mucus feeders as well as haematophagous forms. The specimens studied were *Calicotyle kroyeri*, *Entobdella hippoglossi* and *Udonella caligorum* from the first group and *Polystoma intergerrium*, *Diplozoon paradoxum*, *Discocotyle sagittata*, *Diclidophora merlangi*, *Octodactylus palmata* and *Plectanocotyle gurnardi* from the second group. The oral sucker of *Polystoma sp.* possessed gland cells which produced a highly proteinaceous secretion thought to play a part in attachment, as it was devoid of enzyme activity. Gland cells associated with the oral sucker of *Calicotyle sp.* and with the buccal funnel of *Diplozoon sp.* were assumed to have a similar function.

The nature and characteristics of the pharynx differed between the genera. In *Entobdella sp.* the pharynx was represented by a muscular-glandular feeding organ which could be protruded through the mouth and applied to the surface of the host. The posterior glandular portion of this organ contained about 40–50 large acidophilic gland cells which were capable of inducing liquifaction of a gelatin substrate. The authors suggested that a gelatin-splitting protease was produced by these cells and this corresponded to the interpretation of KEARN (1963) who ascribed a similar activity to this organ. In the other genera studied the pharynx was a purely muscular structure. Oesophageal glands occurred in some species and in *Calicotyle* they gave an acidophilic reaction, a basophilic reaction in *Entobdella* and the corresponding oesophageal region in *Polystoma* exhibited gelatinolytic activity. The oesophageal gland cells in *Polystoma* occurred in two groups—an inner zone which had alkaline phosphatase activity and an outer zone with no enzyme activity but acidophilic in reaction.

The gastrodermal cells of *Calicotyle sp.* showed a strong reaction with non-specific esterase and the intracellular vesicles of the gastrodermis of *Polystoma* were also positive for this type of activity. In other cases, the cells lining the gastrodermis of *Polystoma*, *Diplozoon*, *Discocotyle*, *Diclidophora* and *Octodactylus* showed alkaline phosphatase activity along their distal margins, and the authors suggested that this might be concerned with absorption, see also HALTON (1967a). The changes in the gastrodermis during digestion have been best described for *Polystoma*. The ingested erythrocytes become haemolysed almost immediately and within three hours of ingestion the nuclei have completely disintegrated. The semi-digested material becomes absorbed by the smaller 'younger' cells of the gastrodermis and the cells become distended with this material. At this time the distal edges of the cells exhibit the phosphatase activity mentioned earlier. This process continues until no stainable material can be detected in the lumen and the time for this will depend on the size of the blood meal. Within the cells, digestion takes place in vesicles and as this progresses the granules of haematin appear. Eventually the vesicles are extruded into the lumen which will then contain the haematin plus traces of esterase activity from the contents. In this way, esterase active material may become interspersed with the gut contents and possibly play a part in luminal digestion. Because of this sequence of events the gastrodermis appears discontinuous and becomes represented by collections of cells separated by gaps, or by debris where cells have disintegrated in the course of the egestion of the haematin. This is the picture as it appears under the light microscope and it is possible that a slightly different picture will be resolved under the electron microscope as has been the case with certain digenea. However, all sanguivorous forms exhibit this apparently discontinuous gastrodermis, whereas the mucus feeders exhibit a continuous layer without any differentiation. This histological pattern seems to represent a fundamental difference between the Mono and Polyopisthocotyles. Thus the evidence presented by these workers (and that from earlier papers reviewed by them) suggests that digestion may occur intraluminally and intracellularly and this process may be aided by secretions from the gland cells associated with the oral sucker and pharynx so that extracorporeal digestion is also a possibility in some cases. The apparently deciduous nature of the gastrodermis in the Polyopisthocotylea seems to be associated with the digestion of haemoglobin and the elimination of insoluble haematin.

DIGENEA

The histochemistry of the digenean gut has not been extensively studied and observations are restricted to a few species. *Alkaline phosphatase activity.* In most cases (*F. hepatica* SAITO, 1952;

TARAZONA VILAS, 1958; *Paragonimus westermani*: YAMAO, 1952; *S. mansoni*: ROBINSON, 1961: NIMMO-SMITH and STANDEN, 1963: FRIPP, 1966a; *C. bushiensis*, *H. lühei*, *D. spathaceum* and *A. gracilis minor*—ERASMUS and ÖHMAN, 1963; ÖHMAN, 1965, 66a,b) alkaline phosphatase activity was not detected in caecal cells. Conflicting observations exist in the case of *S. mansoni*. Negative results were obtained by ROBINSON (1961), NIMMO-SMITH and STANDEN (1963) and FRIPP (1966a) but some degree of activity was recorded by DUNSANIC (1959) and LEWERT and DUSANIC (1961). SAITO (1961) found the caecal epithelium of *F. hepatica* positive for alkaline phosphatase activity on ATP and fructose—1, 6-diphosphate substrates only. HALTON (1967a) was unable to detect alkaline phosphatase actvity in the caecum of eight digenean species.

Acid phosphatase activity. In contrast to the observations on the Monogenea, the caecal cells of the digenean species studied constantly exhibit strong acid phosphatase activity (see HALTON, 1967a). TARAZONA VILAS (1958) and SAITO (1961) demonstrated this activity in the caecal cells of *F. hepatica*; YAMAO (1952) in *Paragonimus westermani*; NIMMO-SMITH and STANDEN (1963) and FRIPP (1966a) in both male and female *Schistosoma mansoni*; ERASMUS and ÖHMAN (1963) in the caecum of adult *Cyathocotyle bushiensis*; MA (1964) in adult *Clonorchis sinensis*; ÖHMAN (1965) in adult *Diplostomum spathaceum*, in adult *Apatemon gracilis minor* (ÖHMAN, 1966a) and *Holostephanus lühei* (ÖHMAN, 1966b).

Esterase activity. Esterase activity corresponding to Type B (PEARSE, 1960) was detected in the caecal cells and lumen of *Apatemon gracilis minor* (ÖHMAN, 1966a) no activity in the gut of *Diplostomum sp.* (ÖHMAN, 1965) and a particularly resistant type in the caecum of *Cyathocotyle bushiensis* (ERASMUS and ÖHMAN, 1963) and *Holostephanus lühei* (ÖHMAN, 1966b). A similarly resistant non-specific esterase has also been identified in the gut of *Fasciola hepatica* by HALTON (1967c). This worker (HALTON, 1963) has described non-specific esterase activity in the suckers and associated gland cells of *Haplometra cylindracea* and *F. hepatica*.

Leucine aminopeptidase activity. Faint activity was detected in the caecal cells of *Cyathocotyle bushiensis* (ERASMUS and ÖHMAN, 1963) and THORSELL and BJÖRKMAN (1965) observed that gelatinolytic activity occurred anterior to *F. hepatica* lying on the substrate and that this did not occur with flukes in which the oral sucker was ligated. The oesophageal glands only, were positive in *Schistosoma rodhaini* (FRIPP, 1967).

Glycogen. This reserve carbohydrate does not seem to be present in the caecal cells of *Eurytrema coelomaticum*, *E. lancreatum*, *Paragonimus westermani*, *Dicrocoelium lanceatum*, *F. hepatica* (YAMAO, 1952); *Fasciola hepatica*, (ORTNER-SCHOENBACH, 1913, VON BRAND and MERCADO, 1961); *Haematoloechus medioplexus* (BURTON, 1962), *Cyathocotyle bushiensis* (loc. cit.), *Diplostomum spathaceum*, *Apatemon gracilis minor* and *Holostephanus lühei* (loc. cit.), nor in a wide range of species tested by HALTON, 1967d.

RNA and Protein. Quite distinct, positive reactions for RNA and proteins (mercury-bromophenol blue test) were obtained in the caecal cells of *Cyathocotyle*, *Holostephanus*, *Diplostomum* and *Apatemon* (loc. cit.). This particular result is very significant in relation to the ultrastructural observations to be described later and suggests active protein synthesis by the caecal cells.

UPTAKE OF LABELLED COMPOUNDS AND ENZYME ACTIVITY OF TISSUE EXTRACTS

A few observations exist concerning the uptake of labelled compounds. PANTELOURIS and GRESSON (1960) found that after the injection of Fe^{59} labelled ferric chloride and L-phenylalanine-C^{14} directly into the gut of *Fasciola hepatica* a high concentration of labelled iron localized in the caecal epithelium although the phenylalanine was not detected. The uptake under *in vitro* conditions of radioglucose by *Haematoloechus medioplexus* was investigated by BURTON (1962) and he noted that after 12, 24 and 48 hours hardly any activity was exhibited by the caecal epithelium. The absence of the amino acid was regarded by the former authors as indicating a rapid passage through the cells and the absence of activity in the second case was interpreted as indicating an absence of glycogenesis from the caecal epithelium, and this is supported by the histochemical evidence quoted above. PANTELOURIS (1964a) found that although S^{35} labelled free sulphate was not localized in the caecal epithelium DL-methionine-S^{35} was, and he suggested that this compound was being metabolized in the caecal cells and the resulting sulphur passed into the tissues to be incorporated into yolk, by the vitelline glands, and into mucopolysaccharides by other tissues of the body. More recently, THORSELL and BJÖRKMAN (1965) have shown that after 10 minutes incubation *in vitro*, DL-leucine-2-C^{14}, L-methionine-S^{35}, DL-phenylalanine-3-C^{14}, and DL-tryptophane-3-C^{14} all localized intensely in the caecal cells of *Fasciola hepatica*. In contrast, they were unable to demonstrate the uptake of ferritin after three hours incubation. SENFT (1966) has described the uptake by *Schistosoma mansoni* under *in vitro* conditions of C^{14} labelled L-arginine, L-histidine, D,L-tryptophan, and L-methionine as well as C^{14} glucose and C^{14} leucine (SENFT, 1963) although the route via which these substances entered the worm was not determined. NOLLEN (1968b) found that both routes were involved in the absorption of labelled glucose, tyrosine, leucine and thymidine by adult *Philophthalmus*, although he was of the opinion that tyrosine and leucine entered largely via the gut.

ROGERS (1940) showed that the pigment in the gut of *S. mansoni* was haematin and TIMMS and BUEDING (1959) have described a protease from schistosomes capable of digesting haemoglobin although cellular localization of this activity was not determined. Hydrolytic enzyme activity (Lipase, proteinase, dipeptidase and amylase) has also been detected in tissue extracts of *Fasciola hepatica* (PENNOIT-DE COOMAN and VAN GREMBERGEN, 1942) but their role in caecal digestion must be interpreted with caution as their localization is unknown.

The many attempts at *in vitro* culture of trematodes (reviewed by SILVERMAN, 1965) have indicated that various compounds increase the survival times but the route by which these substances enter the helminth is not apparent. The observations of many workers indicate that, as well as being able to absorb soluble materials as outlined above, the majority of trematodes when viewed alive have cellular debris in their caeca and that this exhibits various degrees of disintegration, suggesting extracellular digestion of host tissue masses. The idea that ingestion of materials in a solid or semi-solid form is significant to the biology of these animals is supported by the observation of BELL and SMYTH (1958) who found that *Diplostomum phoxini* would only develop well *in vitro* when a semi-solid medium was used. This was also seen to enter the gut.

Although the data available is restricted, it is becoming apparent that the trematode is able to take in solid and liquid materials into its gut, that absorption of soluble substances can take place at this site and also that the caecal cells are involved in active enzyme synthesis (high RNA content) with the secretion of these substances into the gut lumen. The careful comparison of the histochemical characteristics of the host tissue, parasite caecal contents and the caecal cells does indicate without any doubt that secretion from the cells into the lumen takes place. REZNIK (1963) has also suggested the possibility of secretory activity by the caecal epithelium of *Fasciola hepatica* and *Dicrocoelium lanceatum* and proposed that it was apocrine in the former and exocrine in the latter.

A comparative study of the histology and histochemistry of several digenean caeca as well as observations on the process of feeding has been made by HALTON (1967b). He found that feeding was suctorial involving the activity of the pharynx and the oral sucker. The abrasive action of the oral sucker was supplemented by enzymatic secretion in *Haplometra cylindracea*. Digestion was predominantly an extracellular process but details varied according to the nature of the food and the species involved. This author was of the opinion that the gastrodermis was both secretory and absorptive in function although specific gland cells were absent from the gastrodermis of all species.

ULTRASTRUCTURE

The histology of the trematode caecum and the changes exhibited by the epithelium have been best described at the optical level by MÜLLER (1923), STEPHENSON (1947b) and DAWES (1962b) for *Fasciola hepatica*. The cells appear to exhibit a secretory cycle which corresponded to the transformation of the tall columnar cell into a stumpy mass of residual cytoplasm. Dawes was unable to distinguish distinct absorptive and secretory cells as described by GRESSON and THREADGOLD (1959). In his account, Dawes mentions the presence of 'fringes' at the periphery of the cells and WOTTON and SOGANDARES-BERNAL (1963) also reported micro-villus-like structures on the caecal cells of certain paramphistomes. Microvilli have been described from the caecal cells of 7 digenean species but not from 5 monogeneans examined by HALTON (1966).

Electron microscope studies on the caecal cells of *Fasciola* have been made by GRESSON and THREADGOLD (1959) and THORSELL and BORKMAN (1965) and on *Schistosoma mansoni* by SENFT *et al* (1961) (Fig. 99). The caecal lining in *Fasciola* consists of an epithelium of cells resting on a basement membrane. The major gross distinction between the individual cells of the epithelium is their size which varies from 0·01–0·02 mm to 0·04–0·05 mm (DAWES, 1962). To the outside (parenchymal side) of the gut is a thin layer of muscles. The cytoplasm of the cells is very dense, and contains masses of parallel endoplasmic reticulum with its associated ribosomes and also mitochondria. The nuclei are generally basal in position. The presence of the abundant endoplasmic reticulum explains the very strong reaction for RNA obtained in the histochemical studies on the caecum of *Cyathocotyle*, *Diplostomum* and *Apatemon* by ERASMUS and ÖHMAN (1963) and ÖHMAN (1965, 66a, 66b). The plasma membrane delimiting the lateral surface of the cells is not always obvious and it may become irregular and folded so that the cells interdigitate with one another. THORSELL and BÖRKMAN (1965) describe the infolding of the luminal plasma membrane into the cytoplasm with the formation of septate desmosomes-like structures and it is possible that this appearance represents the infolded plasma membrane at the apex of adjacent cells or it could be the infolded luminal plasma membrane of a single cell. The major portion of the luminal plasma membrane becomes involved in the formation of numerous microvilli. These vary considerably in length but are uniform in diameter and in *Fasciola* are 0·7 μ thick and consist of a dense core with a triple-layered plasma membrane on either side. All cells, tall or short, possess this covering of microvilli and this corresponds to the 'fringe' described at the optical level (Plate 10-1). The microvilli may extend straight out into the lumen or may become looped and where adjacent microvilli touch they seem to become fused. In this way the gap between the sides of the adjacent cells may become traversed by a meshwork of microvilli and this meshwork corresponds to some of the vacuoles seen in cells at the opical level. Both Gresson and Threadgold and Thorsell and Borkman refer to dense granules in the cytoplasm. These occur also in

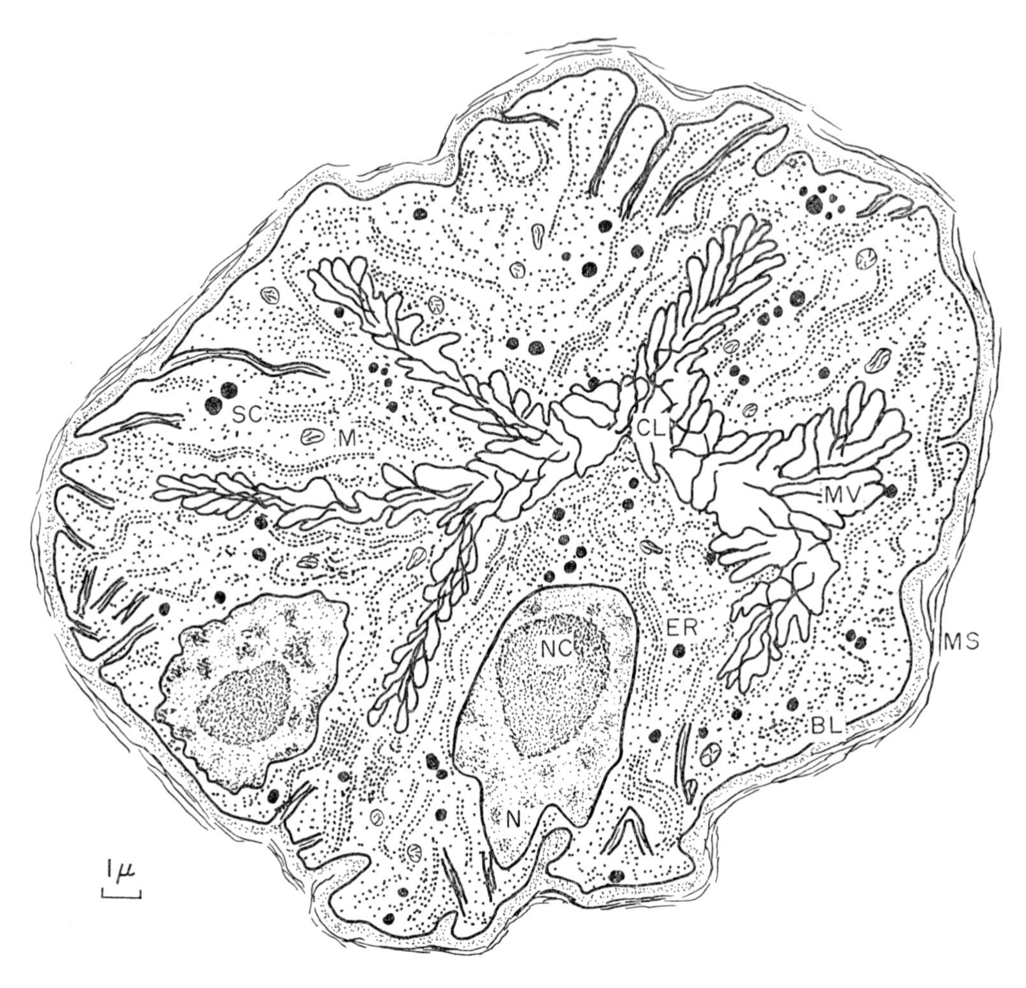

FIG. 99 Semidiagrammatic drawing of a transverse section of the caecum of *Cyathocotyle bushiensis* based on electron-icrographs. (BL: basement layer; CL: lumen of caecum; ER: endoplasmic reticulum; M: mitochondrion; MS: muscle; N: nucleus; NC: nucleolus; SC: secretory body.)

the caecal cytoplasm of *Cyathocotyle bushiensis*. In *Fasciola* the granules are dense and enclosed in a membrane and Thorsell and Borkman suggest they are secretory in function and are discharged from the cell into the caecal lumen. Gresson and Threadgold, however, suggest they may be storage bodies or may represent material absorbed from the lumen, although Thorsell and Borkman were unable to demonstrate the uptake of ferritin by the caecal epithelium. In *Cyathocotyle sp.* the contents, enclosed by a membrane, are not homogeneous and consist of an electron dense matrix in which occurs small vacuoles. Within the cell cytoplasm the bodies vary in size and there does not appear to be any regular distribution associated with a gradual change in size which might be expected if they represented ingested material. The significance (lysosomal and secretory or pinocytotic and storage) of these structures has still to be determined but a study of caecal epithelial where the inclusion of host material into the cells is known to take place, i.e. the sanguiniferous monogenea, might help to resolve the difficulties.

In general terms therefore it may be stated that the caecal epithelium consists of cells with the luminal plasma membrane elevated to form microvilli (loc. cit. and DIKE, 1967) for *Gorgodera amplicava*) or lamellae (DIKE, 1967—*Haematoloechus medioplexus*; DAVIS *et al*, 1968—*H. medioplexus* and MORRIS, 1968—*Schistosoma mansoni*). These elaborations serve to increase the probably absorptive surface area but may also (DIKE, 1967 in *H. medioplexus*) play an important part in enmeshing food particles in the gut lumen. Authors differ in their assessment of the structure of the gastrodermis. Some regard the epithelium as multicellular, whereas others (DIKE, 1967—*G. amplicava* and MORRIS, 1968—*S. mansoni*) interpret it as being a syncytium. Dense endoplasmic reticulum with abundant ribosomes is a common feature and golgi complexes have been described by some (e.g. MORRIS, 1968; DIKE, 1967) but not by others (e.g. DAVIS *et al*, 1968). Common to all descriptions is the presence of membrane bound dense bodies and the illustrations of Davis *et al* suggest that they may discharge to the exterior of the cell into the gut lumen. In *S. mansoni*, MORRIS (1968) provides some evidence suggesting the uptake of droplets containing haematein-like material. All descriptions of digenean gastrodermis refer to the complex infolding of the basal plasma membrane. This structural characteristic presumably will be associated with the same physiological processes (i.e. ion exchange and translocation of dissolved substances) as has been postulated for the elaborate basal plasma membrane of the tegument.

The change in the gross morphology of the caecal cells of *Fasciola* has been correlated with the presence or absence of food in the caecum (GRESSON and THREADGOLD, 1959; DAWES, 1962b). The cells change from the basic columnar form to a stumpy variety in the presence of food. THORSELL and BÖRKMAN (1965) observed that in the cells undergoing secretion, the cytoplasmic organelles are disorganized, the mitochondria swollen and the nuclei slightly pyknotic. The evidence does suggest that there is basically a single type of cells which undergoes a secretory phase. Also suggested by DAVIS *et al* (1968), DIKE (1967) and MORRIS (1968). Whether or not regeneration of the same cell after secretion takes place is not certain, but it does seem likely that the epithelium undergoes a turnover with the formation of new cells. Because of the complex branching exhibited by the gut in *Fasciola hepatica* it is possible to observe branches empty of food where the entire epithelium is intact and columnar. In *Cyathocotyle bushiensis* more typical unbranched caeca occur, and a complete range of morphology from tall columnar cells to a flattened epithelium may be observed in a single transverse section. The same appearance may be observed also in longitudinal section so that the activity of the entire epithelium does not seem to be phased but is irregular, the precise phase of activity varying from place to place. Under the optical microscope, the caecal lining in *C. bushiensis* appears to be completely absent from some regions of the gut but under the electron microscope it may be seen to be present but reduced in height. The ultrastructural features of the epithelium in *Cyathocotyle* resembles that of *Fasciola* although the microvilli seem to enmesh more frequently than suggested by the other workers, and sometimes enclose and come into physical contact with granular material in the gut lumen. Thus the luminal surface of the caecal epithelium in these few trematodes consists of a microvillous surface which must increase the surface area considerably and so play an important part in absorption. In addition the microvilli entangle and form a meshwork which must trap very small particles. In *Cyathocotyle bushiensis* it has been observed that the basal plasma membrane adjacent to the basement membrane becomes much infolded with the folds extending into the cell cytoplasm. This appearance resembles that described for the intestinal cells of vertebrates and of nematodes and it may represent a morphological device for obtaining an increase in surface area of the basal portion of the cell which may be associated with the interchange of substances between the caecum and surrounding tissues. The ultrastructural observations also indicate that considerable protein synthesis associated with secretion takes place in the caecal cell and this fits in with the histochemical observations. The precise nature of the secretory process is not clear but the histochemical demonstration of acid phosphatase and esterase activity in some epithelia and the presence of membrane bounded bodies in the cytoplasm does suggest strongly the possible implication of lysosomes or lysosome-like organelles in the process of digestion.

The basement membrane of the caecum becomes closely associated with a variety of cells (parenchymal cells, flame-cells and vitellaria) and the morphological continuity which exists must play a part in the transference of material from the gut into the surrounding tissues. The experiments with isotopes described above indicate that this transference must take place very rapidly and THREADGOLD and GALLAGHER (1966), in a description of the parenchymal cell of *F. hepatica*, discuss the possibility of transport pathways existing and involving the parenchymal cell. The fact that this cell type interdigitates with various organs as well as with other parenchymal cells provides a morphological basis for this idea. Furthermore the fibrous interstitial material which normally exists between the cells is absent from these areas of intimate contact. Electron microscope studies seem to indicate that descriptions which interprets parenchyma as consisting of a syncytial meshwork enclosing fluid filled spaces must be replaced by one which regards this tissue as consisting of large, individual cells associated with interstitial material (THREADGOLD and GALLAGHER, 1966). This idea also corresponds in general terms to the descriptions produced for planarian parenchyma (PEDERSEN, 1961). Thus it appears as if the idea of diffusion of metabolites through fluid-filled spaces (in the absence of a circulatory system) must be replaced by one involving cellular agencies. Distribution could involve fixed cells which might exhibit differential transport characteristics or might be associated with migratory cells as has often been described in the Mollusca. Particularly significant in relation to this later concept is an observation by CHENG and STREISFELD (1963) which described the presence of cells in *Megalodiscus temperatus* and *Haematoloechus sp.* capable of phagocytising indian ink particles. In the case of *Haematoloechus* they found that phagocytes containing ingested ink particles were present in the tissues adjacent to the caecum, within the caecal epithelium and in the lumen of the caeca. The cells free in the lumen had, in most cases, ruptured and released their contents. The identification of these cells under the electron microscope and a description of their characteristics would be very interesting as it seems possible that cells of this type would also be involved in transport systems associated with nutrient distribution or excretion.

The gut exhibits a diversity of surfaces (inner surface or oral sucker, pharynx, oesophagus and caeca) which do not differ greatly under the optical microscope. The electron microscope does however, reveal considerable differences between these surfaces. The oral sucker is lined with tegument which resembles that of the general body surface but the surface of the pharynx in *C. bushiensis* possesses a tegument which has an irregularly folded plasma membrane and suggests that pinocytosis may take place here. The basement layer of the pharynx is also very thick and this seems to be related to the presence of the well-developed radial muscles characteristic of this structure and their insertion into the basement layer. The changeover from tegumentary surface to the microvillous caecal epithelium takes place very abruptly indeed.

The structure of the caecal surface of larval trematodes has not yet been studied extensively. The non-encysting metacercariae of strigeoid trematodes are active tissue feeders and in *Diplostomulum phoxini* the caecal epithelium is well developed and, although not developed to the same extent as that of the adult, does possess short, widely spaced microvilli on its surface. In the gut of the redia of *Cryptocotyle lingua*, KRUPA *et al* (1967) showed that the gastrodermis possessed on its luminal surface flat, ribbon-like folds. These authors suggested that semi-digested material in the gut lumen becomes entrapped by these folds and the resultant droplets taken into the cytoplasm. The gastrodermis contained nuclei, mitochondria and some glycogen, but golgi complexes, secretory bodies and endoplasmic reticulum were not described.

The trematode gut, in most instances, does not possess an anus and consequently egestion of materials must occur through the mouth. The wall of the caecum is muscular and is able to contract but movement of food is also aided by the expansion and contraction of the body as a whole. The process of egestion has not been studied and the possibility of phased periods of ingestion and egestion remains to be investigated. The various reports of lysis of gelatinous substrates in the region of the oral sucker confirm that lytic material may be extruded in this way.

EXTRACORPOREAL DIGESTION

It seems possible that extracorporeal digestion may take place to some extent in many trematodes· As described above the egestion of material from the parasite's caecum will result in the contamination of the host's tissues with debris containing enzymes. This was demonstrated in *Cyathocotyle bushiensis* (ERASMUS and ÖHMAN, 1963) after allowing adult trematodes to browse on denatured egg white. When the blocks were flooded with the appropriate azodye substrate mixture for the demonstration of acid phosphatase activity, small coloured patches appeared, which corresponded to the extrusion of enzymatically active materials from the gut via the oral sucker. In the same way THORSELL and BORKMAN (1965) obtained lysis of gelatine immediately in front of the oral sucker of *F. hepatica* and KEARN (1963), observing the process of feeding by *Entobdella solae*, described how secretion from the glands of the feeding organ is discharged onto the host skin and confined to a small area after being enclosed and sealed from the surrounding environment by the lip of the feeding organ. In this way digestion in the trematodes may not only be intracellular, and intraluminal but also extracorporeal. However the egestion of lytic material from the gut may not be particularly significant in the feeding of many digeneans as the movement of the parasite away from the site of attachment, or the dilution and removal of the deposit by host fluids, will detract considerably from the efficiency of the process.

The role of extracorporeal digestion becomes much more significant in the strigeoid trematodes. One of the most distinctive features of the strigeoids is the adhesive organ present on the ventral surface (Text-Fig. 90). This specialised structure has associated with it gland cells and the organ itself may be acetabular (e.g. *Holostephanus*, *Cyathocotyle*, *Diplostomum*) or lobular (*Apatemon*, *Cotylurus*) in form. In the acetabular type the organ is capable of partial eversion so that the host mucosa becomes drawn into the lumen whereas in the other type, the cup-shaped forebody encloses a mass of host villi which then comes into intimate contact with the adhesive organ lobes. In most cases the mucosal epithelium becomes destroyed at the point of contact with the adhesive organ or its lobes, and this feature has prompted several workers (BRANDES, 1890; MÜHLING, 1896; LA RUE, 1927, 1932; SZIDAT, 1929; VAN HAITSMA, 1931a,b; BAER, 1933, LEE, 1962 and ERASMUS, 1962a) to suggest that secretion of a histolytic substance to the exterior of the parasite occurs and that the mucosa becomes disrupted and the partially digested material ingested (BAER, 1933; ERASMUS, 1962a). Because of the intimacy of the association between this structure and host tissues many people have suggested an additional absorptive function for the organ.

The role of the strigeoid adhesive organ in feeding has been investigated in selected genera by ERASMUS (1962a), ERASMUS and ÖHMAN (1963) and ÖHMAN (1965, 66a,b). Using a range of histochemical techniques these authors have shown that the gland cells associated with the adhesive organ are rich in protein, RNA, and exhibit enzyme activity of various kinds.

The gland cells of *Cyathocotyle bushiensis* showed acid, alkaline phosphatase, esterase and leucineaminopeptidase activity; *Holostephanus lühei* the same; *Diplostomum spathaceum* acid phosphatase and esterase activity and *Apatemon gracilis minor* acid phosphatase activity only. In *D. spathaceum* and *A. gracilis minor* attachment lappets are present on the forebody and these, with their associated cells showed alkaline phosphatase, leucineaminopeptidase and esterase activity in the case of *Apatemon sp.* and alkaline and acid phosphatase and esterase activity in *Diplostomum sp.* A comparison of the histochemistry of the host mucosa in its normal condition and at points of attachment, and also observations based on the *in vitro* incubation of the parasites on non-living substrates suggested very strongly, that secretion to the exterior of the parasite of enzymatically active material occurs at the sites of attachment. The nature of the adhesive organ in *Cyathocotyle* and *Holostephanus* permitted attachment to a flat substrate and incubation of these parasites on gelatine film resulted in a dissolution of the gelatine in regions which corresponded to the attachment sites of the adhesive organ. A gelatinolytic activity could also be attributed to the secretion from these two species. Histological and histochemical evidence suggests that mucus from the host goblet cells and also the disintegrated host cells are ingested by the attached parasite. In this way extracorporeal digestion plays a significant part in the feeding of these parasites. A further feature is that the removal of the mucosal epithelium enables the surface of the adhesive organ to come into intimate contact with the lamina propria and its capillaries thus facilitating absorption from the host blood system if this does indeed occur. Electron microscope studies of *C. bushiensis* (ERASMUS and ÖHMAN, 1965) have indicated that the gland cells possess the ultrastructural characteristics of secretory structures and that the secretion is complex in morphology. The tegument covering the adhesive organ surface is, as mentioned earlier, markedly different from the rest of the tegument and is elevated into microvilli. This morphological feature increases the surface area in contact with the host mucosa and would be a very significant morphological adaptation if absorption of nutrients from the host does occur. Histochemical tests at the ultrastructural level (ERASMUS, 1966, 1968) demonstrate alkaline phosphatase activity within the membrane of the wall of the microvillus and this again could be correlated with absorption. Secretion from the parasite may follow two routes. It may emerge through the wall of the microvillus or, and this seems the most likely route, collect in the distended tips of the microvilli which then bud off and break down, releasing the secretion. This is a type of secretory process which has been described for certain mammalian cells (KUROSUMI, 1961). The granular component of the secretion exhibits strong acid phosphatase activity and this can be detected in the lumen of the microvillus and in its distended tip (Plate 9-3).

Electron microscope studies of the lappets of *Apatemon sp.* and *Diplostomum sp.* (ERASMUS, 1969b, 1970) indicate that these regions come into particularly intimate contact with the host tissue and that secretion of probably lytic material into or onto the host tissues occurs at this point. It seems quite likely that extracorporeal digestion of host tissues may occur at these regions of attachment by the parasite.

Many trematodes possess gland cells associated with suckers (HALTON, 1967b; HALTON and DERMOTT, 1967), with the pharynx, or with the tegument and it is possible that secretion of lytic substances is more general than realised. Furthermore, the role these biochemically complex secretions play in stimulating the immunological responses of the host must not be underestimated.

These secretions are rich in protein and phospholipid and come into intimate contact with the host tissues and blood system.

IV. The secretory process

It is now apparent that secretory processes have been demonstrated at all levels of the life-cycle and that the processes may be associated with a variety of functions with varying emphasis at different stages. The ultrastructural and histochemical studies described earlier in this and other chapters permit a preliminary assessment of the secretory process as seen in digenetic trematodes.

The observations of many workers have established the significance of gland cells and secretory processes in larval stages such as miracidia and cercariae (STIREWALT, 1963b). In the case of the adult trematode many descriptions are available concerning the presence of gland cells associated with the tegument and various suckers but the precise nature of these cells and the assessment of their role in the biology of the parasite has only begun.

The application of ultrastructural techniques has revealed the considerable association which exists between the host parasite interface and the process of secretion. The epithelial lining of the caecum has been shown to be a region of enzyme synthesis and secretion as well as functioning in absorption. The tegument, with its various specialisation over regions of attachment organs, has associated with it cells which exhibit a variety of secretory processes. Ultrastructural studies have influenced considerably our ideas concerning the process of secretion and KUROSUMI (1961) has produced a revised terminology incorporating the morphological features which are apparent only under the electron microscope. Within the trematodes secretory processes as far as they have been determined appear to fit into the following patterns:

I. HOLOCRINE SECRETION. (Extrusion type I, KUROSUMI, 1961)

The classical interpretation involves the complete conversion or replacement of cell cytoplasm with secretory products so that the gland cells exhibiting this type of secretion have a restricted survival time terminating with the accumulation of secretion. The evidence available indicates that this secretory process is exhibited by the gland cells of cercariae especially those referred to as 'penetration gland cells'. STIREWALT (1965) has noticed that when cercariae of *Schistosoma mansoni*, *S. japonicum*, *S. haematobium* and *Schistosomatium douthitti* emerge from their snail hosts the gland cells are filled with secretion so that the nuclei, in many cases, are pressed against the cell membrane and have begun to degenerate in some individuals. Once the contents of these glands are exhausted no further secretion may be produced. Similar characteristics are exhibited by the 'penetration gland cell' of miracidia and the 'penetration gland cells' of strigeoid cercariae (see pages 144, 153). The significance of holocrine processes is considerable in the case of migrating and penetrating forms as their period of activity may become rapidly curtailed once the formed secretion is utilised, and no replacement of the particular cell type takes place. Holocrine processes also occur in adult trematodes and it seems likely that the gland cells associated with the lappets and adhesive organ of strigeoid trematodes (ERASMUS and ÖHMAN, 1965; ERASMUS, 1969a,b,c; 1970) come into this category as well as the subtegumentary gland cells (not tegumentary cells). described from *Diplostomum* species (ÖHMAN, 1965). These cells exhibit esterase activity as well as containing acid mucopolysaccharide and PAS diastase resistant material.

They are particularly abundant in the fore-body and referred to as 'Hautdrusen' by KRAUSE (1914), have their cytoplasm gradually replaced by secretory bodies. The 'neck' of these cells penetrates the tegument so that the secretion is discharged directly to the exterior. The persistence of these cells through the life-span of this trematode (several weeks) has not been studied and it seems likely that cellular replacement and a cell turnover must occur to maintain the cell population. This phenomenon within the trematoda would repay investigation and must contribute to the factors determining the life-span of adult flukes.

II. MEROCRINE SECRETION. (Extrusion Type II, III, IV. KUROSUMI, 1961)

In this case the cell survives the cycle of secretory activity with a loss of only a small portion of the cell cytoplasm (apocrine) or the loss of secretory products only (eccrine = merocrine of some authors). The fact that cell destruction does not occur results in the association of this type of secretory activity with those processes which must continue for a considerable time and probably for the life of the parasite. The major categories (apocrine and eccrine) have been sub-divided by Kurosumi in the light of information derived from ultrastructural studies.

The presence of a cytoplasmic tegument complicates the process of secretion in that the products may not be discharged directly to the exterior but may be released from the gland cells by one process, followed by an accumulation of products for a greater or lesser time in the tegument, and their subsequent release from the tegument by a quite different secretory process. Very little detailed information is available but what has been described does indicate the following major patterns of the secretory process.

The formation of secretory 'blebs' by the limiting plasma membrane does not seem to have been described from the trematodes so far although this type of apocrine secretion (Extrusion Type II of Kurosumi) has been indicated in the cestode *Moniezia expansa* (HOWELLS, 1967). Microapocrine secretion (Extrusion type III) has been suggested for the release of secretion from the microvilli associated with the adhesive organ of *Cyathocotyle bushiensis* (ERASMUS and ÖHMAN, 1965). This mode of secretion is represented by the pinching off of the tips of microvilli containing secretory products. This type of extrusion has been described from a number of mammalian tissues (KUROSUMI, 1961).

The other secretory and extrusion mechanism likely to be reported from the trematodes is a form of eccrine secretion (Extrusion type IV). In this case membrane bounded bodies approach the outer plasma membrane, they attach to the limiting membrane and dissolution occurs at the point of contact so that the contents of the body are released to the exterior of the parasite. This sequence occurs during the release of material present in the secretory bodies, produced in the tegumentary cells of *C. bushiensis* (ERASMUS, 1967c), through the limiting plasma membrane (see p. 228). A similar process must also occur in the formation of the metacercarial cyst wall by cercarial bodies. The products of 'cystogenous' cells pass from the secretory cells via the cytoplasmic extensions penetrating the basement layer into the tegument. Release of this material to the exterior occurs by the passage of the substances through the limiting plasma membrane (DIXON and MERCER, 1964; BILS and MARTIN, 1966). Data from a variety of sources (see earlier in this chapter) suggest that extensive secretion must take place from the caecal epithelium. The histochemical observations indicate that acid phosphatase and esterase release must take place from this epithelium and THORSELL and BJORKMAN (1965) suggests that the dense, membrane bounded bodies abundant in the caecal epithelium of *F. hepatica* and other trematodes 'empty their content' into the caecal lumen. The secretory process in *F. hepatica* has been described as apocrine (DAWES, 1962b) and this described the process as seen under the optical microscope. If the suggestions of THORSELL and BJORKMAN (loc. cit.) are correct however, it is possible that eccrine extrusion processes are also present, in that the membrane enclosing the secretory body fuses at a restricted point with that limiting the ceacal epithelium allowing the contents to stream out into the lumen. If excretion takes place through the tegument, and there seems no reason to suppose that it does not, it seems likely that this latter process would be involved in the movement of materials through and to the exterior of the tegument. The process has the advantage that the external plasma membrane is perforated presumably to a very limited extent for a very short time.

There is no evidence available so far which demonstrates that the fifth secretory process (Extrusion type V) occurs within the trematodes. In this process Kurosumi suggests that material may pass through the limiting plasma membrane without perforation of the membrane and without loss of cytoplasm. In other words the passage of material must take place at the molecular level. The suggestion that molecules pass into the parasite through the limiting plasma membrane is not new and there is some evidence which indirectly supports this idea (see earlier in this chapter) particularly the uptake of ferritin including transmembranosis in *F. hepatica* (THORSELL and BJORKMAN, 1965). If one accepts an inward passage of molecules there is no reason for not accepting the reverse movement of molecules. Thus it seems likely that the processes of secretion and also excretion may involve extrusion mechanisms resembling types IV and V as outlined by Kurosumi i.e. modified eccrine secretory processes.

It will be apparent from this brief section that the change in our concept of the 'cuticle' of trematodes must influence considerably our ideas concerning secretion and the extrusion mechanisms associated with excretion. The adoption of many of the ideas concerning the behaviour and characteristics of the plasma membrane from current cell biology will undoubtedly allow a more realistic interpretation of the nature of the limiting external surface in trematodes.

References

AGERSBORG, H. P. K. 1924. Studies on the effects of parasitism upon the tissues. I. With special reference to certain gastropod molluscs. *Q. Jl. microsc. Sci.* **68,** 361–401.

AGOSIN, M. and ARAVENA, L. C. 1960. Studies on the metabolism of *Echinococcus granulosus.* IV. Enzymes of the pentose phosphate pathway. *Expl. Parasit.* **10,** 28–38.

ALVARADO, R. 1951. El Tegumento, La musculature y el parénquima de *Fasciola hepatica. Trab. Inst. cienc. nat. Acosta* **3,** 1–90.

ALVEY, C. M. 1936. The morphology and development of the monogenetic trematode *Sphyranura oligorchis* (Alvey, 1933) and the description of *Sphyranura polyorchs* n.sp. *Parasitology,* **28,** 229–53.

AMEEL, D. J. 1934. Paragonimus, its life history and distribution in North America and its taxonomy (Trematode: Troglotrematidae). *Am. J. Hyg.* **19,** 279–317.

AMEEL, D. J., CORT, W. W. and VAN DER WOUDE, A. 1949. Germinal development in the mother sporocyst and redia of *Halipegus eccentricus* Thomas, 1939. *J. Parasit.* **35,** 569–78.

AMEEL, D. J., CORT, W. W. and VAN DER WOUDE, A. 1951. Development of the mother sporocyst and rediae of *Paragonimus kellicoti* Ward, 1908. *J. Parasit.* **37,** 395–404.

ANDRADE, Z. A. and BARKA, T. 1962. Histochemical observations on experimental schistosomiasis of mouse. *Am. J. trop. Med. Hyg.* **11,** 12–16.

ARVY, L. 1954. Distomatose cérébro-rachidienne due à *Diplostomulum phoxini* (Faust), Hughes 1920, chez *Phoxinus laevis* AG. *Annls. Parasit. hum. comp.* **29,** 510–20.

AXMANN, M. C. 1947. Morphological studies on glycogen deposition in schistosomes and other flukes. *J. Morph.* **80,** 321–34.

AZIM, M. A. 1933. *Prohemistomum vivax* and its development from *Cercaria vivax. Z. ParasitKde* **5,** 432–6.

BAER, J. G. 1933. Contribution à l'etude de la faune helminthologique africaine. *Revue suisse Zool.* **40**:31–84.

BAER, J. G. and EUZET, L. 1961. Classe des Monogènes. In Grassé, P. P.; *Traité de Zoologie,* **4**:243–325. Paris: Masson

BAER, J. G. and JOYEUX, C. 1961. Classe des Trématodes. In Grassé, P. P.: *Traité de Zoologie,* **4,** 561–692. Paris: Masson.

BARBOSA, F. S. 1962. Aspects of the ecology of the intermediate hosts of *Schistosoma mansoni* interfering with the transmission of bilharziasis in north-eastern Brazil. pp. 23–35. *Ciba Foundation Symposium Bilharziasis.* Edit. Wolstenholme and O'Connor. J. & A. Churchill Ltd. London. 433 p.

BARBOSA, F. S. and BARRETO, A. C. 1960. Differences in susceptibility of brazilian strains of *Australorbis glabratus* to *Schistosoma mansoni. Expl. Parasit.* **9,** 137–40.

BARBOSA, F. S. and COELHO, M. V. 1956. Resquisa de imunidade adquiri da homologa em *Australorbis glabratus* na infestações par *Schistosoma mansoni. Revta bras. Malar. Doenç. trop.* **8,** 49–56.

BASCH, P. F. 1966. Patterns of transmission of the trematode *Eurytrema pancreaticum* in Malaysia. *Am. J. vet. Res.* **27,** 234–40.

BASCH, P. F. and LIE, J. K. 1966a. Infection of single snails with two different trematodes. I. Simultaneous exposure and early development of a schistosome and an echinostome. *Z. ParasitKde.* **27,** 252–9.

BASCH, P. F. and LIE, J. K. 1966b. Infection of single snails with two different trematodes. II. Dual exposures to a schistosome and an echinostome at staggered intervals. *Z. ParasitKde.* **27,** 260–70.

BAUER, O. N. 1959. *Bulletin of the State Scientific Research Institute of lake and river fisheries.* Vol. 49. Parasites of freshwater fish and the biological basis for their control. Translated by Kochva, L. Israel Program for Scientific Translations, Jerusalem 1962. 236 p.

BAUMAN, P. M., BENNETT, H. J. and INGALLS, J. W. 1948. The molluscan intermediate host and schistosomiasis japonica. Observations on the production and rate of emergence of cercariae of *Schistosoma japonica* from the molluscan intermediate host *Oncomelanica quadrasi. Am. J. trop. Med.* **28,** 567–75.

BAYLIS, H. A. 1939. A larval trematode (*Diplostomulum volvens*) in the lens of the eye of a rainbow trout. *Proc. Linn. Soc. Lond.* **151,** 130.

BAYLIS, H. A. and TAYLOR, E. L. 1930. Observations and experiments on a dermatitis producing cercaria and other cercaria from *Lymnaea stagnalis* in Great Britain. *Trans. R. Soc. trop. Med. Hyg.* **24,** 219–44.

BEČEJAC, Š. and KRVAVICA, S. 1964. On the presence and localisation of alkaline phosphatase activity in the liver fluke (*Fasciola hepatica* L.) *Vet. Arh.* **34,** 56–9.

BEČEJAC, Š. and LUI, A. 1959. The action of temperature and oxygen on the development of the eggs of liver fluke (*Fasciola hepatica* L.) *Vet. Arh. Zagreb.* **29,** 293–301.

BEČEJAC, S., LUI, A., KRVAVICA, S., and KRALT, N. 1964. Acetylcholinesterase and butyrocholinesterase activity in the lancet fluke (*Dicrocoelium lanceatum* Rudolphi). *Vet. Arhiv.* **34,** 87–9.

BECKER, W. 1964. Der Einfluss von Trematoden auf den Gastoffwechsel von *Stagnicola palustris* Müll. *Z. ParasiKde.* **25,** 77–102.

BELL, E. J. and HOPKINS, C. A. 1956. The development of *Diplostomum phoxini* (Strigeida, Trematoda). *Ann. Trop. Med. Parasit.* **50,** 275–82.

BELL, E. J. and SMYTH, J. D. 1958. Cytological and histochemical criteria for evaluating development of trematodes and pseudophyllidean cestodes *in vivo* and *in vitro*. *Parasitology,* **48,** 131–48.

BELTON, C. M. and HARRIS, P. J. 1967. Fine structure of the cuticle of the cercaria of *Acanthatrium oregonense* (Macy). *J. Parasit.* **53,** 715–24.

BENNETT, H. J. 1936. The life history of *Cotylophoron cotylophoron* a trematode from ruminants. *Illinois biol. Monogr.* **14,** No. 4. 119 pp.

BENNETT, H. J. and HUMES, A. G. 1939. Studies on the pre-cercarial development of *Stichorchis subtriquetrus* (Trematoda: Paramphistomidae). *J. Parasit.* **25,** 223–31.

BENNETT, H. J. and JENKINS, L. L. 1950. The longevity of the miracidium of *Cotylophoron cotylophoron*. *Proc. La Acad. Sci.,* **13,** 5–13.

BENNINGTON, E. E. and PRATT, I. 1960. The life history of the salmon-poisoning fluke, *Nanophyetus salmincola* (Chapin). *J. Parasit.* **46,** 91–100.

BERRIE, A. D. 1960. The influence of various definitive hosts on the development of *Diplostomum phoxini* (Strigeida, Trematoda). *J. Helminth.* **34,** 205–10.

BERTHIER, L. 1954. Contribution a l'étude du tegument de *Fasciola hepatica* L. *Arch. Zool. exp. gén.,* **91,** 89–102.

BETTENDORF, H. 1897. Über Muskulatur und Sinneszellen der Trematoden. *Zool. J. ber. Neapel* **10,** 307–58.

BEVELANDER, G. 1933. Response to light in the cercariae of *Bucephalus elegans*. *Physiol Zool.* **6,** 289–305.

BILS, R. F. and MARTIN, W. E. 1966. Fine structure and development of the trematode integument. *Trans. Am. microsc. Soc.* **85,** 78–88.

BISSET, K. A. 1947. Bacterial infection and immunity in lower vertebrates and invertebrates. *J. Hyg. Camb.* **45,** 128–35.

BJÖRKMAN, N. and THORSELL, W. 1964a. On the fine structure and resorptive function of the cuticle of the liver fluke, *Fasciola hepatica* L. *Expl. Cell Res.* **33,** 319–29.

BJÖRKMAN, N. and THORSELL, W. 1964b. On the ultrastructure of the ovary of the liver fluke (*Fasciola hepatica* L.) *Z. b. Zellforsch.,* **63,** 538–49.

BJÖRKMAN, N., THORSELL, W., and LIENERT, E. 1963. Studies on the action of some enzymes on the cuticle of *Fasciola hepatica* L. *Experientia.,* **19,** 3–5.

BLOCHMANN, F. 1910. Sterben von Aquarienfischen durch Einwonderung von *Cercaria fissicauda* La Val. *Zentbl. Bakt. ParasitKde, Abt I* **56,** 47–9.

BOGITSH, B. J. 1962. The chemical nature of metacercarial cysts. I. Histological and histochemical observations on the cyst of *Posthodiplostomum minimum*. *J. Parasit.* **48,** 55–60.

BOGITSH, B. J. 1963. Histochemical observations on the cercariae of *Posthodiplostomum minimum*. *Expl. Parasit.* **14,** 193–202.

BOGITSH, B. J. 1966a. Histochemical observations on *Posthodiplostomum minimum*. II. Esterases in subcuticular cells, holdfast organ cells and the nervous system. *Expl. Parasit.* **19,** 64–70.

BOGITSH, B. J. 1966b. Histochemical observations on *Posthodiplostomum minimum*. III. Alkaline phosphatase activity in metacercariae and adults. *Expl. Parasit.* **19,** 339–47.

BOGITSH, B. J. and ALDRIDGE, F. P. 1967. Histochemical observations on *Posthlodiplostomum minimum* IV. Electron microscopy of tegument and associated structures. *Expl. Parasit.* **21,** 1–8.

BOGOMOLOVA, N. A. 1957. A cytological study of the miracidia of *Fasciola hepatica*. *Dokl. Akad. Nauk SSSR.* **117** (2), 313–15.

BORAY, J. C. 1969. Experimental fascioliasis in Australia. p. 95–210. In *Advances in Parasitology* V 7. Ed. B. Dawes. Academic Press. London and New York. 1967.

BOSMA, N. J. 1934. The life history of the trematode *Alaria mustelae,* sp.nov. *Trans. Am. microsc. Soc.* **53,** 116–53.

BOVIEN, P. 1931. Notes on the cercaria of the liver fluke (*Cercaria Fasciolae hepaticae*). *Vidensk. Meddr. dansk naturh. Foren.* **92,** 223–6.

BRANDES, G. 1890. Die Familie der Holostomiden. *Zool. Jb.*, **5,** 553–604.

BRANDT, B. B. 1936. Parasites of certain North Carolina Salientia. *Ecol. Monogr.*, **6,** 491–532.

BROOKS, C. P. 1953. A comparative study of *Schistosoma mansoni* in *Tropicornis havanensis* and *Australorbis glabratus. J. Parasit.* **39,** 159–65.

BROOKS, F. G. 1930. Studies on the germ cell cycles of trematodes. *Am. J. Hyg.*, **12,** 299–340.

BROWN, F. J. 1933. On the excretory system and life-history of *Lecithodendrium chilostomum* (Mehl.) and other bat trematodes, with a note on the life history of *Dicrocoelium dendriticum* (Rudolphi). *Parasitology*, **25,** 317–28.

BRYANT, C. and WILLIAMS, J. P. G. 1962. Some aspects of the metabolism of the liver fluke, *Fasciola hepatica* L. *Expl. Parasit.* **12,** 372–6.

BUEDING, E. 1949. Metabolism of parasitic helminths. *Physiol. Rev.* **29,** 195–218.

BUEDING, E. 1950. Carbohydrate metabolism of *Schistosoma mansoni. J. gen. Physiol.*, **33,** 457–95.

BUEDING, E. 1962. Comparative biochemistry of parasitic helminths. *Comp. Biochem. Physiol.*, **4,** 343–51.

BUEDING, E. and CHARMS, B. 1951. Respiratory metabolism of parasitic helminths without participation of the cytochrome system. *Nature Lond.*, **167,** 149.

BUEDING, E. and CHARMS, B. 1952. Cytochrome C, cytochrome oxidase and succinoxidase activities of helminths. *J. biol. chem.*, **196,** 615–27.

BUGGE, G. 1902. Zur Kenntniss des Excretionsgefäss—Systems der Cestoden und Trematoden. *Zool. Jb.* (*Anat.*) **16,** 177–234.

BULLOCK, W. L. 1949. Histochemical studies on the Acanthocephala. I. The distribution of lipase and phosphatase. *J. Morph.* **84,** 185–99.

BURNS, W. C. 1961. Penetration and development of *Allassogonoporus vespertilionis* and *Acanthatrium oregonense* (Trematoda: Lecithodendriidae) cercariae in caddis fly larvae. *J. Parasit.* **47,** 927–32.

BURTON, P. R. 1960. Gametogenesis and fertilisation in the frog lung fluke, *Haematoloechus medioplexus* Stafford (Trematoda: Plagiorchiidae). *J. Morph.* **107,** 93–121.

BURTON, P. R. 1962. *In vitro* uptake of radioglucose by a frog lung-fluke and correlation with the histochemical identification of glycogen. *J. Parasit.* **48,** 874–82.

BURTON, P. R. 1963. A histochemical study of vitelline cells, egg capsules and Mehlis' gland in the frog lung fluke, *Haematoloechus medioplexus. J. exp. Zool.* **154,** 247–58.

BURTON, P. R. 1964. The ultrastructure of the integument of the frog lung-fluke, *Haematoloechus medioplexus* (Trematoda: Plagiorchiidae). *J. Morph.* **115,** 305–18.

BURTON, P. R. 1966. The ultrastructure of the integument of the frog bladder fluke, *Gorgoderina sp. J. Parasit.* **52,** 926–34.

BURTON, P. R. 1967. Fine structure of the reproductive system of a frog lung fluke. I. Mehlis' gland and associated ducts. *J. Parasit.* **53,** 540–55.

BYCHOWSKY, B. E. 1957. *Monogenetic Trematodes, their systematics and phylogeny.* English Translation 1961. Ed. W. J. Hargis. *Amer. Instit. Biol. Sci.* 627 p.

BYKHOVSKAJA-PAVLOVSKAJA, I. E. 1964a. *An influence of alimentary factors on the fauna of bird trematodes. Acad. Sci. U.S.S.R.* 1964, 9 pp. Report presented at the First International Congress of Parasitology.

BYKHOVSKAJA-PAVLOVSKAJA, I. E. 1964b. Parasitic Worms and Aquatic Conditions. To the methods and problems of parasitological investigations of animals bound to aquatic environment. Prague & Czechoslovak Academy of Sciences, p. 29–36.

BYRD, E. E. and MAPLES, W. P., 1963. Excretory system of *Chaledocystus pennsylvaniensis* (Cheng, 1961) Byrd & Maples, 1963 (Trematoda) *J. Parasit.* **49,** (No. 5, Sec. 2), 42.

BYRD, E. E. and SCOFIELD, G. F. 1952. Developmental stages in the Digenea. I. Observations on the hatchability and infectivity of ochetosomatid eggs in physid snails. *J. Parasit.* **38,** 532–9.

BYRD, E. E. and SCOFIELD, G. F. 1954. Developmental Stages in the Digenea. III. Observations on the number of daughter sporocysts and cercariae produced in *Physa gyrina* as a result of single and multiple ochetosomatid egg exposures. *J. Parasit.* **40,** 1–21.

CABLE, R. M. 1934. Studies on the germ-cell cycle of *Cryptocotyle lingua.* II. Germinal development in the larval stages. *Q. Jl. microsc. Sci.* **76,** 573–614.

CABLE, R. M. 1956. Marine cercariae of Puerto Rico. *Scient. Surv. P. Rico* **16,** 491–576.

CABLE, R. M. 1965. 'Thereby hangs a Tail'. *J. Parasit.* **51,** 3–12.

CABLE, R. M. and HUNNINEN, A. V. 1942. Studies on *Deropristis inflata* (Molin); its life history and affinities to trematodes of the family Acanthocolpidae. *Biol. Bull. mar. biol. Lab., Woods Hole*, **82,** 292–312.

CABLE, R. M. and NAHLAS, F. M. 1962. *Bivesicula caribbensis* sp.n. (Trematoda: Digenea) and its life history. *J. Parasit.* **48,** 536–8.

CAMPBELL, W. C. and TODD, A. C. 1956. Emission of cercariae and metacercariae in snail faeces. *Trans. Am. microsc. Soc.* **75,** 241–3.

CAPRON, A., BIGUET, J., ROSE, F. and VERNES, A. 1965. Les antigènes des *Schistosoma mansoni*. II. Êtude immuno-électrophorétique comparée de divers stades larvaires et des adultes des deux sexes. Aspects immunologiques des relations hote-parasite de la cercaire et de l'adulte de *S. mansoni*. *Annls. Inst. Pasteur. Paris* **109,** 798.

CARDELL, R. R. 1962. Observations on the ultrastructure of the body of the cercaria of *Himasthla quissetensis* (Miller and Northup, 1926). *Trans. Am. microsc. Soc.* **81,** 124–31.

CARDELL, R. R. and PHILPOTT, D. E. 1960. The ultrastructure of the tail of the cercaria of *Himasthla quissetensis* (Miller and Northup 1926). *Trans. Am. microsc. Soc.* **79,** 442.

CASTELLÁ, E. B. 1952. La anhidrasa carbónica en algunos parásitos intestinales. *An. Inst. Invest. vet.* **4,** 344–50.

CHANDLER, A. C. 1942. The morphology and life cycle of a new strigeid, *Fibricola texensis*, parasitic in raccoons. *Trans. Am. microsc. Soc.* **61,** 156–67.

CHEN, PIN DJI. 1937. The germ cell cycle in the trematode *Paragonimus kellicotti* Ward. *Trans. Am. microsc. Soc.* **56,** 208–36.

CHENG, T. C. 1961. Description, life history and developmental pattern of *Glypthelmins pennsylvaniensis* n.sp. (Trematoda: Brachycoeliidae), new parasite of frogs. *J. Parasit.* **47,** 469–77.

CHENG, T. C. 1963a. The effects of Echinoparyphium larvae on the structure of and glycogen deposition in the hepatopancreas of *Helisoma trivolvis* and glycogenesis in the parasite larvae. *Malacologica* **1,** 291–303.

CHENG, T. C. 1963b. Biochemical requirements of larval trematodes. *Ann. N.Y. Acad. Sci.* **113,** 289–321.

CHENG, T. C. 1963c. Activation of *Gorgodera amplicava* cercariae by molluscan sera. *Expl. Parasit.* **13,** 342–7.

CHENG, T. C. 1964. Studies on phosphatase systems in hepatopancreatic cells of the molluscan host of *Echinoparyphium* sp. and in the redia and cercariae of this trematode. *Parasitology* **54,** 73–9.

CHENG, T. C. 1964. Biology of Animal Parasites. W. B. Saunders Co., Philadelphia, London.

CHENG, T. C. 1965. Histochemical observations on changes in the lipid composition of the American oyster, *Crassostrea virginica* (Gmelin) parasitized by the trematode *Bucephalus* sp. *J. Invert. Pathol.* **7,** 398–407.

CHENG, T. C. and BURTON, R. W. 1965. Relationships between *Bucephalus sp.* and *Crassostrea virginica*: Histopathology and sites of infection. *Chesapeake Sci.*, **6,** 3–16.

CHENG, T. C. and COOPERMAN, J. S. 1964. Studies on host-parasite relationships between larval trematodes and their hosts. V. The invasion of the reproductive system of *Helisoma trivolvis* by the sporocysts and cercariae of *Glypthelmins pennsylvaniensis*. *Trans. Am. microsc. Soc.* **83,** 12–23.

CHENG, T. C. and JAMES, H. A. 1960. The histopathology of *Crepidostomum sp.* infection in the second intermediate host, *Sphaerium striatinum*. *Proc. helminth Soc. Wash.* **27,** 67–8.

CHENG, T. C. and SAUNDERS, B. G. 1962. Internal defense mechanisms in mollusca and an electrophoretic analysis of a naturally occurring serum haemagglutin in *Viviparus molleatus* Reeve. *Proc. Pa. Acad. Sci.* **36,** 72–83.

CHENG, T. C. and SNYDER, R. W. 1962a. Studies on host-parasite relationships between larval trematodes and their hosts. I. A review. II. The utilization of the host's glycogen by the intramolluscan larvae of *Glypthelmins pennsylvaniensis* Cheng, and associated phenomena. *Trans. Am. microsc. Soc.* **81,** 209–28.

CHENG, T. C. and SNYDER, R. W. 1962b. Studies on host-parasite relationships between larval trematodes and their hosts. III. Certain aspects of lipid metabolism in *Helisoma trivolvis* (Say) infected with the larvae of *Glypthelmins pennsylvaniensis* Cheng and related phenomena. *Trans. Am. microsc. Soc.* **81,** 327–31.

CHENG, T. C. and SNYDER, R. W. 1963. Studies on host-parasite relationships between larval trematodes and their hosts. IV. A histochemical determination of glucose and its role in the metabolism of molluscan host and parasite. *Trans. Am. microsc. Soc.* **82,** 343–6.

CHENG, T. C. and STREISFELD, S. D. 1963. Innate phagocytosis in the trematodes *Megalodiscus temperatus* and *Haematoloechus* sp. *J. Morph.* **113,** 375–80.

CHENG, T. C. and YEE, H. W. F. 1968. Histochemical demonstration of amino-peptidase activity associated with the intramolluscan stages of *Philophthalmus gralli* Mathias and Leger. *Parasitology* **58,** 473–80.

CHENG, T. C., SHUSTER, C. N. JNR. and ANDERSON, A. H. 1966a. A comparative study of the susceptibility and response of eight species of marine pelecypods to the trematode *Himasthla quissetensis*. *Trans. Am. microsc. Soc.* **85,** 284–95.

CHENG, T. C., SHUSTER, C. N. JNR. and ANDERSON, A. H. 1966b. Effects of plasma and tissue extracts of marine pelecypods on the cercaria of *Himasthla quissetensis*. *Expl Parasit.* **19,** 9–14.

CHERNIN, E. 1964. Maintenance *in vitro* of larval *Schistosoma mansoni* in tissues from the snail, *Australorbis glabratus. J. Parasit.* **50,** 531–45.

CHERNIN, E. 1966. Transplantation of larval *Schistosoma mansoni* from infected to uninfected snails. *J. Parasit.* **52,** 473–82.

CHERNIN, E. and DUNAVAN, C. A. 1962. The influence of host-parasite dispersion upon the capacity of *Schistosoma mansoni* miracidia to infect *Australorbis glabratus. Am. J. trop. Med. Hyg.* **11,** 455–71.

CHOWANIEC, W. 1961. Influence of environment on the development of liver fluke, and the problem of superinvasion and reinvasion in the intermediate host. *Acta Parasit. Pol.* **9,** 463–79.

CHU, G. W. T. C. 1958. Pacific area distribution of fresh-water and marine cercarial dermatitis. *Pacif. Sci.* **12,** 299–312.

CHU, G. W. T. C. and CUTRESS, C. E. 1954. *Austrobilharzia variglandis* Miller and Northup 1926), Penner, 1953 (Trematoda: Schistosomatidae) in Hawaii with notes on its biology. *J. Parasit.* **40,** 515–24.

CHUBB, J. C. 1963. On the characterization of the parasite fauna of the fish of Llyn Tegid. *Proc. zool. Soc. Lond.* **141,** 609–21.

CHUBB, J. C. 1964. A preliminary comparison of the specific composition of the parasite fauna of the fish of Llyn Padarn, Caernarvonshire, an oligotrophic lake and Llyn Tegid (Bala Lake), Merionethshire, a late oligotrophic or early mesotrophic lake. *Whadomosci Parazytologiczne* **X,** Nr. 4–5, 499–510.

CLEGG, J. A. 1965. Secretion of lipoprotein by the Mehlis' gland in *Fasciola hepatica. Ann. N.Y. Acad. Sci.* **118,** 969–86.

CLEGG, J. A. and MORGAN, J. 1966. The lipid composition of the lipoprotein membranes on the egg-shell of *Fasciola hepatica. Comp. Biochem. Physiol.* **18,** 573–88.

CLEGG, J. A. and SMITHERS, S. R. 1968. Death of schistosome cercariae during penetration of the skin. II. Penetration of mammalian skin by *Schistosoma mansoni. Parasitology* **58,** 111–128.

COELHO, M. V. 1954. Ação des formas larvárias de *Schistosoma mansoni* sõbre a reprodução de *Australorbis glabratus. Publções avals. Inst. Aggeu Magalhães* **3,** 39–53.

COELHO, M. V. 1957. Aspectos do desenvolvimento das formas larvais de *Schistosoma mansoni* em *Australorbis nigricans. Revta. bras. Biol.* **17,** 325–37.

COELHO, M. V. 1962. Susceptibilidade de *Australorbis tenagophilus* a infecção por *Schistosoma mansoni Rev. Inst. Med. trop. S. Paulo* **4,** 289–95.

COIL, W. H. 1958. Alkaline phosphatase in the trematode excretory system. *Proc. helminth Soc. Wash.* **25,** 137–8.

COIL, W. H. 1965. Observations on egg shell formation in *Hydrophitrema gigantica* Sandars, 1960 (Hemiuridae: Digenea). *Z. ParasitKde.* **25,** 510–17.

COIL, W. H. 1966. Egg shell formation in the notocotylid trematode, *Ogmocotyle indica* (Bhalerao, 1942) Ruiz, 1946. *Z. ParasiKtde.* **27,** 205–9.

COIL, W. H. and KUNTZ, R. E. 1963. Observations on the histochemistry of *Syncoelium spathulatum* n.sp. *Proc. helminth Soc. Wash.* **30,** 60–5.

CORT, W. W. 1915. Some North American larval trematodes. *Illinois Biol. Monog.* **1,** (No. 4), 87 pp.

CORT, W. W. 1917. Homologies of the excretory system of the forked-tailed cercariae. *J. Parasit.* **4,** 48–57.

CORT, W. W. 1919. Notes on the eggs and miracidia of the human schistosomes. *Univ. Calif. Publs Zool.* **18,** 509–19.

CORT, W. W. 1922. A study of the escape of cercariae from their snail hosts. *J. Parasit.* **8,** 177–84.

CORT, W. W. 1928. Schistosome dermatitis in the U.S.A. (Michigan). *J. Am. med. Ass.* **90,** 1027.

CORT, W. W. and AMEEL, D. J. 1944. Further studies on the development of the sporocyst stages of plagiorchiid trematodes. *J. Parasit.* **30,** 37–56.

CORT, W. W., AMEEL, D. J. and OLIVIER, L. 1944. An experimental study of the development of *Schistosomatium douthitti* (Cort, 1914) in its intermediate host. *J. Parasit.* **30,** 1–17.

CORT, W. W., AMEEL, D. J. and VAND DER WOUDE, A. 1948. Studies on germinal development in redia of the trematode order *Fasciolatoidea* Szidat, 1936. *J. Parasit.* **34,** 428–51.

CORT, W. W., AMEEL, D. J. and VAN DER WOUDE, A. 1949. Germinal masses in redial embryos of an echinostome and a Psilostome. *J. Parasit.* **35,** 579–82.

CORT, W. W., AMEEL, D. J. and VAN DER WOUDE, A. 1951. Early developmental stages of strigeid mother sporocysts. *Proc. helminth Soc. Wash.*, **18,** 5–9.

CORT, W. W., AMEEL, D. J. and VAN DER WOUDE, A. 1954. Germinal development in the sporocysts and rediae of the digenetic trematodes. *Expl. Parasit.*, **3,** 185–225.

CORT, W. W., BRACKETT, S. and OLIVIER, L. 1944. Lymnaeid snails as second intermediate hosts of the strigeid trematode *Cotylurus flabelliformis* (Faust, 1917). *J. Parasit.* **30,** 309–21.

CORT, W. W., BRACKETT, S., OLIVIER, L. and NOLF, L. O. 1945. Influence of larval trematode infections in snails on their intermediate host relations to the strigeid trematode, *Cotylurus flabelliformis* (Faust, 1917). *J. Parasit.* **31,** 61–78.

CORT, W. W. and OLIVIER, L. 1941. Early developmental stages of strigeid trematodes in the first intermediate host. *J. Parasit.* **27,** 493–504.

CORT, W. W. and OLIVIER, L. 1943a. The development of the larval stages of *Plagiorchis muris* Tanabe, 1922, in the first intermediate host. *J. Parasit.* **29,** 81–99.

CORT, W. W. and OLIVIER, L. 1943b. The development of the sporocysts of a schistosome *Cercaria stagnicolae* Talbot 1936. *J. Parasit.* **29,** 164–74.

CORT, W. W., OLIVIER, L. and BRACKETT, S. 1941. The relation of Physid and Planorbid snails to the life-cycle of the strigeid trematode *Cotylurus flabelliformis*. *J. Parasit.* **27,** 437.

CORT, W. W. and TALBOT, S. B. 1936. Studies on schistosome dermatitis. III. Observations on the behaviour of the dermatitis-producing schistosome cercariae. *Am. J. Hyg.* **23,** 385–96.

CRANDALL, R. B. 1960. The life-history and affinities of the turtle lung fluke *Heronimus chelydrae* MacCallum, 1902. *J. Parasit.* **46,** 289–307.

CUCKLER, A. C. 1940a. The life cycle of *Fibricola cratera* (Barker and Noll, 1915) Dubois, 1932 (Trematoda: Strigeata). *J. Parasit.* **26,** (6) p. 32–3.

CUCKLER, A. C. 1940b. Studies on the migration and development of *Alaria spp.* (Trematoda: Strigeata) in the definitive host. *J. Parasit.* **26,** (6) p. 36.

CULBERTSON, J. T. 1941. *Immunity against animal parasites*. Columbia University Press. 274 pp.

DAMIAN, R. T. 1966. An immunodiffusion analysis of some antigens of *Schistosoma mansoni* adults. *Expl. Parasit.* **18,** 255–65.

DAMIAN, R. T. 1967. Common antigens between adult *Schistosoma mansoni* and the laboratory mouse. *J. Parasit.* **53,** 60.

DAVENPORT, D., WRIGHT, C. A. and CAUSLEY, D. 1962. Technique for the study of the behaviour of mobile micro-organisms. *Science,* **135,** 1059–60.

DAVIS, A. D., BOGITSH, B. J. and NUNNALLY, D. A. 1968. Cytochemical and biochemical observations on the digestive tracts of Digenetic trematodes. I. Ultrastructure of *Haematoloechus medioplexus* gut. *Expl. Parasit.* **22,** 96–106.

DAVIS, D. J. 1936. Report on the preparation of an histolytic ferment present in the bodies of cercariae. *J. Parasit.* **22,** 108–10.

DAVIS, H. S., HOFFMAN, G. L. and SURBER, E. W. 1961. Notes on *Sanguinicola davisi* (Trematoda: Sanguinicolidae) in the gills of trout. *J. Parasit.* **47,** 512–14.

DAWES, B. 1940. Notes on the formation of the egg capsules in the monogenetic trematode, *Hexacotyle extensicauda* Dawes 1940. *Parasitology* **32,** 287–95.

DAWES, B. 1946 and 1953. *The Trematoda.* 1st and 2nd edit. Cambridge University Press. 644 pp.

DAWES, B. 1952. Trematode life-cycle enacted in a London pond. *Nature, Lond.* **170,** 72–3.

DAWES, B. 1959. Penetration of the Liver-fluke *Fasciola hepatica* into the snail *Limnaea truncatula*. *Nature, Lond.* **184,** 1334.

DAWES, B. 1960. A study of the miracidium of *Fasciola hepatica* and an account of the mode of penetration of the sporocyst into *Limnaea truncatula*. *Sobretiro del Libro Homenje al Dr. Eduardo Caballero y Caballero,* 95–111.

DAWES, B. 1961a. On the early stages of *Fasciola hepatica* penetrating into the liver of an experimental host, the mouse: a histological picture. *J. Helminth R. T. Leifer Supplement,* 1961, 41–52.

DAWES, B. 1961b. Juvenile stages of *Fasciola hepatica* in the liver of the mouse. *Nature, Lond.* **190,** 646–7.

DAWES, B. 1962a. On the growth and maturation of *Fasciola hepatica* L. in the mouse. *J. Helminth* **36,** 11–38.

DAWES, B. 1962b. A histological study of the caecal epithelium of *Fasciola hepatica* L. *Parasitology* **52,** 483–93.

DAWES, B. 1963a. The migration of juvenile forms of *Fasciola hepatica* L. through the wall of the intestines in the mouse, with some observations on food and feeding. *Parasitology* **53,** 109–22.

DAWES, B. 1963b. Hyperplasia of the bile duct in fascioliasis and its relation to the problem of motivation in the liver fluke *Fasciola hepatica* L. *Parasitology* **53,** 123–33.

DAWES, B. 1963c. Some observations of *Fasciola hepatica* L. during feeding operations in the hepatic parenchyma of the mouse, with notes on the nature of liver damage in this host. *Parasitology* **53,** 135–43.

DAWES, B. 1964. A preliminary study of the prospect of inducing immunity in fascioliasis by means of infections with X-irradiated metacercarial cysts and subsequent challenge with normal cysts of *Fasciola hepatica* L. *Parasitology* **54,** 369–89.

DAWES, B. and HUGHES, D. L. 1964. Fascioliasis: the invasive stages of *Fasciola hepatica* in mammalian hosts. p. 97–168. In *Advances in Parasitology,* V. 2. Ed. B. Dawes. Academic Press, London and New York, 332 p. 1964.

DEFOREST, A. 1957. Larval trematodes of the Columbia Basin, Washington. *J. Parasit.* **43,** Sect. 2, Suppl. 32.

DEMARTINI, J. D. and PRATT, I. 1964. The life-cycle of *Telorchis pugetensis* Lloyd and Cuberlet, 1932. (Trematoda: Monorchiidae). *J. Parasit.* **50,** 101–5.

DE WITT, W. B. 1955. Influence of temperature on penetration of snail hosts by *Schistosoma mansoni* miracidium. *Expl. Parasit.* **4,** 244–55.

DIKE, S. C. 1967. Ultrastructure of the ceca of the digenetic trematodes *Gorgodera amplicava* and *Haematoloechus medioplexus*. *J. Parasit.* **53,** 1173–85.

DINNIK, J. A. and DINNIK, N. N. 1956. Observations on the succession of redial generations of *Fasciola gigantica* Cobbold in a snail host. *Z. Tropenmed. Parasit.* **7,** 397–419.

DINNIK, J. A. and DINNIK, N. N. 1957. Development of *Paramphistomum sukari* Dinnik, 1954 (Trematoda: Paramphistomidae) in a snail host. *Parasitology* **57,** 209–16.

DIXON, K. E. 1964. Excystment of metacercariae of *Fasciola hepatica* L. *in vitro*. *Nature, Lond.* **202,** 1240.

DIXON, K. E. 1965. The structure and histochemistry of the cyst wall of the metacercariae of *Fasciola hepatica* L. *Parasitology* **55,** 215–26.

DIXON, K. E. 1966a. A morphological and histochemical study of the cystogenic cells of the cercaria of *Fasciola hepatica* L. *Parasitology* **56,** 287–97.

DIXON, K. E. 1966b. The physiology of excystment of the metacercaria of *F. hepatica* L. *Parasitology* **56,** 589–602.

DIXON, K. E. and MERCER, E. H. 1964. The fine structure of the cyst wall of the metacercaria of *Fasciola hepatica*. *Q. Jl microsc. Sci.* **105,** 385–9.

DIXON, K. E. and MERCER, E. H. 1965. The fine structure of the nervous system of the cercaria of the liver fluke *Fasciola hepatica* L. *J. Parasit.* **51,** 967–76.

DIXON, K. E. and MERCER, E. H. 1967. The formation of the cyst wall of the metacercaria of *Fasciola hepatica*. *Z. Zellforsch. mikrosk. Anat.* **77,** 345–60.

DOBROVOLNY, C. G. 1939. Life history of *Plagioporus sinitsini* Mueller and embryology of new cotylocercous cercariae (Trematoda). *Trans. Am. Microsc. Soc.* **58,** 121–55.

DOBROWOLSKI, K. A. 1958. Pasożyty pijawek jeziora Drużno (Parazytofauna biocenozy jeziora Druzno—cześć V). *Acta parasit. polon.* **6,** 179–94.

DOGIEL, V. A. 1962. *General Parasitology*. Revised and enlarged by Yu. I. Polyawski and E. M. Kheisin. Translated by Z. Kabata 1964. Oliver & Boyd, Edn. London, 516 p.

DOGIEL, V. A., PETRUSHEUSKI, G. K. and POLYSANSKI, YU. I. 1958. *Parasitology of Fishes*, Translated by Z. Kabata. Oliver & Boyd, Edinburgh & London 1961.

DÖNGES, J. 1963. Reizphysiologische Untersuchungen an der Cercaria von *Posthodiplostomum cuticola* (v. Nordmann, 1832) Dubois, 1936, dem Erreger des Diplostomatiden—Melanoms der Fische. *Verh. dt. zool. Ges.* 216–23.

DÖNGES, J. 1964. Der Lebenszyklus von *Posthodiplostomum cuticola* (v. Nordmann, 1832) Dubois, 1936 (Trematoda, Diplostomatiidae). *Z. ParasitKde.* **24,** 169–248.

DOUGHERTY, J. W. 1952. Intermediary protein metabolism in helminths. I. Transaminase reactions in *Fasciola hepatica*. *Expl. Parasit.* **1,** 331–8.

DOW, C., ROSS, J. G. and TODD, J. R. 1968. The histopathology of *Fasciola hepatica* infections in sheep. *Parasitology* **58,** 129–35.

DUBOIS, G. 1929. Les cercaires de la région de Neuchâtel. *Bull. Soc. neuchâtel. Sci. nat.* **53,** N.S.*2*, Year 1928, 3–177.

DUBOIS, G. 1938. Monographie des Strigeida (Trematoda). *Mém. Soc. neuchât. Sci. nat.* **6,** 535 pp.

DUBOIS, G. 1953. Systématique des Strigeida Complément de la Monographie. *Mém. Soc. neuchât. Sci. nat.* **8,** 1–141.

DUKE, B. O. L. 1952. On the route of emergence of the cercariae of *Schistosoma mansoni* from *Australorbis glabratus* *J. Helminth.* **26,** 133–46.

DUSANIC, D. G. 1959. Histochemical observations of alkaline phosphatase in *Schistosoma mansoni*. *J. infect. Dis.* **105,** 1–8.

DUSANIC, D. G. and LEWERT, R. M. 1963. Alterations of proteins and free amino acids of *Australorbis glabratus* hemolymph after exposure to *Schistosoma mansoni* miracidia. *J. infect. Dis.* **112,** 243–6.

DUTT, S. C. and SRIVASTAVA, H. D. 1961. On the epidermal structures of the miracidia of six species of mammalian schistosomes, and a new technique of specific diagnosis of animal schistosomiasis. *Indian J. Helminth.* **13,** 100–11.

DUTT, S. C. and SRIVASTAVA, H. D. 1962a. Biological studies on *Orientobilharzia dattai* (Dutt & Srivastava; 1952) Dutt and Srivastava; 1955—a blood fluke of ruminants. *Indian J. vet. Sci.* **32,** 216–28.

DUTT, S. C. and SRIVASTAVA, H. D. 1962b. Studies on the morphology and the history of the mammalian blood fluke *Orientobilharzia dattai* (Dutt and Srivastava) Dutt and Srivastava. II. The molluscan phases of the life-cycle and the intermediate host specificity. *Ind. J. vet. Sci.* **32,** 33–43.

EBRAHIMZADEH, A. 1966. Histologische Untersuchungen über den Feinbau des Oogontop bei digenen Trematoden. *Z. ParasitKde.* **27,** 127–68.

EDNEY, J. M. 1950. Productivity in *Clinostomum marginatum* (Trematoda: Clinostomatidae). *Tr. Am. Microsc. Soc.* **69,** 186–8.

EHRLICH, I., RIJAVEC, M. and KURELEC, B. 1963. *Bull. scient. Cons. Acads. RPF Yougosl.* **8,** 133.

EL MOFTY, A. 1962. Clinical aspects of Bilharziasis; pages 174–197 in *Ciba Foundation Symposium Bilharziasis* Edit. by G. E. W. Wolstenholme and Mawe O'Connor, J. A. Churchill Ltd. London, 1962.

ERASMUS, D. A. 1957a. Studies on the phosphatase systems of Cestodes. Part I. The enzymes present in *Taenia pisiformis* (Cysticercus and adult). *Parasitology* **47,** 70–80.

ERASMUS, D. A. 1957b. The phosphatase systems of Cestodes. Part 2. Studies on *Cysticercus tenuicollis* and *Moniezia expansa* (adult). *Parasitology,* **47,** 81–91.

ERASMUS, D. A. 1958. Studies on the morphology, biology and development of a strigeid cercaria (*Cercaria X* Baylis 1930). *Parasitology* **48,** 312–35.

ERASMUS, D. A. 1959. The migration of *Cercaria X* Baylis (Strigeid) within the fish intermediate host. *Parasitology* **49,** 173–90.

ERASMUS, D. A. 1962a. Studies on the adult and metacercaria of *Holostephanus lühei* Szidat, 1936. *Parasitology* **52,** 353–74.

ERASMUS, D. A. 1962b. Distribution of certain strigeid trematodes in Great Britain. *Nature, Lond.* **195,** 828–9.

ERASMUS, D. A. 1966. Electron microscope and histochemical studies on the cuticle and subcuticular structures in the strigeid trematode *Cyathocotyle bushiensis*. p. 493–494. In *Proceedings of the First International Congress of Parasitology,* Rome 1964. Ed. A. Corradetti. Pergamon Press, Tamburini Editore.

ERASMUS, D. A. 1967a. Histochemical observations on the structure and composition of the cyst wall enclosing the metacercaria of *Cyathocotyle bushiensis* Khan, 1962 (Strigeoidea: Trematoda). *J. Helminth.* **41,** (1) 11–14.

ERASMUS, D. A. 1967b. Ultrastructural observations on the reserve bladder system of *Cyathocotyle bushiensis* Khan, 1962 (Trematoda: Strigeoidea) with special reference to lipid excretion. *J. Parasit.* **53,** 525–36.

ERASMUS, D. A. 1967c. The host-parasite interface of *Cyathocotyle bushiensis* Khan, 1962 (Trematoda: Strigeoidea) II. Electron microscope studies of the tegument. *J. Parasit.* **53,** 703–14.

ERASMUS, D. A. 1968. The host-parasite interface of *Cyathocotyle bushiensis* Khan, 1962 (Trematoda: Strigeoidea). III. Electron microscope observations on non-specific phosphatase activity. *Parasitology* **58,** 371–5.

ERASMUS, D. A. 1969a. Studies on the host-parasite interface of strigeoid trematodes. IV. The ultrastructure of the lappets of *Apatemon gracilis minor* Yamaguti, 1933. *Parasitology* **59,** 193–201.

ERASMUS, D. A. 1969b. Studies on the host-parasite interface of strigeoid trematodes. V. Regional differentiation of the adhesive organ of *Apatemon gracilis minor* Yamaguti, 1933. *Parasitology* **59,** 245–56.

ERASMUS, D. A. 1969c. Studies on the host-parasite interface of strigeoid trematodes. VI. Ultrastructural observations on the lappets of *Diplostomum phoxini* Faust, 1918. *Z. ParasitKde.* **32,** 48–58.

ERASMUS, D. A. 1970. The host-parasite interface of strigeoid trematodes. VII. Ultrastructural observations on the adhesive organ of *Diplostomum phoxini* Faust, 1918. *Z. ParasitKde.* **33,** 211–224.

ERASMUS, D. A. and BENNETT, L. J. 1965. A study of some of the factors affecting excystation *in vitro* of the metacercarial stages of *Holostephanus lühei* Szidat, 1936 and *Cyathocotyle bushiensis* Khan, 1962 (Strigeida: Trematoda). *J. Helminth.* **39,** 185–96.

ERASMUS, D. A. and ÖHMAN, C. 1963. The structure and function of the adhesive organ in strigeid trematodes. *Ann. N.Y. Acad. Sci.* **113,** 7–35.

ERASMUS, D. A. and ÖHMAN, C. 1965. Electron microscope studies of the gland cells and host-parasite interface of the adhesive organ in strigeid trematodes, *J. Parasit.* **51,** 761–9.

ERICKSON, D. G. 1965. The fate of gamma-irradiated *Schistosoma mansoni* cercariae in mice. *Am. J. trop. Med. Hyg.* **14,** 574–8.

ERICKSON, D. G. and CALDWELL, W. L. 1965. Acquired resistance in mice and rats after exposure to gamma-irradiated cercariae. *Am. J. trop. Med. Hyg.* **14,** 566–73.

ETGES, F. J. 1960. On the life-history of *Prosthodendrium* (*Acanthatrium*) *anaplocami* n.sp. (Trematoda: Lecithodendriidae). *J. Parasit.* **46,** 235–40.

ETGES, F. J. and DECKER, C. L. 1963. Chemosensitivity of the miracidium of *Schistosoma mansoni* to *Australorbis glabratus* and other snails. *J. Parasit.* **49,** 114–16.

ETGES, F. J. and GRESSO, W. 1965. Effect of *Schistosoma mansoni* infection on fecundity in *Australorbis glabratus*. *J. Parasit.* **51,** 757–60.

EVANS, H. E. and MACKIEWICZ, J. S. 1958. The incidence and location of metacercarial cysts (Trematoda: Strigeida) on 35 species of Central New York fishes. *J. Parasit.* **44,** 231–5.

EWERS, W. H. 1964a. An analysis of the molluscan hosts of the trematodes of birds and mammals and some speculations on host-specificity. *Parasitology* **54,** 571–8.

EWERS, W. H. 1964b. The influence of the density of snails on the incidence of larval trematodes. *Parasitology* **54,** 579–83.

FAIN, A. 1953. Contribution à l'étude des formes larvaires des Trématodes au Congo belge et spécialement de la larve de *Schistosoma mansoni*. *Mém. Inst. r. colon. belge Section Sci. Naturelles et Médicales. Menaires* **22,** 1–312.

FARLEY, J. 1962. The effect of temperature and pH on the longevity of *Schistosomatium douthitti* miracidia. *Can. J. Zool.* **40,** 615–20.

FAUST, E. C. 1917. Life-history studies on Montana trematodes. *Illinois biol. Monogr.* **4,** No. 1, 1–121.

FAUST, E. C. 1918. Studies on Illinois cercariae. *J. Parasit.* **4,** 93–110.

FAUST, E. C. 1919. The excretory system in Digenea. 1–111. *Biol. Bull. mar. biol. Lab., Woods Hole,* **36,** 315–44.

FAUST, E. C. 1920. Pathological changes in the gastropod liver produced by fluke infections. *Johns Hopkins Hosp. Bull.* **31,** 79–84.

FAUST, E. C. 1924. Notes on larval flukes from China. II. Studies on some larval flukes from the central and south coast provinces of China. *Am. J. Hyg.* **4,** 241–300.

FAUST, E. C. 1932. The excretory system as a method of classification of digenetic trematodes. *Q. Rev. Biol.,* **7,** 458–68.

FAUST, E. C. 1949. *Human Helminthology.* Henry Kimpton, London 744 p.

FAUST, E. C. and HOFFMAN, W. A. 1934. Studies on schistosomiasis mansoni in Puerto Rico. III. Biological studies. I. The extra-mammalian phases of the life cycle. *Puerto Rico J. publ. Hlth trop. Med.* **10,** 1–47.

FAUST, E. C. and MELENEY, H. E. 1924. Studies on schistosomiasis japonica. *Am. J. Hyg. Monogr. Ser. no. 3.* 399 pp.

FERGUSON, M. S. 1940. Excystment and sterilisation of metacercariae of the avian strigeid trematode *Posthodiplostomum minimum* and their development into adult worms in culture. *J. Parasit.* **26,** 359–72.

FERGUSON, M. S. 1943a. Migration and localization of an animal parasite within the host. *J. exp. Zool.,* **93,** 375–401.

FERGUSON, M. S. 1943b. *In vitro* cultivation of trematode metacercariae free from micro-organisms. *J. Parasit.* **29,** 319–23.

FERGUSON, M. S. and HAYFORD, R. A. 1941. The life history and control of an eye fluke. An account of a serious hatchery disease caused by a parasitic worm. *Prog. Fish Cult.* **54,** 1–13.

FILES, V. S. 1951. A study of the vector-parasite relationship in *Schistosoma mansoni*. *Parasitology* **41,** 264–9.

FLURY, F. and LEEB, F. 1926. Zur Chemie und Toxikologie der Distomeen. *Klin. Wschr.* 1926, 2054–5.

FOLGER, H. T. and ALEXANDER, L. E. 1938. The response to mechanical shock by the cercariae of *Bucephalus elegans*. *Physiol. Zoöl.* **11,** 82–8.

FRAENKEL, G. S. and GUNN, D. L. 1961. *The orientation of animals.* Dover Edition, Dover Publications, Inc., N.Y. 376 p.

FRANKLAND, H. M. T. 1955. The life-history and bionomics of *Diclidophora denticulata* (Trematode: Monogenea). *Parasitology* **45,** 313–51.

FRANZEN, A. 1956. On spermiogenesis, morphology of the spermatozoa and biology of fertilization among invertebrates. *Zool. Bidr. Upps.* **31,** 355–482.

FREEMAN, R. F. H. and LLEWELLYN, J. 1958. An adult digenetic trematode from an invertebrate host *Proctoeces subtennis* (Linton) from the lamellibranch *Scrobicularia plana*. *J. mar. biol. Ass. U.K.* **37,** 435–57.

FRIED, B. 1962a. *In vitro* studies on *Philophthalmus* sp. an ocular trematode. *J. Parasit.* **48,** 510.

FRIED, B. 1962b. Growth of *Philophthalmus* sp. (Trematoda) on the chorioallantois of the chick. *J. Parasit.* **48,** 545–50.

FRIEDL, F. E. 1961a. Studies on larval *Fascioloides magna*. I. Observations on the survival of rediae *in vitro*. *J. Parasit.* **47,** 71–5.

FRIEDL, F. E. 1961b. Studies on larval *Fascioloides magna* II. *In vitro* survival of axenic rediae in amino acids and sugars. *J. Parasit.* **47,** 244–7.

FRIPP, P. J. 1966a. The Histochemical Localisation of Hydrolytic enzymes in schistosomes. p. 714–5. *Proceedings of the First International Congress of Parasitology, Roma,* 1964. Ed. A. Corradetti. Pergamon Press, London.

FRIPP, P. J. 1966b. Histochemical localization of β-Glucoronidase in schistosomes. *Expl Parasit.* **19,** 254–63.

FRIPP, P. J. 1967. The histochemical localization of Leucine aminopeptidase in *Schistosoma rodhaim*. *Comp. Biochem. Physiol.* **20,** 207–309.

GAZZINELLI, G. and PELLEGRINO, J. 1964. Elastolytic activity of *Schistosoma mansoni* cercarial extract. *J. Parasit.* **50,** 591–2.

GAZZINELLI, G., RAMALHO-PINTO, F. J. and PELLEGRINO, J. 1966. Purification and characterisation of the proteolytic enzyme complex of cercarial extract. *Comp. Biochem. Physiol.* **18,** 689–700.

GEIMAN, Q. M. and MCKEE, R. W. 1950. Protein Metabolism of Parasites. *J. Parasit.* **36,** 211–26.

GILBERTSON, D. E., ETGES, F. J. and OGLE, J. D. 1967. Free amino acids of *Australorbis glabratus* hemolymph: Comparison of four geographic strains and effect of infection by *Schistosoma mansoni*. *J. Parasit.* **53,** (3), 565–8.

GINETZINSKAJA, T. A. 1950. New data on the mechanism of penetration and migration of cercariae in the tissues of the host. *Dok. Acad. Nauk SSSR.* **72,** 433–5.

GINETZINSKAJA, T. A. 1960. Glycogen in the body of cercariae and the dependence of its distribution upon the peculiarities of their biology (In Russian). *Dokl. Akad. Nauk SSSR.* **135,** 1012–15.

GINETZINSKAJA, T. A. 1961. The dynamics of the storage of fat in the course of the life cycle of trematodes. (In Russian) *Dokl. Acad. Nauk SSSR.* **139,** 1016–19.

GINETZINSKAJA, T. A. 1965. The nature of trematode life-cycles. *Vest. leningr. gos. Univ. Ser. biol.*, **20,** 5–13 (In Russian).

GINETZINSKAJA, T. A. and DOBROVOLSKII, A. A. 1962. Glycogen and fat in the various phases of the life-cycle of trematodes (In Russian). *Vest. leningr. Univ. No. 9. Biol. Ser. fasc. 2.*, 67–81.

GINETZINSKAJA, T. A., MASHANSKI, V. F. and DOBROVOLSKII, A. A. 1966. Ultrastructure of the wall and method of feeding of rediae and sporocysts (trematodes). *Dokl. Akad. Nauk SSSR.* **166,** 1003–4.

GIOVANNOLA, A. 1936a. Inversion of the periodicity of emission of cercariae from their snail hosts by reversal of light and darkness. *J. Parasit.* **22,** 292–5.

GIOVANNOLA, A. 1936b. Some observations on the emission of cercariae of *Schistosoma mansoni* (Trematoda: Schistosomatidae) from *Australorbis glabratus*. *Proc. helminth Soc. Wash.* **3,** 60–1.

GOIL, M. M. 1958a. Fat metabolism in trematode parasites. *Z. ParasitKde.* **18,** 320–3.

GOIL, M. M. 1958b. Protein metabolism in trematode parasites. *J. Helminth.* **32,** 119–24.

GOIL, M. M. 1964. Physiological studies on trematodes-lipid fractions in *Gastrothylax crumenifer*. *Parasitology* **54,** 81–5.

GOIL, M. M. 1966. Physiological studies on trematodes: the osmotic activity of *Gastrothylax crumenifer*. *Parasitology* **56,** 101–4.

GÖNNERT, R. 1962. Histologische untersuchungen über den Feinbau der Eibildungstatte (Oogentop) von *Fasciola hepatica*. *Z. ParasitKde.* **21,** 475–92.

GOODCHILD, C. G. and KIRK, D. E. 1960. The life history of *Spirorchis elegans* Stunkard, 1923 (Trematoda: Spirochiidae) from the painted turtle. *J. Parasit.* **46,** 219–29.

GORSHKOV, P. V. 1964. Independence of ontogenesis and annual reproductive cycle of *P. intergerrimum* from hormonal activity of its host. *Zool. Zh.* **43,** 272–4.

GRASSÉ, P. P. 1961. Traité de Zoologie IV. *Platyhelminths, Mésozoaires, Acanthocéphales, Némertiens*. Mason et Cie. Paris, 1st edit. 1961, 944 p.

GRESSON, R. A. R., 1964. Oogenesis in the hermaphroditic Digenea (Trematoda) *Parasitology* **54,** 409–21.

GRESSON, R. A. R. and PERRY, M. M. 1961. Electron microscope studies of spermateleosis in *Fasciola hepatica* L. *Expl. Cell Res.* **22,** 1–8.

GRESSON, R. A. R. and THREADGOLD, L. T. 1959. A light and electron microscope study of the epithelial cells of the gut of *Fasciola hepatica* L. *J. biophys. biochem. Cytol.* **6,** 157–62.

GRESSON, R. A. R. and THREADGOLD, L. T. 1964. The large neurones and interstitial material of *Fasciola hepatica* L. *Proc. R. Soc. Edinb.* **68,** 261–6.

GUILFORD, H. G. 1958. Observations on the development of the miracidia and the germ cell cycle in *Heronimus chelydrae* MacCullum (Trematoda) *J. Parasit.* **44,** 64–74.

GUILFORD, H. G. 1961. Gametogenesis, egg-capsule formation, and early miracidial development in the digenetic trematode *Halipegus eccentricus* Thomas. *J. Parasit.* **47,** 757–64.

HALL, J. E. 1960. Some lecithodendriid metacercariae from Indiana and Michigan. *J. Parasit.* **46,** 309–15.

HALL, J. E. and GROVES, A. E. 1961. Studies on virgulate cercariae from *Nitocris dilatatus* Conrad and their entry into arthropod second intermediate hosts. *J. Parasit.* **47,** 41–2.

HALTON, D. W. 1963. Some hydrolytic enzymes in two digenetic trematodes. *Proc. XVI. Int. Congr. Zool.* **1,** 29.

HALTON, D. W. 1966. Occurrence of microvilli-like structures in the gut of digenetic trematodes. *Experientia.* **22,** 828–9.

HALTON, D. W. 1967a. Studies on phosphatase activity in Trematoda. *J. Parasit.* **53,** 46–54.

HALTON, D. W. 1967b. Observations on the nutrition of digenetic trematodes. *Parasitology* **57,** 639–60.

HALTON, D. W. 1967c. Histochemical studies of carboxylic esterase activity in *Fasciola hepatica*. *J. Parasit.* **53,** 1210–16.

HALTON, D. W. 1967d. Studies on glycogen deposition in trematodes. *Comp. Biochem. Physiol.* **23,** 113–20.

HALTON, D. W. and DERMOTT, E. 1967. Electron microscopy of certain gland cells in two digenetic trematodes. *J. Parasit.* **53,** 1186–91.

HALTON, D. W. and JENNINGS, J. B. 1965. Observations on the nutrition of monogenetic trematodes. *Biol. Bull. mar. biol. Lab., Woods Hole,* **129,** 257–72.

HANUMANTHA-RAO, K. 1959. Histochemistry of Mehlis' gland and egg-shell formation in the liver fluke *Fasciola hepatica* L. *Experientia.* **15,** 464–5.

HARGIS, W. J. 1959. The host specificity of monogenetic trematodes. *Expl. Parasit.* **6,** 610–25.

HASWELL, W. A. 1903. On two remarkable sporocysts occurring in *Mytilus latus* on the coast of New Zealand. *Proc. Linn. Soc. N.S.W.* 1902. Part 4, 497–515.

HEMENWAY, W. 1948. Studies on the excystment of *Clinostomum* metacercariae by the use of artificial digest. *Iowa. Acad. Sci.* **55,** 375–81.

HENDELBERG, J. 1962. Paired flagella and nucleus migration in the spermatogenesis of *Dicrocoelium* and *Fasciola* (Digenea, Trematoda) *Zool. Bidr. Uppsala*, **35,** 569–88.

HERBER, E. C. 1950. Studies on the biochemistry of cyst envelopes of the fluke, *Notocotylus urbanensis*. *Proc. Pa. Acad. Sci.* **24,** 140–2.

HERSMENOV, B. R., TULLOCH, G. S. and JOHNSTON, A. D. 1966. The fine structure of trematode sperm-tails. *Trans. Amer. Microsc. Soc.* **85,** 480–3.

HEYNEMAN, D. 1960. On the origin of complex life cycles in the Digenetic flukes. Sobretino del Libro Homenaje al Dr. Eduardo Caballero y Caballero. Mexico. 1960, 1933–52.

HEYNEMAN, D. 1966. Successful infection with larval echinostomes surgically implanted into the body cavity of the normal snail host. *Expl. Parasit.* 18, 220–23.

HOCKLEY, D. J. 1968. Small spines on the egg shells of *Schistosoma*. *Parasitology* **58,** 367–70.

HOEPPLI, R., FENG, L. C. and CHU, H. J. 1938. Attempts to culture helminths of vertebrates in artificial media. Chinese Med. J. (Suppl. 11): 343–74.

HOFF, C. C. 1941. A case of correlation between infection of snail hosts with *Cryptocotyle lingua* and the habits of gulls. *J. Parasit.* **27,** 539.

HOFFMAN, G. L. 1955. Notes on the life cycle of *Fibricola cratera* (Trematoda: Strigeida). *J. Parasit.* **41,** 327.

HOFFMAN, G. L. 1956. The life-cycle of *Crassiphiala bulboglossa* (Trematoda Strigeida). Development of the metacercariae and cyst, and effect on the fish hosts. *J. Parasit.* **42,** 435–444.

HOFFMAN, G. L. 1958a. Experimental studies on the cercaria and metacercaria of a strigeoid trematode, *Posthodiplostomum minimum*. *Expl. Parasit.* **7,** 23–50.

HOFFMAN, G. L. 1958b. Studies on the life-cycle of *Ornithodiplostomum ptychocheilus* (Faust) (Trematoda: Strigeoidea) and the "self-cure" of infected fish. *J. Parasit.* **44,** 416–21.

HOFFMAN, G. L. 1959. Studies on the life cycle of *Apatemon gracilis pellucidus* (Yamag.) *Trans. Am. Fish. Soc.* **88,** 96–9.

HOFFMAN, G. L. 1960. Synopsis of Strigeoidea (Trematoda) of fishes and their life-cycles. Fishery Bulletin 175. Vol. 60. Fishery Bulletin of the Fish and Wild Life Service. 469 pp. U.S. Dept. of the Interior.

HOFFMAN, G. L. and DUNBAR, C. E. 1963. Studies on *Neogogatea kentuckiensis* (Cable, 1935) n.comb. (Trematoda: Strigeoidea: Cyathocotylidae). *J. Parasit.* **49,** 737–44.

HOFFMAN, G. L. and HOYME, J. B. 1958. The experimental histopathology of the "Tumor" on the brain of the stickleback caused by *Diplostomum baeri eucaliae* Hoffman and Hundley, 1957. (Trematoda: Strigeoidea) *J. Parasit.* **44,** 374–8.

HOFFMAN, G. L. and HUNDLEY, J. B. 1957. The life-cycle of *Diplostomum baeri eucaliae* n.subsp. (Trematoda: Strigeida). *J. Parasit.* **43,** 613–27.

HOLL, F. 1932. The ecology of certain fishes and amphibians with special reference to their helminth and linguatulid parasites. *Ecol. Monogr.* **2,** 85–107.

HORAK, I. G. 1965. Studies on paramphistomiasis. VII. The immunisation of sheep, goats and cattle. Preliminary report. *Jl. S. Afr. vet. med. Ass.* **36,** 361–3.

HORSTMANN, H. J. 1962. Sauerstoffverbrauch und Glykogengehalt der Eier von *Fasciola hepatica* wahrend der Entwicklung der Miracidien. *Ztsch. Parasitenk.* **21,** 437–45.

HOWELL, R. M. 1966. Collagenase activity of immature *Fasciola hepatica*. *Nature, Lond.* **209,** 713–4.

HOWELLS, R. E. 1967. *Histochemical and electron microscope studies on the cuticle and associated structures of certain cestode parasites*. Ph.D. Thesis. Univ. Wales. 1967.

HOWELLS, R. E. 1969. Observations on the nephridial system of the cestode *Moniezia expansa* (Rud., 1805). *Parasitology* **59,** 449–59.

HSÜ, H. F. and HSÜ, S. Y. LI. 1962. *Schistosoma japonicum* in Formosa: A critical review. *Expl. Parasit.* **12,** 459–65.

HUANG, F. Y. and CHU, C. H. 1962. Sheng Wu Hua Hsuen Yu Shing Wu Wu Li Hseum Pao 2, 286–294. *Chem. Abst.* **59,** 14333 (1963).

HUFF, C. G. 1940. Immunity in invertebrates. *Physiol. Rev.* **20,** 68–88.

HUGGHINS, E. J. 1956. Ecological studies on a trematode of bullheads and cormorants at Spring Lake, Illinois. *Trans. Am. microsc. Soc.* **75,** 281–9.

HUGHES, D. H. 1962. Observations on the immunology of *Fasciola hepatica* infections in mice and rabbits. *Parasitology* **52,** 4 p.

HUGHES, R. C. and HALL, L. J. 1929. Studies on the trematode family Strigeidae (Holostomidae). No. XVI. *Diplostomulum huronense* (La Rue) *Pap. Mich. Acad. Sci.* **10,** 489–94.

HUNTER, G. W. 1960. Studies on Schistosomiasis. XIII Schistosome Dermatitis in Colorado. *J. Parasit.* **46,** 231–4.

HUNTER, G. W. III and BIRKENHOLZ, D. E. 1961. Notes on larval trematodes of Gunnison County, Colorado. *Trans. Am. microsc. Soc.* **80,** 358–64.

HUNTER, G. W. and HAMILTON, J. M. 1941. Studies on host-parasite reactions to larval parasites. IV. The cyst of *Uvulifer amploblitis. Trans. Am. microsc. Soc.* **60,** 498–507.

HUNTER, G. W. and HUNTER, W. S. 1937. Studies on host reactions to larval parasites. III. An histolytic ferment from the cercariae of *Cryptocotyle lingua J. Parasit.* **23,** (Suppl): 572.

HUNTER, G. W. III and HUNTER, W. S. 1940. Studies on the development of the metacercaria and the nature of the cyst of *Posthodiplostomum minimum* (MacCallum, 1921) (Trematoda; Strigeata). *Trans. Am. microsc. Soc.* **59,** 52–63.

HUNTER, W. R. 1961. Annual variations in growth and density in natural populations of freshwater snails in the west of Scotland. *Proc. zool. Soc. Lond.* **136,** 219–53.

HUNTER, W. S. and CHAIT, D. C. 1952. Notes on excystment and culture *in vitro* of the micro-phallid trematode *Gynaecotyle adunca* (Linton, 1945). *J. Parasit.* **33,** 79–84.

HUNTER, W. S. and VERNBERG, W. B. 1955a. Studies on oxygen consumption in digenetic trematodes I. *Expl. Parasit.* **4,** 54–61.

HUNTER, W. S. and VERNBERG, W. B. 1955b. Studies on oxygen consumption in digenetic trematodes. II. Effects of two extremes in oxygen tension. *Expl. Parasit.* **4,** 427–34.

HURST, C. T. 1927. Structural and functional changes produced in the gastropod mollusk, *Physa occidentalis,* in the case of parasitism by the larvae of *Echinostoma revolutum. Univ. Calif. Publs. Zool.* **29,** 321–404.

HUSSEY, K. L. 1941. Comparative enbryological development of the excretory system in digenetic trematodes. *Trans. Am. microsc. Soc.* **60,** 171–210.

HUSSEY, K. L. 1943. Further studies on the comparative embryological development of the excretory system in digenetic trematodes. *Trans. Am. microsc. Soc.* **62,** 271–9.

HUSSEY, K. L., CORT, W. W. and AMEEL, D. J. 1958. The production of cercariae by a strigeid mother sporocyst. *J. Parasit.* **44,** 289–91.

HYMAN, L. H. 1951. *The Invertebrates: Platyhelminthes and Rhynchocoela. The acoelomate Bilateria.* VII + 550 pp. New York: McGraw-Hill 1951.

ILES, C. 1960. The larval trematodes of certain fresh-water molluscs. II. Experimental studies on the life-cycle of two species of Furcocercariae. *Parasitology* **50,** 401–17.

INGALLS, J. W., HUNTER, G. W., MCMULLEN, D. B. and BAUMAN, P. M. 1949. The molluscan intermediate host and *Schistosomiasis Japonica.* I. Observations on the conditions governing the hatching of the eggs of *Schistosoma japonicum. J. Parasit.* **35,** 147–51.

INGERSOLL, E. M. 1956. *In vitro* survival of rediae of *Cyclocoelum microstomum. Expl. Parasit.* **5,** 231–7.

ISHII, Y. 1934. Studies on the development of *Fasciolopsis buski.* Part I. Development of the eggs outside the host. (Japanese text). (English summary, Suppl. pp. 29–30) *J. med. Ass. Formosa* **33,** 349–412.

ISSEROF, H. 1963. Ultrastructure of the eyespot and excretory system in the miracidia of a philophthalmid trematode. *J. Parasit.* **49,** (5) Suppl., 41–2, 58.

ISSEROFF, H. 1964. Fine structure of the eyespot in the miracidia of *Philophthalmus megalurus* (Cort, 1914). *J. Parasit.* **50,** 549–54.

ITO, J. and WATANABE, K. 1959. Studies on mucoid glands in the cercaria of *Notocotylus magniovatus* Yamaguti, 1934 (Notocotylidae: Trematoda). *Jap. J. med. Sci. Biol.* **12,** 139–143.

JAMES, B. L. 1964. The life-cycle of *Parvatrema homoeotecnum* sp.nov. (Trematoda: Digenea) and a review of the family Gymnophallidae Morozov, 1955. *Parasitology* **54,** 1–41.

JAMES, B. L. 1965. The effects of parasitism by larval Digenea on the digestive gland of the intertidal prosobranch *Littorina saxatilis* (Olivi) subsp. *tenebrosa* (Montagu). *Parasitology* **55,** 93–115.

JAMES, B. L. 1968a. The occurrence of *Parvatrema homoeotecnum* James, 1964 (Trematoda: Gymnophallidae) in a population of *Littorina saxitalis tenebrosa* (Mont.). *J. nat. Hist.* **2,** 21–37.

JAMES, B. L. 1968b. The occurrence of larval digenea in ten species of intertidal prosobranch molluscs in Cardigan Bay. *J. nat. Hist.* **2,** 329–43.

JAMES, B. L. and BOWERS, E. A. 1967a. The effects of parasitism by the daughter sporocyst, of *Cercaria bucephalopsis haimaena* Lacaze-Duthiers, 1854 on the digestive tubules of the cockle *Cardium edule* L. *Parasitology* **57,** 67–77.

JAMES, B. L. and BOWERS, E. A. 1967b. Histochemical observations on the occurrence of carbohydrates, lipids and enzymes in the daughter sporocyst of *Cercaria bucephalopsis haimaena* Lacaze-Duthiers, 1854 (Digenea: Bucephalidae.) *Parasitology* **57,** 79–86.

JAMES, B. L. and BOWERS, E. A. 1967c. Reproduction in the daughter sporocyst of *Cercaria bucephalopsis haimaena* (Lacaze-Duthiers, 1854) (Bucephalidae) and *Cercaria dichotoma* Lebour, 1911 (non Müller) (Gymnocephalidae). *Parasitology* **57,** 607–25.

JAMES, B. L., BOWERS, E. A. and RICHARDS, J. G. 1966. The ultrastructure of *Cercaria bucephalopsis haimaena* Lacaze-Duthiers, 1854 (Digenea: Bucephalidae) from the edible cockle, *Cardium edule* L. *Parasitology* **56,** 753–62.

JARECKA, L. 1958. Plankton crustaceans in the life-cycle of tapeworms occurring at Drużno lake (Parasitofauna of the biocoenosis of Drużno Lake—part II. *Acta Parasit. Polon.* **6,** 65–109.

JENKINSON, J. 1967. *The ultrastructure of the tegument of a strigeoid sporocyst and cercaria.* Personal communication.

JENNINGS, J. B. 1956. A technique for the detection of *Polystoma integerrimum* in the common frog (*Rana temporaria*) *J. Helminth.* **30,** 119-20.

JENNINGS, J. B. 1959. Studies on digestion in the Monogenetic Trematode *Polystoma intergerrimum. J. Helminth.* **33,** 197–204.

JOHNSTON, T. H. and BECKWITH, A. C. 1947. Larval trematodes from Australian freshwater molluscs Part XII. *Trans. Roy. Soc. S. Aust.* **71,** 324–33.

KAGAN, I. G. 1951. Aspects in the life history of *Neoleucochloridium problematicum* (Magath, 1920) new comb. and *Leucochloridium cyanocittae,* McIntosh 1932 (Trematoda: Brachylaemidae). *Trans. Am. microsc. Soc.* **70,** 281–318.

KAGAN, I. G. 1952. Further contributions to the life-history of *Neoleucochloridium problematicum* (Magath, 1921) new comb. (Trematoda: Brachylaemidae) *Trans. Am. microsc. Soc.* **71,** 20–44.

KAGAN, I. G. 1966. Mechanisms of immunity in trematode infection p. 277–299. In *Biology of Parasites,* ed. E. J. L. Soulsby. Acad. Press. 1966, 354 pp.

KAGAN, I. G. and GEIGER, S. J. 1965. The susceptibility of three strains of *Australorbis glabratus* to *Schistosoma mansoni* from Brazil and Puerto Rico. *J. Parasit.* **51,** 622–7.

KAGAN, I. G. and NORMAN, L. 1963. Analysis of helminth antigens (*Echinococcus granulosus* and *Schistosoma manosni*) by agar gel methods. *Ann. N.Y. Acad. Sci.* **113,** 130–53.

KEARN, G. C. 1963. Feeding in some Monogenean skin parasites: *Entobdella solea* on *Solea solea* and *Acanthocotyle sp.* on *Raia clavata. J. mar. biol. Ass. U.K.* **43,** 749–66.

KEARN, G. C. 1967. Experiments on host-finding and host-specificity in the monogenean skin parasite *Entobdella soleae. Parasitology* **57,** 585–605.

KENDALL, S. B. 1949a. Bionomics in *Limnaea truncatula* and the parthenitae of *Fasciola hepatica* under drought conditions. *J. Helminth.* **23,** 57–68.

KENDALL, S. B. 1949b. Nutritional factors affecting the rate of development of *Fasciola hepatica* in *Limnaea truncatula. J. Helminth.* **23,** 179–90.

KENDALL, S. B. 1950. Snail hosts of *Fasciola hepatica* in Britain. *J. Helminth.* **24,** 63–74.

KENDALL, S. B. 1964. Some factors influencing the development and behaviour of trematodes in the molluscan hosts, p. 51–73 in *Host-Parasite Relationships in Invertebrate Hosts.* Ed. A. R. Taylor. Second Symposium of the British Society for Parasitology. Blackwell Scientific Publications, Oxford 134 p.

KENDALL, S. B. 1965. Relationships between the species of *Fasciola* and their molluscan hosts. 59–98. In *Advances in Parasitology,* V. 3. Ed. B. Dawes. Academic Press, N. York.

KENDALL, S. B. and MCCULLOUGH, F. S. 1951. The emergence of *Fasciola hepatica* cercariae from the snail *Limnaea truncatula. J. Helminth.* **25,** 77–92.

KENT, N. H. 1963. Seminar on immunity to Parasitic Helminths. V. Antigens. *Expl. Parasit.* **13,** 45–56.

KHAN, D. 1960a. Studies on larval trematodes infecting freshwater snails in London (U.K.) and some adjoining areas Part I. Echinostome cercariae. *J. Helminth.* **34,** 277–304.

KHAN, D. 1960b. Studies on larval trematodes infecting freshwater snails in London (U.K.) and some adjoining areas. Part II. Gymnocephalous cercariae. *J. Helminth.* **34,** 305–18.

KHAW, O. K. 1935. *In vitro* experiments on the viability and excystment of *Paragonimus* cyst. *Proc. Soc. exp. Biol. Med.* **32,** 1003–5.

KLOETZEL, K. 1958. Observacões sôbre o tropismo do miracidio de *Schistosoma mansoni* pelo molusco *Australorbis glabratus. Rev. Brasil. Biol.* **18,** 223–32.

KNOX, B. E. 1965a. *Metabolic functions of the integument in Fasciola hepatica L.* Ph.D. thesis, Queen's University of Belfast, 1965.

KNOX, B. E. 1965b. Uptake of nutrients by *Fasciola hepatica* (Abstract). *Parasitology* **55,** 17–18.

KNOX, B. E. and PANTELOURIS, E. M. 1966. Osmotic behaviour of *Fasciola hepatica* L. in modified Hedon-Fleig Media. *Comp. Biochem. Physiol.* **18,** 609–15.

KOMIYA, Y. 1938. Die Entwichlung des exkretions-systems einiger trematodenlarven aus Alster und Elbe, nebst. bemerkungen uber ihren entwicklungszyklus. *Ztsch. Parasitenk.* **10,** 16–385.

KOMIYA, Y. 1961. *The excretory system of digenetic trematodes.* Committee of Jubilee publications in the commemoration of Dr. Yoshitaka Komiya at the 10th anniversary as a chief of department of Parasitology, National Institute of Health, Tokyo, Japan 1961.

KOMIYA, Y. and TAJIMI, T. 1940a. Study on *Clonorchis sinensis* in the District of Shanghai. 5. The cercaria and metacercaria of *Clonorchis sinensis* with special reference to their excretory system. *J. Shanghai Science Inst. S.IV.* **5,** 91–106.

KOMIYA, Y. and TAJIMI, T. 1940b. Study of *Clonorchis sinensis* in the District of Shanghai. 6. The life cycle of *Exorchis oviformis* with special reference to the similarity of its larval forms to that of *Clonorchis sinensis. J. Shanghai Science Inst., S.IV.* **5,** 109–123.

KOMIYA, Y. and TAJIMI, T. 1941. Metacercariae from chinese *Pseudorasbora parva* Temminck and Schlegel with special reference to their excretory system. I. (Metacercariae from Chinese Freshwaters No. 1.) *J. Shanghai Sci. Inst.* **1,** 69–106.

KOURI, P. and NAUSS, R. W. 1938. Formation of the egg shell in *Fasciola hepatica* as demonstrated by histological methods. *J. Parasit.* **24,** 291–310.

KOZICKA, J. 1958. Diseases of fishes of Drużno lake (Parasitofauna of the biocoenosis of Drużno lake—part VII). *Acta Parasit. Polon.* **6,** 393–432.

KOZICKA, J. 1959. Parasites of fishes of Drużno lake (Parisitofauna of the biocoenosis of Drużno lake—part VIII). *Acta. Parasit. Polon.* **7,** 1–72.

KRAUSE, R. K. L. 1914. Beitrage zur Kenntnis der Hemistominen. *Ztschr. Wissensch. Zool.* **112,** 93–238.

KRUIDENIER, F. J. 1949. Mucoid glands in *Fasciola hepatica* cercariae. *J. Parasit.* **35,** (6) Suppl. p. 20, no. 34.

KRUIDENIER, F. J. 1951a. Studies in the use of mucoids by *Clinostomum marginatum. J. Parasit.* **37,** (5 sec 2) 25–26.

KRUIDENIER, F. J. 1951b. The formation and function of mucoids in virgulate cercariae including a study of the virgula organ. *Am. Midl. Nat.* **46,** 660–83.

KRUIDENIER, F. J. 1953a. Studies on the function and formation of mucoid glands in cercariae: opisthorchoid cercariae. *J. Parasit.* **39,** 385–91.

KRUIDENIER, F. J. 1953b. Studies on mucoid secretion and function in the cercariae of *Paragonimus kellicotti* Ward (Trematoda: Troglotrematidae). *J. Morph.* **92,** 531–43.

KRUIDENIER, F. J. 1953c. The formation and function of mucoids in cercariae: monostome cercariae. *Trans. Am. microsc. Soc.* **72,** 57–67.

KRUIDENIER, F. J. 1953d. Studies on the formation and function of mucoids in cercariae: non-virgulate xiphidiocercariae. *Am. Midl. Nat.* **50,** 382–96.

KRUIDENIER, F. J. 1959. Ultrastructure of the excretory system of cercariae. *J. Parasit.* 45 (supp.): 59 No. 144.

KRUIDENIER, F. J. 1960a. Observations on the ultrastructure and histochemistry of cercarial glands. *J. Parasit.* **46,** (Suppl.) p. 19, No. 52.

KRUIDENIER, F. J. 1960b. Ultrastructure in the tails of furcocercous cercariae. *J. Parasit.* **46,** No. 5. Sect. 2 (Suppl.) No. 92, p. 32.

KRUIDENIER, F. J. and MEHRA, K. N. 1957. Mucosubstances in Plagiorchoid and monostomate cercariae (Trematoda: Digenea). *Trans. Illin. State Acad. Sci.* **50,** 267–78.

KRUIDENIER, F. J. and VATTER, A. E. 1958. Ultrastructure at the surface of cercariae of *S. mansoni* and of a Plagiorchioid (*Tetrapapillatrema concavocorpa*?). *J. Parasit.* **44,** (4) p. 42, No. 101.

KRUIDENIER, F. J. and VATTER, A. E. 1960. Microstructure of muscles in cercariae of the digenetic trematodes *Schistosoma mansoni* and *Tetrapapillatrema concavocorpa.* Fourth International Conference on Electron Microscopy, 1960.

KRULL, W. H. 1933. New snail and rabbit hosts for *Fasciola hepatica* Linn. *J. Parasit.* **20,** 49–52.

KRULL, W. H. 1935. Studies on the life history of *Halipegus occidualis* 1905. *Am. Midl. Nat.* **16,** 129–43.

KRUPA, P. L., BAL, A. K. and COUSINEAU, G. H. 1967. Ultrastructure of the redia of *Cryptocotyle lingua. J. Parasit.* **53,** 725–34.

KRVAVICA, S. and MARTINCIC, T. 1964. The content of coenzyme A in the liver fluke (*Fasciola hepatica* L.). *Veterinarski Arhib.*, **34,** 60–2.

KUBLITSKENE, O. A. 1963. Age changes in the microstructure of the cuticle of *Fasciola hepatica* (In Russian). *Zoologicheski Zhurnal* **42,** 1613–18.

KÜMMEL, G. 1958. Das terminalongan der Protonephridien, Feinstruktur und Deutung der Funktion. *Z. Naturf.* **136,** 676–9.

KÜMMEL, G. 1960a. Feinstruktur der Wimperflamme in den Protonephridien. *Protoplasma* **51,** 371–6.

KÜMMEL, G. 1960b. Die Feinstruktur des Pigmenthecherocells bei Miracidien von *Fasciola hepatica* L. *Zool. Beit.* **5,** 345–54.

KUNTZ, R. E. 1948. Abnormalities in development of helminth parasites with a description of several anomalies in cercariae of digenetic trematodes. *Proc. helminth. Soc. Wash.* **15,** 73–7.

KUNTZ, R. E. 1950. Embryonic development of the excretory system in fork-tailed cercariae of the schistosomes and in a blunt-tailed brachylaemid cercaria. *Trans. Am. microsc. Soc.* **69,** 1–20.

KUNTZ, R. E. 1951. Embryonic development of the excretory system in a psilostome cercaria, a gymnocephalous (fasciolid) cercariae and in three monostome cercariae. *Trans. Am. microsc. Soc.* **70,** 95–118.

KUNTZ, R. E. 1952. Embryonic development of the excretory system in a pleurolophocercous cercaria, three stylet cercariae and in a microcaudate eucotylid cercaria. *Trans. Am. microsc. Soc.* **71,** 45–81.

KUPRIIANOVA-SHAKHMATOVA, R. A. 1957. Study on the trematode larvae of the freshwater molluscs of the central region of the Volgo. (In Russian). *Helminthologia Bratislava,* **2,** 67–76.

KURELEC, B. 1966. The enzymes of the ornithine cycle in the liver fluke *Fasciola hepatica* L. Proceedings of the First International Congress of Parasitology, Roma, 1964. Ed. A. Corradetti, Pergamon Press, London. 63–64.

KURELEC, B. and EHRLICH, I. 1963. Uber die Natur der von *Fasciola hepatica* L. *in vitro* ausgeschiedenen amino und ketosäuren. *Expl. Parasit.* **13,** 113–17.

KUROSUMI, K. 1961. Electron microscopic analysis of the secretion mechanism. *Int. Rev. Cytol.* **11,** 1–124.

LAL, M. B. 1953. A new trematode metacercaria from the eyes of Trout. *Nature, Lond.* **171,** 130.

LAL, M. B. and PREMVATI, 1955. Studies in histopathology. Changes induced by larval monostome in the digestive gland of the snail *Melanoides tuberculatus* (Müller). *Proc. Indian Acad. Sci.* **42,** Sec. B, No. 2, 293–9.

LAL, M. B. and SHRIVASTAVA, S. C. 1960. Some histochemical observations on the cuticle of *Fasciola indica* Verma, 1953. *Experientia* **16,** 185–6.

LAMPE, P. H. J. 1927. The development of *Schistosoma mansoni. Proc. R. Soc. Med.* **20,** 1510–16.

LANGERON, M. 1924. Recherches sur les Cercaires des piscines de Gafsa et enquête sur la bilharziose tunisienne (Septembre–Octobre, 1920). *Arch. Inst. Pasteur, Tunis* **13,** 19–67.

LARSON, O. R. 1961. Larval trematodes of freshwater snails of Lake Itasca, Minnesota. *Proc. Minnes. Acad. Sci.* **29,** 252–254.

LA RUE, G. R. 1927. A new species of Strigea from the herring gull, *Larus argentatus* (Pont.) with remarks on the function of the holdfast organ. *J. Parasit.* **13,** 226.

LA RUE, G. R. 1932. Morphology of *Cotylurus communis* Hughes (Trematoda: Strigeida). *Trans. Amer. Micr. Soc.* **51,** 28–47.

LA RUE, G. R. 1938. Life history studies and their relation to problems in taxonomy of digenetic trematodes. *J. Parasit.* **24,** 1–11.

LA RUE, G. R. 1951. Host-parasite relationships among the digenetic trematodes. *J. Parasit.* **37,** 333–42.

LA RUE, G. R. 1957. The classification of digenetic trematoda: A review and a new system. *Expl. Parasit.* **6,** 306–49.

LA RUE, G. R., BUTLER, E. P. and BERKHOUT, P. G. 1926. Studies on the trematode family Strigeidae (Holostomidae). No. IV. The eye of fishes, an important habitat for larval Strigeidae. *Trans. Am. microsc. Soc.* **45,** 282–8.

LAURIE, J. S. 1961. Carbohydrate absorption in cestodes from elasmobranch fishes. *Comp. Biochem. Physiol.* **4,** 63–71.

LAUTENSCHLAGER, E. W. 1961. Ultrastructure of the cuticular region and flame-cell system of the metacercaria *Diplostomulum trituri. J. Parasit.* **47,** (4) Suppl. p. 46, No. 106.

LAUTENSCHLAGER, E. W. and CARDELL, R. R. 1959. Ultrastructure of the surface layers of a strigeid metacercaria, *Diplostomulum trituri* and an echinostome cercaria, *Himasthla quissetensis. J. Parasit.* **45,** (4) Suppl. p. 18.

LAUTENSCHLAGER, E. W. and CARDELL, R. R. 1961. Ultrastructure of the cuticular region and flame-cell system of the metacercaria *Diplostomulum trituri. J. Parasit.* **47,** (4) Suppl. p. 46.

LAZARUS, M. 1950. The respiratory metabolism of helminths. *Aust. J. Sci. Res.* **3,** 245–50.

LEBOUR, M. V. 1916. Medusae as hosts for larval trematodes. *J. mar. biol. Ass. U.K.* **11,** 57–9.

LEBOUR, M. R. 1917. Some parasites of *Sagitta bipunctata. J. mar. biol. Ass. U.K.* **11,** 201–6.

LEE, D. L. 1962. Studies on the function of the pseudosuckers and holdfastorgan of *Diplostomum phoxini* Faust (Strigeida, Trematoda). *Parasitology* **52,** 103–12.

LEE, D. L. 1966. The structure and composition of the helminth cuticle. p. 187–254 In *Advances in Parasitology* V. 4. Ed. B. Dawes, Academic Press, New York.

LEE, D. L. and SMITH, M. H. 1965. Hemoglobins of parasitic animals. *Expl. Parasit.* **16,** 392–424.

LEE, C. L. and LEWERT, R. M. 1957. Studies on the presence of mucopolysaccharidase in penetrating helminth larvae. *J. infect. Dis.* **101,** 287–94.

LEE, H. F. 1962. Life history of *Heterobilharzia americana* Price 1929, a schistosome of the racoon and other mammals in south-eastern united states. *J. Parasit.* **48,** 728–39.

LEES, E. 1962. The incidence of helminth parasites in a particular frog population. *Parasitology* **52,** 95–102.

LENGY, J. 1960. Study on *Paramphistomum microbothrium* Fischoeder, 1901, a rumen parasite of cattle in Israel. *Bull. Res. Coun., Israel, Sect.* 9b, 71–130.

LENHOFF, H. M., SCHROEDER, R. and LEIGH, W. H. 1960. The collagen-like nature of metacercarial cysts of a new species of *Ascocotyle*. *J. Parasit.* **46,** (5), Suppl., No. 106, p. 36.

LENTZ, T. L. 1966. *The cell biology of Hydra.* North-Holland Publishing Company, Amsterdam, 199 pp.

LEVINE, M. D., GARZOLI, R. E., KUNTZ, R. E. and KILLOUGH, J. H. 1948. On the demonstration of hyaluronidase in the cercaria of *Schistosoma mansoni*. *J. Parasit.* **34,** 158–61.

LEWERT, R. M. 1958. Invasiveness of helminth larvae. Rice Institute Pamphlet, **45** (1) 97–113.

LEWERT, R. M. and DUSANIC, D. G. 1961. Effects of a symmetrical diaminodibenzylalkane on alkaline phosphatase of *Schistosoma mansoni*. *J. infect. Dis.* **109,** 85–9.

LEWERT, R. M. and HOPKINS, D. R. 1964. Histochemical demonstration of calcium in pre-acetabular glands of cercariae and the role of calcium ions in invasiveness. *J. Parasit.* **50** (3) No. 44, p. 30.

LEWERT, R. M. and HOPKINS, D. R. 1965. Cholinesterase activity in *Schistosoma mansoni* cercariae. *J. Parasit.* **51,** 616.

LEWERT, R. M. and LEE, C. L. 1954. Studies on the passage of helminth larvae through host tissues. I. Histochemical studies on extracellular changes caused by penetrating larvae. II. Enzymatic activity of larvae *in vitro* and *in vivo*. *J. infect. Dis.* **95,** 13–51.

LEWERT, R. M. and LEE, C. L. 1956. Quantitative studies of the collagenase-like enzymes of cercariae of *Schistosoma mansoni* and the larvae of *Strongyloides ratti*. *J. infect. Dis.* **99,** 1–14.

LEWERT, R. M. and PARA, B. J. 1965. Carbon-14 labelled *S. mansoni*. *J. Parasit.* **51,** (2) Suppl. No. 93, p. 42–3.

LE ZOTTE, L. A. 1954. Studies on marine digenetic trematodes of Puerto Rico: the family Bivesiculidae, its biology and affinities. *J. Parasit.* **40,** 148–162.

LICHTENBERG, F. V. and SADUN, E. H. 1963. Parasite migration and host reaction in mice exposed to irradiated cercariae of *Schistosoma mansoni*. *Expl. Parasit.* **13,** 256–65.

LIE, J. K. 1966a. Antagonistic interaction between *Schistosoma mansoni* sporocysts and Echinostome rediae in the snail *Australorbis glabratus*. *Nature, Lond.* **211,** 1213–15.

LIE, J. K. 1966b. Studies on Echinostomatidae (Trematoda) in Malaya. XIII. Integumentary papillae on six species of echinostome cercariae. *J. Parasit.* **52,** 1041–8.

LIE, J. K., BASCH, P. F. and UMATHEVY, T. 1965. Antagonism between two species of larval trematodes in the same snail. *Nature, Lond.* **206,** 422–3.

LIE, J. K., BASCH, P. F. and UMATHEVY, T. 1966. Studies on Echinostomatidae (Trematoda) in Malaya. XLL. Antagonism between two species of echinostome trematodes in the same lymnaeid snail. *J. Parasit.* **52,** 454–7.

LINCICOME, D. R. 1962. Frontiers in Research in Parasitism: 1. Cellular and Humoral Reactions in experimental Schistosomiasis. *Expl. Parasit.* **12,** 211–40.

LLEWELYN, C. 1957. The morphology, biology and incidence of the larval digenea parasitic in certain fresh water molluscs. Ph.D. thesis, University of Wales, 1–302.

LLEWELLYN, J. 1954. Observations on the food and gut pigment of the Polyopisthocotylea (Trematoda: Monogenea). *Parasitology* **44,** 428–37.

LLEWELLYN, J. 1956a. The host specificity, microecology, adhesive attitudes and comparative morphology of some trematode gill parasites. *J. mar. biol. Ass. U.K.* **35,** 113–27.

LLEWELLYN, J. 1956b. The adhesive mechanisms of monogenetic trematodes: the attachment of *Plectanocotyle gurnardi* (v. Ben. and Hesse) to the gills of *Trigla*. *J. mar. biol. Ass. U.K.* **35,** 507–14.

LLEWELLYN, J. 1957a. The larvae of some monogenetic trematode parasites of Plymouth fishes. *J. mar. biol. Ass. U.K.* **36,** 243–59.

LLEWELLYN, J. 1957b. The mechanism of the attachment of *Kuhnia scombri* (Kuhn, 1829) (Trematoda: Monogenea) to the gills of its host *Scomber scombrusi* including a note on the taxonomy of the parasite. *Parasitology* **47,** 30–9.

LLEWELLYN, J. 1957c. Host-specificity in monogenetic trematodes: In First Symposium on host specificity among parasites of vertebrates, Neuchâtel. Univ. of Neuchâtel, 199–211.

LLEWELLYN, J. 1958. The adhesive mechanisms of monogenetic trematodes; the attachment of species of the Diclidophoridae to the gills of gadoid fishes. *J. mar. biol. Ass. U.K.* **37,** 67–79.

LLEWELLYN, J. 1960. Amphibdellid (Monogenean) Parasites of Electric Rays (Torpedinidae). *J. mar. biol. Ass. U.K.* **39,** 561–89.

LLEWELLYN, J. 1962a. The life histories and population dynamics of monogenean gill parasites of *Trachurus trachurus* (L.). *J. mar. biol. Ass. U.K.* **42,** 587–600.

LLEWELLYN, J. 1962b. The effects of the host and its habits on the morphology and life-cycle of a monogenean parasite. In: Symposium on Helminths Bound to Aquatic Conditions (1962), Prague: Parasitologicky Ustav Cskoslovenske Akademie Ved.

LLEWELLYN, J. 1963. Larvae and larval development of Monogeneans. p. 287–326. *Advances in Parasitology*, Vol. I. Ed. B. Dawes, Academic Press, London and New York, pp. 347.

LLEWELLYN, J. 1965. The evolution of Parasitic Platyhelminthes (p. 47–78). Evolution of Parasites, Ed. A. E. R. Taylor, Third Symposium of the British Society for Parasitology, Blackwell Scientific Publications, Oxford, 1965. pp. 133.

LLEWELLYN, J. 1968. Larvae and larval development of Monogenea. p. 373–383. In *Advances in Parasitology*, Vol. 6 Ed. B. Dawes, Academic Press, London and New York, pp. 416.

LLEWELLYN, J. and EUZET, L. 1964. Spermatophores in the monogenean *Entobdella diadema* Monticelli from the skin of sting-rays, with a note on the taxonomy of the parasite. *Parasitology* **54,** 337–44.

LLEWELLYN, J. and OWEN, I. L. 1960. The attachment of the monogenean *Discocotyle sagitta* Leuchart to the gills of *Salmo trutto* L. *Parasitology* **50,** 51–9.

LOWE, C. Y. 1966. Comparative studies of the lymphatic system of four species of Amphistomes. *Z. Parasitenk.* **27,** 169.

LÜHE, M. 1909. Trematodes: In Die Süsswasserfauna Deutschlands. Heft, 17, 217 pp.

LUMSDEN, R. D. 1965. Macromolecular structure of glycogen in some Cyclophyllidean and Trypanorhynch Cestodes. *J. Parasit.* **51,** 501–15.

LUMSDEN, R. D. 1966. Fine structure of the medullary parenchymal cells of a Trypanorhynch Cestode, *Lacistorhyncus tenuis* (v. Beneden, 1858), with emphasis on specializations for glycogen metabolism. *J. Parasit.* **52,** 417–27.

LUMSDEN, R. D., GONZALEZ, G., MILLS, R. R. and VILES, J. M. 1968. Cytological studies on the absorptive surfaces of cestodes III. Hydrolysis of phosphate esters. *J. Parasit.* **54,** 524–35.

LUMSDEN, R. D. and HARRINGTON, G. W. 1966. Incorporation of linoleic acid by the cestode *Hymenolepis diminuta*, *J. Parasit.* **52,** 695–700.

LUTTA, A. S. 1939. The dynamics of the glycogen and fat in parasitic worms. Uchen. Zapiski Leningrad. Gos. Univ. Ser. 11, pp. 129–171 (In Russian).

LYAIMAN, E. M. 1949. Kurs boleznei ryb (A manual of Fish Diseases). Moskva, Pishchepromizdat. 1949.

LYNCH, J. E. 1933. The miracidium of *Heronimus chelydrae* MacCullum. *Q. Jl. Microsc. Sci.* **76,** 13–33.

LYNCH, D. L. and BOGITSH, B. J. 1962. The chemical nature of metacercarial cysts. II. Biochemical investigations on the cyst of *Posthodiplostomum minimum*. *J. Parasit.* **48,** 241–3.

LYONS, K. M. 1966. The chemical nature and evolutionary significance of monogenean attachment sclerites. *Parasitology* **56,** 63–100.

LYSAGHT, A. M. 1941. The biology and trematode parasites of the gastropod *Littorina neritoides* (L.) on the Plymouth breakwater. *J. mar. biol. Ass. U.K.* **25,** 41–67.

MA, L. 1963. Trace elements and polyphenol oxidase in *Clonorchis sinensis*. *J. Parasit.* **49,** 197–203.

MA, L. 1964. Acid Phosphatase in *Clonorchis sinensis*. *J. Parasit.* **50,** 235–40.

MACCULLUM, G. A. 1927. A new ectoparasitic trematode *Epibdella melleni* sp. nov. *Zoopathologica.* **1,** 291–300.

MACINNIS, A. J. 1965. Responses of *Schistosoma mansoni* miracidia to chemical attractants. *J. Parasit.* **51,** 731–46.

MACY, R. W. 1960. The life cycle of *Plagiorchis vespertilionis parorchis*, n. ssp., (Trematoda: Plagiorchiidae), and observations on the effects of light on the emergence of the cercaria. *J. Parasit.* **46,** 337–45.

MAGALHAẼS, FILMO, A., KRUPP, I. M. and MALEK, E. A. 1965. Localization of antigen and presence of antibody in tissues of mice infected with *Schistosoma mansoni* as indicated by fluorescent antibody technics. *Am. J. trop. Med. Hyg.* **14,** 84–99.

MAGATH, T. B. and MATHIESON, D. R. 1946. Factors affecting the hatching of ova of *Schistosoma japonicum*. *J. Parasit.* **32,** 64–8.

MANDLOWITZ, S., DUSANIC, D. and LEWERT, R. M. 1960. Peptidase and lipase activity of extracts of *Schistosoma mansoni* cercariae. *J. Parasit.* **46,** 89–90.

MANSON-BAHR, P. H. and FAIRLEY, N. H. 1920. Observations on bilharziasis amongst the Egyptian expeditionary force. *Parasitology* **12,** 33–71.

MANSOUR, T. E. 1959. Studies on the carbohydrate metabolism of the liver fluke, *Fasciola hepatica*. *Biochem. Biophys. Acta.* **34,** 456–464.

MANTER, H. W. 1926. Some North American fish trematodes. *Illinois biol. Monogr.* **10,** 1–138.

MARTIN, W. E. 1945. Two new species of marine cercariae. *Trans. Am. Microsc. Soc.* **64,** 203–12.

MARTIN, W. E. and BILS, R. F. 1964. Trematode excretory concretions: Formation and fine structure. *J. Parasit.* **50,** 337–44.

MATHIAS, P. 1925. Recherches expérimentales sur le cycle évolutif de quelques trématodes. *Bull. Biol.* **49,** 1–123.

MATTES, O. 1926. Zur Biologie der Larven Entwichlung von *Fasciola hepatica*, Besonders uber den Einfluss der Wasserstoffionenkonzentration Auf Das Ausschlupfen Der Miracidien. *Zool. Anz.* **69,** 138–56.

MCDANIEL, J. S. 1966. Excystment of *Cryptocotyle lingua* metacercariae. *Biol. Bull. mar. biol. Lab. Woods Hole,* **130,** 369–77.

MCDANIEL, J. S. and DIXON, K. E. 1967. Utilization of exogenous glucose by the rediae of *Parorchis acanthus* (Digenea: Philophthalmidae) and *Cryptocotyle lingua* (Digenea: Heterophyidae). *Biol. Bull. nar. biol. Lab.* **133,** 591–9.

MCQUAY, R. M. 1953. Studies on variability in the susceptibility of an American snail *Tropicorbis havanensis* to infection with the Puerto Rican strain of *Schistosoma mansoni. Trans. R. Soc. trop. Med. Hyg.* **47,** 56–61.

MERCER, E. H. and DIXON, K. E. 1967. The fine structure of the cystogenic cells of the cercaria of *Fasciola hepatica* L. *Z. Zellforsch.* **77,** 331–44.

MEYERHOF, E. and ROTHSCHILD, M. 1940. A prolific trematode. *Nature, Lond.* **146,** 367.

MICHELSON, E. H. 1963. Development and specificity of miracidial immobilizing substances in extracts of the snail *Australorbis glabratus* exposed to various agents. *Ann. N.Y. Acad. Sci.* **113,** 486–91.

MICHELSON, E. W. 1964. Miracidia immobilizing substances in extracts prepared from snails infected with *Schistosoma mansoni. Am. J. trop. Med. Hyg.* **13,** 36–42.

MILLER, H. M. 1926a. Behaviour studies on Tortugas larval trematodes with notes on the morphology of two additional species. Carnegie Inst., Washington. Year Book. 1925–26, 243–7.

MILLER, H. M. 1926b. Comparative studies on furcocercous cercariae. *Illinois biol. Monogr.* **10,** 1–112.

MILLER, H. M. 1928. Variety of behaviour of larval trematodes. *Science* **68,** 117–118.

MILLER, H. M. 1930. Continuation of study on behaviour and reactions of marine cercariae from Tortugas. Carnegie Inst., Washington. Year Book 28, 292–294.

MILLER, H. M. and MAHAFFY, E. E. 1930. Reactions of *Cercaria hamata* to light and to mechanical stimuli. *Biol. Bull. mar. biol. Lab., Woods Hole,* **59,** 95–103.

MILLER, H. M. and MCCOY, O. R. 1929. An experimental study of the behaviour of *Cercaria floridensis* in relation to its fish intermediate host. Carnegie Inst., Washington. Year book. 1928–29, **28,** 295–7.

MILLER, H. M. and MCCOY, O. R. 1930. An experimental study of the behaviour of *Cercaria floridensis* in relation to its fish intermediate host. *J. Parasit.* **16,** 185–97.

MILLER, H. M. and NORTHUP, F. E. 1926. The seasonal infestation of *Nassa obsoleta* (Say) with larval trematodes. *Biol. Bull. mar. biol. Lab., Woods Hole,* **50,** 490–508.

MONNÉ, L. 1959. On the external cuticles of various helminths and their role in the host-parasite relationship. *Ark. Zool.* **12,** 343–58.

MOORE, D. V. 1964. Efficacy of mass exposure of *Australorbis glabratus* to *Schistosoma mansoni. J. Parasit.* **50,** 798–9.

MOOSE, J. W. 1963. Growth Inhibition of young *Oncomelania nosophora* exposed to *Schistosoma japonicum. J. Parasit.* **49,** 151–2.

MOOSE, J. W. and WILLIAMS, J. E. 1964. The susceptibility of geographical races of *Oncomelania formosana* to infection with human strains of *Schistosoma japonicum.* 406th *Med. Lab. Res. Report.* 30th June, 1964.

MORITA, M. 1965. Electron microscope studies on Planaria. 1. Fine structure of muscle fiber in the head of the planarian *Dugesia dorotocephala. J. Ultrastruct. Res.* **13,** 383–95.

MORITA, M. and BEST, J. B. 1965. Electron microscopic studies on Planaria. II. Fine structure of the neurosecretory system in the planarian *Dugesia dorotocephala. J. Ultrastruct. Res.* **13,** 396–408.

MORRIS, G. P. 1968. Fine structure of the gut epithelium of *Schistosoma mansoni. Experientia* **24,** 480–2.

MORRIS, G. P. and THREADGOLD, L. T. 1967. A presumed sensory structure associated with the tegument of *Schistosoma mansoni. J. Parasit.* **53,** 537–9.

MORRIS, G. P. and THREADGOLD, L. T. 1968. Ultrastructure of adult *Schistosoma mansoni. J. Parasit.* **54,** 15–27.

MOULDER, J. W. 1950. The oxygen requirement of parasites. *J. Parasit.* **36,** 193–200.

MUFTIC, M. 1969. Metamorphosis of miracidia into cercariae of *Schistosoma mansoni in vitro. Parasitology* **59,** 365–71.

MÜHLING, P. 1896. Beiträge zur Kenntnis einiger Trematoden. *Zentbl. Bakt. Parasitkde Abt. I,* **20,** 588–90.

MULLER, R. 1966. Glycogen metabolism in the frog lung fluke *Haplometra cylindracea* (Zeder, 1800). *J. Parasit.* **52,** 50–3.

MÜLLER, W. 1923. Die Nahrung von *Fasciola hepatica* und ihre Verdauung. *Zool. Anz.* **57,** 273–81.

MYER, D. G. 1960. On the life history of *Mesostephanus kentuckiensis* (Cable, 1935) N. Comb. (Trematoda: Cyathocotylidae). *J. Parasit.* **46,** 819–32.

NADAKAL, A. M. 1960a. Chemical nature of cercariae eye-spot and other tissue pigments. *J. Parasit.* **46,** 475–81.

NADAKAL, A. M. 1960b. Types and sources of pigments in certain species of larval trematodes. *J. Parasit.* **46,** 777–86.

NADAKAL, A. M. 1960c. Biochromes in certain species of larval trematodes. *J. Parasit.* **46,** (5) Suppl. No. 89, p. 31.

NAJARIAN, H. H. 1953. The life history of *Echinoparyphium flexum* (Linton, 1892) (Dietz, 1910) (Trematoda: Echinostomatidae) *Science, N.Y.* **117,** 564–5.

NAJARIAN, H. H. 1961a. The life cycle of *Plagiorchis goodmani* N. Comb. (Trematoda: Plagiorchiidae). *J. Parasit.* **47,** 625–34.

NAJARIAN, H. H. 1961b. Egg-laying capacity of the snail *Bulinus truncatus* in relation to infection with *Schistosoma haematobium*. *Tex. Rep. Biol. Med.* **19,** 327–31.

NASIR, P. 1960a. Trematode parasites of snails from Edgbaston Pool: the life history of the strigeoid *Cotylurus brevis* Dubois and Rausch, 1950. *Parasitology* **50,** 551–75.

NASIR, P. 1960b. Studies on the life history of *Echinostoma nudicaudatum* n. sp. (Echinostomatidae: Trematoda). *J. Parasit.* **46,** 833–47.

NASIR, R. and ERASMUS, D. A. 1964. A key to the cercaria from British Freshwater Molluscs. *J. Helminth.* **38,** 245–68.

NEGUS, M. R. S. 1968. The nutrition of sporocysts of the trematode *Cercaria doricha* Rothschild, 1935 in the molluscan host *Turritella communis* Risso. *Parasitology* **58,** 355–66.

NEUHAUS, W. 1953. Ueber den chemischen Sinn der Miracidien von *Fasciola hepatica*. *Ztsch. Parasitenk.* **15,** 476–490.

NEWTON, W. L. 1952. The comparative tissue reaction of two strains of *Australorbis glabratus* to infection with *Schistosoma mansoni*. *J. Parasit.* **38,** 362–6.

NEWTON, W. L. 1953. The inheritance of susceptibility to infection with *Schistosoma mansoni* in *Australorbis glabratus* *Expl. Parasit.* **2,** 242–57.

NEWTON, W. L. 1954. Tissue response to *Schistosoma mansoni* in second generation snails from a cross between two strains of *Australorbis glabratus*. *J. Parasit.* **40,** 352–5.

NIMMO-SMITH, R. H. and STANDEN, O. D. 1963. Phosphomonoesterases of *Schistosoma mansoni*. *Expl. Parasit.*, **13,** 305–22.

NOBLE, G. A. 1967. A forty-foot fluke. *J. Parasit.* **53,** 645.

NOLF, L. O. and CORT, W. W. 1933. On immunity reactions of snails to the penetration of the cercariae of the strigeid trematode, *Cotylurus flabelliformis* (Faust) *J. Parasit.* **20,** 38–48.

NOLLEN, P. M. 1968a. Autoradiographic studies on reproduction in *Philophthalmus megalurus* (Cort, 1914) (Trematoda). *J. Parasit.* **54,** 43–8.

NOLLEN, P. M. 1968b. Uptake and incorporation of glucose, tyrosine, leucine and thymidine by adult *Philophthalmus megalurus* (Cort, 1914) (Trematoda) as determined by autoradiography. *J. Parasit.* **54,** 295–304.

ÖHMAN, C. 1965. The structure and function of the adhesive organ in strigeid trematodes. Part II. *Diplostomum spathaceum* Braun, 1893. *Parasitology* **55,** 481–502.

ÖHMAN, C. 1966a. The structure and function of the adhesive organ in strigeid trematodes Part III. *Apatemon gracilis minor* Yamaguti, 1933. *Parasitology* **56,** 209–26.

ÖHMAN, C. 1966b. The structure and function of the adhesive organ in strigeid trematodes. Part IV. *Holostephanu lühei* Szidat, 1936. *Parasitology* **56,** 481–91.

OLIVER, J. H. JR. and SHORT, R. B. 1956. Longevity of miracidia of *Schistosomatium douthitti*. *Expl. Parasit.* **5,** 238–49.

OLIVIER, L. J. 1940. Life history studies on two strigeid trematodes of the Douglas Lake Region, Michigan. *J. Parasit.* **26,** 447–77.

OLIVIER, L. J. 1951. The influence of light on the emergence of *Schistosomatium douthitti* cercariae from their snail host. *J. Parasit.* **37,** 201–4.

OLIVIER, L. J. and MAO, C. P. 1949. The early larval stage of *Schistosoma mansoni* Sambon, 1907 in the snail host, *Australorbis glabratus* (Say, 1818). *J. Parasit.* **35,** 267–75.

OLIVIER, L., VON BRAND, T., and MEHLMAN, B. 1953. The influence of lack of oxygen on *Schistosoma mansoni* cercariae and on infected *Australorbis glabratus*. *Expl. Parasit.* **2,** 258–70.

OLLERENSHAW, C. B. 1958a. Climate and liver fluke disease in Anglesey. *Trans. R. Soc. trop. Med. Hyg.* **52,** 303.

OLLERESHAW, C. B. 1958b. Climate and liver flukes. *Agriculture, Lond.* **65,** 231–5.

OLLERENSHAW, C. B. 1959. Ecology of liver fluke (*Fasciola hepatica*). *Vet. Rec.* **71,** 957–63.

OLLERENSHAW, C. B. 1964. The effect of temperature on the development of *Fasciola hepatica* in the intermediate host *Lymnaea truncatula* and its influence on the epidemiology of the disease. Paper delivered at the First International Congress of Parasitology, Rome, 1964.

OLLERENSHAW, C. B. and ROWLANDS, W. T. 1959. A method of forecasting the incidence of fascioliasis in Anglesey. *Vet. Rec.* **71,** 591–598.

OLSON, R. E. 1966. Some experimental fish hosts of the strigeid trematode *Bolbophorus confusus* and effects of temperature on the cercaria and metacercaria. *J. Parasit.* **52,** 327–34.

ONORATO, A. R. and STUNKARD, H. W. 1931. The effect of certain environmental factors on the development and hatchings of the eggs of blood flukes. *Biol. Bull. mar. biol. Lab., Woods Hole,* **61,** 120–32.

ORTIGOZA, R. O. and HALL, J. E. 1963a. Studies on the glandular apparatus and secretions of virgulate Xiphidiocercariae. I. Intravital and Histochemical data. *Expl. Parasit.* **14,** 160–77.

ORTIGOZA, R. O. and HALL, J. E. 1963b. Studies on the glandular apparatus and secretions of virgulate Xiphidiocercariae. II. Qualitative chemical analysis of secretions. *Expl. Parasit.* **14,** 178–85.

ORTNER-SCHÖNBACH, P. 1913. Zut Morphologie des Glykogens bei Trematoden und Cestoden. *Arch. Zellforsch.* **11,** 413–449.

OSHIMA, T., YOSHIDA, Y., and KIHATA, M. 1958a. Studies on the excystation of *Paragonimus westermani.* I. Especially on the effect of bile salts. *Bull. Inst. Pub. Health, Tokyo,* **7,** 256–269.

OSHIMA, T. and KIHATA, M. 1958b. Studies of the excystation of the metacercaria of *Paragonimus westermani* II. Influence of pepsin pretreatment on the effect of bile salts. *Bull. Inst. Pub. Health, Tokyo,* **7,** 270–274.

PALING, J. E. 1965. The population dynamics of the monogenean gill parasite *Discocotyle sagittata* Leuckart on Windermere trout, *Salmo trutta.* L. *Parasitology* **55,** 607–94.

PALING, J. E. 1966. The functional morphology of the genitalia of the spermatophore-producing monogenean parasite (*Diplectanum aequans*) (Wagener) Diesing, with a note on the copulation of the parasite. *Parasitology* **56,** 367–83.

PALM, V. 1962a. Glykogenund Fettstoffwechsel bei *Cercaria limnaea ovata. Ztsch. Parasitenk.* **22,** 261–266.

PALM, V. 1962b. Glykogen und Fett bei Trematodenlarvenstadien am Beispiel von *Dolichosaccus rastellus* und *Haplometra cylindraceum* (Plagiorchiida). *Acta Parasitol. Polon.* **10,** 117–123.

PALMER, E. D. 1939. Diplostomiasis, a hatchery disease of fresh-water fishes new to North America. *Prog. Fish Cult* **45,** 41–47.

PAN, C. 1963. Generalized and focal tissue responses in the snail *Australorbis glabratus,* infected with *Schistosoma mansoni. Ann. N.Y. Acad. Sci.,* **113,** 475–85.

PANITZ, E. and KNAPP, S. E. 1967. Acetylcholinesterase activity in *Fasciola hepatica* miracidia. *J. Parasit.* **53,** 354.

PANTELOURIS, E. M. 1964a. Sulfur uptake by *Fasciola hepatica* L. *Life Sci.* **3,** 1–5.

PANTELOURIS, E. M. 1964b. Localization of glycogen in *Fasciola hepatica* L. and an effect of Insulin. *J. Helminth* **38,** 283–6.

PANTELOURIS, E. M. 1965a. Utilization of methionine by the liver fluke, *Fasciola hepatica. Res. Vet. Sci.* **6,** 334–336.

PANTELOURIS, E. M. 1965b. The common Liver Fluke. Pergamon Press, London, 259 p.

PANTELOURIS, E. M. 1967. Esterases of *Fasciola hepatica* L. *Res. vet. Sci.* **8,** 157–9.

PANTELOURIS, E. M. and GRESSON, R. A. R. 1960. Autoradiographic studies on *Fasciola hepatica* L. *Parasitology* **50,** 165–9.

PANTELOURIS, E. M. and HALE, P. 1962. Iron and Vitamin C in *Fasciola hepatica* L. *Res. vet. Sci.* **3,** 300–303.

PANTELOURIS, E. M. and THREADGOLD, L. T. 1963. The excretory system of the adult *Fasciola hepatica* L. La Cellule, LXIV, 63–67.

PAPERNA, I. 1963. *Enterogyrus cichlidarum* n. gen. n. sp., a monogenetic trematode parasitic in the intestine of a fish. *Bull. Res. Coun. Israel, Sect. B,* 183–7.

PARENSE, W. L. and CORRÉA, L. R. 1963. Variation in susceptibility of populations of *Australorbis glabratus* to a strain of *Schistosoma mansoni. Rev. Inst. Med. Trop. São Paulo,* **5,** 15–22.

PAUL, D. 1934. Beobachtungen über die Darmparasiten silesischer Anuren. *Ztsch. Parasitenk.* **7,** 172–197.

PEARSE, A. G. E. 1960. *Histochemistry, theoretical and applied.* 2nd ed. 998 pp. London. J. and A. Churchill Ltd.

PEARSON, J. C. 1956. Studies on the life cycles and morphology of the larval stages of *Alaria arisaemoides* Augustine and Uribe, 1927 and *Alaria canis* La Rue and Fallis, 1936 (Trematoda: Diplostomidae). *Can. J. Zool.* **34,** 295–387.

PEARSON, J. C. 1959. Observations on the morphology and life-cycle of *Strigea elegans* Chandler and Rausch, 1947 (Trematoda: Strigeidae). *J. Parasit.* **45,** 155–74.

PEARSON, J. C. 1961. Observations on the morphology and life-cycle of *Neodiplostomum intermedium* (Trematoda: Diplostomatidae). *Parasitology* **51,** 133–72.

PEDERSEN, K. J. 1961. Studies on the nature of planarian connective tissue. *Zeit. fur. Zellforch.* **53,** 569–608.

PELSENEER, P. 1906. Trématodes parasites de mollusques marins. *Bull. scient. Fr. Belg.* **40,** 161–86.

PELSENEER, P. 1928. Les parasites des mollusques et les mollusques parasites. *Bull. Soc. zool. Fr.* **53,** 158–89.

PENNOIT-DE COOMAN, E. and VAN GREMBERGEN, G. 1942. Vergelijkend onderzoek van het fermentensysteem by vrÿlevende en parasitaire Plathelminthen. *Verband. Koninkl. Vlaamsche Akad., Wetensch.* **4,** 7–77.

PEPLER, W. J. 1958. Histochemical demonstration of an acetylcholinesterase in the ova of *Schistosoma mansoni. J. Histochem. Cytochem.* **6,** 139–41.

PESIGAN, T. P., HAIRSTON, N. G., JAUREGUI, J. J., GARCIA, E. G., SANTOS, A. T., SANTOS, B. C. and BESA, A. A. 1958. Studies on *Schistosoma japonicum* infections in the Phillipines. 2. The molluscan host. *Bull. Wld. Hlth. Org.* **18,** 481–578.

PIEPER, M. B. 1953. The life history and germ cell cycle of *Spirorchis artericola* (Ward, 1921). *J. Parasit.* **39,** 310–25.

PIKE, A. W. 1965. The morphology, histology and ecology of certain trematode and cestode parasites in the fauna of selected fresh-water environments. Ph.D. thesis, Univ. of Wales.

PIKE, A. W. and ERASMUS, D. A. 1967. The formation, structure and histochemistry of the metacercarial cyst of three species of digenetic trematode. *Parasitology* **57,** (4), 683–94.

PINO, E. CONDE-DEL., PEREZ-VILAR, M., CINTRON-RIVERA, A. A. and SEÑERIZ, R. 1966. Studies in *Schistosoma mansoni*. I. Malic and Lactic Dehydrogenase of Adult Worms and Cercariae. *Expl. Parasit.* **18,** 320–6.

PIRILÄ, V. and WIKGREN, B. 1957. Cases of swimmers itch in Finland. *Acta, Derm-Vener.* **37,** 140–148.

POND, G. G. 1964. Comparative ultrastructure of photoreceptors in three types of ocellate cercariae. *J. Parasit.* **50,** (3), Suppl. No. 93, p. 43–4.

POND, G. G. and CABLE, R. M. 1966. Fine structure of photoreceptors in three types of Ocellate cercariae. *J. Parasit.* **52,** 483–93.

PRATT, I. and BARTON, G. D. 1941. The effects of four species of larval trematodes upon the liver and ovotestis of the snail *Stagnicola emarginata angulata* (Sowerby). *J. Parasit.* **27,** 283–8.

PRATT, I. and LINDQUIST, W. D. 1943. The modification of the digestive gland tubules in the snail *Stagnicola* following parasitization. *J. Parasit.*, **29,** 176–81.

PREMVATI, 1955. *Cercaria duplicata* n. sp. from the snail *Melanoides tuberculatis* (Muller). *J. Zoool. Soc., India,* **7,** 13–24.

PRENENT, M. 1922. Recherches sur le parenchyme des platyhelminthes, essai d'histologie comparée. *Arch. Morphol.* **5,** 1–474.

PRICE, E. W. 1931. Life history of *Schistosomatium douthitti* (Cort). *Am. J. Hyg.*, **13,** 685–727.

PROBERT, A. J. 1963. Biological studies on some helminths parasitic in the fresh water fauna of South Wales. Ph.D. thesis, Univ. of Wales.

PROBERT, A. J. 1966a. Histochemical studies on the rediae and cercariae of *Echinoparyphium recurvatum* Linstow. *Nature. Lond.* **210,** 550–551.

PROBERT, A. J. 1966b. Studies on the incidence of larval trematodes infecting the fresh water molluscs of Llangorse Lake, South Wales. *J. Helminth* **49,** 115–30.

PROBERT, A. J. and ERASMUS, D. A. 1965. The migration of *Cercaria X* Baylis (Strigeida) within the molluscan intermediate host *Limnaea stagnalis*. *Parasitology* **55,** 77–92.

RADKE, M. G., RITCHIE, L. S. and ROWAN, W. B. 1961. Effects of water velocities on worm burdens of animals exposed to *Schistosoma mansoni* cercariae released under laboratory and field conditions. *Expl. Parasit.* **11,** 323–31.

RANKIN, J. 1937. An ecological study of parasites of some North Carolina salamanders. *Ecol. Monogr.* **6,** 171–269.

RANKIN, J. S. 1939. Ecological studies on larval trematodes from W. Massachussetts. *J. Parasit.* **25,** 309–29.

RANZOLI, F. 1956. Cellule vitelline e ovociti in *Fasciola hepatica*. *Boll. Zool.* **23,** 559–63.

READ, C. P. 1966. Nutrition of Intestinal Helminths p. 101–126. In *Biology of Parasites* Ed. E. J. L. Soulsby, Academic Press, 1966.

READ, C. P. and YOGORE, M. 1955. Respiration of *Paragonimus westermani*. *J. Parasit.* **41,** (6), p. 28.

REES, F. G. 1931. Some observations and experiments on the biology of larval trematodes. *Parasitology* **23,** 428–40.

REES, F. G. 1932. An investigation into the occurrence, structure and life-histories of the trematode parasites of four species of Lymnaea (*L. truncatula* (Müll.), *L. pereger* (Müll.), *L. palustris* (Müll.) and *L. stagnalis* (Linne) and *Hydrobia jenkinsi* (Smith) in Glamorgan and Monmouth. *Proc. zool. Soc. Lond.* **1932,** (1) 1–32.

REES, F. G. 1934. *Cercaria patellae* Lebour, 1911, and its effects on the digestive gland and gonads of *Patella vulgata*. *Proc. zool. Soc. Lond.* 1935, 45–53.

REES, F. G. 1937. The anatomy and encystment of *Cercaria purpurae* Lebour, 1911. *Proc. zool. Soc. Lond.* **107,** 65–73.

REES, F. G. 1940. Studies on the germ cell cycle of the digenetic trematode *Parorchis acanthus* Nicoll. Part II. Structure of the miracidium and germinal development in the larval stages. *Parasitology* **32,** 372–91.

REES, F. G. 1947. A study of the effect of light, temperature and salinity on the emergence of *Cercaria purpurae* Lebour from *Nucella lapillus* (L.). *Parasitology* **38,** 228–42.

REES, F. G. 1955. The adult and diplostomulum stage (*Diplostomulum phoxini* (Faust) of *Diplostomum pelmatoides* Dubois and an experimental demonstration of part of the life-cycle. *Parasitology* **45,** 295–312.

REES, F. G. 1957. *Cercaria Diplostomi phoxini* (Faust) a furcocercaria which develops into the *Diplostomulum phoxini* in the brain of the minnow. *Parasitology* **47,** 126–37.

REES, F. G. 1966. Light and electron microscope studies of the redia of *Parorchis acanthus* Nicoll. *Parasitology* **56,** 589–602.

REES, F. G. 1967. The histochemistry of the cystogenous gland cells and cyst wall of *Parorchis acanthus* Nicoll, and some details of the morphology and fine structure of the cercaria. *Parasitology* **57,** 87–110.

REES, W. J. 1936. The effects of parasitism by larval trematodes on the tissues of *Littorina littorea* (Linne). *Proc. zool. Soc. Lond.* 1936, 357–72.

REISINGER, E. and GRAAK, B. 1962. Untersuchungen an *Codonocephalus* (Trematoda digenea: Strigeidae), Nervensystem und Paranephridialer plexus. *Ztsch. Parasitenk.* **22,** 1–42.

RENNISON, B. D. 1953. A morphological and histochemical study of egg-shell formation in *Diclidophora merlangi.* M.Sc. thesis, Univ. of Dublin.

REZNIK, G. K. 1963. Histological and histochemical investigations of the alimentary tract of *Fasciola hepatica* and *Dicrocoelium lanceatum. Tr. Vses, Inst. Gel'mintol.* **10,** 238–44. *Chem. Abstracts* 1964, **60,** 14881d.

RICHARDS, C. S. 1961. Emergence of cercariae of *Schistosoma mansoni* from *Australorbis glabratus. J. Parasit.* **47,** 428.

RIJAVEC, M. 1966. On the presence of arginase and ornithine transcarbamylase in some parasitic platyhelminthes and in different developmental stages of the liver fluke *Fasciola hepatica* L. Proceedings of the First International Congress of Parasitology, Roma, 1964. Ed. A Corradetti. Pergamon Press, London, p. 62–63.

RIJAVEC, M., KURELEC, B. and EHRLICH, I. 1962. Uber den Verbrauch der Serumalbumine *in vitro* durch *Fasciola hepatica. Biol. Glasnik.* **15,** 103–107.

ROBINSON, D. L. H. 1961. Phosphatase in *Schistosoma mansoni. Nature, Lond.* **191,** 473–4.

ROBINSON, E. J. 1949. The life history of *Postharmostomum helicis* (Leida, 1847) n.comb. (Trematoda: Brachylaemidae). *J. Parasit.* **35,** 513–33.

ROEWER, C. P. 1906. Beitrage zur Histogenese von *Cercariaeum helicis. Jena. Z. Naturw.* **41,** 185–228.

ROGERS, W. 1940. Hematological studies on the gut contents of certain nematode and trematode parasites. *J. Helminth.* **18,** 53–62.

ROGERS, W. P. 1949. On the relative importance of aerobic metabolism in small nematode parasites of the alimentary tract. 1. Oxygen tensions in the normal environment of the parasites. *Aust. J. sci. Res. Ser. B,* **2,** 157–65.

ROGERS, W. P. 1962. *The Nature of parasitism. The Relationship of some metazoan parasites to their hosts,* 287 pp. Academic Press, New York.

ROHDE, K. 1965. The cell types found in the pharynx of *Polystomoides malayi* Rohe, 1963. *Med. J. Malaya,* 20, (1), 55.

ROHDE, K. 1966. Sense receptors of *Multicotyle purvisi* Dawes 1941 (Trematoda, Aspidobothria). *Nature, Lond.* **211** 820–2.

ROHRBACHER, G. H. 1957. Observations on the survival *in vitro* of bacteria-free adult common liver fluke, *Fasciola hepatica* Linn. 1758. *J. Parasit.* **43,** 9–18.

ROSS, O. A. and BUEDING, E. 1950. Survival of *Schistosoma mansoni in vitro. Proc. Soc. exp. Biol. Med.* **73,** 179–82.

ROSS, I. C. and MCKAY, A. C. 1929. The bionomics of *Fasciola hepatica* in New South Wales and of the intermediate host *Lymnaea brazieri. Bull. Counc. sci. industr. Res. Aust.* **43,** 1–62.

ROSS, J. G., DOW, C. and TODD, J. R. 1967a. A study of *Fasciola hepatica* infections in sheep. *Vet. Rec.* **80,** 543–6.

ROSS, J. G., DOW, C. and TODD, J. R. 1967b. The pathology of *Fasciola hepatica* infection in pigs: comparison of the infection in pigs and other hosts. *Br. vet. J.* **123,** 317–22.

ROSS, J. G., TODD, J. R. and DOW, C. 1966. Single infections of calves with the liver fluke *Fasciola hepatica* (Linnaeus, 1758), *J. comp. Path.* **76,** 67–81.

ROTHMAN, A. H. 1968. Enzyme localization and colloid transport in *Haematoloechus medioplexus. J. Parasit.* **54,** 286–94.

ROTHMAN, A. H. and LEE, D. L. 1963. Histochemical demonstration of dehydrogenase activity in the cuticle of cestodes. *Expl. Parasit.* **14,** 333–6.

ROTHSCHILD, M. 1936a. Gigantism and variation in *Peringia ulvae* Pennant, 1777, caused by infection with larval trematodes. *J. mar. biol. Ass. U.K.* **30,** 537–46.

ROTHSCHILD, M. 1936b. The process of encystment of a cercaria parasitic in *Lymnaea tenera euphratica. Parasitology* **28,** 56–62.

ROTHSCHILD, M. 1938. Further observations on the effect of trematode parasites on *Peringia ulvae* (Pennant, 1777). *Novit. Zool.* **41,** 84–102.

ROTHSCHILD, M. 1941a. The effect of trematode parasites on the growth of *Littorina neritoides* (L.). *J. mar. biol. Ass. U.K.* **25,** 84–102.

ROTHSCHILD, M. 1941b. Observations on the growth and trematode infections of *Peringia ulvae* (Pennant, 1777) in a pool in the Tamar Saltings, Plymouth. *Parasitology* **33,** 406–15.

ROWAN, W. B. 1956. The mode of hatching of the egg of *Fasciola hepatica. Expl. Parasit.* **5,** 118–37.

ROWAN, W. B. 1957. The mode of hatching of the egg of *Fasciola hepatica.* II. Colloidal nature of the viscous cushion. *Expl. Parasit.* **6,** 131–42.

ROWAN, W. B. 1958. Daily periodicity of *Schistosoma mansoni* cercariae in Puerto Rican waters. *Amer. J. trop. Med. Hyg.* **7,** 374–81.

ROWAN, W. B. and GRAM, A. L. 1959. Relation of water velocity to *Schistosoma mansoni* infection in mice. *Amer. J. trop. Med. Hyg.* **8,** 630–4.

ROWCLIFFE, S. A. and OLLERENSHAW, C. B. 1960. Observations on the bionomics of the egg of *Fasciola hepatica*. *Ann. trop. Med. Parasit.* **54,** 172.

RUSHTON, W. 1937. Blindness in fresh-water fish. *Nature, Lond.* **140,** 1114.

RUSHTON, W. 1938. Blindness in fresh-water fish. *Nature, Lond.* **141,** 289.

RUSSELL, C. M. 1954. The effects of various environmental factors and the hatching of eggs of *Plagitura salamandra* Hall (Trematoda: Plagiorchiidae). *J. Parasit.* **40,** 461–4.

RUSZKOWSKI, J. S. 1931. Sur la decouverte d'un ectoparasite *Amphibdella torpedinis* dans le coeur des torpilles. *Publ. Staz. zool. Napoli,* **11,** 161–7.

RYBICKA, K. 1958. Tasiemce ptaków (excl. Anseriformes) jeziora Drużno (Parazytofauna biocenozy jeziora Drużno-częćś IV.) *Acta Parasit. Polon,* **6,** 143–178.

RYBICKA, K. 1966. Embryogenesis in cestodes p. 107–187. In *Advances in Parasitology* v. 4. Ed. B. Dawes, Academic Press, London.

SADUN, E. H. 1963. Immunization in schistosomiasis by previous exposure to homologous and heterologous cercariae by inoculation of preparations from schistosomes and by exposure to irradiated cercariae. *Ann. N.Y. Acad. Sci.* **113,** 418–39.

SADUN, E. H. and WILLIAMS, J. S. 1966. Biochemical aspects of schistosomiasis mansoni in mice in relation to worm burdens and duration of infection. *Expl. Parasit.* **18,** 266–73.

SAITO, A. 1961. Histochemical study on the digestive tract of *Fasciola hepatica*. J. Tokyo Med. Coll., **19,** 1487–1497.

SALT, G. 1963. The defence reactions of insects to metazoan parasites. *Parasitology* **53,** 527–642.

SAWADA, T., *et al* 1956. Cytochemical studies on the hepatic tissue of mice following infections with *Schistosoma japonicum*. *Am. J. trop. Med. Hyg.* **5,** 847–59.

SCHELL, S. C. 1960. A case of conjoined twin Rediae. *J. Parasit.* **46,** 448.

SCHELL, S. C. 1961. Development of mother and daughter sporocysts of *Haplometrana intestinalis* Lucher, a Plagiorchioid trematode of frogs. *J. Parasit.* **47,** 493–500.

SCHELL, S. C. 1962. Development of the sporocyst generations of *Glypthelmins quieta* (Stafford, 1900) (Trematoda: Plagiorchioidea), a parasite of frogs. *J. Parasit.* **48,** 387–94.

SCHELL, S. C. 1965. The life history of *Haematoloechus breviplexus* Stafford, 1902 (Trematoda: Haplometridae McMullen, 1937) with emphasis on the development of the sporocysts. *J. Parasit.* **51,** 587–93.

SCHREIBER, F. G. and SCHUBERT, M. 1949a. Experimental infection of the snail *Australorbis glabratus* with the trematode *Schistosoma mansoni* and the production of cercariae. *J. Parasit.* **35,** 91–100.

SCHREIBER, F. G. and SCHUBERT, M. 1949b. Results of exposure of the snail *Australorbis glabratus* to varying numbers of miracidia of *Schistosoma mansoni*. *J. Parasit.* **35,** 590–2.

SEKARDI, L. 1966. The effect of insulin on glucose uptake and glycogen synthesis in the liver fluke *Fasciola hepatica* L. p. 61–62. In Proceedings of the 1st International Congress of Parasitology, Rome 1964. Ed. A. Corradetti. Pergamon Press, Tamburini Editore.

SENFT, A. W. 1959. Electron microscope study of the integument, flame cells and gut of the human parasite, *Schistosoma mansoni*. *Biol. Bull. mar. biol. Lab. Woods Hole,* **117,** 387.

SENFT, A. W. 1963. Observations on amino acid metabolism of *Schistosoma mansoni* in a chemically defined medium. *Ann. N.Y. Acad. Sci.* **113,** 272–88.

SENFT, A. W. 1965. Recent development in the understanding of amino-acid and protein metabolism by *Schistosoma mansoni in vitro*. *Ann. trop. Med. Parasit.* **59,** 164–8.

SENFT, A. W. 1966. Studies in arginine metabolism by Schistosomes. I. Arginine uptake and lysis by *Schistosoma mansoni*. *Comp. Biochem. Physiol.* **18,** 209–16.

SENFT, A. W. 1967. Studies in arginine metabolism by schistosomes. II. Arginine depletion in mammals and snails infected with *S. mansoni* or *S. haematobium*. *Comp. Biochem. Physiol.* **21,** 299–306.

SENFT, A. W., PHILPOTT, D. E. and PELOFSKY, A. H. 1961. Electron microscope observations of the integument, flame cells, and gut of *Schistosoma mansoni*. *J. Parasit.* **47,** 217–29.

SENGER, C. M. 1954. Notes on the growth, development and survival of two echinostome trematodes. *Expl. Parasit.* **3,** 491–6.

SERKOVA, O. P. and BYCHOWSKY, B. E. 1940. *Asymphylodora progenetica* n.sp. nebst einigen Angaben uber ihre Morphologie und Entwicklungsgeschichte. *Mag. Parasitol. Inst. Zool. Acad. Sci. U.R.S.S.* **8,** 162–75.

SEWELL, M. M. H. 1963. The immunology of fasciolasis I. Autofixation of guinea-pig complement. *Immunology* **6,** 453–61.

SEWELL, M. M. H. 1966. The pathogenesis of fasciolasis. *Vet. Rec.* **78,** 98–105.

SEWELL, R. B. S. 1922. Cercariae Indicae. *Indian J. med. Res.* **10,** (Sppl.): 1–370.

SHAPIRO, J. E., HERSHENOV, B. R. and TULLOCH, G. S. 1961. The fine structure of *Haematoloechus* spermatozoon tail. *J. biophys. biochem. Cytol.* **9,** 211–17.

SHAW, J. N. 1932. Studies of the liver fluke (*Fasciola hepatica*). *J. Amer. vet. med. Ass.* **81,** 76–82.

SHORT, R. B. and MENZEL, M. Y. 1959. Chromosomes in parthenogenetic miracidia and embryonic cercariae of *Schistomatium douthitti*. *Expl. Parasit.* **8,** 249–64.

SIDDIQI, A. H. and LUTZ, P. L. 1966. Osmotic and ionic regulation in *Fasciola gigantica* (Trematoda: Digenea). *Expl. Parasit.* **19,** 348–57.

SILVERMAN, P. H. 1965. *In vitro* cultivation procedures for parasitic helminths p. 159–222. In *Advances in Parasitology* 3. Ed. B. Dawes, Academic Press, London.

SIMMS, B. T. 1932. Pathogenicity of metacercaria of *Nanophyetus salmincola* Chapin for fish hosts. *J. Parasit.* **19,** 160.

SINDEMANN, C., ROSENFIELD, A. and STROM, L. 1957. The ecology of marine dermatitis-producing schistosomes. II. Effects of certain environmental factors on emergence of cercariae of *Australobilharzia variglandis*. *J. Parasit.* **43,** 382.

SINGH, R. N. 1959. Seasonal infestation of *Indoplanorbis exustus* (Deshayes) with furcocercous cercariae. *Proc. natn. Acad. Sci. India*, **29** (2), 62–72.

SINGH, K. S. and LEWERT, R. M. 1959. Observations on the formation and chemical nature of metacercarial cysts of *Notocotylus urbanensis*. *J. infect. Dis.* **104,** 138–141.

SINITSÎN, D. F. 1911. Le generation parthénogénétique des trématodes et sa descendance dans les mollusques de la Mer Noire. *Mém. Acad. Sci., St. Pétersb.* (8) **30,** 1–127.

SMITH, J. C. 1965. Bibliography on the metabolism of Endoparasites exclusive of Arthropods, 1951–1962. *Expl. Parasit.* **16,** 236–90.

SMITHERS, S. R. 1962a. Stimulation of acquired resistance to *Schistosoma mansoni* in monkeys: Role of eggs and worms. *Expl. Parasit.* **12,** 263–73.

SMITHERS, S. R. 1962b. Acquired resistance to Bilharziasis page 239–258 in *Ciba Foundation Symposium Bilharziasis* Ed. by G. E. W. Wolstenholme and Mawe O'Connor. J. A. Churchill Ltd., London 1962.

SMITHERS, S. R. 1967. The induction and nature of antibody response to parasites, p. 43–49. In, 'Immunological aspects of parasitic infections'. PAHO. Special Session. Publication No. 150. WHO. 1967.

SMITHERS, S. R. 1968. Immunity to blood helminths, in *Immunity to Parasites*, pp. 55–66. Sixth symposium of the British Society for Parasitology. Ed. A. E. R. Taylor, Blackwell, Oxford.

SMITHERS, S. R. and OGLIVIE, B. M. 1965. Reagin-like antibodies and the passive transfer of resistance in experimental schistosomiasis. *Parasitology* **55,** 2 p.

SMITHERS, S. R., ROODYN, D. B. and WILSON, R. J. M. 1965. Biochemical and morphological characteristics of subcellular fractions of male *Schistosoma mansoni*. *Expl. Parasit.* **16,** 195–206.

SMITHERS, S. R. and TERRY, R. J. 1967. Resistance to experimental infection with *Schistosoma mansoni* in rhesus monkeys induced by the transfer of adult worms. *Trans. R. Soc. trop. Med. Hyg.* **61,** 517–33.

SMYTH, J. D. 1954. A technique for the histochemical demonstration of polyphenol oxidase and its application to egg-shell formation in Helminths and Byssus formation. *Q. Jl. microsc. Sci.* **95,** 139–52.

SMYTH, J. D. 1959. Maturation of larval pseudophyllidean cestodes and strigeid trematodes under axenic conditions; the significance of nutritional levels in platyhelminth development. *Ann. New York Acad. Sci.* **77,** 102–125.

SMYTH, J. D. 1963. The biology of cestode life-cycles. Technical Communication No. 34. Commonwealth Bureau of Helminthology, pp. 38.

SMYTH, J. D. 1966a. Genetic aspects of speciation in trematodes and cestodes: some speculations. Proceedings of the First International Congress of Parasitology (Rome, 1964). Vol. 1. Ed. A. Corradetti, Pergamon Press, p. 473–474.

SMYTH, J. D. 1966b. The Physiology of Trematodes. University Reviews in Biology No. 7, pp. 256, Oliver and Boyde, London.

SMYTH, J. D., BINGLEY, W. J. and HILL, G. R. 1945. The distribution of vitamin C in *Nyctotherus cordiformis* Ehrenberg, *Opisthioglyphe ranae* Frolich, and *Toxocara canis* Werner. *J. exp. Biol.* **21,** 13–16.

SMYTH, J. D. and CLEGG, J. A. 1959. Egg shell formation in trematodes and cestodes. *Expl. Parasit.* **8,** 286–323.

SNYDER, R. W. JR. and CHENG, T. C. 1961. The effect of the larvae of *Glypthelmins pennsylvaniensis* (Trematoda: Brachycoeliidae) on glycogen deposition in the hepatopancreas of *Helisoma trivolvis* (Say). *J. Parasit.* **47,** 52.

SOULSBY, E. J. L. 1967. Lymphocyte, macrophage and other cell reactions to parasites, p. 66–84. In 'Immunological aspects of parasitic infections'. PAHO. Special Session. Publication No. 150. WHO, 1967.

SOUTHWELL, T. and KIRSHNER, A. 1937. Parasitic infections in a swan and in a brown trout. *Ann. trop. Med. Parasit.* **31,** 427–34.

STANDEN, O. D. 1951. The effects of temperature, light and salinity upon the hatching of the ova of *Schistosoma mansoni*. *Trans. R. Soc. trop. Med. Hyg.* **45,** 225–41.

STANDEN, O. D. 1952. Experimental infection of *Australorbis glabratus* with *Schistosoma mansoni* I—Individual and mass infection of snails and the relationship of infection to temperature and season. *Ann. trop. Med. Parasit.* **46,** 48–53.

STAUBER, L. A. 1961. Immunity in invertebrates, with special reference to the oyster. *Proc. natn. Shellfish Assoc.* **50,** 7–20.

STEPHENSON, W. 1947a. Physiological and histochemical observations on the adult liver fluke *Fasciola hepatica* L. Survival *in vitro*. *Parasitology* **38,** 116–122.

STEPHENSON, W. 1947b. Physiological and histochemical observations on the adult liver fluke, *Fasciola hepatica* L. II. Feeding. *Parasitology* **38,** 123–7.

STEPHENSON, W. 1947c. Physiological and histochemical observations on the adult liver fluke *Fasciola hepatica* L. III. Eggshell formation. *Parasitology* **38,** 128–39.

STEPHENSON, W. 1947d. Physiological and histochemical observations on the adult liver fluke, *Fasciola hepatica* L IV. The excretory system. *Parasitology* **38,** 140–4.

STIREWALT, M. A. 1954. Effect of snail maintenance temperatures on development of *Schistosoma mansoni*. *Expl. Parasit.* **3,** 504–16.

STIREWALT, M. A. 1959a. Isolation and characterization of deposits of secretion from the acetabular gland complex of cercariae of *Schistosoma mansoni*. *Expl. Parasit.* **8,** 199–214.

STIREWALT, M. A. 1959b. Chronological analysis, pattern and rate of migration of cercariae of *Schistosoma mansoni* in body, ear and tail skin of mice. *Ann. trop. Med. Parasit.* **53,** 400–13.

STIREWALT, M. A. 1963a. Seminar on immunity to parasitic Helminths. IV. Schistosome Infections. *Expl. Parasit.* **13,** 18–44.

STIREWALT, M. A. 1963b. Chemical Biology of secretions of larval helminths. *Ann. N.Y. Acad. Sci.* **113,** 36–53.

STIREWALT, M. A. 1965. Mucus in schistosome cercariae. *Ann. N.Y. Acad. Sci.* **118,** 966–8.

STIREWALT, M. A. 1966. Skin penetration mechanisms of helminths, p. 41–59. In *Biology of Parasites*: ed. E. J. L. Soulsby. *Acad. Press*, 1966, 354 pp.

STIREWALT, M. A. and EVANS, A. S. 1952. Demonstration of an enzymatic factor in cercariae of *Schistosoma mansoni* by the streptococcal decapsulation test. *J. infect. Dis.* **91,** 191–7.

STIREWALT, M. A. and EVANS, A. S. 1960. Chromatographic analysis of secretions from the acetabular glands of cercariae of *Schistosoma mansoni*. *Expl. Parasit.* **10,** 75–80.

STIREWALT, M. A. and FREGEAU, W. A. 1965. Effect of selected experimental conditions on penetration and maturation of *Schistosoma mansoni* in mice. I. Environmental. *Expl. Parasit.* **17,** 168–79.

STIREWALT, M. A. and FREGEAU, W. A. 1966. An invasive enzyme system present in cercariae but absent in schistosomules of *Schistosoma mansoni*. *Expl. Parasit.* **19,** 206–15.

STIREWALT, M. A. and FREGEAU, W. A. 1968. Effect of selected environmental conditions on penetration and maturation of *Schistosoma mansoni* in mice. II. Parasite—related conditions. *Expl. Parasit.* **22,** 73–95.

STIREWALT, M. A. and HACKEY, J. R. 1957. Penetration of host skin by cercariae of *Schistosoma mansoni*. I. Observed entry into skin of mouse, hamster, rat, monkey and man. *J. Parasit.* **42,** 565–80.

STIREWALT, M. A. and KRUIDENIER, F. J. 1961. Activity of the acetabular secretory apparatus of cercariae of *Schistosoma mansoni* under experimental conditions. *Expl. Parasit.* **11,** 191–211.

STIREWALT, M. A. and WALTERS, M. 1964. Histochemical assay of glands of cercariae of *Schistosoma mansoni*. *J. Parasit.* **50** (3), Suppl. No. 94, p. 44.

STUNKARD, H. W. 1923. Studies on North American blood flukes. *Bull. Amer. Mus. nat. Hist.* **48,** 165–221.

STUNKARD, H. W. 1930. The life cycle of *Cryptocotyle lingua* (Creplin) with notes on the physiology of the metacercaria. *J. Morph. Physiol.* **50,** 143–83.

STUNKARD, H. W. 1931. Further observations on the occurrence of anal openings in digenetic trematodes. *Ztsch. Parasitenk.* **3,** 713–25.

STUNKARD, H. W. 1934. The life history of *Typhlocoelum cymbium* (Diesing 1850) Kossack 1911 (Trematoda, Cyclocoelidae). A contribution to the phylogeny of the monostomes. *Bull. Soc. zool. Fr.* **59,** 447–66.

STUNKARD, H. W. 1946. Interrelationships and taxonomy of the digenetic trematodes. *Biol. Rev.* **21,** 148–58.

STUNKARD, H. W. 1959. Progenetic maturity and phylogeny of Digenetic trematodes. *J. Parasit.* **45,** 15.

STUNKARD, H. W. 1963. Systematics, taxonomy and nomenclature of the Trematoda. *Q. Rev. Biol.* **38,** 221–33.

STUNKARD, H. W. and SHAW, C. R. 1931. The effect of dilution of sea-water on the activity and longevity of certain marine cercariae, with descriptions of two new species. *Biol. Bull. mar. biol. Lab. Woods Hole,* **61,** 242–71.

STYCZYŃSKA, E. 1958a. Acanthocephala of the biocoenosis of Drużo lake (Parasytofauna biocoenosis of Drużno lake—part VI). *Acta. Parasit. Polon.* **6,** 195–211.

STYCZYŃSKA, E. 1958b. Some observationson the development and bionomics of larvae of *Filicollis anatis* Shrank (Parasitofauna of the biocoenosis of Drużno lake—Part VII). *Acta Parasit. Polon.* **6,** 213–24.

STYCZYŃSKA-JUREWICZ, E. 1961. On the geotaxis, invasivity and span of life of *Opisthoglyphe ranae* Duj. cercariae. *Bull. Acad. pol. Sci. Cl. ll, Sér. Sci. biol.* **9,** 31–5.

STYCZYŃSKA-JUREWICZ, E. 1962. Behaviour of cercariae of *Opisthoglyphe ranae* Duj. as an adaptation to the behaviour of tadpoles in the oxygen conditions of small water bodies. *Polskie Archwm Hydrobiol.* **10,** 197–214.

SUDDS, R. H. 1960. Observations on schistosome miracidial behaviour in the presence of normal and abnormal snail hosts and subsequent studies of these hosts. *J. Elisha Mitchell Scient. Soc.* **76,** 121–33.

SUGIURA, S., SASAKI, T., HOSAKA, Y. and ONO, R. 1954. A study of several factors influencing hatching of *Schistosoma japonicum* eggs. *J. Parasit.* **40,** 381–6.

SULGOSTOWSKA, T. 1958. Flukes of birds of Drużno lake (Parasitofauna of the biocoenosis of Drużno lake—part III). *Acta Parasit. Polon.* **6,** 111–142.

SZIDAT, L. 1924. Beiträge zur Entwickelungsgeschichte der Holostomiden II. *Zool. Anz.* **61,** 249–66.

SZIDAT, L. 1927. Über ein Fischsterben in Kurischen Haff und seine Ursachen. *Z. Fisch.* **25,** 83–90.

SZIDAT, L. 1929. Beiträge zur Kenntinis der Gattung Strigeida, I. *Ztsch. Parasitenk.* **1,** 612–74.

SZIDAT, L. 1932. Ueber cysticerke Riesencercarien, insbesondere *Cercaria mirabilis* M. Braun und *Cercaria splendens* n.sp., und ihre Entwicklung im Magen von Raubfischen zu Trematoden der Gattung Azygia Looss. *Ztsch. Parasitenk.* **4,** 477–505.

SZIDAT, L. 1956. Über den Entwicklungszyklus mit progenetischen Larvenstadien (*Cercariaeen*) von *Generchella genarchella Travassos* 1928 (Trematoda, Hemiuridae) und die Möglichkeit einer hormonalen Beeinflussung der Parasiten durch ihre Wirtstiere. *Z. Tropen med. Parasit.* **7,** 132–53.

SZIDAT, L. and NANI, A. 1951. Diplostomiasis cerebralis del pejerrey Una grave epizootia que afecta la economia nacional producida par de trematodes que destruyyen el cerebro de los pejeurreyes. *Proc. Inst. Nac. Invest. Cien. Nat.* (Cien Zool.) **1,** 323–84.

TAKAHASHI, T., MORI, K. and SHIGETA, Y. 1961. Phototactic, thermotactic and geotactic responses of miracidia of *Schistosoma japonicum*. *Jap. J. Parasit.* **10,** 686–91.

TANDON, R. S. 1960a. Studies on the lymphatic system of Amphistomes of Ruminants. I. *Carmyerius spatiosus* (Stiles and Goldberger, 1910). *Zool. Anz.* **164,** 213–17.

TANDON, R. S. 1960b. Studies on the lymphatic system of Amphistomes of ruminants. 2. The genera *Gastrothylax* and *Fischoederius*. *Zool. Anz.* **164,** 217–21.

TARAZONA VILLAS, J. M. 1958. Contribution al estudio de las gliceromonofosfatases en Los plathehelmintos parasitos. *Rev. Ibér. Parasit.*, **18,** 233–242. *Helm. Abs.* 1958, **27.** 287c.

TAYLOR, A. E. R. and BAKER, J. R. 1968. *The cultivation of parasites in vitro*. Oxford: Blackwell Scientific Publications, xii + 377 pp.

TAYLOR, E. L. 1964. Fascioliasis and the liver fluke. *F.A.O. agric. Stud.* No. 64, XXIX 234 pp.

TAYLOR, E. L. and PARFITT, J. W. 1957. Mouse test for the infectivity of metacercariae with particular reference to metacercariae in snail faeces. *Trans. Am. microsc. Soc.* **76,** 327–8.

TERRY, R. J. 1968. Applications of immunology to helminth disease, in "Immunity to Parasites" pp. 91–103. Sixth Symposium of the British Society for Parasitology. ed. A. E. R. Taylor, Blackwell, Oxford.

THOMAS, A. P. 1883. The life history of the liver fluke. *Q. Jl. microsc. Soc.* **23,** 90–133.

THOMAS, J. D. 1958. Studies on the structure, life-history and ecology of the trematode *Phyllodistomum simile* Nybelin, 1926 (Gorgoderidae: Gorgoderinae) from the urinary bladder of Brown Trout *Salmo trutta* L. *Proc. zool. Soc. Lond.* **130,** 397–435.

THOMAS, J. D. 1964. A comparison between the helminth burdens of male and female brown trout, *Salmo trutta* L. from a natural population in the River Teify, West Wales. *Parasitology* **54,** 263–72.

THOMAS, L. J. 1939. Life cycle of a fluke *Halipegus eccentricus* n.sp. found in the ears of frogs. *J. Parasit.* **25,** 207–22.

THORSELL, W. and BJORKMAN, N. 1965. Morphological and biochemical studies on absorption and secretion in the alimentary tract of *Fasciola hepatica* L. *J. Parasit.* **51,** 217–23.

THORSON, R. E. 1963. Seminar on immunity to parasitic helminths II. Physiology of immunity to helminth infections. *Expl. Parasit.* **13,** 3–12.

THREADGOLD, L. T. 1963a. The ultrastructure of the 'cuticle' of *Fasciola hepatica*. *Expl. Cell Res.* **30,** 240–2.

THREADGOLD, L. T. 1963b. The tegument and associated structures of *Fasciola hepatica*. *Q. Jl. microsc. Sci.* **104,** 505–12.

THREADGOLD, L. T. 1968. Electron microscope studies of *Fasciola hepatica*, VI. The ultrastructural localization of phosphatases. *Expl. Parasit.* **23,** 264–76.

THREADGOLD, L. T. and GALLAGHER, S. S. E. 1966. Electron microscope studies of *Fasciola hepatica*. 1. The ultrastructure and interrelationship of the parenchymal cells. *Parasitology* **56,** 299–304.

THURSTON, J. P. and LAWS, R. M. 1965. *Oculotrema hippopotami* (Trematoda: Monogenea), in Uganda. *Nature* **205,** 1127.

TIMMS, A. and BUEDING, E. 1959. Studies of a proteolytic enzyme from *Schistosoma mansoni*. *Brit. J. Pharmacol.* **14,** 68–73.

TIMON-DAVID, J. 1938. On parasitic trematodes in Echinoderms. *Livro. Jub. Prof. L. Travassos, Rio de Janeiro*, pp. 467–73.

TODD, J. R. and ROSS, J. G. 1966. Origin of hemoglobin in the cecal contents of *Fasciola hepatica*. *Expl. Parasit.* **19,** 151–4.

TRIPP, M. R. 1961. The fate of foreign materials experimentally introduced into the snail *Australorbis glabratus*. *J. Parasit.* **47,** 745–51.

TRIPP, M. R. 1963. Cellular responses of mollusks. *Ann. N. Y. Acad. Sci.* **113,** 467–74.

TROMBA, F. G. 1957. A technique for hatching miracidia in dilute formalin. *J. Parasit.* **40,** 698.

TULLOCH, G. S. and SHAPIRO, J. E. 1957. The ultrastructure of the vitelline cells of *Haematoloechus*. *J. Parasit.* **43,** 628–32.

UDE, J. 1962. Neurosekretorische Zellen im Cerebralganglion von *Dicrocoelium lanceatum* St. u. M. (Trematoda-Digenea) *Zool. Anz.* **169,** 455–7.

UJIIE, N. 1936. On the process of egg-shell formation of *Clonorchis sinensis*, a liver fluke. English abstract. *J. Med. Assoc. Formosa*, **35,** 1894–6.

ULMER, M. J. 1951. *Postharmostomum helicis* (Leidy, 1847) Robinson 1949 (Trematoda) its life history and a revision of the subfamily Brachylaeminae. Part I. *Trans. Amer. Micr. Soc.* **70,** 189–238.

ULMER, M. J. 1957. Notes on the development of *Cotylurus flabelliformis tetracotyles* in the second intermediate host (Trematoda: Strigeidae). *Trans. Amer. Micr. Soc.* **76,** 321–7.

UZMANN, J. R. 1953. *Cercaria milfordensis* nov. sp., a microcercous trematode larva from the marine bivalve, *Mytilus edulis* L. with special reference to its effect on the host. *J. Parasit.* **39,** 445–51.

VAN GREMBERGEN, G. 1949. Le métabolisme respiratoire du trematode *Fasciola hepatica* Linn. *Enzymologia* **13,** 241–57.

VAN GREMBERGEN, G. and PENNOIT DE COOMAN, E. 1944. Natuurw Tijdschr. (Belg) **26,** 91–7.

VAN HAITSMA, J. P. 1931a. Studies on the trematode family Strigeidae (Holostomidae). No. XXII: *Cotylurus flabelliformis* (Faust) and its life-history. *Pap. Mich. Acad. Sci.* **13,** 447–82.

VAN HAITSMA, J. P. 1931b. Studies on the trematode family Strigeidae (Holostomidae). No. XXIII. *Diplostomum flexicaudum* (Cort & Brooks) and stages in its life-history. *Pap. Mich. Acad. Sci.* **13,** 483–516.

VERNBERG, W. B. 1961. Studies on oxygen consumption in the digenetic trematodes VI. The influence of temperature on larval trematodes. *Expl. Parasit.* **11,** 270–5.

VERNBERG, W. B. 1963. Respiration of digenetic trematodes. *Ann. N. Y. Acad. Sci.* **113,** 261–71.

VERNBERG, W. B. and HUNTER, W. S. 1956. Quantitative determinations of the glycogen content of *Gynaecotyla adunca* (Linton, 1905). *Expl. Parasit.* **5,** 441–8.

VERNBERG, W. B. and HUNTER, W. S. 1959. Studies on oxygen consumption in digenetic trematodes III. The relationship of body nitrogen to oxygen uptake. *Expl. Parasit.* **8,** 76–82.

VERNBERG, W. B. and HUNTER, W. S. 1961. Studies on oxygen consumption in digenetic trematodes. V. The influence of temperature on three species of adult trematodes. *Expl. Parasit.* **11,** 34–8.

VERNBERG, W. B. and HUNTER, W. S. 1963. Utilization of certain substrates by larval and adult stages of *Himasthla quissetensis*. *Expl. Parasit.* **14,** 311–15.

VERNBERG, W. B. and VERNBERG, F. J. 1963. Influence of parasitism on thermal resistance of the mud-flat snail. *Nassarius absoleta* Say. *Expl. Parasit.* **14,** 330–2.

VERNBERG, F. J. and VERNBERG, W. B. 1966. Interrelationship between parasites and their hosts. II. Comparative metabolic patterns of thermal acclimation of the trematode *Lintonium vibex* with the host *Spheroides maculatus* *Expl. Parasit.* **18,** 244–50.

VOGEL, H. and BRAND, VON T. 1933. Ueber das verhalten des Fettes in den einzelnen Entwicklungsstadien von *Fasciola hepatica* und seine Beziehungen zum Exkretionssystem. *Ztsch. Parasitenk.* **5,** 425–31.

VON BRAND, T. 1928. Beitrag zur Kenntnis der Zusammensetzung des Fettes von *Fasciola hepatica*. *Z. vergl. Physiol.* **8,** 613–24.

VON BRAND, T. 1950. The carbohydrate metabolism of parasites. *J. Parasit.* **36,** 178–92.

VON BRAND, T. 1952. *Chemical Physiology of Endoparasitic Animals*, 339 pp. Academic Press, New York.

VON BRAND, T. 1960. Recent advances in carbohydrate biochemistry of helminths. *Helminth. Abstr.* **29,** 97–111.

VON BRAND, T. 1966. *Biochemistry of Parasites*, 429 pp. Academic Press, New York.

VON BRAND, T. and FILES, V. S. 1947. Chemical and histological observations on the influence of *Schistosoma mansoni* infection on *Australorbis glabratus*. *J. Parasit.* **33,** 476–82.

VON BRAND, T. and MERCADO, T. I. 1961. Histochemical glycogen studies on *Fasciola hepatica*. *J. Parasit.* **47,** 459–63.

VON BRAND, T. and WEINLAND, E. 1924. Ueber tröpfchenförmige Ausscheidungen bei *Fasciola hepatica* (*Distomum. hepaticum*) *Z. vergl. Physiol.* **2,** 209–14.

VON BRAND, T., WEINBACH, E. C. and CLAGGETT, C. E. 1965. Incorporation of phosphate into the soft tissues and calcareous corpuscles of larval *Taenia taeniaeformis*. *Comp. Biochem. Physiol.* **14,** 11–20.

WAGNER, A. 1959. Stimulation of *Schistosomatium douthitti* cercariae to penetrate their host. *J. Parasit.* **45** (4), 16.

WAITZ, J. A. and SCHARDEIN, J. L. 1964. Histochemical studies of four cyclophyllidean cestodes. *J. Parasit.* **50,** 271–7.

WALL, L. D. 1941a. Life history of *Spirorchis elephantis* (Cort, 1917) a new blood fluke from *Chrysemys picta*. *Amer. Mid. Nat.* **25,** 402–12.

WALL, L. D. 1941b. *Spirorchis parvus* (Stunkard) its life history and the development of its excretory system (Trematoda: Spirorchiidae). *Trans. Amer. Micr. Soc.* **60,** 221–60.

WALL, L. D. 1951. The life history of *Vasotrema robustum* (Stunkard, 1928) (Trematoda: Spirorchiidae). *Trans. Amer. Micr. Soc.* **70,** 173–84.

WALTON, A. C. 1959. Some parasites and their chromosomes. *J. Parasit.* **45,** 1–20.

WARD, H. B. 1916. Notes on two free-living larval trematodes from North America. *J. Parasit.* **3,** 10–20.

WEINSTEIN, P. P. 1966. The *in vitro* cultivation of helminths with reference to morphogenesis. In: *Biology of Parasites*, ed. E. J. L. Soulsby, Acad. Press, 1966.

WESENBERG-LUND, C. 1931. Contributions to the development of the Trematoda Digenea. Part I. The biology of *Leucochloridium paradoxum*. *D. Kgl. Dansk. Vidensk. Selsk. Skrifter, Naturw. Math. Afd. Raekke*, **9,** 4, 90–142.

WESENBERG-LUND, C. 1934. Development of the Trematoda *Digenea* Part II. (Biology of fresh water cercaria in Danish waters). *Mem. de Acad. Ray. Sci. et Lettres de Danemark, Sect. des Sci.* 3: 1–223.

WHARTON, G. W. 1939. Studies on *Lophotaspis vallei* (Stossich, 1899). (Trematoda: Aspidogastridae). *J. Parasit.* **25,** 83–6.

WHARTON, G. W. 1941. The function of respiratory pigments of certain turtle parasites. *J. Parasit.* **27,** 81–87.

WIKGREN, BO-JUNGAR, 1956. Studies on Finnish larval flukes with a list of known Finnish adult flukes (Trematoda: Malacotylea) *Acta Zoologica Fennica* **91,** 1–106.

WILLEY, C. H. and GROSS, P. R. 1957. Pigmentation in the foot of *Littorina littorea* as a means of recognition of infection with trematode larvae. *J. Parasit.* **43,** 324–327.

WILLIAMS, C. O. 1942. Observations on the life history and taxonomic relationships of the trematode *Aspidogaster conchicola*. *J. Parasit.* **28,** 467–75.

WILLIAMS, H. H. 1967. Helminth diseases of fish. *Helm. Abst.* **36** (3), 261–295.

WILLIAMS, J. P. G. and BRYANT, C. 1963. Intermediary metabolism in the immature liver fluke, *Fasciola hepatica* L. *Nature* **200,** 489.

WILLIAMS, M. O., HOPKINS, C. A. and WYLLIE, M. R. 1961. The *in vitro* cultivation of strigeid trematodes. III. Yeast as a medium constituent. *Expl. Parasit.* **11,** 121–127.

WILLMOTT, S. 1952. The development and morphology of the miracidium of *Paramphistomum hiberniae* Willmott, 1950. *J. Helminth.* **26,** 123–32.

WILSON, R. A. 1967a. The protonephridial system in the miracidium of the liver fluke, *Fasciola hepatica* L. *Comp. Biochem. Physiol.* **20,** 337–42.

WILSON, R. A. 1967b. The structure and permeability of the shell and vitelline membrane of the egg of *Fasciola hepatica*. *Parasitology* **57,** 47–58.

WILSON, R. A. 1967c. A physiological study of the development of the egg of *Fasciola hepatica* L. the common liver fluke. *Comp. Biochem. Physiol.* **21,** 307–20.

WILSON, R. A. 1967d. Personal communication.

WILSON, R. A. 1968. The hatching mechanism of the egg of *Fasciola hepatica*. *Parasitology* **58,** 79–89.

WILSON, R. A. 1969. The fine structure of the protonephridial system in the miracidia of *Fasciola hepatica*. *Parasitology* **59,** 461–467.

WINFIELD, G. F. 1932. On the immunity of snails infested with the sporocyst of the strigeid *Cotylurus flabelliformis* to the penetration of its cercaria. *J. Parasit.* **19,** 130–33.

WIŠNIEWSKI, L. W. 1937. Über die Ausschwarmung des Cercarien aus den Schencken. *Zoologica Poloniae* **2,** 67–97.

WIŠNIEWSKI, W. L. 1958a. Characterization of the parasitofauna of an entrophic lake. (Parasito-fauna of the biocoenosis of Družno lake—part I). *Acta parasit. pol.* **6,** 1–64.

WOOTTON, D. M. 1966a. Comparative studies on the life-cycle of two cyclocoelids (Cyclocoelidae-Trematoda). Proc. First Internat. Congress of Parasitology (1964). Vol. 1. Ed. A. Corradetti. Pergamon Press, p. 524–525.

WOOTTON, D. M. 1966. The cotylocidium larva of *Cotylogasteroides occidentalis* (Nickerson, 1902) Yamaguti 1963 (Aspidogasteridae Aspidocotylea—Trematoda). *Proceedings of the First International Congress of Parasitology* (1964). Vol. 1, ed. A. Corradetti, Pergamon Press, p. 547–8.

WOOTTON, R. M. and SOGANDARES-BERNAL, F. 1963. A report on the occurrence of microvillus-like structures in the caeca of certain trematodes (Paramphistomatidae). *Parasitology* **53,** 157–61.

WORLD HEALTH ORGANIZATION 1965. W.H.O. Technical Report Series No. 315. Immunology and Parasitic Diseases. Report of a W.H.O. Expert Committee. pp. 64. World Health Organization, Geneva.

WOUDE, A. VAN DER 1954. Germ cell cycle of *Megalodiscus temperatus* (Stafford, 1905). Harwood, 1932 (Paramphistomidae: Trematoda). *Amer. Mid. Nat.* **51,** 172–202.

WRIGHT, C. 1959. Host location by trematode miracidia. *Amer. Trop. Med. and Parasitol.* **53,** 288–292.

WRIGHT, C. A. 1966a. The pathogenesis of helminths in the Mollusca. *Helm. Abstracts* **35,** 207–224.

WRIGHT, C. A. 1966b. Relationships between Schistosomes and their molluscan hosts in Africa. *J. Helminth.* **40,** 403–412.

WUNDER, W. 1924. Die Schwimmbewung von *Bucephalus polymorphus* v.Baer. *Z. vergl. Physiol.* **1,** 289–296.

WUNDER, W. 1932. Untersuchungen über Pigmentierung und Encystierung von Cercarien. *Z. Morph. Ökal. Thiere* **25,** 336–352.

WYKOFF, D. E. and LEPES, T. J. 1957. Studies on *Clonorchis sinensis* I. Observations on the route of migration in the definitive host. *Am. J. trop. Med. Hyg.* **6,** 1061–5.

WYLLIE, R. M.; WILLIAMS, O. M. and HOPKINS, C. A. 1960. The *in vitro* cultivation of strigeid trematodes. II. Replacement of a yolk medium. *Exp. Parasit.* **10,** 51–57.

YAMAGUTI, S. 1940. Vergleichend—anatomische studien der Miracidien. *Ztsch. Parasitenk.* **11,** 657–68.

YAMAGUTI, S. 1958. *Systema Helminthum* Vol. I. parts 1 and 2. The Digenetic Trematodes of Vertebrates. Interscience, New York 1958, 1575 pp.

YAMAGUTI, S. 1963. *Systema Helminthum* Vol. IV. Monogenea and Aspidocotylea. Interscience Publishers, New York 1963, pp. 699.

YAMAO, Y. 1952. Histochemical studies on endoparasites IX. On the distribution of glycogen. *Dobutsugaku Zasshi,* **61,** 317–22.

YOGORE, M. G., LEWERT, R. M. and MADRASO, E. D. 1965. Immunodiffusion studies on paragonimiasis. *Am. J. trop. Med. Hyg.* **14,** 586–91.

YOKOGAWA, M. 1965. Paragonimus and Paragonimiasis p. 99–158. *In Advances in Parasitology,* Vol. 3, ed. B. Dawes, Academic Press, London, 315 p. 1965.

YOKOGAWA, S., CORT, W. W. and YOKOGAWA, M. 1960. Paragonimus and Paragonimiasis. *Expl. Parasit.* **10,** 81–137, 139–205.

YOKOGAWA, M., YOSHIMURA, H., SANO, M., OKURA, T. and TSUJI, M. 1962. The route of migration of the larva of *Paragonimus westermani* in the final hosts. *J. Parasit.* **48,** 525–31.

ŽDÁRSKÁ, Z. 1964. Contributions to the knowledge of metabolic and morphological changes in the metacercaria of *Echinostoma revolutum. Věst. čsl. Spol. Zool.* **28,** 285–9.

Appendix

Key[1] to the Families of the Subclass Digenea Based on the Method of Developing the Excretory Bladder and on other Larval Character.

X Primitive excretory bladder retained, i.e. not replaced by cells from mesoderm, hence definitive excretory bladder not epithelial;

Cercariae with forked or single tails; caudal excretory vessels present in developing cercariae (except perhaps in certain species of Renicolidae); stylet always absent: Superorder: Anepitheliocystidia N.N.

I Cercariae usually fork-tailed; miracidia with one or two pairs of flamecells: Order: Strigeatoidea LA RUE, 1926

A Cercariae fork-tailed; usually distomate; excretory bladder V-shaped; protonephridia mesostomate or stenostomate; penetration glands present; active penetration into next host; eggs large; miracidia large, with two pairs of flame-cells: Suborder: Strigeata LA RUE, 1926

(i) Cercariae usually longifurcate, tail-stem usually slender, oral sucker well developed; acetabulum usually present, two to four pairs of large penetration glands located in acetabular zone (Strigeidae, Diplostomatidae), or many glands near cecal bifurcation (Cyathocotylidae); protonephridia mesostomate: development in filiform sporocysts; three-host life cycle:
Superfamily: Strigeoidea RAILLIET, 1919
Family: Strigeidae* RAILLIET, 1919
Diplostomatidae* POIRIER, 1886
Cyathocotylidae* POCHE, 1926
Proterodiplostomatidae DUBOIS, 1937
Bolbocephalodidae STRAND, 1935
Brauninidae BOSMA, 1931

(ii) Cercariae brevifurcate; pharyngeate; oral sucker replaced by extensible penetration organ as in Schistosomatidae; acetabulum rudimentary; penetration glands as in Strigeidae and Diplostomatidae; eyespots pigmented; development in rediae, three host life cycle:
Superfamily: Clinostomatoidea DOLLFUS, 1931
Family: Clinostomatidae* LÜHE, 1901

(iii) Cercariae brevifurcate, apharyngeate; oral sucker replaced by extensible penetration organ; six or seven pairs of penetration glands; with or without pigmented eyespots; development in simple sporocysts; cercariae penetrating into final host, hence two-host life cycle:
Superfamily: Schistosomatoidea STILES and HASSALL, 1926

[1] Includes certain families for which no life histories are known. However, their relationships are indicated by the comparative morphology of the adults.

Family: Schistosomatidea* LOOSS, 1899
Spirorchiidae* STUNKARD, 1921
Aporocotylidae* ODHNER, 1912

B Cercariae fork-tailed or variously modified from that condition; distomate (i–iii) or gasterostomate (iv); eggs large or small; miracidium with one pair of flamecells, or not observed:

(i) Cercariae usually furcocystocercous; distomate or monostomate; protonephridia stenostomate, with flamecell groups in the tail, development in redia:
Suborder: Azygiata N.N.
(a) Cercariae furcocystocercous:
Superfamily Azygioidea SKRJABIN and GUSCHANSKAJA, 1956
Cercariae distomate; pharyngeata; bladder Y-shaped; eyespots lacking; usually progenetic; two-host life cycle:
Family: Azygiidae* ODHNER, 1911
Cercariae monostomate; bivesiculate (vessels remaining unfused); eyespots pigmented:
Family: Bivesiculidae*† YAMAGUTI, 1939
(b) Cercariae brevifurcate; tail stem bearing pair of anteriorly placed appendages; body leaf-like; distomate; apharyngeate; eyespots pigmented; branches of gut fused posteriorly, genital pore anterior to sub-terminal oral sucker.
Superfamily: Transversotrematoidea N.N.
Family: Transversotrematidae**† YAMAGUTI, 1953

(ii) Cercariae distomate; tail very short and bilobed (*Pseudhyptiasmus*) or lacking; excretory bladder V-shaped; developing in rediae; encysting in or near rediae;
Suborder: Cyclocoelata N.N.
Superfamily: Cyclocoeloidea NICHOLL, 1934
Family: Cyclocoelidae* KOSSACK, 1911
Typhlocoelidae BITTNER and SPREHN, 1928
Bothrigastridae DOLLFUS, 1948

(iii) Cercariae distomate, tail forked, of moderate size, greatly reduced, or lacking; protonephridia stenostomate; development in sporocysts:
Suborder: Brachylaimata N.N.
(a) Tail functional, rudimentary, or lacking; excretory vesicle V-shaped with short arms; development in branching sporocysts in aquatic or terrestrial snails; life cycle with two or three hosts; those with three hosts provided with penetration glands near acetabulum and in oral sucker:
Superfamily: Brachylaimoidea ALLISON, 1943 (*orthog. emend.*)
Family: Brachylaimidae*† JOYEUX and FOLEY, 1930 (*orthog. emend.*)

* One or more life histories are known in the family.
† Number of flamecells in miracidium is not known.
** Life histories are incompletely known.

(b) Cercariae with tail forked, modified to single tail, or lacking; tail stem with or without paired multiple setae; excretory vesicle U- or lyre-shaped, with short stem and long broad arms; protonephridia stenostomate; penetration glands numerous and far anterior; development in simple sporocysts in marine lamellibranchs; three-host life cycle:
Superfamily: Fellodistomatoidea N.N.
Family: Fellodistomatidae*† NICOLL, 1913

(iv) Cercariae gasterostomate; tail stem short and bulbous; furcae very long and active; excretory vesicle cylindrical; protonephridia mesostomate; development in branched sporocysts in lamellibranchs of fresh and brackish waters; life cycle with three hosts:
Superfamily: Bucephaloidea LA RUE, 1926
Family: Bucephalidae*† POCHE, 1907

II Cercariae with large bodies and strong single tails; cystogenous glands numerous; protonephridia stenostomate; miracidia with one pair of flamecells; development in rediae:
Order: Echinostomida N.N.

A Cercariae echinostomate or exhibiting modifications therefrom in time of appearance of collar and collarspines and in degree of development of these structures; development in collared rediae with stumpy appendages; life cycle usually involving three hosts:
Suborder: Echinostomata SZIDAT, 1939
Superfamily: Echinostomatoidea FAUST, 1929
Family: Echinostomatidae* LOOSS, 1902
Cathaemasiidae* FUHRMANN, 1929
Campulidae ODHNER, 1926
Fasciolidae* RAILLIET, 1895
Psilostomatidae* ODHNER, 1911
Philophthalmidae* TRAVASSOS, 1918
Rhopaliidae LOOSS, 1898
Haplosplanchnidae**† POCHE, 1926
? Rhytidodidae ODHNER, 1926.

B Cercariae amphistomate or monostomate; penetration apparatus lacking; bodies heavily pigmented; two or three pigmented eyespots; development in rediae lacking collar and usually lacking stumpy appendages; cercariae emerging from rediae before completing growth; two-host life cycle; encystment on substrate:
Suborder: Paramphistomata SZIDAT, 1936

(i) Cercariae typically amphistomate; pharynx present and often replacing oral sucker; eggs medium to large size, without filaments:
Superfamily: Paramphistomatoidea STILES and GOLDBERGER, 1910
Family: Paramphistomatidae* FISCHOEDER, 1901
Gastrothylacidae* STILES and GOLDBERGER, 1910
Cladorchiidae SOUTHWELL and KIRSCHNER, 1937
Diplodiscidae* SKRJABIN, 1949

Brumptidae SKRJABIN, 1949
Gastrodiscidae* STILES and GOLDBERGER, 1910
Stephanopharyngidae SKRJABIN, 1949
Heronimidae* WARD, 1918
Microscaphidiidae TRAVASSOS, 1922
? Mesometridae POCHE, 1926

(ii) Cercariae monostomate; pharynx lacking; main collecting vessels fused anteriorly; protrusible cup-shaped attaching structures situated postero-laterally; eggs small, with polar filaments:
Superfamily: Notocotyloidea N.N.
Family: Notocotylidae* LÜHE, 1909
Pronocephalidae* LOOSS, 1902
Rhabdopoeidae POCHE, 1926

C Cercariae of rhodometopa type (distomate; pharyngeate; body large; two or four groups of small penetration glands anterior to ventral sucker; tail large, frequently provided with dorsal, ventral, and lateral fins; excretory bladder large, Y-shaped, with lateral fins; excretory bladder large, Y-shaped, with lateral diverticula arising from stem and arms, post-acetabular commissure present or lacking; protonephridia mesostomate; caudal vessels usually present; in developing cercariae; development in simple sporocysts in marine gastropods; three-host life cycle):
Order: Renicolida N.N.
Suborder: Renicolata N.N.
Superfamily: Renicoloidea N.N.
Family: Renicolidae**† DOLLFUS, 1939

XX Primitive excretory bladder surrounded by, and then replaced by, layer of cells derived from mesoderm, hence definitive bladder thick-walled and epithelial:
Cercarial tail single, reduced in size, or lacking; caudal excretory vessels present or lacking; miracidium with one pair of flame-cells:
Superorder: Epitheliocystidia N.N.

I Cercariae completely lacking caudal excretory vessels at any stage of development; stylet present or lacking:
Order: Plagiorchiida N.N.

A Cercariae typically distomate and pharyngeate; of various xiphidiocercarial types (armatae, ornatae, virgulae, microcotylae, or tailless); stylet horizontal; protonephridia mesostomate; encystment in invertebrates (chiefly arthropods), rarely in vertebrates:
Suborder: Plagiorchiata N.N.
Superfamily: Plagiorchioidea DOLLFUS, 1930
Family: Dicrocoeliidae* ODHNER, 1910
Eucotylidae* SKRJABIN, 1924
Haplometridae* MCMULLEN, 1937
Lecithodendriidae* ODHNER, 1910
Lissorchiidae* POCHE, 1926
Macroderoididae* MCMULLEN, 1937
Microphallidae* TRAVASSOS, 1921

Ochetosomatidae* LEAO, 1944
Plagiorchiidae* LÜHE, 1901
? Stomylotrematidae POCHE, 1926
? Cephalogonimidae NICOLL, 1915
? Urotrematidae POCHE, 1926
? Cephaloporidae TRAVASSOS, 1934
? Collyriclidae WARD, 1918
? Mesotretidae POCHE, 1926

B Cercariae of various types (ophthalmoxiphidiocercariae, microcercous, cotylomicrocercous, macrocercous, rhopalocercous, ophthalmotrichocercous, tailless, or of megaperid type with muscular tail having lateral and ventral fins); stylet usually nor horizontal, if present; protonephridia usually mesostomate; excretory bladder saccate or Y-shaped; development in rediae or sporocysts, in snails or lamellibranchs; encystment in invertebrates (chiefly arthropods), rarely in vertebrates; usually three-host life cycle:
Superfamily: Allocreadioidea NICOLL, 1934
Family: Acanthocolpidae* LÜHE, 1909
Allocreadiidae* STOSSICH, 1903
Lepocreadiidae* NICOLL, 1934
Megaperidae**† MANTER, 1934
Monorchiidae* ODHNER, 1911
Opecoelidae* OZAKI, 1925
Opistholebetidae FUKUI, 1929
Provisionally assigned to this superfamily:
Gorgoderidae* LOOSS, 1901
Gyliauchenidae (Goto and Matsudaira) in OZAKI, 1933
Troglotrematidae* ODHNER, 1914
Zoogonidae* ODHNER, 1911

II Cercariae with caudal excretory vessel during development; stylet always lacking:
Order: Opisthorchiida N.N.

A Primary excretory pores on margins of tail near body-tail furrow; bodies of opisthorchioid type; oral sucker protrusible, with large spines and openings of penetration glands in crypts anterior to subterminal mouth; ventral sucker usually rudimentary; tails pleuro- or parapleuro-lophocercous, magna- or even zygo-cercous; protonephridia mesostomate or stenostomate; bladder V-shaped or globular; development in sporocysts or rediae; encystment in lower vertebrates; eggs small:
Suborder: Opisthorchiata N.N.
Superfamily: Opisthorchioidea FAUST, 1929
Family: Opisthorchiidae* BRAUN, 1901
Heterophyidae* ODHNER, 1914
Acanthostomatidae* POCHE, 1926
Cryptogonimidae* CIUREA, 1933
Pachytrematidae BAER, 1944
Ratziidae BAER, 1944

B Primary excretory pores on tail distant from body-tail furrow; cercariae of

cystophorous type or modified therefrom; bladder saccate or cylindrical; protonephridia stenostomate; main collecting vessels fused anteriorly; eggs small to medium size, with or without filament; miracidia non-ciliate, but with spinose anterior tip; development in rediae; second intermediate host a copepod:
Suborder: Hemiurata SKRJABIN and GUSCHANSKAJA, 1954
Superfamily: Hemiuroidea FAUST, 1929
Family: Hemiuridae* LÜHE, 1901
Halipegidae* POCHE, 1926
Dinuridae SKRJABIN and GUSCHANSKAJA, 1954
Lechithasteridae SKRJABIN and GUSCHANSKAJA, 1954
Lecithochiridae SKRJABIN and GUSCHANSKAJA, 1954
Bathycotylidae DOLLFUS, 1932
Isoparorchiidae POCHE, 1926
Ptychogonimidae DOLLFUS, 1937
Didymozoidae**† POCHE, 1907

INDEX TO GENERA

Numbers in bold type refer to figures and tables.

AUTHOR INDEX

SUBJECT INDEX